THE SCIENTIFIC PAPERS

OF

J. WILLARD GIBBS, Ph.D., LL.D.

THE
SCIENTIFIC PAPERS

OF

J. WILLARD GIBBS, Ph.D., LL.D.

FORMERLY PROFESSOR OF MATHEMATICAL PHYSICS IN YALE UNIVERISTY

IN TWO VOLUMES

VOL. I

THERMODYNAMICS

OX BOW PRESS
WOODBRIDGE, CONNECTICUT

1993 Reprint published by:
OX BOW PRESS
P.O. Box 4045
Woodbridge, Connecticut, U.S.A. 06525
(203) 387-5900   FAX (203) 387-0035

Library of Congress Cataloging-in-Publication Data

Gibbs, J. Willard (Josiah Willard), 1839-1903
        The scientific papers of J. Willard Gibbs
        p.   cm.
        Originally published:London; New York;
        Longmans, Green, 1906.
        ISBN 0-918024-77-3 (pbk.   : alk. paper)
        1. Science.   2. Thermodynamics.   I. Title
Q113.G44   1993
530-dc20

                                                93-19391
                                                CIP

The paper in this book meets the guidelines for performance and durability of the Committee on Production Guidelines for Book Longevity of the Council on Library Resources.

Printed in the United States of America

PREFACE.

WITH the exception of Professor J. Willard Gibbs's last work, *Elementary Principles in Statistical Mechanics*, and of his lectures upon Vector Analysis, adapted for use as a text-book by his pupil Dr. E. B. Wilson, and printed like the former as a volume of the Yale Bicentennial Series, none of his contributions to mathematical and physical science were published in separate form, but appeared in the transactions of learned societies and in various scientific journals.

These scattered papers, which constitute the larger and perhaps the more important part of his published work, are here presented in a collected edition, from which, so far as known to the editors, no printed paper has been omitted. A small amount of hitherto unpublished matter has also been included. Permission for the present reprint of the different papers contained in these volumes has in every case been granted by the authorities in charge of the publications in which they originally appeared, a courtesy for which the editors desire here to make due acknowledgment.

In the arrangement of the papers a grouping by subject has been adopted in preference to a strict chronological order. Within the separate groups, however, the chronological order has in general been preserved.

The papers on Thermodynamics, which form somewhat more than one half of the whole, constitute the first volume. Among these is the well-known memoir *On the Equilibrium of Heterogeneous Substances*, which has proved to be of such fundamental importance to Physical Chemistry and has been translated into German by Professor Ostwald, and into French by Professor Le Chatelier.

Shortly before the author's death he had yielded to numerous requests for a republication of his thermodynamic papers, and had arranged for a volume which was to contain the *Equilibrium of Heterogeneous Substances* and the two earlier papers, *Graphical Methods in the Thermodynamics of Fluids*, and *A Method of Geometrical Representation of the Thermodynamic Properties of Substances by means of Surfaces*. To these he proposed to add some supplementary chapters, the preparation of which he had hardly more than commenced when he was overtaken by his last illness. The manuscript of a portion of this additional material (evidently a first draft) was found among the author's papers and has been printed at the end of the first volume. It is believed that it will be of interest and value in spite of its unfinished and somewhat fragmentary condition.

The remaining papers, which compose the second volume, are divided between mathematical and physical science. Most of them naturally fall under one of the following heads: Dynamics, Vector Analysis and Multiple Algebra, the Electromagnetic Theory of Light, and are so grouped in the volume in the order named. A fourth section is made up of the unclassified papers.

In the first section the short abstract of a paper read before the American Association for the advancement of Science is worthy of notice as showing that the fundamental ideas and methods of the treatise on Statistical Mechanics were well developed in the author's mind at least seventeen years before the publication of that work.

The second section includes the *Elements of Vector Analysis*, privately printed in 1881-1884 for the use of the author's classes, but never published. It contains in a very condensed form all the essential features of Professor Gibbs's system of Vector Analysis, but without the illustrations and applications which he was accustomed to give in his lectures on this subject. Copies of this pamphlet have been for many years past practically unobtainable. Here is also placed a hitherto unpublished letter to the editor of Klinkerfues' *Theoretische Astronomie*, on the use of the author's vector method for the determination of orbits.

Five papers on the Electromagnetic Theory of Light constitute the third section. The fourth and last is composed of miscellaneous papers, including biographical sketches of Clausius and of the author's colleague Hubert A. Newton.

The editors have spared no pains to make the reprint typographically accurate. In a few cases slight corrections had been made by Professor Gibbs in his own copies of the papers. These changes, together with the correction of obvious misprints in the originals, have been incorporated in the present edition without comment.

Where for the sake of clearness it has seemed desirable to the editors to insert a word or two in a footnote or in the text itself, the addition has been indicated by enclosing it within square brackets [ ], a sign which is otherwise used only in the formulæ.

A sketch of the life and estimate of the work of Professor Gibbs, by one of the editors, is placed at the beginning of the first volume. It is taken, with some additions, from the *American Journal of Science*, September 1903.

HENRY ANDREWS BUMSTEAD.
RALPH GIBBS VAN NAME.

YALE UNIVERSITY,
NEW HAVEN,
October 1906.

CONTENTS OF VOLUME I.

THERMODYNAMICS.

JOSIAH WILLARD GIBBS.

[Reprinted with some additions from the *American Journal of Science*, ser. 4, vol. XVI., September, 1903.]

JOSIAH WILLARD GIBBS was born in New Haven, Connecticut, February 11, 1839, and died in the same city, April 28, 1903. He was descended from Robert Gibbs, the fourth son of Sir Henry Gibbs of Honington, Warwickshire, who came to Boston about 1658. One of Robert Gibbs's grandsons, Henry Gibbs, in 1747 married Katherine, daughter of the Hon. Josiah Willard, Secretary of the Province of Massachusetts, and of the descendants of this couple, in various parts of the country, no fewer than six have borne the name Josiah Willard Gibbs.

The subject of this memorial was the fourth child and only son of Josiah Willard Gibbs, Professor of Sacred Literature in the Yale Divinity School from 1824 to 1861, and of his wife, Mary Anna, daughter of Dr. John Van Cleve of Princeton, N.J. The elder Professor Gibbs was remarkable among his contemporaries for profound scholarship, for unusual modesty, and for the conscientious and painstaking accuracy which characterized all of his published work. The following brief extracts from a discourse commemorative of his life, by Professor George P. Fisher, can hardly fail to be of interest to those who are familiar with the work of his distinguished son: "One who should look simply at the writings of Mr. Gibbs, where we meet only with naked, laboriously classified, skeleton-like statements of scientific truth, might judge him to be devoid of zeal even in his favorite pursuit. But there was a deep fountain of feeling that did not appear in these curiously elaborated essays. . . . Of the science of comparative grammar, as I am informed by those most competent to judge, he is to be considered in relation to the scholars of this country as the leader." Again, in speaking of his unfinished translation of Gesenius's *Hebrew Lexicon*: "But with his wonted thoroughness, he could not leave a word until he had made the article upon it perfect, sifting what the author had written by independent investigations of his own."

The ancestry of the son presents other points of interest. On his father's side we find an unbroken line of six college graduates. Five of these were graduates of Harvard,—President Samuel Willard, his son Josiah Willard, the great grandfather, grandfather and father of the elder Professor Gibbs, who was himself a graduate of Yale. Among his mother's ancestors were two more Yale graduates, one of whom, Rev. Jonathan Dickinson, was the first President of the College of New Jersey.

Josiah Willard Gibbs, the younger, entered Yale College in 1854 and was graduated in 1858, receiving during his college course several prizes for excellence in Latin and Mathematics; during the next five years he continued his studies in New Haven, and in 1863 received the degree of doctor of philosophy and was appointed a tutor in the college for a term of three years. During the first two years of his tutorship he taught Latin and in the third year Natural Philosophy, in both of which subjects he had gained marked distinction as an undergraduate. At the end of his term as tutor he went abroad with his sisters, spending the winter of 1866-67 in Paris and the following year in Berlin, where he heard the lectures of Magnus and other teachers of physics and of mathematics. In 1868 he went to Heidelberg, where Kirchhoff and Helmholtz were then stationed, returning to New Haven in June, 1869. Two years later he was appointed Professor of Mathematical Physics in Yale College, a position which he held until the time of his death.

It was not until 1873, when he was thirty-four years old, that he gave to the world, by publication, evidence of his extraordinary powers as an investigator in mathematical physics. In that year two papers appeared in the *Transactions of the Connecticut Academy*, the first being entitled "Graphical Methods in the Thermodynamics of Fluids," and the second "A Method of Geometrical Representation of the Thermodynamic Properties of Substances by Means of Surfaces." These were followed in 1876 and 1878 by the two parts of the great paper "On the Equilibrium of Heterogeneous Substances," which is generally, and probably rightly, considered his most important contribution to physical science, and which is unquestionably among the greatest and most enduring monuments of the wonderful scientific activity of the nineteenth century. The first two papers of this series, although somewhat overshadowed by the third, are themselves very remarkable and valuable contributions to the theory of thermodynamics; they have proved useful and fertile in many direct ways, and, in addition, it is difficult to see how, without them, the third could have been written. In logical development the three are very closely connected, and methods first brought forward in the earlier papers are used continually in the third.

Professor Gibbs was much inclined to the use of geometrical

illustrations, which he employed as symbols and aids to the imagination, rather than the mechanical models which have served so many
great investigators; such models are seldom in complete correspondence
with the phenomena they represent, and Professor Gibbs's tendency
toward rigorous logic was such that the discrepancies apparently
destroyed for him the usefulness of the model. Accordingly he usually
had recourse to the geometrical representation of his equations, and
this method he used with great ease and power. With this inclination,
it is probable that he made much use, in his study of thermodynamics,
of the volume-pressure diagram, the only one which, up to that time,
had been used extensively. To those who are acquainted with the
completeness of his investigation of any subject which interested him,
it is not surprising that his first published paper should have been a
careful study of all the different diagrams which seemed to have any
chance of being useful. Of the new diagrams which he first described
in this paper, the simplest, in some respects, is that in which entropy
and temperature are taken as coordinates; in this, as in the familiar
volume-pressure diagram, the work or heat of any cycle is proportional
to its area in any part of the plane; for many purposes it is far more
perspicuous than the older diagram, and it has found most important
practical applications in the study of the steam engine. The diagram,
however, to which Professor Gibbs gave most attention was the
volume-entropy diagram, which presents many advantages when the
properties of bodies are to be studied, rather than the work they do or
the heat they give out. The chief reason for this superiority is that
volume and entropy are both proportional to the quantity of substance,
while pressure and temperature are not; the representation of coexistent states is thus especially clear, and for many purposes the gain in
this direction more than counter-balances the loss due to the variability
of the scale of work and heat. No diagram of constant scale can, for
example, adequately represent the triple state where solid, liquid and
vapor are all present; nor, without confusion, can it represent the
states of a substance which, like water, has a maximum density; in
these and in many other cases the volume-entropy diagram is superior
in distinctness and convenience.

In the second paper the consideration of graphical methods in
thermodynamics was extended to diagrams in three dimensions.
James Thomson had already made this extension to the volume-pressure
diagram by erecting the temperature as the third coordinate, these
three immediately cognizable quantities giving a surface whose interpretation is most simple from elementary considerations, but which,
for several reasons, is far less convenient and fertile of results than
one in which the coordinates are thermodynamic quantities less directly
known. In fact, if the general relation between the volume, entropy

and energy of any body is known, the relation between the volume, pressure and temperature may be immediately deduced by differentiation; but the converse is not true, and thus a knowledge of the former relation gives more complete information of the properties of a substance than a knowledge of the latter. Accordingly Gibbs chooses as the three coordinates the volume, entropy and energy and, in a masterly manner, proceeds to develop the properties of the resulting surface, the geometrical conditions for equilibrium, the criteria for its stability or instability, the conditions for coexistent states and for the critical state; and he points out, in several examples, the great power of this method for the solution of thermodynamic problems. The exceptional importance and beauty of this work by a hitherto unknown writer was immediately recognized by Maxwell, who, in the last years of his life, spent considerable time in carefully constructing, with his own hands, a model of this surface, a cast of which, very shortly before his death, he sent to Professor Gibbs.

One property of this three dimensional diagram (analogous to that mentioned in the case of the plane volume-entropy diagram) proved to be of capital importance in the development of Gibbs's future work in thermodynamics; the volume, entropy and energy of a mixture of portions of a substance in different states (whether in equilibrium or not), are the sums of the volumes, entropies and energies of the separate parts, and, in the diagram, the mixture is represented by a single point which may be found from the separate points, representing the different portions, by a process like that of finding centers of gravity. In general this point is not in the surface representing the stable states of the substance, but within the solid bounded by this surface, and its distance from the surface, taken parallel to the axis of energy, represents the available energy of the mixture. This possibility of representing the properties of mixtures of different states of the same substance immediately suggested that mixtures of substances differing in chemical composition, as well as in physical state, might be treated in a similar manner; in a note at the end of the second paper the author clearly indicates the possibility of doing so, and there can be little doubt that this was the path by which he approached the task of investigating the conditions of chemical equilibrium, a task which he was destined to achieve in such' a magnificent manner and with such advantage to physical science.

In the discussion of chemically homogeneous substances in the first two papers, frequent use had been made of the principle that such a substance will be in equilibrium if, when its energy is kept constant, its entropy cannot increase; at the head of the third paper the author puts the famous statement of Clausius: "Die Energie der Welt ist constant. Die Entropie der Welt strebt einem Maximum zu." He proceeds to show that the above condition for equilibrium, derived from the two laws of thermodynamics, is of universal application, carefully removing one restriction after another, the first to go being that the substance shall be chemically homogeneous. The important analytical step is taken of introducing as variables in the fundamental differential equation, the masses of the constituents of the heterogeneous body; the differential coefficients of the energy with respect to these masses are shown to enter the conditions of equilibrium in a manner entirely analogous to the "intensities," pressure and temperature, and these coefficients are called potentials. Constant use is made of the analogies with the equations for homogeneous substances, and the analytical processes are like those which a geometer would use in extending to n dimensions the geometry of three.

It is quite out of the question to give, in brief compass, anything approaching an adequate outline of this remarkable work. It is universally recognized that its publication was an event of the first importance in the history of chemistry, that in fact it founded a new department of chemical science which, in the words of M. Le Chatelier, is becoming comparable in importance with that created by Lavoisier. Nevertheless it was a number of years before its value was generally known; this delay was due largely to the fact that its mathematical form and rigorous deductive processes make it difficult reading for any one, and especially so for students of experimental chemistry whom it most concerns; twenty-five years ago there was relatively only a small number of chemists who possessed sufficient mathematical knowledge to read easily even the simpler portions of the paper. Thus it came about that a number of natural laws of great importance which were, for the first time, clearly stated in this paper were subsequently, during its period of neglect, discovered by others, sometimes from theoretical considerations, but more often by experiment. At the present time, however, the great value of its methods and results are fully recognized by all students of physical chemistry. It was translated into German in 1891 by Professor Ostwald and into French in 1899 by Professor Le Chatelier; and, although so many years had passed since its original publication, in both cases the distinguished translators give, as their principal reason for undertaking the task, not the historical interest of the memoir, but the many important questions which it discusses and which have not even yet been worked out experimentally. Many of its theorems have already served as starting points or guides for experimental researches of fundamental consequence; others, such as that which goes under the name of the "Phase Rule," have served to classify and explain, in a simple and logical manner, experimental facts of much apparent complexity; while still others, such as the theories of catalysis, of solid solutions, and of the action of semi-permeable diaphragms and osmotic pressure, showed that many facts, which had previously seemed mysterious and scarcely capable of explanation, are in fact simple, direct and necessary consequences of the fundamental laws of thermodynamics. In the discussion of mixtures in which some of the components are present only in very small quantity (of which the most interesting cases at present are dilute solutions) the theory is carried as far as is possible from à *priori* considerations; at the time the paper was written the lack of experimental facts did not permit the statement, in all its generality, of the celebrated law which was afterward discovered by van't Hoff; but the law is distinctly stated for solutions of gases as a direct consequence of Henry's law and, while the facts at the author's disposal did not permit a further extension, he remarks that there are many indications "that the law expressed by these equations has a very general application."

It is not surprising that a work containing results of such consequence should have excited the profoundest admiration among students of the physical sciences; but even more remarkable than the results, and perhaps of even greater service to science, are the methods by which they were attained; these do not depend upon special hypotheses as to the constitution of matter or any similar assumption, but the whole system rests directly upon the truth of certain experiential laws which possess a very high degree of probability. To have obtained the results embodied in these papers in any manner would have been a great achievement; that they were reached by a method of such logical austerity is a still greater cause for wonder and admiration. And it gives to the work a degree of certainty and an assurance of permanence, in form and matter, which is not often found in investigations so original in character.

In lecturing to students upon mathematical physics, especially in the theory of electricity and magnetism, Professor Gibbs felt, as so many other physicists in recent years have done, the desirability of a vector algebra by which the more or less complicated space relations, dealt with in many departments of physics, could be conveniently and perspicuously expressed; and this desire was especially active in him on account of his natural tendency toward elegance and conciseness of mathematical method. He did not, however, find in Hamilton's system of quaternions an instrument altogether suited to his needs, in this respect sharing the experience of other investigators who have, of late years, seemed more and more inclined, for practical purposes, to reject the quaternionic analysis, notwithstanding its beauty and logical completeness, in favor of a simpler and more direct treatment of the subject. For the use of his students, Professor Gibbs privately printed in 1881 and 1884 a very concise account of the vector analysis which he had developed, and this pamphlet was to some extent circulated among those especially interested in the subject. In the development of this system the author had been led to study deeply the *Ausdehnungslehre* of Grassmann, and the subject of multiple algebra in general; these investigations interested him greatly up to the time of his death, and he has often remarked that he had more pleasure in the study of multiple algebra than in any other of his intellectual activities. His rejection of quaternions, and his championship of Grassmann's claim to be considered the founder of modern algebra, led to some papers of a somewhat controversial character, most of which appeared in the columns of *Nature*. When the utility of his system as an instrument for physical research had been proved by twenty years' experience of himself and of his pupils, Professor Gibbs consented, though somewhat reluctantly, to its formal publication in much more extended form than in the original pamphlet. As he was at that time wholly occupied with another work, the task of preparing this treatise for publication was entrusted to one of his students, Dr. E. B. Wilson, whose very successful accomplishment of the work entitles him to the gratitude of all who are interested in the subject.

The reluctance of Professor Gibbs to publish his system of vector analysis certainly did not arise from any doubt in his own mind as to its utility, or the desirability of its being more widely employed; it seemed rather to be due to the feeling that it was not an original contribution to mathematics, but was rather an adaptation, for special purposes, of the work of others. Of many portions of the work this is of course necessarily true, and it is rather by the selection of methods and by systematization of the presentation that the author has served the cause of vector analysis. But in the treatment of the linear vector function and the theory of dyadics to which this leads, a distinct advance was made which was of consequence not only in the more restricted field of vector analysis, but also in the broader theory of multiple algebra in general.

The theory of dyadics* as developed in the vector analysis of 1884 must be regarded as the most important published contribution of Professor Gibbs to pure mathematics. For the vector analysis as an *algebra* does not fulfil the definition of the linear associative algebras of Benjamin Peirce, since the scalar product of vectors lies outside the vector domain; nor is it a geometrical analysis in the sense of

---

* The three succeeding paragraphs are by Professor Percey F. Smith; they form part of a sketch of Professor Gibbs's work in pure mathematics, which Professor Smith contributed to the *Bulletin of the American Mathematical Society*, vol. x, p. 34 (October, 1903).

Grassmann, the vector product satisfying the combinatorial law, but yielding a vector instead of a magnitude of the second order. While these departures from the systems mentioned testify to the great ingenuity and originality of the author, and do not impair the utility of the system as a tool for the use of students of physics, they nevertheless expose the discipline to the criticism of the pure algebraist. Such objection falls to the ground, however, in the case of the theory mentioned, for dyadics yield, for $n = 3$, a linear associative algebra of nine units, namely nonions, the general nonion satisfying an identical equation of the third degree, the Hamilton-Cayley equation.

It is easy to make clear the precise point of view adopted by Professor Gibbs in this matter. This is well expounded in his vice-presidential address on multiple algebra, before the American Association for the Advancement of Science, in 1886, and also in his warm defense of Grassmann's priority rights, as against Hamilton's, in his article in *Nature* "Quaternions and the *Ausdehnungslehre.*" He points out that the key to matricular algebras is to be found in the open (or indeterminate) product (i.e., a product in which no equations subsist between the factors), and, after calling attention to the brief development of this product in Grassmann's work of 1844, affirms that Sylvester's assignment of the date 1858 to the "second birth of Algebra" (this being the year of Cayley's *Memoir on Matrices*) must be changed to 1844. Grassmann, however, ascribes very little importance to the open product, regarding it as offering no useful applications. On the contrary, Professor Gibbs assigns to it the first place in the three kinds of multiplication considered in the *Ausdehnungslehre,* since from it may be derived the algebraic and the combinatorial products, and shows in fact that both of them may be expressed in terms of indeterminate products. Thus the multiplication rejected by Grassmann becomes, from the standpoint of Professor Gibbs, the key to all others. The originality of the latter's treatment of the algebra of dyadics, as contrasted with the methods of other authors in the allied theory of matrices, consists exactly in this, that Professor Gibbs regards a matrix of order n as a multiple quantity in n^2 units, each of which is an indeterminate product of two factors. On the other hand, C. S. Peirce, who was the first to recognize (1870) the quadrate linear associative algebras identical with matrices, uses for the units a *letter pair*, but does not regard this combination as a product. In addition, Professor Gibbs, following the spirit of Grassmann's system, does not confine himself to one kind of multiplication of dyadics, as do Hamilton and Peirce, but considers two sorts, both originating with Grassmann. Thus it may be said that quadrate, or matricular algebras, are brought entirely within the wonderful system expounded by Grassmann in 1844.

As already remarked, the exposition of the theory of dyadics given in the vector analysis is not in accord with Grassmann's system. In a footnote to the address referred to above, Professor Gibbs shows the slight modification necessary for this purpose, while the subject has been treated in detail and in all generality in his lectures on multiple algebra delivered for some years past at Yale University.

Professor Gibbs was much interested in the application of vector analysis to some of the problems of astronomy, and gave examples of such application in a paper, "On the Determination of Elliptic Orbits from Three Complete Observations" (*Mem. Nat. Acad. Sci.*, vol. iv, pt. 2, pp. 79-104). The methods developed in this paper were afterwards applied by Professors W. Beebe and A. W. Phillips * to the computation of the orbit of Swift's comet (1880 V) from three observations, which gave a very critical test of the method. They found that Gibbs's method possessed distinct advantages over those of Gauss and Oppolzer; the convergence of the successive approximations was more rapid and the labor of preparing the fundamental equations for solution much less. These two papers were translated by Buchholz and incorporated in the second edition of Klinkerfues' *Theoretische Astronomie.*

Between the years 1882 and 1889, five papers appeared in *The American Journal of Science* upon certain points in the electromagnetic theory of light and its relations to the various elastic theories. These are remarkable for the entire absence of special hypotheses as to the connection between ether and matter, the only supposition made as to the constitution of matter being that it is fine-grained with reference to the wave-length of light, but not infinitely fine-grained, and that it does disturb in some manner the electrical fluxes in the ether. By methods whose simplicity and directness recall his thermodynamic investigations, the author shows in the first of these articles that, in the case of perfectly transparent media, the theory not only accounts for the dispersion of colors (including the "dispersion of the optic axes" in doubly refracting media), but also leads to Fresnel's laws of double refraction for any particular wave-length without neglect of the small quantities which determine the dispersion of colors. He proceeds in the second paper to show that circular and elliptical polarization are explained by taking into account quantities of a still higher order, and that these in turn do not disturb the explanation of any of the other known phenomena; and in the third paper he deduces, in a very rigorous manner, the general equations of monochromatic light in media of every degree of transparency, arriving

---

* *Astronomical Journal*, vol. ix, pp. 114-117, 121-124, 1889.

at equations somewhat different from those of Maxwell in that they do not contain explicitly the dielectric constant and conductivity as measured electrically, thus avoiding certain difficulties (especially in regard to metallic reflection) which the theory as originally stated had encountered; and it is made clear that "a point of view more in accordance with what we know of the molecular constitution of bodies will give that part of the ordinary theory which is verified by experiment, without including that part which is in opposition to observed facts." Some experiments of Professor C. S. Hastings in 1888 (which showed that the double refraction in Iceland spar conformed to Huyghens's law to a degree of precision far exceeding that of any previous verification) again led Professor Gibbs to take up the subject of optical theories in a paper which shows, in a remarkably simple manner, from elementary considerations, that this result and also the general character of the facts of dispersion are in strict accord with the electrical theory, while no one of the elastic theories which had, at that time, been proposed could be reconciled with these experimental results. A few months later upon the publication of Sir William Thomson's theory of an infinitely compressible ether, it became necessary to supplement the comparison by taking account of this theory also. It is not subject to the insuperable difficulties which beset the other elastic theories, since its equations and surface conditions for perfectly homogeneous and transparent media are identical in form with those of the electrical theory, and lead in an equally direct manner to Fresnel's construction for doubly-refracting media, and to the proper values for the intensities of the reflected and refracted light. But Gibbs shows that, in the case of a fine-grained medium, Thomson's theory does not lead to the known facts of dispersion without unnatural and forced hypotheses, and that in the case of metallic reflection it is subject to similar difficulties; while, on the other hand, "it may be said for the electrical theory that it is not obliged to invent hypotheses, but only to apply the laws furnished by the science of electricity, and that it is difficult to account for the coincidences between the electrical and optical properties of media unless we regard the motions of light as electrical." Of all the arguments (from theoretical grounds alone) for excluding all other theories of light except the electrical, these papers furnish the simplest, most philosophical, and most conclusive with which the present writer is acquainted; and it seems likely that the considerations advanced in them would have sufficed to firmly establish this theory even if the experimental discoveries of Hertz had not supplied a more direct proof of its validity.

In his last work, *Elementary Principles in Statistical Mechanics*,

Professor Gibbs returned to a theme closely connected with the subjects of his earliest publications. In these he had been concerned with the development of the consequences of the laws of thermodynamics which are accepted as given by experience; in this empirical form of the science, heat and mechanical energy are regarded as two distinct entities, mutually convertible of course with certain limitations, but essentially different in many important ways. In accordance with the strong tendency toward unification of causes, there have been many attempts to bring these two things under the same category; to show, in fact, that heat is nothing more than the purely mechanical energy of the minute particles of which all sensible matter is supposed to be made up, and that the extra-dynamical laws of heat are consequences of the immense number of independent mechanical systems in any body,—a number so great that, to human observation, only certain averages and most probable effects are perceptible. Yet in spite of dogmatic assertions, in many elementary books and popular expositions, that "heat is a mode of molecular motion," these attempts have not been entirely successful, and the failure has been signalized by Lord Kelvin as one of the clouds upon the history of science in the nineteenth century. Such investigations must deal with the mechanics of systems of an immense number of degrees of freedom and (since we are quite unable in our experiments to identify or follow individual particles), in order to compare the results of the dynamical reasoning with observation, the processes must be statistical in character. The difficulties of such processes have been pointed out more than once by Maxwell, who, in a passage which Professor Gibbs often quoted, says that serious errors have been made in such inquiries by men whose competency in other branches of mathematics was unquestioned.

On account, then, of the difficulties of the subject and of the profound importance of results which can be reached by no other known method, it is of the utmost consequence that the principles and processes of statistical mechanics should be put upon a firm and certain foundation. That this has now been accomplished there can be no doubt, and there will be little excuse in the future for a repetition of the errors of which Maxwell speaks; moreover, theorems have been discovered and processes devised which will render easier the task of every future student of this subject, as the work of Lagrange did in the case of ordinary mechanics.

The greater part of the book is taken up with this general development of the subject without special reference to the problems of rational thermodynamics. At the end of the twelfth chapter the author has in his hands a far more perfect weapon for attacking such problems than any previous investigator has possessed, and its triumphant use in the last three chapters shows that such purely mechanical systems as he has been considering will exhibit, to human perception, properties in all respects analogous to those which we actually meet with in thermodynamics.   No one can understandingly read the thirteenth chapter without the keenest delight, as one after another of the familar formulæ of thermodynamics appear almost spontaneously, as it seems, from the consideration of purely mechanical systems.   But it is characteristic of the author that he should be more impressed with the limitations and imperfections of his work than with its successes; and he is careful to say (p. 166): "But it should be distinctly stated, that if the results obtained when the numbers of degrees of freedom are enormous coincide sensibly with the general laws of thermodynamics, however interesting and significant this coincidence may be, we are still far from having explained the phenomena of nature with respect to these laws.   For, as compared with the case of nature, the systems which we have considered are of an ideal simplicity.   Although our only assumption is that we are considering conservative systems of a finite number of degrees of freedom, it would seem that this is assuming far too much, so far as the bodies of nature are concerned.   The phenomena of radiant heat, which certainly should not be neglected in any complete system of thermodynamics, and the electrical phenomena associated with the combination of atoms, seem to show that the hypothesis of a finite number of degrees of freedom is inadequate for the explanation of the properties of bodies."   While this is undoubtedly true, it should also be remembered that, in no department of physics have the phenomena of nature been explained with the completeness that is here indicated as desirable.   In the theories of electricity, of light, even in mechanics itself, only certain phenomena are considered which really never occur alone.   In the present state of knowledge, such partial explanations are the best that can be got, and, in addition, the problem of rational thermodynamics has, historically, always been regarded in this way.   In a matter of such difficulty no positive statement should be made, but it is the belief of the present writer that the problem, as it has always been understood, has been successfully solved in this work; and if this belief is correct, one of the great deficiencies in the scientific record of the nineteenth century has been supplied in the first year of the twentieth.

In methods and results, this part of the work is more general than any preceding treatment of the subject; it is in no sense a treatise on the kinetic theory of gases, and the results obtained are not the properties of any one form of matter, but the general equations of thermodynamics which belong to all forms alike.   This corresponds to the generality of the hypothesis in which nothing is assumed as to the mechanical nature of the systems considered, except that they are mechanical and obey Lagrange's or Hamilton's equations. In this respect it may be considered to have done for thermodynamics what Maxwell's treatise did for electromagnetism, and we may say (as Poincaré has said of Maxwell) that Gibbs has not sought to give a mechanical explanation of heat, but has limited his task to demonstrating that such an explanation is possible. And this achievement forms a fitting culmination of his life's work.

The value to science of Professor Gibbs's work has been formally recognized by many learned societies and universities both in this country and abroad. The list of societies and academies of which he was a member or correspondent includes the Connecticut Academy of Arts and Sciences, the National Academy of Sciences, the American Academy of Arts and Sciences, the American Philosophical Society, the Dutch Society of Sciences, Haarlem, the Royal Society of Sciences, Göttingen, the Royal Institution of Great Britain, the Cambridge Philosophical Society, the London Mathematical Society, the Manchester Literary and Philosophical Society, the Royal Academy of Amsterdam, the Royal Society of London, the Royal Prussian Academy of Berlin, the French Institute, the Physical Society of London, and the Bavarian Academy of Sciences. He was the recipient of honorary degrees from Williams College, and from the universities of Erlangen, Princeton, and Christiania. In 1881 he received the Rumford Medal from the American Academy of Boston, and in 1901 the Copley Medal from the Royal Society of London.

Outside of his scientific activities, Professor Gibbs's life was uneventful; he made but one visit to Europe, and with the exception of those three years, and of summer vacations in the mountains, his whole life was spent in New Haven, and all but his earlier years in the same house, which his father had built only a few rods from the school where he prepared for college and from the university in the service of which his life was spent. His constitution was never robust—the consequence apparently of an attack of scarlet fever in early childhood—but with careful attention to health and a regular mode of life his work suffered from this cause no long or serious interruption until the end, which came suddenly after an illness of only a few days. He never married, but made his home with his sister and her family. Of a retiring disposition, he went little into general society and was known to few outside the university; but by those who were honoured by his friendship, and by his students, he was greatly beloved. His modesty with regard to his work was proverbial among all who knew him, and it was entirely real and unaffected. There was never any doubt in his mind, however, as to the accuracy of anything which he published, nor indeed did he underestimate its importance; but he seemed to regard it in an entirely impersonal way and never doubted, apparently, that what he had accomplished could have been done equally well by almost anyone who might have happened to give his attention to the same problems. Those nearest him for many years are constrained to believe that he never realized that he was endowed with most unusual powers of mind; there was never any tendency to make the importance of his work an excuse for neglecting even the most trivial of his duties as an officer of the college, and he was never too busy to devote, at once, as much time and energy as might be necessary to any of his students who privately sought his assistance.

Although long intervals sometimes elapsed between his publications his habits of work were steady and systematic; but he worked alone and, apparently, without need of the stimulus of personal conversation upon the subject, or of criticism from others, which is often helpful even when the critic is intellectually an inferior. So far from publishing partial results, he seldom, if ever, spoke of what he was doing until it was practically in its final and complete form. This was his chief limitation as a teacher of advanced students; he did not take them into his confidence with regard to his current work, and even when he lectured upon a subject in advance of its publication (as was the case for a number of years before the appearance of the *Statistical Mechanics*) the work was really complete except for a few finishing touches. Thus his students were deprived of the advantage of seeing his great structures in process of building, of helping him in the details, and of being in such ways encouraged to make for themselves attempts similar in character, however small their scale. But on the other hand, they owe to him a debt of gratitude for an introduction into the profounder regions of natural philosophy such as they could have obtained from few other living teachers. Always carefully prepared, his lectures were marked by the same great qualities as his published papers and were, in addition, enriched by many apt and simple illustrations which can never be forgotten by those who heard them. No necessary qualification to a statement was ever omitted, and, on the other hand, it seldom failed to receive the most general application of which it was capable; his students had ample opportunity to learn what may be regarded as known, what is guessed at, what a proof is, and how far it goes. Although he disregarded many of the shibboleths of the mathematical rigorists, his logical processes were really of the most severe type; in power of deduction, of generalization, in insight into hidden relations, in critical acumen, utter lack of prejudice, and in the philosophical breadth of his view of the object and aim of physics, he has had few superiors in the history of the science; and no student could come in contact with this serene and impartial mind without feeling profoundly its influence in all his future studies of nature.

In his personal character the same great qualities were apparent. Unassuming in manner, genial and kindly in his intercourse with his fellow-men, never showing impatience or irritation, devoid of personal ambition of the baser sort or of the slightest desire to exalt himself, he went far toward realizing the ideal of the unselfish, Christian gentleman. In the minds of those who knew him, the greatness of his intellectual achievements will never overshadow the beauty and dignity of his life.

H. A. Bumstead.

Bibliography.

1873.   Graphical methods in the thermodynamics of fluids. *Trans. Conn. Acad.*, vol. ii, pp. 309–342.

A method of geometrical representation of the thermodynamic properties of substances by means of surfaces. *Ibid.*, pp. 382–404.

1875–1878.   On the equilibrium of heterogeneous substances. *Ibid.*, vol. iii, pp. 108–248; pp. 343–524. Abstract, *Amer. Jour. Sci.* (3), vol. xvi, pp. 441–458.

(A German translation of the three preceding papers by W. Ostwald has been published under the title, "Thermodynamische Studien," Leipzig, 1892; also a French translation of the first two papers by G. Roy, with an introduction by B. Brunhes, under the title "Diagrammes et surfaces thermodynamiques," Paris, 1903, and of the first part of the *Equilibrium of Heterogeneous Substances* by H. Le Chatelier under the title "Équilibre des Systèmes Chimiques," Paris, 1899.)

1879.   On the fundamental formulae of dynamics. *Amer. Jour. Math.*, vol. ii, pp. 49–64.

On the vapor-densities of peroxide of nitrogen, formic acid, acetic acid, and perchloride of phosphorus. *Amer. Jour. Sci.* (3), vol. xviii, pp. 277–293; pp. 371–387.

1881 and 1884.   Elements of vector analysis arranged for the use of students in physics. New Haven, 8°, pp. 1–36 in 1881, and pp. 37–83 in 1884. (Not published.)

1882–1883.   Notes on the electromagnetic theory of light. I. On double refraction and the dispersion of colors in perfectly transparent media. *Amer. Jour. Sci.* (3), vol. xxiii, pp. 262–275. II. On double refraction in perfectly transparent media which exhibit the phenomena of circular polarization. *Ibid.*, pp. 460–476. III. On the general equations of monochromatic light in media of every degree of transparency. *Ibid.*, vol. xxv, pp. 107–118.

1883.   On an alleged exception to the second law of thermodynamics. *Science*, vol. i, p. 160.

1884.   On the fundamental formula of statistical mechanics, with applications to astronomy and thermodynamics. (Abstract.) *Proc. Amer. Assoc. Adv. Sci.*, vol. xxxiii, pp. 57, 58.

1886.   Notices of Newcomb and Michelson's "Velocity of light in air and refracting media" and of Ketteler's "Theoretische Optik." *Amer. Jour. Sci.* (3), vol. xxxi, pp. 62–67.

On the velocity of light as determined by Foucault's revolving mirror. *Nature*, vol. xxxiii, p. 582.

On multiple algebra. (Vice-president's address before the section of mathematics and astronomy of the American Association for the Advancement of Science.) *Proc. Amer. Assoc. Adv. Sci.*, vol. xxxv, pp. 37–66.

1887 and 1889.   Electro-chemical thermodynamics. (Two letters to the secretary of the electrolysis committee of the British Association.) *Rep. Brit. Assoc. Adv. Sci.* for 1886, pp. 388–389, and for 1888, pp. 343-346.

1888.   A comparison of the elastic and electrical theories of light, with respect to the law of double refraction and the dispersion of colors. *Amer. Jour. Sci.* (3), vol. xxxv, pp. 467–475.

1889.   A comparison of the electric theory of light and Sir William Thomson's theory of a quasi-labile ether. *Amer. Jour. Sci.*, vol. xxxvii, pp. 129–144.

 Reprint, *Phil. Mag.* (5), vol. xxvii, pp. 238–253.

 On the determination of elliptic orbits from three complete observations. *Mem. Nat. Acad. Sci.*, vol. iv, pt. 2, pp. 79–104.

 Rudolf Julius Emanuel Clausius. *Proc. Amer. Acad.*, new series, vol. xvi, pp. 458–465.

1891.   On the rôle of quaternions in the algebra of vectors. *Nature*, vol. xliii, pp. 511–513.

 Quaternions and the *Ausdehnungslehre*. *Nature*, vol. xliv, pp. 79–82.

1893.   Quaternions and the algebra of vectors. *Nature*, vol. xlvii, pp. 463, 464.

1893.   Quaternions and vector analysis. *Nature*, vol. xlviii, pp. 364–367.

1896.   Velocity of propagation of electrostatic force. *Nature*, vol. liii, p. 509.

1897.   Semi-permeable films and osmotic pressure. *Nature*, vol. lv, pp. 461, 462.

 Hubert Anson Newton. *Amer. Jour. Sci.* (4), vol. iii, pp. 359–376.

1898-99.   Fourier's series. *Nature*, vol. lix, pp. 200, 606.

1901.   Vector analysis, a text book for the use of students of mathematics and physics, founded upon the lectures of J. Willard Gibbs, by E. B. Wilson. Pp. xviii + 436. Yale Bicentennial Publications. C. Scribner's Sons.

1902.   Elementary principles in statistical mechanics developed with especial reference to the rational foundation of thermodynamics. Pp. xviii + 207. Yale Bicentennial Publications. C. Scribner's Sons.

1906.   Unpublished fragments of a supplement to the "Equilibrium of Heterogeneous Substances." *Scientific Papers*, vol. i, pp. 418-434.

 On the use of the vector method in the determination of orbits. Letter to Dr. Hugo Buchholz, editor of Klinkerfues' *Theoretische Astronomie*. *Scientific Papers*, vol. ii, pp. 149-154.

THE SCIENTIFIC PAPERS

OF

J. WILLARD GIBBS, Ph.D., LL.D.

I.

GRAPHICAL METHODS IN THE THERMODYNAMICS OF FLUIDS.

[*Transactions of the Connecticut Academy*, II., pp. 309–342, April–May, 1873.]

ALTHOUGH geometrical representations of propositions in the thermo-dynamics of fluids are in general use, and have done good service in disseminating clear notions in this science, yet they have by no means received the extension in respect to variety and generality of which they are capable. So far as regards a general graphical method, which can exhibit at once all the thermodynamic properties of a fluid concerned in reversible processes, and serve alike for the demonstration of general theorems and the numerical solution of particular problems, it is the general if not the universal practice to use diagrams in which the rectilinear co-ordinates represent volume and pressure. The object of this article is to call attention to certain diagrams of different construction, which afford graphical methods co-extensive in their applications with that in ordinary use, and prefer-able to it in many cases in respect of distinctness or of convenience.

Quantities and Relations which are to be represented by the Diagram.

We have to consider the following quantities :—

v, the volume,
p, the pressure,
t, the (absolute) temperature, ⎫ of a given body in any state,
ϵ, the energy,
η, the entropy, ⎭

also W, the work done, ⎫ by the body in passing from one state
and H, the heat received,* ⎭ to another.

---

* Work spent upon the body is as usual to be considered as a negative quantity of work done by the body, and heat given out by the body as a negative quantity of heat received by it.

It is taken for granted that the body has a uniform temperature throughout, and that the pressure (or expansive force) has a uniform value both for all points in the body and for all directions. This, it will be observed, will exclude irreversible processes, but will not entirely exclude solids, although the condition of equal pressure in all directions renders the case very limited, in which they come within the scope of the discussion.

These are subject to the relations expressed by the following differential equations:—

$$dW = ap\,dv, \tag{a}$$

$$d\epsilon = \beta\,dH - dW, \tag{b}$$

$$d\eta = \frac{dH^*}{t}, \tag{c}$$

where a and β are constants depending upon the units by which v, p, W and H are measured. We may suppose our units so chosen that $a = 1$ and $\beta = 1$,[†] and write our equations in the simpler form,

$$d\epsilon = dH - dW, \tag{1}$$

$$dW = p\,dv, \tag{2}$$

$$dH = t\,d\eta. \tag{3}$$

Eliminating dW and dH, we have

$$d\epsilon = t\,d\eta - p\,dv. \tag{4}$$

The quantities v, p, t, ϵ and η are determined when the state of the body is given, and it may be permitted to call them *functions of the state of the body*. The state of a body, in the sense in which the term is used in the thermodynamics of fluids, is capable of two independent variations, so that between the five quantities v, p, t, ϵ and η there exist relations expressible by three finite equations, different in general for different substances, but always such as to be in harmony with the differential equation (4). This equation evidently signifies that if ϵ be expressed as function of v and η, the partial differential co-efficients of this function taken with respect to v and to η will be equal to $-p$ and to t respectively.[‡]

---

* Equation (a) may be derived from simple mechanical considerations. Equations (b) and (c) may be considered as defining the energy and entropy of any state of the body, or more strictly as defining the differentials $d\epsilon$ and $d\eta$. That functions of the state of the body exist, the differentials of which satisfy these equations, may easily be deduced from the first and second laws of thermodynamics. The term *entropy*, it will be observed, is here used in accordance with the original suggestion of Clausius, and not in the sense in which it has been employed by Professor Tait and others after his suggestion. The same quantity has been called by Professor Rankine the *Thermodynamic function*. See Clausius, *Mechanische Wärmetheorie*, Abhnd. ix. § 14; or *Pogg. Ann.*, Bd. cxxv. (1865), p. 390; and Rankine, *Phil. Trans.*, vol. 144, p. 126.

† For example, we may choose as the unit of volume, the cube of the unit of length,—as the unit of pressure the unit of force acting upon the square of the unit of length,—as the unit of work the unit of force acting through the unit of length,—and as the unit of heat the thermal equivalent of the unit of work. The units of length and of force would still be arbitrary as well as the unit of temperature.

‡ An equation giving ϵ in terms of η and v, or more generally any finite equation between ϵ, η and v for a definite quantity of any fluid, may be considered as the fundamental thermodynamic equation of that fluid, as from it by aid of equations (2), (3) and (4) may be derived all the thermodynamic properties of the fluid (so far as reversible processes are concerned), viz.: the fundamental equation with equation (4) gives the three relations existing between v, p, t, ϵ and η, and these relations being known, equations (2) and (3) give the work W and heat H for any change of state of the fluid.

On the other hand W and H are not functions of the state of the body (or functions of any of the quantities v, p, t, ϵ and η), but are determined by the whole series of states through which the body is supposed to pass.

Fundamental Idea and General Properties of the Diagram.

Now if we associate a particular point in a plane with every separate state, of which the body is capable, in any continuous manner, so that states differing infinitely little are associated with points which are infinitely near to each other,* the points associated with states of equal volume will form lines, which may be called *lines of equal volume*, the different lines being distinguished by the numerical value of the volume (as lines of volume 10, 20, 30, etc.). In the same way we may conceive of *lines of equal pressure, of equal temperature, of equal energy, and of equal entropy*. These lines we may also call *isometric, isopiestic, isothermal, isodynamic, isentropic*,† and if necessary use these words as substantives.

Suppose the body to change its state, the points associated with the states through which the body passes will form a line, which we may call the *path* of the body. The conception of a path must include the idea of direction, to express the order in which the body passes through the series of states. With every such change of state there is connected in general a certain amount of work done, W, and of heat received, H, which we may call the *work* and the *heat* of the *path*.‡ The value of these quantities may be calculated from equations (2) and (3),

$$dW = pdv,$$
$$dH = td\eta,$$

i.e.,
$$W = \int pdv, \tag{5}$$
$$H = \int td\eta, \tag{6}$$

---

* The method usually employed in treatises on thermodynamics, in which the rectangular co-ordinates of the point are made proportional to the volume and pressure of the body, is a single example of such an association.

† These lines are usually known by the name given them by Rankine, *adiabatic*. If, however, we follow the suggestion of Clausius and call that quantity *entropy*, which Rankine called the *thermodynamic function*, it seems natural to go one step farther, and call the lines in which this quantity has a constant value *isentropic*.

‡ For the sake of brevity, it will be convenient to use language which attributes to the diagram properties which belong to the associated states of the body. Thus it can give rise to no ambiguity, if we speak of the volume or the temperature of a point in the diagram, or of the work or heat of a line, instead of the volume or temperature of the body in the state associated with the point, or the work done or the heat received by the body in passing through the states associated with the points of the line. In like manner also we may speak of the body moving along a line in the diagram, instead of passing through the series of states represented by the line.

the integration being carried on from the beginning to the end of the path. If the direction of the path is reversed, W and H change their signs, remaining the same in absolute value.

If the changes of state of the body form a cycle, i.e., if the final state is the same as the initial, the path becomes a *circuit*, and the work done and heat received are equal, as may be seen from equation (1), which when integrated for this case becomes $0 = H - W$.

The circuit will enclose a certain area, which we may consider as positive or negative according to the direction of the circuit which circumscribes it. The direction in which areas must be circumscribed in order that their value may be positive, is of course arbitrary. In other words, if x and y are the rectangular co-ordinates, we may define an area either as $\int y\,dx$, or as $\int x\,dy$.

If an area be divided into any number of parts, the work done in the circuit bounding the whole area is equal to the sum of the work done in all the circuits bounding the partial areas. This is evident from the consideration, that the work done in each of the lines which separate the partial areas appears twice and with contrary signs in the sum of the work done in the circuits bounding the partial areas. Also the heat received in the circuit bounding the whole area is equal to the sum of the heat received in all the circuits bounding the partial areas.*

If all the dimensions of a circuit are infinitely small, the ratio of the included area to the work or heat of the circuit is independent of the shape of the circuit and the direction in which it is described, and varies only with its position in the diagram. That this ratio is independent of the direction in which the circuit is described, is evident from the consideration that a reversal of this direction simply changes the sign of both terms of the ratio. To prove that the ratio is independent of the shape of the circuit, let us suppose the area ABCDE (fig. 1) divided up by an infinite number of isometrics v_1v_1, v_2v_2, etc., with equal differences of volume dv, and an infinite number of isopiestics p_1p_1, p_2p_2, etc., with equal differences of pressure dp. Now from the

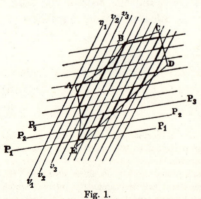

Fig. 1.

---

principle of continuity, as the whole figure is infinitely small, the ratio of the area of one of the small quadrilaterals into which the figure is divided to the work done in passing around it is approximately the same for all the different quadrilaterals. Therefore the area of the figure composed of all the complete quadrilaterals which fall within the given circuit has to the work done in circumscribing this figure the same ratio, which we will call γ. But the area of this figure is approximately the same as that of the given circuit, and the work done in describing this figure is approximately the same as that done in describing the given circuit (eq. 5). Therefore the area of the given circuit has to the work done or heat received in that circuit this ratio γ, which is independent of the shape of the circuit.

Now if we imagine the systems of equidifferent isometrics and isopiestics, which have just been spoken of, extended over the whole diagram, the work done in circumscribing one of the small quadrilaterals, so that the increase of pressure directly precedes the increase of volume, will have in every part of the diagram a constant value, viz., the product of the differences of volume and pressure $(dv \times dp)$, as may easily be proved by applying equation (2) successively to its four sides. But the area of one of these quadrilaterals, which we could consider as constant within the limits of the infinitely small circuit, may vary for different parts of the diagram, and will indicate proportionally the value of γ, which is equal to the area divided by $dv \times dp$.

In like manner, if we imagine systems of isentropics and isothermals drawn throughout the diagram for equal differences $d\eta$ and dt, the heat received in passing around one of the small quadrilaterals, so that the increase of t shall directly precede that of η, will be the constant product $d\eta \times dt$, as may be proved by equation (3), and the value of γ, which is equal to the area divided by the heat, will be indicated proportionally by the areas.*

---

* The indication of the value of γ by systems of equidifferent isometrics and isopiestics, or isentropics and isothermals, is explained above, because it seems in accordance with the spirit of the graphical method, and because it avoids the extraneous consideration of the co-ordinates. If, however, it is desired to have analytical expressions for the value of γ based upon the relations between the co-ordinates of the point and the state of the body, it is easy to deduce such expressions as the following, in which x and y are the rectangular co-ordinates, and it is supposed that the sign of an area is determined in accordance with the equation $A = \int y\,dx$:—

$$\frac{1}{\gamma} = \frac{dv}{dx} \cdot \frac{dp}{dy} - \frac{dp}{dx} \cdot \frac{dv}{dy} = \frac{d\eta}{dx} \cdot \frac{dt}{dy} - \frac{dt}{dx} \cdot \frac{d\eta}{dy},$$

where x and y are regarded as the independent variables ;—or

$$\gamma = \frac{dx}{dv} \cdot \frac{dy}{dp} - \frac{dy}{dv} \cdot \frac{dx}{dp},$$

This quantity γ, which is the ratio of the area of an infinitely small circuit to the work done or heat received in that circuit, and which we may call the scale on which work and heat are represented by areas, or more briefly, the *scale of work and heat,* may have a constant value throughout the diagram or it may have a varying value. The diagram in ordinary use affords an example of the first case, as the area of a circuit is everywhere proportional to the work or heat. There are other diagrams which have the same property, and we may call all such *diagrams of constant scale.*

In any case we may consider the scale of work and heat as known for every point of the diagram, so far as we are able to draw the isometrics and isopiestics or the isentropics and isothermals. If we write δW and δH for the work and heat of an infinitesimal circuit, and δA for the area included, the relations of these quantities are thus expressed :—*

$$\delta W = \delta H = \frac{1}{\gamma}\, \delta A. \tag{7}$$

We may find the value of W and H for a circuit of finite dimensions by supposing the included area A divided into areas δA infinitely small in all directions, for which therefore the above equation will hold, and taking the sum of the values of δH or δW for the various areas δA. Writing W^c and H^c for the work and heat of the circuit C, and Σ^c for a summation or integration performed within the limits of this circuit, we have

where v and p are the independent variables ;—or

$$\gamma = \frac{dx}{d\eta} \cdot \frac{dy}{dt} - \frac{dy}{d\eta} \cdot \frac{dx}{dt},$$

where η and t are the independent variables ;—or

$$\frac{1}{\gamma} = \frac{-\dfrac{d^2\epsilon}{dv\, d\eta}}{\dfrac{dx}{dv} \cdot \dfrac{dy}{d\eta} - \dfrac{dy}{dv} \cdot \dfrac{dx}{d\eta}},$$

where v and η are the independent variables.

These and similar expressions for $\frac{1}{\gamma}$ may be found by dividing the value of the work or heat for an infinitely small circuit by the area included. This operation can be most conveniently performed upon a circuit consisting of four lines, in each of which one of the independent variables is constant. E.g., the last formula can be most easily found from an infinitely small circuit formed of two isometrics and two isentropics.

* To avoid confusion, as dW and dH are generally used and are used elsewhere in this article to denote the work and heat of an infinite short path, a slightly different notation, δW and δH, is here used to denote the work and heat of an infinitely small circuit. So δA is used to denote an element of area which is infinitely small in all directions, as the letter d would only imply that the element was infinitely small in one direction. So also below, the integration or summation which extends to all the elements written with δ is denoted by the character Σ, as the character $\int$ naturally refers to elements written with d.

$$W^c = H^c = \Sigma^c \frac{1}{\gamma} \delta A \qquad (8)$$

We have thus an expression for the value of the work and heat of a circuit involving an integration extending over an area instead of one extending over a line, as in equations (5) and (6).

Similar expressions may be found for the work and the heat of a path which is not a circuit. For this case may be reduced to the preceding by the consideration that $W=0$ for a path on an isometric or on the line of no pressure (eq. 2), and $H=0$ for a path on an isentropic or on the line of absolute cold. Hence the work of any path S is equal to that of the circuit formed of S, the isometric of the final state, the line of no pressure and the isometric of the initial state, which circuit may be represented by the notation $[S, v'', p^0, v']$. And the heat of the same path is the same as that of the circuit $[S, \eta'', t^0, \eta']$. Therefore using W^S and H^S to denote the work and heat of any path S, we have

$$W^S = \Sigma^{[S,\, v'',\, p^0,\, v']} \frac{1}{\gamma} \delta A, \qquad (9)$$

$$H^S = \Sigma^{[S,\, \eta'',\, t^0,\, \eta']} \frac{1}{\gamma} \delta A, \qquad (10)$$

where as before the limits of the integration are denoted by the expression occupying the place of an index to the sign Σ.* These equations evidently include equation (8) as a particular case.

It is easy to form a material conception of these relations. If we imagine, for example, mass inherent in the plane of the diagram with a varying (superficial) density represented by $\frac{1}{\gamma}$, then $\Sigma \frac{1}{\gamma} \delta A$ will

---

* A word should be said in regard to the sense in which the above propositions should be understood. If beyond the limits, within which the relations of v, p, t, ϵ and η are known and which we may call the limits of the known field, we continue the isometrics, isopiestics, &c., in any way we please, only subject to the condition that the relations of v, p, t, ϵ and η shall be consistent with the equation $d\epsilon = t\,d\eta - p\,dv$, then in calculating the values of quantities W and H determined by the equations $dW = p\,dv$ and $dH = t\,d\eta$ for paths or circuits in any part of the diagram thus extended, we may use any of the propositions or processes given above, as these three equations have formed the only basis of the reasoning. We will thus obtain values of W and H, which will be identical with those which would be obtained by the immediate application of the equations $dW = p\,dv$ and $dH = t\,d\eta$ to the path in question, and which in the case of any path which is entirely contained in the known field will be the true values of the work and heat for the change of state of the body which the path represents. We may thus use lines outside of the known field without attributing to them any physical signification whatever, without considering the points in the lines as representing any states of the body. If however, to fix our ideas, we choose to conceive of this part of the diagram as having the same physical interpretation as the known field, and to enunciate our propositions in language based upon such a conception, the unreality or even the impossibility of the states represented by the lines outside of the known field cannot lead to any incorrect results in regard to paths in the known field.

evidently denote the mass of the part of the plane included within the limits of integration, this mass being taken positively or negatively according to the direction of the circuit.

Thus far we have made no supposition in regard to the nature of the law, by which we associate the points of a plane with the states of the body, except a certain condition of continuity. Whatever law we may adopt, we obtain a method of representation of the thermodynamic properties of the body, in which the relations existing between the functions of the state of the body are indicated by a net-work of lines, while the work done and the heat received by the body when it changes its state are represented by integrals extending over the elements of a line, and also by an integral extending over the elements of certain areas in the diagram, or, if we choose to introduce such a consideration, by the mass belonging to these areas.

The different diagrams which we obtain by different laws of association are all such as may be obtained from one another by a process of *deformation*, and this consideration is sufficient to demonstrate their properties from the well-known properties of the diagram in which the volume and pressure are represented by rectangular co-ordinates. For the relations indicated by the net-work of isometrics, isopiestics etc., are evidently not altered by deformation of the surface upon which they are drawn, and if we conceive of mass as belonging to the surface, the mass included within given lines will also not be affected by the process of deformation. If, then, the surface upon which the ordinary diagram is drawn has the uniform superficial density 1, so that the work and heat of a circuit, which are represented in this diagram by the included area, shall also be represented by the mass included, this latter relation will hold for any diagram formed from this by deformation of the surface on which it is drawn.

The choice of the method of representation is of course to be determined by considerations of simplicity and convenience, especially in regard to the drawing of the lines of equal volume, pressure, temperature, energy and entropy, and the estimation of work and heat. There is an obvious advantage in the use of diagrams of constant scale, in which the work and heat are represented simply by areas. Such diagrams may of course be produced by an infinity of different methods, as there is no limit to the ways of deforming a plane figure without altering the magnitude of its elements. Among these methods, two are especially important,—the ordinary method in which the volume and pressure are represented by rectilinear co-ordinates, and that in which the entropy and temperature are so represented. A diagram formed by the former method may be called, for the sake of distinction, a *volume-pressure* diagram,—one formed by the latter, an *entropy-temperature* diagram. That the latter as well as the former satisfies the condition that $\gamma = 1$ throughout the whole diagram, may be seen by reference to page 5.

The Entropy-temperature Diagram compared with that in ordinary use.

Considerations independent of the nature of the body in question.

As the general equations (1), (2), (3) are not altered by interchanging v, $-p$ and $-W$ with η, t and H respectively, it is evident that, so far as these equations are concerned, there is nothing to choose between a volume-pressure and an entropy-temperature diagram. In the former, the work is represented by an area bounded by the path which represents the change of state of the body, two ordinates and the axis of abscissas. The same is true of the heat received in the latter diagram. Again, in the former diagram, the heat received is represented by an area bounded by the path and certain lines, the character of which depends upon the nature of the body under consideration. Except in the case of an ideal body, the properties of which are determined by assumption, these lines are more or less unknown in a part of their course, and in any case the area will generally extend to an infinite distance. Very much the same inconveniences attach themselves to the areas representing work in the entropy-temperature diagram.* There is, however, a consideration of a

---

* In neither diagram do these circumstances create any serious difficulty in the estimation of areas representing work or heat. It is always possible to divide these areas into two parts, of which one is of finite dimensions, and the other can be calculated in the simplest manner. Thus in the entropy-temperature diagram the work done in a path AB (fig. 2) is represented by the area included by the path AB, the isometric BC, the line of no pressure and the isometric DA. The line of no pressure and the adjacent parts of the isometrics in the case of an actual gas or vapor are more or less undetermined in the present state of our knowledge, and are likely to remain so; for an ideal gas the line of no pressure coincides with the axis of abscissas, and is an asymptote to the isometrics. But, be this as it may, it is not necessary to examine the form of the remoter parts of the diagram.

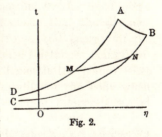

Fig. 2.

If we draw an isopiestic MN, cutting AD and BC, the area MNCD, which represents the work done in MN, will be equal to $p(v'' - v')$, where p denotes the pressure in MN, and v'' and v' denote the volumes at B and A respectively (eq. 5). Hence the work done in AB will be represented by $ABNM + p(v'' - v')$. In the volume-pressure diagram, the areas representing heat may be divided by an isothermal, and treated in a manner entirely analogous.

Or we may make use of the principle that, for a path which begins and ends on the same isodynamic, the work and heat are equal, as appears by integration of equation (1). Hence, in the entropy-temperature diagram, to find the work of any path, we may extend it by an isometric (which will not alter its work), so that it shall begin and end

general character, which shows an important advantage on the side of the entropy-temperature diagram. In thermodynamic problems, heat received at one temperature is by no means the equivalent of the same amount of heat received at another temperature. For example, a supply of a million calories at 150^c is a very different thing from a supply of a million calories at 50^c. But no such distinction exists in regard to work. This is a result of the general law, that heat can only pass from a hotter to a colder body, while work can be transferred by mechanical means from one fluid to any other, whatever may be the pressures. Hence, in thermodynamic problems, it is generally necessary to distinguish between the quantities of heat received or given out by the body at different temperatures, while as far as work is concerned, it is generally sufficient to ascertain the total amount performed. If, then, several heat-areas and one work-area enter into the problem, it is evidently more important that the former should be simple in form, than that the latter should be so. Moreover, in the very common case of a circuit, the work-area is bounded entirely by the path, and the form of the isometrics and the line of no pressure are of no especial consequence.

It is worthy of notice that the simplest form of a perfect thermodynamic engine, so often described in treatises on thermodynamics, is

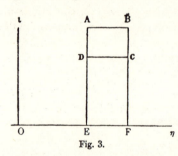

Fig. 3.

represented in the entropy-temperature diagram by a figure of extreme simplicity, viz: a rectangle of which the sides are parallel to the co-ordinate axes. Thus in figure 3, the circuit ABCD may represent the series of states through which the fluid is made to pass in such an engine, the included area representing the work done, while the area ABFE represents the heat received from the heater at the highest temperature AE, and the area CDEF represents the heat transmitted to the cooler at the lowest temperature DE.

There is another form of the perfect thermodynamic engine, viz: one with a perfect regenerator as defined by Rankine, *Phil. Trans.* vol. 144, p. 140, the representation of which becomes peculiarly simple in the entropy-temperature diagram. The circuit consists of two equal straight lines AB and CD (fig. 4) parallel to the axis of abscissas, and two precisely similar curves of any form BC and AD.

---

on the same isodynamic, and then take the heat (instead of the work) of the path thus extended. This method was suggested by that employed by Cazin, *Théorie élémentaire des machines à air chaud*, p. 11, and Zeuner, *Mechanische Wärmetheorie*, p. 80, in the reverse case, viz: to find the heat of a path in the volume-pressure diagram.

The included area ABCD represents the work done, and the areas ABba and CDdc represent respectively the heat received from the heater and that transmitted to the cooler. The heat imparted by the fluid to the regenerator in passing from B to C, and afterward restored to the fluid in its passage from D to A, is represented by the areas BCcb and DAad.

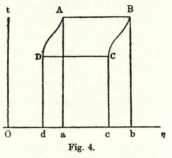

Fig. 4.

It is often a matter of the first importance in the study of any thermo-dynamic engine, to compare it with a perfect engine. Such a comparison will obviously be much facilitated by the use of a method in which the perfect engine is represented by such simple forms.

The method in which the co-ordinates represent volume and pressure has a certain advantage in the simple and elementary character of the notions upon which it is based, and its analogy with Watt's indicator has doubtless contributed to render it popular. On the other hand, a method involving the notion of *entropy*, the very existence of which depends upon the second law of thermodynamics, will doubtless seem to many far-fetched, and may repel beginners as obscure and difficult of comprehension. This inconvenience is perhaps more than counter-balanced by the advantages of a method which makes the second law of thermodynamics so prominent, and gives it so clear and elementary an expression. The fact, that the different states of a fluid can be represented by the positions of a point in a plane, so that the ordi-nates shall represent the temperatures, and the heat received or given out by the fluid shall be represented by the area bounded by the line representing the states through which the body passes, the ordinates drawn through the extreme points of this line, and the axis of abscissas,—this fact, clumsy as its expression in words may be, is one which presents a clear image to the eye, and which the mind can readily grasp and retain. It is, however, nothing more nor less than a geometrical expression of the second law of thermodynamics in its application to fluids, in a form exceedingly convenient for use, and from which the analytical expression of the same law can, if desired, be at once obtained. If, then, it is more important for purposes of instruction and the like to familiarize the learner with the second law, than to defer its statement as long as possible, the use of the entropy-temperature diagram may serve a useful purpose in the popularizing of this science.

The foregoing considerations are in the main of a general character, and independent of the nature of the substance to which the graphical method is applied. On this, however, depend the forms of the isometrics, isopiestics and isodynamics in the entropy-temperature diagram, and of the isentropics, isothermals and isodynamics in the volume-pressure diagram. As the convenience of a method depends largely upon the ease with which these lines can be drawn, and upon the peculiarities of the fluid which has its properties represented in the diagram, it is desirable to compare the methods under consideration in some of their most important applications. We will commence with the case of a perfect gas.

Case of a perfect gas.

A perfect or ideal gas may be defined as such a gas, that for any constant quantity of it the product of the volume and the pressure varies as the temperature, and the energy varies as the temperature, i.e.,

$$pv = at, \tag{A}*$$

$$\epsilon = ct. \tag{B}$$

The significance of the constant a is sufficiently indicated by equation (A). The significance of c may be rendered more evident by differentiating equation (B) and comparing the result

$$d\epsilon = cdt$$

with the general equations (1) and (2), viz:

$$d\epsilon = dH - dW, \quad dW = pdv.$$

If $dv = 0$, $dW = 0$, and $dH = cdt$, i.e.,

$$\left(\frac{dH}{dt}\right)_v = c, \dagger \tag{C}$$

i.e., c is the quantity of heat necessary to raise the temperature of the body one degree under the condition of constant volume. It will be observed, that when different quantities of the same gas are considered, a and c both vary as the quantity, and $c \div a$ is constant; also, that the value of $c \div a$ for different gases varies as their specific heat determined for equal volumes and for constant volume.

With the aid of equations (A) and (B) we may eliminate p and t from the general equation (4), viz:

$$d\epsilon = td\eta - pdv,$$

---

* In this article, all equations which are designated by arabic numerals subsist for any body whatever (subject to the condition of uniform pressure and temperature), and those which are designated by small capitals subsist for any quantity of a perfect gas as defined above (subject of course to the same conditions).

† A subscript letter after a differential co-efficient is used in this article to indicate the quantity which is made constant in the differentiation.

which is then reduced to $$\frac{d\epsilon}{\epsilon}=\frac{1}{c}d\eta-\frac{a}{c}\frac{dv}{v},$$

and by integration to $$\log\epsilon=\frac{\eta}{c}-\frac{a}{c}\log v.* \tag{D}$$

The constant of integration becomes 0, if we call the entropy 0 for the state of which the volume and energy are both unity.

Any other equations which subsist between v, p, t, ϵ and η may be derived from the three independent equations (A), (B) and (D). If we eliminate ϵ from (B) and (D), we have

$$\eta=a\log v+c\log t+c\log c. \tag{E}$$

Eliminating v from (A) and (E), we have

$$\eta=(a+c)\log t-a\log p+c\log c+a\log a. \tag{F}$$

Eliminating t from (A) and (E), we have

$$\eta=(a+c)\log v+c\log p+c\log\frac{c}{a}. \tag{G}$$

If v is constant, equation (E) becomes

$$\eta=c\log t+\text{Const.},$$

i.e., the isometrics in the entropy-temperature diagram are logarithmic curves identical with one another in form,—a change in the value of v having only the effect of moving the curve parallel to the axis of η. If p is constant, equation (F) becomes

$$\eta=(a+c)\log t+\text{Const.},$$

so that the isopiestics in this diagram have similar properties. This identity in form diminishes greatly the labour of drawing any considerable number of these curves. For if a card or thin board be cut in the form of one of them, it may be used as a pattern or ruler to draw all of the same system.

The isodynamics are straight in this diagram (eq. B).

To find the form of the isothermals and isentropics in the volume-pressure diagram, we may make t and η constant in equations (A) and (G) respectively, which will then reduce to the well-known equations of these curves:—

$$pv=\text{Const.},$$
and $$p^{c}v^{a+c}=\text{Const.}$$

---

* If we use the letter e to denote the base of the Naperian system of logarithms, equation (D) may also be written in the form

$$\epsilon=e^{\frac{\eta}{c}}.v^{-\frac{a}{c}}.$$

This may be regarded as the fundamental thermodynamic equation of an ideal gas. See the last note on page 2. It will be observed, that there would be no real loss of generality if we should choose, as the body to which the letters refer, such a quantity of the gas that one of the constants a and c should be equal to unity.

The equation of the isodynamics is of course the same as that of the isothermals. None of these systems of lines have that property of identity of form, which makes the systems of isometrics and isopiestics so easy to draw in the entropy-temperature diagram.

Case of condensable vapors.

The case of bodies which pass from the liquid to the gaseous condition is next to be considered. It is usual to assume of such a body, that when sufficiently superheated it approaches the condition of a perfect gas. If, then, in the entropy-temperature diagram of such a body we draw systems of isometrics, isopiestics and isodynamics, as if for a perfect gas, for proper values of the constants a and c, these will be asymptotes to the true isometrics, etc., of the vapor, and in many cases will not vary from them greatly in the part of the diagram which represents vapor unmixed with liquid, except in the vicinity of the line of saturation. In the volume-pressure diagram of the same body, the isothermals, isentropics and isodynamics, drawn for a perfect gas for the same values of a and c, will have the same relations to the true isothermals, etc.

In that part of any diagram which represents a mixture of vapor and liquid, the isopiestics and isothermals will be identical, as the pressure is determined by the temperature alone. In both the diagrams which we are now comparing, they will be straight and parallel to the axis of abscissas. The form of the isometrics and isodynamics in the entropy-temperature diagram, or that of the isentropics and isodynamics in the volume-pressure diagram, will depend upon the nature of the fluid, and probably cannot be expressed by any simple equations. The following property, however, renders it easy to construct equidifferent systems of these lines, viz: any such system will divide any isothermal (isopiestic) into equal segments.

It remains to consider that part of the diagram which represents the body when entirely in the condition of liquid. The fundamental characteristic of this condition of matter is that the volume is very nearly constant, so that variations of volume are generally entirely inappreciable when represented graphically on the same scale on which the volume of the body in the state of vapor is represented, and both the variations of volume and the connected variations of the connected quantities may be, and generally are, neglected by the side of the variations of the same quantities which occur when the body passes to the state of vapor.

Let us make, then, the usual assumption that v is constant, and see how the general equations (1), (2), (3) and (4) are thereby affected.

We have first,
$$dv = 0,$$
then
$$dW = 0,$$
and
$$d\epsilon = t\,d\eta.$$
If we add
$$dH = t\,d\eta,$$

these four equations will evidently be equivalent to the three independent equations (1), (2) and (3), combined with the assumption which we have just made. For a liquid, then, ϵ, instead of being a function of two quantities v and η, is a function of η alone,—t is also a function of η alone, being equal to the differential co-efficient of the function ϵ; that is, the value of one of the three quantities t, ϵ and η, is sufficient to determine the other two. The value of v, moreover, is fixed without reference to the values of t, ϵ and η (so long as these do not pass the limits of values possible for liquidity); while p does not enter into the equations, i.e., p may have any value (within certain limits) without affecting the values of t, ϵ, η or v. If the body change its state, continuing always liquid, the value of W for such a change is 0, and that of H is determined by the values of any one of the three quantities t, ϵ and η. It is, therefore, the relations between t, ϵ, η and H, for which a graphical expression is to be sought; a method, therefore, in which the co-ordinates of the diagram are made equal to the volume and pressure, is totally inapplicable to this particular case; v and p are indeed the only two of the five functions of the state of the body, v, p, t, ϵ and η, which have no relations either to each other, or to the other three, or to the quantities W and H, to be expressed.* The values of v and p do not really determine the state of an incompressible fluid,—the values of t, ϵ and η are still left undetermined, so that through every point in the volume-pressure diagram which represents the liquid there must pass (in general) an infinite number of isothermals, isodynamics and isentropics. The character of this part of the diagram is as follows:—the states of liquidity are represented by the points of a line parallel to the axis of pressures, and the isothermals, isodynamics and isentropics, which cross the field of partial vaporization and meet this line, turn upward and follow its course.†

In the entropy-temperature diagram the relations of t, ϵ and η are

---

* That is, v and p have no such relations to the other quantities, as are expressible by equations; p, however, cannot be *less* than a certain function of t.

† All these difficulties are of course removed when the differences of volume of the liquid at different temperatures are rendered appreciable on the volume-pressure diagram. This can be done in various ways,—among others, by choosing as the body to which v, etc., refer, a sufficiently large quantity of the fluid. But, however we do it, we must evidently give up the possibility of representing the body in the state of vapor in the same diagram without making its dimensions enormous.

distinctly visible. The line of liquidity is a curve AB (fig. 5) determined by the relation between t and η. This curve is also an isometric. Every point of it has a definite volume, temperature, entropy and energy. The latter is indicated by the isodynamics E_1E_1, E_2E_2, etc., which cross the region of partial vaporization and terminate in the line of liquidity. (They do not in this diagram turn and follow the line.) If the body pass from one state to another, remaining liquid, as from M to N in the figure, the heat received is represented as usual by the area MNnm. That the work done is nothing, is indicated by the fact that the line AB is an isometric. Only the isopiestics in this diagram are superposed in the line of fluidity, turning downward where they meet this line and following its course, so that for any point in this line the pressure is undetermined. This is, however, no inconvenience in the diagram, as it simply expresses the fact of the case, that when all the quantities v, t, ϵ and η are fixed, the pressure is still undetermined.

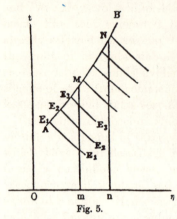

Fig. 5.

Diagrams in which the Isometrics, Isopiestics, Isothermals, Isodynamics and Isentropics of a Perfect Gas are all Straight Lines.

There are many cases in which it is of more importance that it should be easy to draw the lines of equal volume, pressure, temperature, energy and entropy, than that work and heat should be represented in the simplest manner. In such cases it may be expedient to give up the condition that the scale (γ) of work and heat shall be constant, when by that means it is possible to gain greater simplicity in the form of the lines just mentioned.

In the case of a perfect gas, the three relations between the quantities v, p, t, ϵ and η are given on pages 12, 13, equations (A), (B) and (D). These equations may be easily transformed into the three

$$\log p + \log v - \log t = \log a, \qquad \text{(H)}$$

$$\log \epsilon - \log t = \log c, \qquad \text{(I)}$$

$$\eta - c \log \epsilon - a \log v = 0; \qquad \text{(J)}$$

so that the three relations between the quantities $\log v$, $\log p$, $\log t$, $\log \epsilon$ and η are expressed by linear equations, and it will be possible to make the five systems of lines all rectilinear in the same diagram, the distances of the isometrics being proportional to the differences of the logarithms of the volumes, the distances of the isopiestics being proportional to the differences of the logarithms of the pressures, and so with the isothermals and the isodynamics,—the distances of the isentropics, however, being proportional to the differences of entropy simply.

The scale of work and heat in such a diagram will vary inversely as the temperature. For if we imagine systems of isentropics and isothermals drawn throughout the diagram for equal small differences of entropy and temperature, the isentropics will be equidistant, but the distances of the isothermals will vary inversely as the temperature, and the small quadrilaterals into which the diagram is divided will vary in the same ratio: $\therefore \ \gamma \backsim 1 \div t$. (See p. 5.)

So far, however, the form of the diagram has not been completely defined. This may be done in various ways: e.g., if x and y be the rectangular co-ordinates, we may make

$$\begin{cases} x = \log v, \\ y = \log p; \end{cases} \quad \text{or} \quad \begin{cases} x = \eta, \\ y = \log t; \end{cases} \quad \text{or} \quad \begin{cases} x = \log v, \\ y = \eta; \end{cases} \text{etc.}$$

Or we may set the condition that the logarithms of volume, of pressure and of temperature, shall be represented in the diagram on the same scale. (The logarithms of energy are necessarily represented on the same scale as those of temperature.) This will require that the isometrics, isopiestics and isothermals cut one another at angles of 60°.

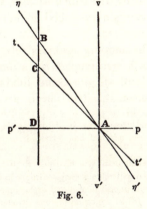

Fig. 6.

The general character of all these diagrams, which may be derived from one another by projection by parallel lines, may be illustrated by the case in which $x = \log v$, and $y = \log p$.

Through any point A (fig. 6) of such a diagram let there be drawn the isometric vv′, the isopiestic pp′, the isothermal tt′ and the isentropic $\eta\eta'$. The lines pp′ and vv′ are of course parallel to the axes. Also by equation (H)

$$\tan tAp = \left(\frac{dy}{dx}\right)_t = \left(\frac{d \log p}{d \log v}\right)_t = -1,$$

and by (G) $\qquad \tan \eta Ap = \left(\frac{dy}{dx}\right)_\eta = \left(\frac{d \log p}{d \log v}\right)_\eta = -\frac{c+a}{c}.$

Therefore, if we draw another isometric, cutting $\eta\eta'$, tt′, and pp′ in B, C and D,

$$\frac{BD}{CD} = \frac{c+a}{c}, \quad \frac{BC}{CD} = \frac{a}{c}, \quad \frac{CD}{BC} = \frac{c}{a}.$$

Hence, in the diagrams of different gases, CD÷BC will be proportional to the specific heat determined for equal volumes and for constant volume.

As the specific heat, thus determined, has probably the same value for most simple gases, the isentropics will have the same inclination in diagrams of this kind for most simple gases. This inclination may easily be found by a method which is independent of any units of measurement, for

$$\text{BD} : \text{CD} :: \left(\frac{d \log p}{d \log v}\right)_{\eta} : \left(\frac{d \log p}{d \log v}\right)_{t} :: \left(\frac{dp}{dv}\right)_{\eta} : \left(\frac{dp}{dv}\right)_{t},$$

i.e., BD÷CD is equal to the quotient of the co-efficient of elasticity under the condition of no transmission of heat, divided by the co-efficient of elasticity at constant temperature. This quotient for a simple gas is generally given as 1·408 or 1·421. As

$$\text{CA} \div \text{CD} = \sqrt{2} = 1\cdot414,$$

BD is very nearly equal to CA (for simple gases), which relation it may be convenient to use in the construction of the diagram.

In regard to compound gases the rule seems to be, that the specific heat (determined for equal volumes and for constant volume) is to the specific heat of a simple gas inversely as the volume of the compound is to the volume of its constituents (in the condition of gas); that is, the value of BC÷CD for a compound gas is to the value of BC÷CD for a simple gas, as the volume of the compound is to the volume of its constituents. Therefore, if we compare the diagrams (formed by this method) for a simple and a compound gas, the distance DA and therefore CD being the same in each, BC in the diagram of the compound gas will be to BC in the diagram of the simple gas as the volume of the compound is to the volume of its constituents.

Although the inclination of the isentropics is independent of the quantity of gas under consideration, the rate of increase of η will vary with this quantity. In regard to the rate of increase of t, it is evident that if the whole diagram be divided into squares by isopiestics and isometrics drawn at equal distances, and isothermals be drawn as diagonals to these squares, the volumes of the isometrics, the pressures of the isopiestics and the temperatures of the isothermals will each form a geometrical series, and in all these series the ratio of two contiguous terms will be the same.

The properties of the diagrams obtained by the other methods mentioned on page 17 do not differ essentially from those just described. For example, in any such diagram, if through any point we draw an isentropic, an isothermal and an isopiestic, which cut any isometric not passing through the same point, the ratio of the segments of the isometric will have the value which has been found for BC : CD.

In treating the case of vapors also, it may be convenient to use diagrams in which $x = \log v$ and $y = \log p$, or in which $x = \eta$ and $y = \log t$; but the diagrams formed by these methods will evidently be radically different from one another. It is to be observed that each of these methods is what may be called a *method of definite scale* for work and heat; that is, the value of γ in any part of the diagram is independent of the properties of the fluid considered. In the first method $\gamma = \dfrac{1}{e^{x+y}}$, in the second $\gamma = \dfrac{1}{e^y}$. In this respect these methods have an advantage over many others. For example, if we should make $x = \log v$, $y = \eta$, the value of γ in any part of the diagram would depend upon the properties of the fluid, and would probably not vary in any case, except that of a perfect gas, according to any simple law.

The conveniences of the entropy-temperature method will be found to belong in nearly the same degree to the method in which the co-ordinates are equal to the entropy and the logarithm of the temperature. No serious difficulty attaches to the estimation of heat and work in a diagram formed on the latter method on account of the variation of the scale on which they are represented, as this variation follows so simple a law. It may often be of use to remember that such a diagram may be reduced to an entropy-temperature diagram by a vertical compression or extension, such that the distances of the isothermals shall be made proportional to their differences of temperature. Thus if we wish to estimate the work or heat of the circuit ABCD (fig. 7), we may draw a number of equidistant ordinates (isentropics) as if to estimate the included area, and for each of the ordinates take the differences of temperature of the points where it cuts the circuit; these differences of temperature will

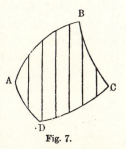

Fig. 7.

be equal to the lengths of the segments made by the corresponding circuit in the entropy-temperature diagram upon a corresponding system of equidistant ordinates, and may be used to calculate the area of the circuit in the entropy-temperature diagram, i.e., to find the work or heat required. We may find the work of any path by applying the same process to the circuit formed by the path, the isometric of the final state, the line of no pressure (or any isopiestic; see note on page 9), and the isometric of the initial state. And we may find the heat of any path by applying the same process to a circuit formed by the path, the ordinates of the extreme points and the line of absolute cold. That this line is at an infinite distance occasions no difficulty. The lengths of the ordinates in the entropy-temperature diagram which we desire are given by the temperature of points in the path determined (in either diagram) by equidistant ordinates.

The properties of the part of the entropy-temperature diagram representing a mixture of vapor and liquid, which are given on page 14, will evidently not be altered if the ordinates are made proportional to the logarithms of the temperatures instead of the temperatures simply.

The representation of specific heat in the diagram under discussion is peculiarly simple. The specific heat of any substance at constant volume or under constant pressure may be defined as the value of

$$\left(\frac{dH}{dt}\right)_v \text{ or } \left(\frac{dH}{dt}\right)_p, \text{ i.e., } \left(\frac{d\eta}{d\log t}\right)_v \text{ or } \left(\frac{d\eta}{d\log t}\right)_p,$$

for a certain quantity of the substance. Therefore, if we draw a diagram, in which $x = \eta$ and $y = \log t$, for that quantity of the substance which is used for the determination of the specific heat, the tangents of the angles made by the isometrics and the isopiestics with the ordinates in the diagram will be equal to the specific heat of the substance determined for constant volume and for constant pressure respectively. Sometimes, instead of the condition of constant volume or constant pressure, some other condition is used in the determination of specific heat. In all cases, the condition will be represented by a line in the diagram, and the tangent of the angle made by this line with an ordinate will be equal to the specific heat as thus defined. If the diagram be drawn for any other quantity of the substance, the specific heat for constant volume or constant pressure, or for any other condition, will be equal to the tangent of the proper angle in the diagram, multiplied by the ratio of the quantity of the substance for which the specific heat is determined to the quantity for which the diagram is drawn.*

The Volume-entropy Diagram.

The method of representation, in which the co-ordinates of the point in the diagram are made equal to the volume and entropy of the body, presents certain characteristics which entitle it to a somewhat detailed consideration, and for some purposes give it substantial advantages over any other method. We might anticipate some of these advantages from the simple and symmetrical form of the general equations of thermodynamics, when volume and entropy are chosen as independent variables, viz :—†

---

* From this general property of the diagram, its character in the case of a perfect gas might be immediately deduced.

† See page 2, equations (2), (3) and (4).

In general, in this article, where differential coefficients are used, the quantity which is constant in the differentiation is indicated by a subscript letter. In this discussion of the volume-entropy diagram, however, v and η are uniformly regarded as the independent variables, and the subscript letter is omitted.

$$p = -\frac{d\epsilon}{dv}, \tag{11}$$

$$t = \frac{d\epsilon}{d\eta}, \tag{12}$$

$$dW = p\,dv,$$

$$dH = t\,d\eta.$$

Eliminating p and t we have also

$$dW = -\frac{d\epsilon}{dv}\,dv, \tag{13}$$

$$dH = \frac{d\epsilon}{d\eta}\,d\eta. \tag{14}$$

The geometrical relations corresponding to these equations are in the volume-entropy diagram extremely simple. To fix our ideas, let the axes of volume and entropy be horizontal and vertical respectively, volume increasing toward the right and entropy upward. Then the pressure taken negatively will equal the ratio of the difference of energy to the difference of volume of two adjacent points in the same horizontal line, and the temperature will equal the ratio of the difference of energy to the difference of entropy of two adjacent points in the same vertical line. Or, if a series of isodynamics be drawn for equal infinitesimal differences of energy, any series of horizontal lines will be divided into segments inversely proportional to the pressure, and any series of vertical lines into segments inversely proportional to the temperature. We see by equations (13) and (14), that for a motion parallel to the axis of volume, the heat received is 0, and the work done is equal to the decrease of the energy, while for a motion parallel to the axis of entropy, the work done is 0, and the heat received is equal to the increase of the energy. These two propositions are true either for elementary paths or for those of finite length. In general, the work for any element of a path is equal to the product of the pressure in that part of the diagram into the horizontal projection of the element of the path, and the heat received is equal to the product of the temperature into the vertical projection of the element of the path.

If we wish to estimate the value of the integrals $\int p\,dv$ and $\int t\,d\eta$, which represent the work and heat of any path, by means of measurements upon the diagram, or if we wish to appreciate readily by the eye the approximate value of these expressions, or if we merely wish to illustrate their meaning by means of the diagram; for any of these purposes the diagram which we are now considering will have the advantage that it represents the differentials dv and $d\eta$ more simply and clearly than any other.

But we may also estimate the work and heat of any path by means of an integration extending over the elements of an area, viz: by the formulæ of page 7,

$$W^c = H^c = \Sigma^c \frac{1}{\gamma} \delta A,$$

$$W^s = \Sigma^{[S,\, v'',\, p^0,\, v']} \frac{1}{\gamma} \delta A,$$

$$H^s = \Sigma^{[S,\, \eta'',\, t^0,\, \eta']} \frac{1}{\gamma} \delta A.$$

In regard to the limits of integration in these formulæ, we see that for the work of any path which is not a circuit, the bounding line is composed of the path, the line of no pressure and two vertical lines, and for the heat of the path, the bounding line is composed of the path, the line of absolute cold and two horizontal lines.

As the sign of γ, as well as that of δA, will be indeterminate until we decide in which direction an area must be circumscribed in order to be considered positive, we will call an area positive which is circumscribed in the direction in which the hands of a watch move. This choice, with the positions of the axes of volume and entropy which we have supposed, will make the value of γ in most cases positive, as we shall see hereafter.

The value of γ, in a diagram drawn according to this method, will depend upon the properties of the body for which the diagram is drawn. In this respect, this method differs from all the others which have been discussed in detail in this article. It is easy to find an expression for γ depending simply upon the variations of the energy, by comparing the area and the work or heat of an infinitely small circuit in the form of a rectangle having its sides parallel to the two axes.

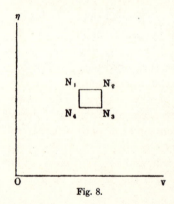

Fig. 8.

Let $N_1 N_2 N_3 N_4$ (fig. 8) be such a circuit, and let it be described in the order of the numerals, so that the area is positive. Also let ϵ_1, ϵ_2, ϵ_3, ϵ_4 represent the energy at the four corners. The work done in the four sides in order commencing at N_1, will be $\epsilon_1 - \epsilon_2$, 0, $\epsilon_3 - \epsilon_4$, 0. The total work, therefore, for the rectangular circuit is

$$\epsilon_1 - \epsilon_2 + \epsilon_3 - \epsilon_4.$$

Now as the rectangle is infinitely small, if we call its sides dv and $d\eta$, the above expression will be equivalent to

$$- \frac{d^2 \epsilon}{dv\, d\eta} dv\, d\eta.$$

Dividing by the area $dv\, d\eta$, and writing $\gamma_{v,\eta}$ for the scale of work and heat in a diagram of this kind, we have

$$\frac{1}{\gamma_{v,\eta}} = -\frac{d^2\epsilon}{dv\, d\eta} = \frac{dp}{d\eta} = -\frac{dt}{dv}. \qquad (15)$$

The two last expressions for the value of $1 \div \gamma_{v,\eta}$ indicate that the value of $\gamma_{v,\eta}$ in different parts of the diagram will be indicated proportionally by the segments into which vertical lines are divided by a system of equidifferent isopiestics, and also by the segments into which horizontal lines are divided by a system of equidifferent isothermals. These results might also be derived directly from the propositions on page 5.

As, in almost all cases, the pressure of a body is increased when it receives heat without change of volume, $\frac{dp}{d\eta}$ is in general positive, and the same will be true of $\gamma_{v,\eta}$ under the assumptions which we have made in regard to the directions of the axes (page 21) and the definition of a positive area (page 22).

In the estimation of work and heat it may often be of use to consider the deformation necessary to reduce the diagram to one of constant scale for work and heat. Now if the diagram be so deformed that each point remains in the same vertical line, but moves in this line so that all isopiestics become straight and horizontal lines at distances proportional to their differences of pressure, it will evidently become a volume-pressure diagram. Again, if the diagram be so deformed that each point remains in the same horizontal line, but moves in it so that isothermals become straight and vertical lines at distances proportional to their differences of temperature, it will become an entropy-temperature diagram. These considerations will enable us to compute numerically the work or heat of any path which is given in a volume-entropy diagram, when the pressure and temperature are known for all points of the path, in a manner analogous to that explained on page 19.

The ratio of any element of area in the volume-pressure or the entropy-temperature diagram, or in any other in which the scale of work and heat is unity, to the corresponding element in the volume-entropy diagram is represented by $\frac{1}{\gamma_{v,\eta}}$ or $-\frac{d^2\epsilon}{dv\, d\eta}$. The cases in which this ratio is 0, or changes its sign, demand especial attention, as in such cases the diagrams of constant scale fail to give a satisfactory representation of the properties of the body, while no difficulty or inconvenience arises in the use of the volume-entropy diagram.

As $-\frac{d^2\epsilon}{dv\, d\eta} = \frac{dp}{d\eta}$, its value is evidently zero in that part of the diagram which represents the body when in part solid, in part liquid, and in part vapor. The properties of such a mixture are very simply and clearly exhibited in the volume-entropy diagram.

Let the temperature and the pressure of the mixture, which are independent of the proportions of vapor, solid and liquid, be denoted by t' and p'. Also let V, L and S (fig. 9) be points of the diagram which indicate the volume and entropy of the body in three perfectly defined states, viz: that of a vapor of temperature t' and pressure p', that of a liquid of the same temperature and pressure, and that of a solid of the same temperature and pressure. And let v_V, η_V, v_L, η_L, v_S, η_S denote the volume and entropy of these states. The position of the point which represents the body, when part is vapor, part liquid, and part solid, these parts being as μ, ν, and $1 - \mu - \nu$, is determined by the equations

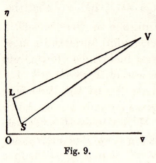

Fig. 9.

$$v = \mu v_V + \nu v_L + (1 - \mu - \nu)v_S,$$

$$\eta = \mu \eta_V + \nu \eta_L + (1 - \mu - \nu)\eta_S,$$

where v and η are the volume and entropy of the mixture. The truth of the first equation is evident. The second may be written

$$\eta - \eta_S = \mu(\eta_V - \eta_S) + \nu(\eta_L - \eta_S),$$

or multiplying by t',

$$t'(\eta - \eta_S) = \mu\, t'(\eta_V - \eta_S) + \nu\, t'(\eta_L - \eta_S).$$

The first member of this equation denotes the heat necessary to bring the body from the state S to the state of the mixture in question under the constant temperature t', while the terms of the second member denote separately the heat necessary to vaporize the part μ, and to liquefy the part ν of the body.

The values of v and η are such as would give the center of gravity of masses μ, ν and $1 - \mu - \nu$ placed at the points V, L and S.* Hence the part of the diagram which represents a mixture of vapor, liquid and solid, is the triangle VLS. The pressure and temperature are constant for this triangle, i.e., an isopiestic and also an isothermal here expand to cover a space. The isodynamics are straight and equidistant for equal differences of energy. For $\dfrac{d\epsilon}{dv} = -p'$ and $\dfrac{d\epsilon}{d\eta} = t'$, both of which are constant throughout the triangle.

---

* These points will not be in the same straight line unless

$$t'(\eta_V - \eta_S) : t'(\eta_L - \eta_S) :: v_V - v_S : v_L - v_S,$$

a condition very unlikely to be fulfilled by any substance. The first and second terms of this proportion denote the heat of vaporization (from the solid state) and that of liquefaction.

This case can be but very imperfectly represented in the volume-pressure, or in the entropy-temperature diagram. For all points in the same vertical line in the triangle VLS will, in the volume-pressure diagram, be represented by a single point, as having the same volume and pressure. And all the points in the same horizontal line will be represented in the entropy-temperature diagram by a single point, as having the same entropy and temperature. In either diagram, the whole triangle reduces to a straight line. It must reduce to a line in any diagram whatever of constant scale, as its area must become 0 in such a diagram. This must be regarded as a defect in these diagrams, as essentially different states are represented by the same point. In consequence, any circuit within the triangle VLS will be represented in any diagram of constant scale by two paths of opposite directions superposed, the appearance being as if a body should change its state and then return to its original state by inverse processes, so as to repass through the same series of states. It is true that the circuit in question is like this combination of processes in one important particular, viz: that $W = H = 0$, i.e., there is no transformation of heat into work. But this very fact, that a circuit without transformation of heat into work is possible, is worthy of distinct representation.

A body may have such properties that in one part of the volume-entropy diagram $\frac{1}{\gamma_{v,\eta}}$, i.e., $\frac{dp}{d\eta}$ is positive and in another negative. These parts of the diagram may be separated by a line, in which $\frac{dp}{d\eta} = 0$, or by one in which $\frac{dp}{d\eta}$ changes abruptly from a positive to a negative value.* (In part, also, they may be separated by an area in which $\frac{dp}{d\eta} = 0$.) In the representation of such cases in any diagram of constant scale, we meet with a difficulty of the following nature.

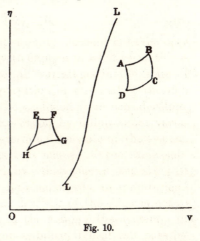

Fig. 10.

Let us suppose that on the right of the line LL (fig. 10) in a volume-entropy diagram, $\frac{dp}{d\eta}$ is positive, and on the left negative. Then, if we draw any circuit ABCD on the right side of LL, the direction

---

* The line which represents the various states of water at its maximum density for various constant pressures is an example of the first case. A substance which as a liquid has no proper maximum density for constant pressure, but which expands in solidifying, affords an example of the second case.

being that of the hands of a watch, the work and heat of the circuit
will be positive. But if we draw any circuit EFGH in the same
direction on the other side of the line LL, the work and heat will
be negative. For

$$W = H = \Sigma \frac{1}{\gamma_{v,\eta}} \delta A = \Sigma \frac{dp}{d\eta} \delta A,$$

and the direction of the circuits makes the areas positive in both
cases. Now if we should change this diagram into any diagram of
constant scale, the areas of the circuits, as representing proportionally
the work done in each case, must necessarily have opposite signs,
i.e., the direction of the circuits must be opposite. We will suppose
that the work done is positive in the diagram of constant scale, when
the direction of the circuit is that of the hands of a watch. Then, in

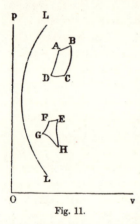

Fig. 11.

that diagram, the circuit ABCD would have
that direction, and the circuit EFGH the contrary direction, as in figure 11. Now if we
imagine an indefinite number of circuits on
each side of LL in the volume-entropy diagram, it will be evident that to transform
such a diagram into one of constant scale, so
as to change the direction of all the circuits
on one side of LL, and of none on the other
the diagram must be *folded over* along that
line; so that the points on one side of LL in
a diagram of constant scale do not represent
any states of the body, while on the other
side of this line, each point, for a certain
distance at least, represents two different states of the body, which in
the volume-entropy diagram are represented by points on opposite
sides of the line LL. We have thus in a part of the field two diagrams
superposed, which must be carefully distinguished. If this be done,
as by the help of different colors, or of continuous and dotted lines,
or otherwise, and it is remembered that there is no continuity between
these superposed diagrams, except along the bounding line LL, all the
general theorems which have been developed in this article can be
readily applied to the diagram. But to the eye or to the imagination,
the figure will necessarily be much more confusing than a volume-
entropy diagram.

If $\frac{dp}{d\eta} = 0$ for the line LL, there will be another inconvenience in

the use of any diagram of constant scale, viz: in the vicinity of the

line LL, $\frac{dp}{d\eta}$, i.e., $1 \div \gamma_{v,\eta}$ will have a very small value, so that areas

will be very greatly reduced in the diagram of constant scale, as compared with the corresponding areas in the volume-entropy diagram. Therefore, in the former diagram, either the isometrics, or the isentropics, or both, will be crowded together in the vicinity of the line LL, so that this part of the diagram will be necessarily indistinct.

It may occur, however, in the volume-entropy diagram, that the same point must represent two different states of the body. This occurs in the case of liquids which can be vaporized. Let MM (fig. 12) be the line representing the states of the liquid bordering upon vaporization. This line will be near to the axis of entropy, and nearly parallel to it. If the body is in a state represented by a point of the line MM, and is compressed without addition or subtraction of heat, it will remain of course liquid. Hence, the points of the space immediately on the left of MM represent simple liquid. On the other hand, the body being in the original state, if its volume should be increased without addition or subtraction of heat, and if the conditions necessary for vaporization are present (conditions relative to the body enclosing the liquid in question, etc.), the liquid will become partially vaporized,

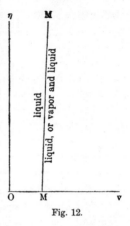

Fig. 12.

but if these conditions are not present, it will continue liquid. Hence, every point on the right of MM and sufficiently near to it represents two different states of the body, in one of which it is partially vaporized, and in the other it is entirely liquid. If we take the points as representing the mixture of vapor and liquid, they form one diagram, and if we take them as representing simple liquid, they form a totally different diagram superposed on the first. There is evidently no continuity between these diagrams except at the line MM; we may regard them as upon separate sheets united only along MM. For the body cannot pass from the state of partial vaporization to the state of liquid except at this line. The reverse process is indeed possible; the body can pass from the state of superheated liquid to that of partial vaporization, if the conditions of vaporization alluded to above are supplied, or if the increase of volume is carried beyond a certain limit, but not by gradual changes or reversible processes. After such a change, the point representing the state of the body will be found in a different position from that which it occupied before, but the change of state cannot be properly represented by any path, as during the change the body does not satisfy that condition of uniform temperature and pressure which has been assumed throughout this article, and which is necessary for the graphical methods under discussion. (See note on page 1.)

Of the two superposed diagrams, that which represents simple liquid is a continuation of the diagram on the left of MM. The isopiestics, isothermals and isodynamics pass from one to the other without abrupt change of direction or curvature. But that which represents a mixture of vapor and liquid will be different in its character, and its isopiestics and isothermals will make angles in general with the corresponding lines in the diagram of simple liquid. The isodynamics of the diagram of the mixture, and those of the diagram of simple liquid, will differ in general in curvature at the line MM, but not in direction, for $\frac{d\epsilon}{dv} = -p$ and $\frac{d\epsilon}{d\eta} = t$.

The case is essentially the same with some substances, as water, for example, about the line which separates the simple liquid from a mixture of liquid and solid.

In these cases the inconvenience of having one diagram superposed upon another cannot be obviated by any change of the principle on which the diagram is based. For no distortion can bring the three sheets, which are united along the line MM (one on the left and two on the right), into a single plane surface without superposition. Such cases, therefore, are radically distinguished from those in which the superposition is caused by an unsuitable method of representation.

To find the character of a volume-entropy diagram of a perfect gas, we may make ϵ constant in equation (D) on page 13, which will give for the equation of an isodynamic and isothermal

$$\eta = a \log v + \text{Const.},$$

and we may make p constant in equation (G), which will give for the equation of an isopiestic

$$\eta = (a+c) \log v + \text{Const.}$$

It will be observed that all the isodynamics and isothermals can be drawn by a single pattern and so also with the isopiestics.

The case will be nearly the same with vapors in a part of the diagram. In that part of the diagram which represents a mixture of liquid and vapor, the isothermals, which of course are identical with the isopiestics, are straight lines. For when a body is vaporized under constant pressure and temperature, the quantities of heat received are proportional to the increments of volume; therefore, the increments of entropy are proportional to the increments of volume. As $\frac{d\epsilon}{dv} = -p$ and $\frac{d\epsilon}{d\eta} = t$, any isothermal is cut at the same angle by all the isodynamics, and is divided into equal segments by equidifferent isodynamics. The latter property is useful in drawing systems of equidifferent isodynamics.

Arrangement of the Isometric, Isopiestic, Isothermal and Isentropic about a Point.

The arrangement of the isometric, the isopiestic, the isothermal and the isentropic drawn through any same point, in respect to the order in which they succeed one another around that point, and in respect to the sides of these lines toward which the volume, pressure, temperature and entropy increase, is not altered by any deformation of the surface on which the diagram is drawn, and is therefore independent of the method by which the diagram is formed.* This arrangement is determined by certain of the most characteristic thermodynamic properties of the body in the state in question, and serves in turn to indicate these properties. It is determined, namely, by the value of $\left(\dfrac{dp}{d\eta}\right)_v$ as positive, negative, or zero, i.e., by the effect of heat as increasing or diminishing the pressure when the volume is maintained constant, and by the nature of the internal thermodynamic equilibrium of the body as stable or neutral,—an unstable equilibrium, except as a matter of speculation, is of course out of the question.

Let us first examine the case in which $\left(\dfrac{dp}{d\eta}\right)_v$ is positive and the equilibrium is stable. As $\left(\dfrac{dp}{d\eta}\right)_v$ does not vanish at the point in question, there is a definite isopiestic passing through that point, on one side of which the pressures are greater, and on the other less, than on the line itself. As $\left(\dfrac{dt}{dv}\right)_\eta = -\left(\dfrac{dp}{d\eta}\right)_v$, the case is the same with the isothermal. It will be convenient to distinguish the sides of the isometric, isopiestic, etc., on which the volume, pressure, etc., increase, as the *positive* sides of these lines. The condition of stability requires that, when the pressure is constant, the temperature shall increase with the heat received,—therefore with the entropy. This may be written $[dt:d\eta]_p > 0.$† It also requires that, when there is no transmission of heat, the pressure should increase as the volume diminishes, i.e., that $[dp:dv]_\eta < 0$. Through the point in question,

---

*It is here assumed that, in the vicinity of the point in question, each point in the diagram represents only one state of the body. The propositions developed in the following pages cannot be applied to points of the line where two superposed diagrams are united (see pages 25–28) without certain modifications.

† As the notation $\dfrac{dt}{d\eta}$ is used to denote the limit of the ratio of dt to $d\eta$, it would not be quite accurate to say that the condition of stability requires that $\left(\dfrac{dt}{d\eta}\right)_p > 0$. This condition requires that the ratio of the differences of temperature and entropy between the point in question and any other infinitely near to it and upon the same isopiestic should be positive. It is not necessary that the limit of this ratio should be positive.

A (fig. 13), let there be drawn the isometric vv' and the isentropic $\eta\eta'$, and let the positive sides of these lines be indicated as in the figure. The conditions $\left(\dfrac{dp}{d\eta}\right)_v > 0$ and $[dp:dv]_\eta < 0$ require that the pressure at v and at η shall be greater than at A, and hence, that the isopiestic shall fall as pp' in the figure, and have its positive side turned as indicated. Again, the conditions $\left(\dfrac{dt}{dv}\right)_\eta < 0$ and $[dt:d\eta]_p > 0$ require that the temperature at η and at p shall be greater than at A, and hence, that the isothermal shall fall as tt' and have its positive side turned as indicated. As it is not necessary that $\left(\dfrac{dt}{d\eta}\right)_p > 0$, the lines pp' and tt' may be tangent to one another at A, provided that they cross one another, so as to have the same order about the point A as is represented in the figure; i.e., they may have a contact of the second (or any even) order.* But the condition that $\left(\dfrac{dp}{d\eta}\right)_v > 0$, and hence $\left(\dfrac{dt}{dv}\right)_\eta < 0$, does not allow pp' to be tangent to vv', nor tt' to $\eta\eta'$.

If $\left(\dfrac{dp}{d\eta}\right)_v$ be still positive, but the equilibrium be neutral, it will be

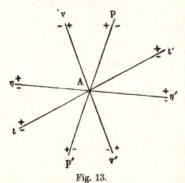

possible for the body to change its state without change either of temperature or of pressure; i.e., the isothermal and isopiestic will be identical. The lines will fall as in figure 13, except that the isothermal and isopiestic will be superposed.

In like manner, if $\left(\dfrac{dp}{d\eta}\right)_v < 0$, it may be proved that the lines will fall as in figure 14 for stable equilibrium, and in the same way for neutral

Fig. 13.

equilibrium, except that pp' and tt' will be superposed.†

---

* An example of this is doubtless to be found at the critical point of a fluid. See Dr. Andrews "On the continuity of the gaseous and liquid states of matter." *Phil. Trans.*, vol. 159, p. 575.

If the isothermal and isopiestic have a simple tangency at A, on one side of that point they will have such directions as will express an unstable equilibrium. A line drawn through all such points in the diagram will form a boundary to the *possible* part of the diagram. It may be that the part of the diagram of a fluid, which represents the superheated liquid state, is bounded on one side by such a line.

† When it is said that the arrangement of the lines in the diagram must be like that in figure 13 or in figure 14, it is not meant to exclude the case in which the figure (13 or 14) must be turned over, in order to correspond with the diagram. In the case, however, of diagrams formed by any of the methods mentioned in this article, if the

The case that $\left(\dfrac{dp}{d\eta}\right)_v = 0$ includes a considerable number of conceivable cases, which would require to be distinguished. It will be sufficient to mention those most likely to occur.

In a field of stable equilibrium it may occur that $\left(\dfrac{dp}{d\eta}\right)_v = 0$ along a line, on one side of which $\left(\dfrac{dp}{d\eta}\right)_v > 0$, and on the other side $\left(\dfrac{dp}{d\eta}\right)_v < 0$. At any point in such a line the isopiestics will be tangent to the isometrics and the isothermals to the isen-

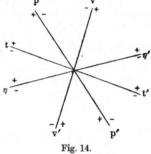

Fig. 14.

tropics. (See, however, note on page 29.)

In a field of neutral equilibrium representing a mixture of two different states of the substance, where the isothermals and isopiestics are identical, a line may occur which has the threefold character of an isometric, an isothermal and an isopiestic. For such a line $\left(\dfrac{dp}{d\eta}\right)_v = 0$. If $\left(\dfrac{dp}{d\eta}\right)_v$ has opposite signs on opposite sides of this line, it will be an isothermal of maximum or minimum temperature.*

The case in which the body is partly solid, partly liquid and partly vapor has already been sufficiently discussed. (See pages 23, 24.)

The arrangement of the isometric, isopiestic, etc., as given in figure 13, will indicate directly the sign of any differential co-efficient of the form $\left(\dfrac{du}{dw}\right)_z$, where u, w and z may be any of the quantities v, p, t, η (and ϵ, if the isodynamic be added in the figure). The value of such a differential co-efficient will be indicated, when the rates of increase of v, p, etc., are indicated, as by isometrics, etc., drawn both for the values of v, etc., at the point A, and for values differing from these by a small quantity. For example, the value of $\left(\dfrac{dp}{dv}\right)_\eta$ will be indicated by the ratio of the segments intercepted upon an isentropic by a pair of isometrics and a pair of isopiestics, of which the differences of volume and pressure have the same numerical value. The case in which W or H appears in the numerator or denominator instead of a

---

directions of the axes be such as we have assumed, the agreement with figure 13 will be *without inversion*, and the agreement with fig. 14 will also be *without inversion* for volume-entropy diagrams, but *with inversion* for volume-pressure or entropy-temperature diagrams, or those in which $x = \log v$ and $y = \log p$, or $x = \eta$ and $y = \log t$.

* As some liquids expand and others contract in solidifying, it is possible that there are some which will solidify either with expansion, or without change of volume, or with contraction, according to the pressure. If any such there are, they afford examples of the case mentioned above.

function of the state of the body, can be reduced to the preceding by the substitution of pdv for dW, or that of $td\eta$ for dH.

In the foregoing discussion, the equations which express the fundamental principles of thermodynamics in an analytical form have been assumed, and the aim has only been to show how the same relations may be expressed geometrically. It would, however, be easy, starting from the first and second laws of thermodynamics as usually enunciated, to arrive at the same results without the aid of analytical formulæ,—to arrive, for example, at the conception of energy, of entropy, of absolute temperature, in the construction of the diagram without the analytical definitions of these quantities, and to obtain the various properties of the diagram without the analytical expression of the thermodynamic properties which they involve. Such a course would have been better fitted to show the independence and sufficiency of a graphical method, but perhaps less suitable for an examination of the comparative advantages or disadvantages of different graphical methods.

The possibility of treating the thermodynamics of fluids by such graphical methods as have been described evidently arises from the fact that the state of the body considered, like the position of a point in a plane, is capable of two and only two independent variations. It is, perhaps, worthy of notice, that when the diagram is only used to demonstrate or illustrate general theorems, it is not necessary, although it may be convenient, to assume any particular method of forming the diagram; it is enough to suppose the different states of the body to be represented continuously by points upon a sheet.

II

A METHOD OF GEOMETRICAL REPRESENTATION OF THE THERMODYNAMIC PROPERTIES OF SUBSTANCES BY MEANS OF SURFACES.

[*Transactions of the Connecticut Academy*, II. pp. 382–404, Dec. 1873.]

THE leading thermodynamic properties of a fluid are determined by the relations which exist between the volume, pressure, temperature, energy, and entropy of a given mass of the fluid in a state of thermodynamic equilibrium. The same is true of a solid in regard to those properties which it exhibits in processes in which the pressure is the same in every direction about any point of the solid. But all the relations existing between these five quantities for any substance (three independent relations) may be deduced from the single relation existing for that substance between the volume, energy, and entropy. This may be done by means of the general equation,

$$de = t\,d\eta - p\,dv, \tag{1}*$$

that is,

$$p = -\left(\frac{de}{dv}\right)_\eta, \tag{2}$$

$$t = \left(\frac{de}{d\eta}\right)_v, \tag{3}$$

where v, p, t, ϵ, and η denote severally the volume, pressure, absolute temperature, energy, and entropy of the body considered. The subscript letter after the differential coefficient indicates the quantity which is supposed constant in the differentiation.

Representation of Volume, Entropy, Energy, Pressure, and Temperature.

Now the relation between the volume, entropy, and energy may be represented by a surface, most simply if the rectangular coordinates of the various points of the surface are made equal to the volume, entropy, and energy of the body in its various states. It may be interesting to examine the properties of such a surface, which

---

* For the demonstration of this equation, and in regard to the units used in the measurement of the quantities, the reader is referred to page 2.

we will call the thermodynamic surface of the body for which it is formed.*

To fix our ideas, let the axes of v, η, and ϵ have the directions usually given to the axes of X, Y, and Z (v increasing to the right, η forward, and ϵ upward). Then the pressure and temperature of the state represented by any point of the surface are equal to the tangents of the inclinations of the surface to the horizon at that point, as measured in planes perpendicular to the axes of η and of v respectively. (Eqs. 2 and 3.) It must be observed, however, that in the first case the angle of inclination is measured upward from the direction of *decreasing* v, and in the second, upward from the direction of *increasing* η. Hence, the tangent plane at any point indicates the temperature and pressure of the state represented. It will be convenient to speak of a plane as representing a certain pressure and temperature, when the tangents of its inclinations to the horizon, measured as above, are equal to that pressure and temperature.

Before proceeding farther, it may be worth while to distinguish between what is essential and what is arbitrary in a surface thus formed. The position of the plane $v = 0$ in the surface is evidently fixed, but the position of the planes $\eta = 0$, $\epsilon = 0$ is arbitrary, provided the direction of the axes of η and ϵ be not altered. This results from the nature of the definitions of entropy and energy, which involve each an arbitrary constant. As we may make $\eta = 0$ and $\epsilon = 0$ for any state of the body which we may choose, we may place the origin of co-ordinates at any point in the plane $v = 0$. Again, it is evident from the form of equation (1) that whatever changes we may make in the units in which volume, entropy, and energy are measured, it will always be possible to make such changes in the units of temperature and pressure, that the equation will hold true in its present form, without the introduction of constants. It is easy to see how a change of the units of volume, entropy, and energy would affect the surface. The projections parallel to any one of the axes of distances between points of the surface would be changed in the ratio inverse to that in which the corresponding unit had been changed. These considerations enable us to foresee to a certain extent the nature of the general properties of the surface which we are to investigate. They

---

*Professor J. Thomson has proposed and used a surface in which the co-ordinates are proportional to the volume, pressure, and temperature of the body. (*Proc. Roy. Soc.*, Nov. 16, 1871, vol. xx, p. 1; and *Phil. Mag.*, vol. xliii, p. 227.) It is evident, however, that the relation between the volume, pressure, and temperature affords a less complete knowledge of the properties of the body than the relation between the volume, entropy, and energy. For, while the former relation is entirely determined by the latter, and can be derived from it by differentiation, the latter relation is by no means determined by the former.

must be such, namely, as shall not be affected by any of the changes mentioned above. For example, we may find properties which concern the plane $v = 0$ (as that the whole surface must necessarily fall on the positive side of this plane), but we must not expect to find properties which concern the planes $\eta = 0$, or $\epsilon = 0$, in distinction from others parallel to them. It may be added that, as the volume, entropy, and energy of a body are equal to the sums of the volumes, entropies, and energies of its parts, if the surface should be constructed for bodies differing in quantity but not in kind of matter, the different surfaces thus formed would be similar to one another, their linear dimensions being proportional to the quantities of matter.

Nature of that Part of the Surface which represents States which are not Homogeneous.

This mode of representation of the volume, entropy, energy, pressure, and temperature of a body will apply as well to the case in which different portions of the body are in different states (supposing always that the whole is in a state of thermodynamic equilibrium), as to that in which the body is uniform in state throughout. For the body taken as a whole has a definite volume, entropy, and energy, as well as pressure and temperature, and the validity of the general equation (1) is independent of the uniformity or diversity in respect to state of the different portions of the body.* It is evident, therefore, that

---

* It is, however, supposed in this equation that the variations in the state of the body, to which dv, $d\eta$, and $d\epsilon$ refer, are such as may be produced *reversibly* by expansion and compression or by addition and subtraction of heat. Hence, when the body consists of parts in different states, it is necessary that these states should be such as can pass either into the other without sensible change of pressure or temperature. Otherwise, it would be necessary to suppose in the differential equation (1) that the proportion in which the body is divided into the different states remains constant. But such a limitation would render the equation as applied to a compound of different states valueless for our present purpose. If, however, we leave out of account the cases in which we regard the states as chemically different from one another, which lie beyond the scope of this paper, experience justifies us in assuming the above condition (that either of the two states existing in contact can pass into the other without sensible change of the pressure or temperature), as at least approximately true, when one of the states is fluid. But if both are solid, the necessary mobility of the parts is wanting. It must therefore be understood, that the following discussion of the compound states is not intended to apply without limitation to the exceptional cases, where we have two different solid states of the same substance at the same pressure and temperature. It may be added that the thermodynamic equilibrium which subsists between two such solid states of the same substance differs from that which subsists when one of the states is fluid, very much as in statics an equilibrium which is maintained by friction differs from that of a frictionless machine in which the active forces are so balanced, that the slightest change of force will produce motion in either direction.

Another limitation is rendered necessary by the fact that in the following discussion the magnitude and form of the bounding and dividing surfaces are left out of account ;

the thermodynamic surface, for many substances at least, can be divided into two parts, of which one represents the homogeneous states, the other those which are not so. We shall see that, when the former part of the surface is given, the latter can readily be formed, as indeed we might expect. We may therefore call the former part the primitive surface, and the latter the derived surface.

To ascertain the nature of the derived surface and its relations to the primitive surface sufficiently to construct it when the latter is given, it is only necessary to use the principle that the volume, entropy, and energy of the whole body are equal to the sums of the volumes, entropies, and energies respectively of the parts, while the pressure and temperature of the whole are the same as those of each of the parts. Let us commence with the case in which the body is in part solid, in part liquid, and in part vapor. The position of the point determined by the volume, entropy, and energy of such a compound will be that of the center of gravity of masses proportioned to the masses of solid, liquid, and vapor placed at the three points of the primitive surface which represent respectively the states of complete solidity, complete liquidity, and complete vaporization, each at the temperature and pressure of the compound. Hence, the part of the surface which represents a compound of solid, liquid, and vapor is a plane triangle, having its vertices at the points mentioned. The fact that the surface is here plane indicates that the pressure and temperature are here constant, the inclination of the plane indicating the value of these quantities. Moreover, as these values are the same for the compound as for the three different homogeneous states corresponding to its different portions, the plane of the triangle is tangent at each of its vertices to the primitive surface, viz: at one vertex to that part of the primitive surface which represents solid, at another to the part representing liquid, and at the third to the part representing vapor.

When the body consists of a compound of two different homogeneous states, the point which represents the compound state will be at the center of gravity of masses proportioned to the masses of the parts of the body in the two different states and placed at the points of the primitive surface which represent these two states (i.e., which represent the volume, entropy, and energy of the body, if its whole mass were supposed successively in the two homogeneous states which occur in its parts). It will therefore be found upon the straight line

so that the results are in general strictly valid only in cases in which the influence of these particulars may be neglected. When, therefore, two states of the substance are spoken of as in contact, it must be understood that the surface dividing them is plane. To consider the subject in a more general form, it would be necessary to introduce considerations which belong to the theories of capillarity and crystallization.

which unites these two points. As the pressure and temperature are evidently constant for this line, a single plane can be tangent to the derived surface throughout this line and at each end of the line tangent to the primitive surface.* If we now imagine the temperature and pressure of the compound to vary, the two points of the primitive surface, the line in the derived surface uniting them, and the tangent

---

* It is here shown that, if two different states of the substance are such that they can exist permanently in contact with each other, the points representing these states in the thermodynamic surface have a common tangent plane. We shall see hereafter that the converse of this is true,—that, if two points in the thermodynamic surface have a common tangent plane, the states represented are such as can permanently exist in contact; and we shall also see what determines the direction of the discontinuous change which occurs when two different states of the same pressure and temperature, for which the condition of a common tangent plane is not satisfied, are brought into contact.

It is easy to express this condition analytically. Resolving it into the conditions, that the tangent planes shall be parallel, and that they shall cut the axis of ϵ at the same point, we have the equations

$$p' = p'', \tag{a}$$

$$t' = t'', \tag{β}$$

$$\epsilon' - t'\eta' + p'v' = \epsilon'' - t''\eta'' + p''v'', \tag{γ}$$

where the letters which refer to the different states are distinguished by accents. If there are three states which can exist in contact, we must have for these states,

$$p' = p'' = p''',$$

$$t' = t'' = t''',$$

$$\epsilon' - t'\eta' + p'v' = \epsilon'' - t''\eta'' + p''v'' = \epsilon''' - t'''\eta''' + p'''v'''.$$

These results are interesting, as they show us how we might foresee whether two given states of a substance of the same pressure and temperature, can or cannot exist in contact. It is indeed true, that the values of ϵ and η cannot like those of v, p, and t be ascertained by mere measurements upon the substance while in the two states in question. It is necessary, in order to find the value of $\epsilon'' - \epsilon'$ or $\eta'' - \eta'$, to carry out measurements upon a process by which the substance is brought from one state to the other, *but this need not be by a process in which the two given states shall be found in contact*, and in some cases at least it may be done by processes in which the body remains always homogeneous in state. For we know by the experiments of Dr. Andrews, *Phil. Trans.*, vol. 159, p. 575, that carbonic acid may be carried from any of the states which we usually call liquid to any of those which we usually call gas, without losing its homogeneity. Now, if we had so carried it from a state of liquidity to a state of gas of the same pressure and temperature, making the proper measurements in the process, we should be able to foretell what would occur if these two states of the substance should be brought together,—whether evaporation would take place, or condensation, or whether they would remain unchanged in contact,—although we had never seen the phenomenon of the coexistence of these two states, or of any other two states of this substance.

Equation (γ) may be put in a form in which its validity is at once manifest for two states which can pass either into the other at a constant pressure and temperature. If we put p' and t' for the equivalent p'' and t'', the equation may be written

$$\epsilon'' - \epsilon' = t'(\eta'' - \eta') - p'(v'' - v').$$

Here the left hand member of the equation represents the difference of energy in the two states, and the two terms on the right represent severally the heat received and

plane will change their positions, maintaining the aforesaid relations. We may conceive of the motion of the tangent plane as produced by rolling upon the primitive surface, while tangent to it in two points, and as it is also tangent to the derived surface in the lines joining these points, it is evident that the latter is a developable surface and forms a part of the envelop of the successive positions of the rolling plane. We shall see hereafter that the form of the primitive surface is such that the double tangent plane does not cut it, so that this rolling is physically possible.

From these relations may be deduced by simple geometrical considerations one of the principal propositions in regard to such

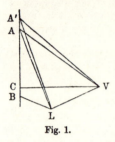

Fig. 1.

compounds. Let the tangent plane touch the primitive surface at the two points L and V (fig. 1), which, to fix our ideas, we may suppose to represent liquid and vapor; let planes pass through these points perpendicular to the axes of v and η respectively, intersecting in the line AB, which will be parallel to the axis of ϵ. Let the tangent plane cut this line at A, and let LB and VC be drawn at right angles to AB and parallel to the axes of η and v. Now the pressure and temperature represented by the tangent plane are evidently $\dfrac{AC}{CV}$ and $\dfrac{AB}{BL}$ respectively, and if we suppose the tangent plane in rolling upon the primitive surface to turn about its instantaneous axis LV an infinitely small angle, so as to meet AB in A', dp and dt will be equal to $\dfrac{AA'}{CV}$ and $\dfrac{AA'}{BL}$ respectively. Therefore,

$$\frac{dp}{dt} = \frac{BL}{CV} = \frac{\eta'' - \eta'}{v'' - v'},$$

where v' and η' denote the volume and entropy for the point L, and v'' and η'' those for the point V. If we substitute for $\eta'' - \eta'$ its equivalent $\dfrac{r}{t}$ (r denoting the heat of vaporization), we have the equation in its usual form, $\dfrac{dp}{dt} = \dfrac{r}{t(v'' - v')}$.

---

the work done when the body passes from one state to the other. The equation may also be derived at once from the general equation (1) by integration.

It is well known that when the two states being both fluid meet in a curved surface, instead of (a) we have $p'' - p' = T\left(\dfrac{1}{r} + \dfrac{1}{r'}\right)$,

where r and r' are the radii of the principal curvatures of the surface of contact at any point (positive, if the concavity is toward the mass to which p'' refers), and T is what is called the *superficial tension*. Equation (β), however, holds good for such cases, and it might easily be proved that the same is true of equation (γ). In other words, the tangent planes for the points in the thermodynamic surface representing the two states cut the plane $v=0$ in the same line.

Properties of the Surface relating to Stability of Thermodynamic Equilibrium.

We will now turn our attention to the geometrical properties of the surface, which indicate whether the thermodynamic equilibrium of the body is stable, unstable, or neutral. This will involve the consideration, to a certain extent, of the nature of the processes which take place when equilibrium does not subsist. We will suppose the body placed in a medium of constant pressure and temperature; but as, when the pressure or temperature of the body at its surface differs from that of the medium, the immediate contact of the two is hardly consistent with the continuance of the initial pressure and temperature of the medium, both of which we desire to suppose constant, we will suppose the body separated from the medium by an envelop which will yield to the smallest differences of pressure between the two, but which can only yield very gradually, and which is also a very poor conductor of heat. It will be convenient and allowable for the purposes of reasoning to limit its properties to those mentioned, and to suppose that it does not occupy any space, or absorb any heat except what it transmits, i.e., to make its volume and its specific heat 0. By the intervention of such an envelop, we may suppose the action of the body upon the medium to be so retarded as not sensibly to disturb the uniformity of pressure and temperature in the latter.

When the body is not in a state of thermodynamic equilibrium, its state is not one of those which are represented by our surface. The body, however, as a whole has a certain volume, entropy, and energy, which are equal to the sums of the volumes, etc., of its parts.[*] If, then, we suppose points endowed with mass proportional to the masses of the various parts of the body, which are in different thermodynamic states, placed in the positions determined by the states and motions of these parts, (i.e., so placed that their co-ordinates are equal to the volume, entropy, and energy of the whole body supposed successively in the same states and endowed with the same velocities as the different parts), the center of gravity of such points thus placed will evidently represent by its co-ordinates the volume, entropy, and energy of the whole body. If all parts of the body are at rest, the point representing its volume, entropy, and energy will be the center of gravity of a number of points upon the primitive surface. The effect of motion in the parts of the body will be to move the corresponding points parallel to the axis of ϵ, a distance equal in each case to the *vis viva* of the whole body, if endowed with the

---

[*] As the discussion is to apply to cases in which the parts of the body are in (sensible) motion, it is necessary to define the sense in which the word *energy* is to be used. We will use the word as *including the vis viva of sensible motions.*

velocity of the part represented;—the center of gravity of points thus determined will give the volume, entropy, and energy of the whole body.

Now let us suppose that the body having the initial volume, entropy, and energy, v', η', and ϵ', is placed (enclosed in an envelop as aforesaid) in a medium having the constant pressure P and temperature T, and by the action of the medium and the interaction of its own parts comes to a final state of rest in which its volume, etc., are v'', η'', ϵ'';—we wish to find a relation between these quantities. If we regard, as we may, the medium as a very large body, so that imparting heat to it or compressing it within moderate limits will have no appreciable effect upon its pressure and temperature, and write V, H, and E, for its volume, entropy, and energy, equation (1) becomes
$$dE = T\,dH - P\,dV,$$
which we may integrate regarding P and T as constants, obtaining
$$E'' - E' = TH'' - TH' - PV'' + PV', \tag{a}$$
where E', E'', etc., refer to the initial and final states of the medium. Again, as the sum of the energies of the body and the surrounding medium may become less, but cannot become greater (this arises from the nature of the envelop supposed), we have
$$\epsilon'' + E'' \leqq \epsilon' + E'. \tag{b}$$
Again as the sum of the entropies may increase but cannot diminish
$$\eta'' + H'' \geqq \eta' + H'. \tag{c}$$
Lastly, it is evident that
$$v'' + V'' = v' + V'. \tag{d}$$
These four equations may be arranged with slight changes as follows:
$$-E'' + TH'' - PV'' = -E' + TH' - PV'$$
$$\epsilon'' + E'' \leqq \epsilon' + E'$$
$$-T\eta'' - TH'' \leqq -T\eta' - TH'$$
$$Pv'' + PV'' = Pv' + PV'.$$
By addition we have
$$\epsilon'' - T\eta'' + Pv'' \leqq \epsilon' - T\eta' + Pv'. \tag{e}$$
Now the two members of this equation evidently denote the vertical distances of the points (v'', η'', ϵ') and (v', η', ϵ') above the plane passing through the origin and representing the pressure P and temperature T. And the equation expresses that the ultimate distance is less or at most equal to the initial. It is evidently immaterial whether the distances be measured vertically or normally, or that the fixed plane representing P and T should pass through the origin; but distances must be considered negative when measured from a point below the plane.

It is evident that the sign of inequality holds in (e) if it holds in either (b) or (c), therefore, it holds in (e) if there are any differences of pressure or temperature between the different parts of the body or between the body and the medium, or if any part of the body has sensible motion. (In the latter case, there would be an increase of entropy due to the conversion of this motion into heat.) But even if the body is initially without sensible motion and has throughout the same pressure and temperature as the medium, the sign < will still hold if different parts of the body are in states represented by points in the thermodynamic surface at different distances from the fixed plane representing P and T. For it certainly holds if such initial circumstances are followed by differences of pressure or temperature, or by sensible velocities. Again, the sign of inequality would necessarily hold if one part of the body should pass, without producing changes of pressure or temperature or sensible velocities, into the state of another part represented by a point not at the same distance from the fixed plane representing P and T. But these are the only suppositions possible in the case, unless we suppose that equilibrium subsists, which would require that the points in question should have a common tangent plane (page 37), whereas by supposition the planes tangent at the different points are parallel but not identical.

The results of the preceding paragraph may be summed up as follows:—Unless the body is initially without sensible motion, and its state, if homogeneous, is such as is represented by a point in the primitive surface where the tangent plane is parallel to the fixed plane representing P and T, or, if the body is not homogeneous in state, unless the points in the primitive surface representing the states of its parts have a common tangent plane parallel to the fixed plane representing P and T, such changes will ensue that the distance of the point representing the volume, entropy, and energy of the body from that fixed plane will be diminished (distances being considered negative if measured from points beneath the plane). Let us apply this result to the question of the stability of the body when surrounded, as supposed, by a medium of constant temperature and pressure.

The state of the body in equilibrium will be represented by a point in the thermodynamic surface, and as the pressure and temperature of the body are the same as those of the surrounding medium, we may take the tangent plane at that point as the fixed plane representing P and T. If the body is not homogeneous in state, although in equilibrium, we may, for the purposes of this discussion of stability, either take a point in the derived surface as representing its state, or we may take the points in the primitive surface which represent the states of the different parts of the body. These points, as we have seen (page 37), have a common tangent plane, which is identical with the tangent plane for the point in the derived surface.

Now, if the form of the surface be such that it falls above the tangent plane except at the single point of contact, the equilibrium is necessarily stable; for if the condition of the body be slightly altered, either by imparting sensible motion to any part of the body, or by slightly changing the state of any part, or by bringing any small part into any other thermodynamic state whatever, or in all of these ways, the point representing the volume, entropy, and energy of the whole body will then occupy a position *above* the original tangent plane, and the proposition above enunciated shows that processes will ensue which will diminish the distance of this point from that plane, and that such processes cannot cease until the body is brought back into its original condition, when they will necessarily cease on account of the form supposed of the surface.

On the other hand, if the surface have such a form that any part of it falls below the fixed tangent plane, the equilibrium will be unstable. For it will evidently be possible by a slight change in the original condition of the body (that of equilibrium with the surrounding medium and represented by the point or points of contact) to bring the point representing the volume, entropy, and energy of the body into a position *below* the fixed tangent plane, in which case we see by the above proposition that processes will occur which will carry the point still farther from the plane, and that such processes cannot cease until all the body has passed into some state entirely different from its original state.

It remains to consider the case in which the surface, although it does not anywhere fall below the fixed tangent plane, nevertheless meets the plane in more than one point. The equilibrium in this case, as we might anticipate from its intermediate character between the cases already considered, is neutral. For if any part of the body be changed from its original state into that represented by another point in the thermodynamic surface lying in the same tangent plane, equilibrium will still subsist. For the supposition in regard to the form of the surface implies that uniformity in temperature and pressure still subsists, nor can the body have any necessary tendency to pass entirely into the second state or to return into the original state, for a change of the values of T and P less than any assignable quantity would evidently be sufficient to reverse such a tendency if any such existed, as either point at will could by such an infinitesimal variation of T and P be made the nearer to the plane representing T and P.

It must be observed that in the case where the thermodynamic surface at a certain point is concave upward in both its principal curvatures, but somewhere falls below the tangent plane drawn through that point, the equilibrium although unstable in regard to *discontinuous* changes of state is stable in regard to *continuous* changes, as appears on restricting the test of stability to the vicinity of the point in question; that is, if we suppose a body to be in a state represented by such a point, although the equilibrium would show itself unstable if we should introduce into the body a small portion of the same substance in one of the states represented by points below the tangent plane, yet if the conditions necessary for such a discontinuous change are not present, the equilibrium would be stable. A familiar example of this is afforded by liquid water when heated at any pressure above the temperature of boiling water at that pressure.*

Leading Features of the Thermodynamic Surface for Substances which take the forms of Solid, Liquid, and Vapor.

We are now prepared to form an idea of the general character of the primitive and derived surfaces and their mutual relations for a substance which takes the forms of solid, liquid, and vapor. The primitive surface will have a triple tangent plane touching it at the three points which represent the three states which can exist in contact. Except at these three points, the primitive surface falls entirely above the tangent plane. That part of the plane which forms a triangle having its vertices at the three points of contact, is the derived surface which represents a compound of the three states of the substance. We may now suppose the plane to roll on the under side of the surface, continuing to touch it in two points without cutting it. This it may do in three ways, viz: it may commence by turning about any one of the sides of the triangle aforesaid. Any pair of points which the plane touches at once represent states which can exist permanently in contact. In this way six lines are traced upon the surface. These lines have in general a common property, that a tangent plane at any point in them will also touch the surface in another point. We must say *in general*, for, as we shall see hereafter, this statement does not hold good for the critical point. A tangent plane at any point of the surface *outside* of these lines has the surface

---

* If we wish to express in a single equation the necessary and sufficient condition of thermodynamic equilibrium for a substance when surrounded by a medium of constant pressure P and temperature T, this equation may be written

$$\delta (\epsilon - T\eta + Pv) = 0,$$

when δ refers to the variation produced by any variations in the state of the parts of the body, and (when different parts of the body are in different states) in the proportion in which the body is divided between the different states. The condition of stable equilibrium is that the value of the expression in the parenthesis shall be a minimum.

entirely above it, except the single point of contact. A tangent plane at any point of the primitive surface *within* these lines will cut the surface. These lines, therefore, taken together may be called the *limit of absolute stability*, and the surface outside of them, the *surface of absolute stability*. That part of the envelop of the rolling plane, which lies between the pair of lines which the plane traces on the surface, is a part of the derived surface, and represents a mixture of two states of the substance.

The relations of these lines and surfaces are roughly represented in horizontal projection* in figure 2, in which the full lines represent lines on the primitive surface, and the dotted lines those on the derived surface. S, L, and V are the points which have a common tangent

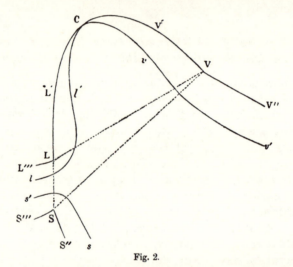

Fig. 2.

plane and represent the states of solid, liquid, and vapor which can exist in contact. The plane triangle SLV is the derived surface representing compounds of these states. LL' and VV' are the pair of lines traced by the rolling double tangent plane, between which lies the derived surface representing compounds of liquid and vapor. VV″ and SS″ are another such pair, between which lies the derived surface representing compounds of vapor and solid. SS‴ and LL‴ are the third pair, between which lies the derived surface representing a compound of solid and liquid. L‴LL′, V′VV″ and S″SS‴ are the boundaries of the surfaces which represent respectively the absolutely stable states of liquid, vapor, and solid.

The geometrical expression of the results which Dr. Andrews,

---

* A horizontal projection of the thermodynamic surface is identical with the diagram described on pages 20–28 of this volume, under the name of the volume-entropy diagram.

Phil. Trans., vol. 159, p. 575, has obtained by his experiments with carbonic acid is that, in the case of this substance at least, the derived surface which represents a compound of liquid and vapor is terminated as follows: as the tangent plane rolls upon the primitive surface, the two points of contact approach one another and finally fall together. The rolling of the double tangent plane necessarily comes to an end. The point where the two points of contact fall together is the *critical point*. Before considering farther the geometrical characteristics of this point and their physical significance, it will be convenient to investigate the nature of the primitive surface which lies between the lines which form the limit of absolute stability.

Between two points of the primitive surface which have a common tangent plane, as those represented by L′ and V′ in figure 2, if there is no gap in the primitive surface, there must evidently be a region where the surface is concave toward the tangent plane in one of its principal curvatures at least, and therefore represents states of unstable equilibrium in respect to continuous as well as discontinuous changes (see pages 42, 43).* If we draw a line upon the primitive surface, dividing it into parts which represent respectively stable and unstable equilibrium, in respect to continuous changes, i.e., dividing the surface which is concave upward in both its principal curvatures from that which is concave downward in one or both, this line, which may be called the *limit of essential instability*, must have a form somewhat like that represented by *ll′Cvv′ss′* in figure 2. It touches the limit of absolute stability at the critical point C. For we may take a pair of points in LC and VC having a common tangent plane as near to C as we choose, and the line joining them upon the primitive surface made by a plane section perpendicular to the tangent plane, will pass through an area of instability.

The geometrical properties of the critical point in our surface may be made more clear by supposing the lines of curvature drawn upon the surface for one of the principal curvatures, that one, namely, which has different signs upon different sides of the limit of essential instability. The lines of curvature which meet this line will in general cross it. At any point where they do so, as the sign of their curvature changes, they evidently cut a plane tangent to the surface, and therefore the surface itself cuts the tangent plane. But where one of these lines of curvature touches the limit of essential instability without crossing it, so that its curvature remains always positive (curvatures being considered positive when the concavity is on the upper side of the surface), the surface evidently does not cut the

---

* This is the same result as that obtained by Professor J. Thomson in connection with the surface referred to in the note on page 34.

tangent plane, but has a contact of the third order with it in the section of least curvature. The critical point, therefore, must be a point where the line of that principal curvature which changes its sign is tangent to the line which separates positive from negative curvatures.

From the last paragraphs we may derive the following physical property of the critical state:—Although this is a limiting state between those of stability and those of instability in respect to continuous changes, and although such limiting states are in general unstable in respect to such changes, yet the critical state is stable in regard to them. A similar proposition is true in regard to absolute stability, i.e., if we disregard the distinction between continuous and discontinuous changes, viz: that although the critical state is a limiting state between those of stability and instability, and although the equilibrium of such limiting states is in general neutral (when we suppose the substance surrounded by a medium of constant pressure and temperature), yet the critical point is stable.

From what has been said of the curvature of the primitive surface at the critical point, it is evident, that if we take a point in this surface infinitely near to the critical point, and such that the tangent planes for these two points shall intersect in a line perpendicular to the section of least curvature at the critical point, the angle made by the two tangent planes will be an infinitesimal of the same order as the cube of the distance of these points. Hence, at the critical point

$$\left(\frac{dp}{dv}\right)_t = 0, \qquad \left(\frac{dp}{d\eta}\right)_t = 0, \qquad \left(\frac{dt}{dv}\right)_p = 0, \qquad \left(\frac{dt}{d\eta}\right)_p = 0,$$

$$\left(\frac{d^2p}{dv^2}\right)_t = 0, \qquad \left(\frac{d^2p}{d\eta^2}\right)_t = 0, \qquad \left(\frac{d^2t}{dv^2}\right)_p = 0, \qquad \left(\frac{d^2t}{d\eta^2}\right)_p = 0,$$

and if we imagine the isothermal and isopiestic (line of constant pressure) drawn for the critical point upon the primitive surface, these lines will have a contact of the second order.

Now the elasticity of the substance at constant temperature and its specific heat at constant pressure may be defined by the equations,

$$e = -v\left(\frac{dp}{dv}\right)_t, \quad s = t\left(\frac{d\eta}{dt}\right)_p;$$

therefore at the critical point

$$e = 0, \qquad \tfrac{1}{s} = 0,$$

$$\left(\frac{de}{dv}\right)_t = 0, \qquad \left(\frac{de}{d\eta}\right)_t = 0, \qquad \left(\frac{d\frac{1}{s}}{dv}\right)_p = 0, \qquad \left(\frac{d\frac{1}{s}}{d\eta}\right)_p = 0.$$

The last four equations would also hold good if p were substituted for t, and vice versa.

We have seen that in the case of such substances as can pass continuously from the state of liquid to that of vapor, unless the primitive surface is abruptly terminated, and that in a line which passes through the critical point, a part of it must represent states which are essentially unstable (i.e., unstable in regard to continuous changes), and therefore cannot exist permanently unless in very limited spaces. It does not necessarily follow that such states cannot be realized at all. It appears quite probable, that a substance initially in the critical state may be allowed to expand so rapidly that, the time being too short for appreciable conduction of heat, it will pass into some of these states of essential instability. No other result is possible on the supposition of no transmission of heat, which requires that the points representing the states of all the parts of the body shall be confined to the isentropic (adiabatic) line of the critical point upon the primitive surface. It will be observed that there is no instability in regard to changes of state thus limited, for this line (the plane section of the primitive surface perpendicular to the axis of η) is concave upward, as is evident from the fact that the primitive surface lies entirely above the tangent plane for the critical point.

We may suppose waves of compression and expansion to be propagated in a substance initially in the critical state. The velocity of propagation will depend upon the value of $\left(\dfrac{dp}{dv}\right)_\eta$, i.e., of $-\left(\dfrac{d^2\epsilon}{d^2v}\right)_\eta$. Now for a wave of compression the value of these expressions is determined by the form of the isentropic on the primitive surface. If a wave of expansion has the same velocity approximately as one of compression, it follows that the substance when expanded under the circumstances remains in a state represented by the primitive surface, which involves the realization of states of essential instability. The value of $\left(\dfrac{d^2\epsilon}{dv^2}\right)_\eta$ in the derived surface is, it will be observed, totally different from its value in the primitive surface, as the curvature of these surfaces at the critical point is different.

The case is different in regard to the part of the surface between the limit of absolute stability and the limit of essential instability. Here, we have experimental knowledge of some of the states represented. In water, for example, it is well known that liquid states can be realized beyond the limit of absolute stability,—both beyond the part of the limit where vaporization usually commences (LL′ in figure 2), and beyond the part where congelation usually commences (LL‴). That vapor may also exist beyond the limit of absolute stability, i.e., that it may exist at a given temperature at pressures greater than that of equilibrium between the vapor and its liquid meeting in a plane surface at that temperature, the considerations adduced by Sir

W. Thomson in his paper "On the equilibrium of a vapor at the
curved surface of a liquid" (*Proc. Roy. Soc. Edinb.*, Session 1869-1870,
and *Phil. Mag.*, vol. xlii, p. 448), leave no room for doubt. By experiments like that suggested by Professor J. Thomson in his paper
already referred to, we may be able to carry vapors farther beyond
the limit of absolute stability.* As the resistance to deformation
characteristic of solids evidently tends to prevent a discontinuous
change of state from commencing within them, substances can doubtless exist in solid states very far beyond the limit of absolute stability.

The surface of absolute stability, together with the triangle representing a compound of three states, and the three developable surfaces
which have been described representing compounds of two states,
forms a continuous sheet, which is everywhere concave upward
except where it is plane, and has only one value of ϵ for any given
values of v and η. Hence, as t is necessarily positive, it has only one
value of η for any given values of v and ϵ If vaporization can take
place at every temperature except 0, p is everywhere positive, and
the surface has only one value of v for any given values of η and ϵ.
It forms the *surface of dissipated energy*. If we consider all the
points representing the volume, entropy, and energy of the body in
every possible state, whether of equilibrium or not, these points will
form a solid figure unbounded in some directions, but bounded in
others by this surface. †

---

* If we experiment with a fluid which does not wet the vessel which contains it,
we may avoid the necessity of keeping the vessel hotter than the vapor, in order to
prevent condensation. If a glass bulb with a stem of sufficient length be placed vertically
with the open end of the stem in a cup of mercury, the stem containing nothing but
mercury and its vapor, and the bulb nothing but the vapor, the height at which the
mercury rests in the stem, affords a ready and accurate means of determining the
pressure of the vapor. If the stem at the top of the column of liquid should be made
hotter than the bulb, condensation would take place in the latter, if the liquid were one
which would wet the bulb. But as this is not the case, it appears probable, that if
the experiment were conducted with proper precautions, there would be no condensation within certain limits in regard to the temperatures. If condensation should take
place, it would be easily observed, especially if the bulb were bent over, so that the
mercury condensed could not run back into the stem. So long as condensation does
not occur, it will be easy to give any desired (different) temperatures to the bulb and
the top of the column of mercury in the stem. The temperature of the latter will
determine the pressure of the vapor in the bulb. In this way, it would appear, we
may obtain in the bulb vapor of mercury having pressures greater for the temperatures than those of saturated vapor.

† This description of the surface of dissipated energy is intended to apply to a substance capable of existing as solid, liquid, and vapor, and which presents no anomalies
in its thermodynamic properties. But, whatever the form of the primitive surface
may be, if we take the parts of it for every point of which the tangent plane does
not cut the primitive surface, together with all the plane and developable derived
surfaces which can be formed in a manner analogous to those described in the preceding
pages, by fixed and rolling tangent planes which do not cut the primitive surface,—

The lines traced upon the primitive surface by the rolling double tangent plane, which have been called the limit of absolute stability, do not end at the vertices of the triangle which represents a mixture of those states. For when the plane is tangent to the primitive surface in these three points, it can commence to roll upon the surface as a double tangent plane not only by leaving the surface at one of these points, but also by a rotation in the opposite direction. In the latter case, however, the lines traced upon the primitive surface by the points of contact, although a continuation of the lines previously described, do not form any part of the limit of absolute stability. And the parts of the envelops of the rolling plane between these lines, although a continuation of the developable surfaces which have been described, and representing states of the body, of which some at least may be realized, are of minor interest, as they form no part of the surface of dissipated energy on the one hand, nor have the theoretical interest of the primitive surface on the other.

Problems relating to the Surface of Dissipated Energy.

The surface of dissipated energy has an important application to a certain class of problems which refer to the results which are theoretically possible with a given body or system of bodies in a given initial condition.

For example, let it be required to find the greatest amount of mechanical work which can be obtained from a given quantity of a certain substance in a given initial state, without increasing its total volume or allowing heat to pass to or from external bodies, except

---

such surfaces taken together will form a continuous sheet, which, if we reject the part, if any, for which $p < 0$, forms the surface of dissipated energy and has the geometrical properties mentioned above.

There will, however, be no such part in which $p < 0$, if there is any assignable temperature t' at which the substance has the properties of a perfect gas except when its volume is less than a certain quantity v'. For the equations of an isothermal line in the thermodynamic surface of a perfect gas are (see equations (B) and (E) on pages 12-13)

$$\epsilon = C$$
$$\eta = a \log v + C'.$$

The isothermal of t' in the thermodynamic surface of the substance in question must therefore have the same equations in the part in which v exceeds the constant v'. Now if at any point in this surface $p < 0$ and $t > 0$ the equation of the tangent plane for that point will be

$$\epsilon = m\eta + nv + C'',$$

where m denotes the temperature and $-n$ the pressure for the point of contact, so that m and n are both positive. Now it is evidently possible to give so large a value to v in the equations of the isothermal that the point thus determined shall fall below the tangent plane. Therefore, the tangent plane cuts the primitive surface, and the point of the thermodynamic surface for which $p < 0$ cannot belong to the surfaces mentioned in the last paragraph as forming a continuous sheet.

such as at the close of the processes are left in their initial condition. This has been called the *available energy* of the body. The initial state of the body is supposed to be such that the body can be made to pass from it to states of dissipated energy by reversible processes.

If the body is in a state represented by any point of the surface of dissipated energy, of course no work can be obtained from it under the given conditions. But even if the body is in a state of thermodynamic equilibrium, and therefore in one represented by a point in the thermodynamic surface, if this point is not in the surface of dissipated energy, because the equilibrium of the body is unstable in regard to discontinuous changes, a certain amount of energy will be available under the conditions for the production of work. Or, if the body is solid, even if it is uniform in state throughout, its pressure (or tension) may have different values in different directions, and in this way it may have a certain available energy. Or, if different parts of the body are in different states, this will in general be a source of available energy. Lastly, we need not exclude the case in which the body has sensible motion and its *vis viva* constitutes available energy. In any case, we must find the initial volume, entropy, and energy of the body, which will be equal to the sums of the initial volumes, entropies, and energies of its parts. ('Energy' is here used to include the *vis viva* of sensible motions.) These values of v, η, and ϵ will determine the position of a certain point which we will speak of as representing the initial state.

Now the condition that no heat shall be allowed to pass to external bodies, requires that the final entropy of the body shall not be less than the initial, for it could only be made less by violating this condition. The problem, therefore, may be reduced to this,—to find the amount by which the energy of the body may be diminished without increasing its volume or diminishing its entropy. This quantity will be represented geometrically by the distance of the point representing the initial state from the surface of dissipated energy measured parallel to the axis of ϵ.

Let us consider a different problem. A certain initial state of the body is given as before. No work is allowed to be done upon or by external bodies. Heat is allowed to pass to and from them only on condition that the algebraic sum of all heat which thus passes shall be 0. From both these conditions any bodies may be excepted, which shall be left at the close of the processes in their initial state. Moreover, it is not allowed to increase the volume of the body. It is required to find the greatest amount by which it is possible under these conditions to diminish the entropy of an external system. This will be, evidently, the amount by which the entropy of the body can be increased without changing the energy of the body or increasing its volume, which is represented geometrically by the distance of the point representing the initial state from the surface of dissipated energy, measured parallel to the axis of η. This might be called the capacity for entropy of the body in the given state.*

---

* It may be worth while to call attention to the analogy and the difference between this problem and the preceding. In the first case, the question is virtually, how great a weight does the state of the given body enable us to raise a given distance, no other permanent change being produced in external bodies? In the second case, the question is virtually, what amount of heat does the state of the given body enable us to take from an external body at a fixed temperature, and impart to another at a higher fixed temperature? In order that the numerical values of the available energy and of the capacity for entropy should be identical with the answers to these questions, it would be necessary in the first case, if the weight is measured in units of force, that the given distance, measured vertically, should be the unit of length, and in the second case, that the difference of the reciprocals of the fixed temperatures should be unity. If we prefer to take the freezing and boiling points as the fixed temperatures, as $\frac{1}{273} - \frac{1}{373} = 0 \cdot 00098$, the capacity for entropy of the body in any given condition would be $0 \cdot 00098$ times the amount of heat which it would enable us to raise from the freezing to the boiling point (i.e., to take from the body of which the temperature remains fixed at the freezing point, and impart to another of which the temperature remains fixed at the boiling point).

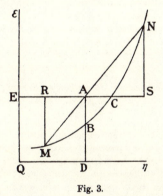

Fig. 3.

The relations of these quantities to one another and to the surface of dissipated energy are illustrated by figure 3, which represents a plane perpendicular to the axis of v and passing through the point A, which represents the initial state of the body. MN is the section of the surface of dissipated energy. Qϵ and Qη are sections of the planes $\eta = 0$ and $\epsilon = 0$, and therefore parallel to the axes of ϵ and η respectively. AD and AE are the energy and entropy of the body in its initial state, AB and AC its available energy and its capacity for entropy respectively. It will be observed that when either the available energy or the capacity for entropy of the body is 0, the other has the same value. Except in this case, either quantity may be varied without affecting the other. For, on account of the curvature of the surface of dissipated energy, it is evidently possible to change the position of the point representing the initial state of the body so as to vary its distance from the surface measured parallel to one axis without varying that measured parallel to the other.

As the different sense in which the word *entropy* has been used by different writers is liable to cause misunderstanding, it may not be out of place to add a

Thirdly.   A certain initial condition of the body is given as before. No work is allowed to be done upon or by external bodies, nor any heat to pass to or from them, from which conditions bodies may be excepted, as before, in which no permanent changes are produced. It is required to find the amount by which the volume of the body can be diminished, using for that purpose, according to the conditions, only the force derived from the body itself.   The conditions require that the energy of the body shall not be altered nor its entropy diminished.   Hence the quantity sought is represented by the distance of the point representing the initial state from the surface of dissipated energy, measured parallel to the axis of volume.

Fourthly.   An initial condition of the body is given as before.   Its volume is not allowed to be increased.   No work is allowed to be done upon or by external bodies, nor any heat to pass to or from them, except a certain body of given constant temperature t'.   From the latter conditions may be excepted as before bodies in which no permanent changes are produced.   It is required to find the greatest amount of heat which can be imparted to the body of constant temperature, and also the greatest amount of heat which can be taken from it, under the supposed conditions.   If through the point of the

---

few words on the terminology of this subject.   If Professor Clausius had defined *entropy* so that its value should be determined by the equation

$$dS = -\frac{dQ}{T},$$

instead of his equation (*Mechanische Wärmetheorie*, Abhand. ix. § 14 ; *Pogg. Ann.* July, 1865)

$$dS = \frac{dQ}{T},$$

where S denotes the entropy and T the temperature of a body and dQ the element of heat imparted to it, that which is here called *capacity for entropy* would naturally be called *available entropy*, a term the more convenient on account of its analogy with the term *available energy*.   Such a difference in the definition of *entropy* would involve no difference in the form of the thermodynamic surface, nor in any of our geometrical constructions, if only we suppose the direction in which entropy is measured to be reversed.   It would only make it necessary to substitute $-\eta$ for η in our equations, and to make the corresponding change in the verbal enunciation of propositions. Professor Tait has proposed to use the word entropy " in the opposite sense to that in which Clausius has employed it " (*Thermodynamics*, § 48.   See also § 178), which appears to mean that he would determine its value by the first of the above equations. He nevertheless appears subsequently to use the word to denote available energy (§ 182, 2d theorem).   Professor Maxwell uses the word entropy as synonymous with available energy, with the erroneous statement that Clausius uses the word to denote the part of the energy which is not available (*Theory of Heat*, pp. 186 and 188).   The term entropy, however, as used by Clausius does not denote a quantity of the same kind (i.e., one which can be measured by the same unit) as energy, as is evident from his equation, cited above, in which Q (heat) denotes a quantity measured by the unit of energy, and as the unit in which T (temperature) is measured is arbitrary, S and Q are evidently measured by different units.   It may be added that entropy as defined by Clausius is synonymous with the thermodynamic function as defined by Rankine.

initial state a straight line be drawn in the plane perpendicular to the axis of v, so that the tangent of the angle which it makes with the direction of the axis of η shall be equal to the given temperature t', it may easily be shown that the vertical projections of the two segments of this line made by the point of the initial state and the surface of dissipated energy represent the two quantities required.*

These problems may be modified so as to make them approach more nearly the economical problems which actually present themselves, if we suppose the body to be surrounded by a medium of constant pressure and temperature, and let the body and the medium together take the place of the body in the preceding problems. The results would be as follows:

If we suppose a plane representing the constant pressure and temperature of the medium to be tangent to the surface of dissipated energy of the body, the distance of the point representing the initial state of the body from this plane measured parallel to the axis of ϵ will represent the available energy of the body and medium, the distance of the point to the plane measured parallel to the axis of η will represent the capacity for entropy of the body and medium, the distance of the point to the plane measured parallel to the axis of v will represent the magnitude of the greatest vacuum which can be produced in the body or medium (all the power used being derived from the body and medium); if a line be drawn through the point in a plane perpendicular to the axis of v, the vertical projection of the segment of this line made by the point and the tangent plane will represent the greatest amount of heat which can be given to or taken from another body at a constant temperature equal to the tangent of the inclination of the line to the horizon. (It represents the greatest amount which can be given to the body of constant temperature, if this temperature is greater than that of the medium; in the reverse case, it represents the greatest amount which can be withdrawn from that body.) In all these cases, the point of contact between the plane and the surface of dissipated energy represents the final state of the given body.

If a plane representing the pressure and temperature of the medium be drawn through the point representing any given initial state of the body, the part of this plane which falls within the surface of dissipated energy will represent in respect to volume, entropy, and energy all the states into which the body can be brought by reversible processes, without producing permanent changes in external bodies (except in the medium), and the solid figure included between

---

* Thus, in figure 3, if the straight line MAN be drawn so that tan NAC$=t'$, MR will be the greatest amount of heat which can be given to the body of constant temperature and NS will be the greatest amount which can be taken from it.

this plane figure and the surface of dissipated energy will represent all the states into which the body can be brought by any kind of processes, without producing permanent changes in external bodies (except in the medium).*

---

* The body under discussion has been supposed throughout this paper to be homogeneous in substance. But if we imagine any material system whatever, and suppose the position of a point to be determined for every possible state of the system, by making the co-ordinates of the point equal to the total volume, entropy, and energy of the system, the points thus determined will evidently form a solid figure bounded in certain directions by the surface representing the states of dissipated energy. In these states, the temperature is necessarily uniform throughout the system; the pressure may vary (e.g., in the case of a very large mass like a planet), but it will always be possible to maintain the equilibrium of the system (in a state of dissipated energy) by a uniform normal pressure applied to its surface. This pressure and the uniform temperature of the system will be represented by the inclination of the surface of dissipated energy according to the rule on page 34. And in regard to such problems as have been discussed in the last five pages, this surface will possess, relatively to the system which it represents, properties entirely similar to those of the surface of dissipated energy of a homogeneous body.

III.

ON THE EQUILIBRIUM OF HETEROGENEOUS SUBSTANCES.

[*Transactions of the Connecticut Academy*, III. pp. 108–248, Oct. 1875–May, 1876, and pp. 343–524, May, 1877–July, 1878.]

" Die Energie der Welt ist constant.
Die Entropie der Welt strebt einem Maximum zu."
CLAUSIUS.*

THE comprehension of the laws which govern any material system is greatly facilitated by considering the energy and entropy of the system in the various states of which it is capable. As the difference of the values of the energy for any two states represents the combined amount of work and heat received or yielded by the system when it is brought from one state to the other, and the difference of entropy is the limit of all the possible values of the integral $\int \frac{dQ}{t}$, (*dQ* denoting the element of the heat received from external sources, and *t* the temperature of the part of the system receiving it,) the varying values of the energy and entropy characterize in all that is essential the effects producible by the system in passing from one state to another. For by mechanical and thermodynamic contrivances, supposed theoretically perfect, any supply of work and heat may be transformed into any other which does not differ from it either in the amount of work and heat taken together or in the value of the integral $\int \frac{dQ}{t}$. But it is not only in respect to the external relations of a system that its energy and entropy are of predominant importance. As in the case of simply mechanical systems, (such as are discussed in theoretical mechanics,) which are capable of only one kind of action upon external systems, viz., the performance of mechanical work, the function which expresses the capability of the system for this kind of action also plays the leading part in the theory of equilibrium, the condition of equilibrium being that the variation of this function shall vanish, so in a thermodynamic system, (such as all material systems actually are,) which is capable of

---

* *Pogg. Ann.* Bd. cxxv. (1865), S. 400 ; or *Mechanische Wärmetheorie*, Abhand. ix. S. 44.

two different kinds of action upon external systems, the two functions which express the twofold capabilities of the system afford an almost equally simple criterion of equilibrium.

Criteria of Equilibrium and Stability.

The criterion of equilibrium for a material system which is isolated from all external influences may be expressed in either of the following entirely equivalent forms:—

I. *For the equilibrium of any isolated system it is necessary and sufficient that in all possible variations of the state of the system which do not alter its energy, the variation of its entropy shall either vanish or be negative.* If ϵ denote the energy, and η the entropy of the system, and we use a subscript letter after a variation to indicate a quantity of which the value is not to be varied, the condition of equilibrium may be written

$$(\delta\eta)_\epsilon \leqq 0. \tag{1}$$

II. *For the equilibrium of any isolated system it is necessary and sufficient that in all possible variations in the state of the system which do not alter its entropy, the variation of its energy shall either vanish or be positive.* This condition may be written

$$(\delta\epsilon)_\eta \geqq 0. \tag{2}$$

That these two theorems are equivalent will appear from the consideration that it is always possible to increase both the energy and the entropy of the system, or to decrease both together, viz., by imparting heat to any part of the system or by taking it away. For, if condition (1) is not satisfied, there must be some variation in the state of the system for which

$$\delta\eta > 0 \quad \text{and} \quad \delta\epsilon = 0;$$

therefore, by diminishing both the energy and the entropy of the system *in its varied state,* we shall obtain a state for which (considered as a variation from the original state)

$$\delta\eta = 0 \quad \text{and} \quad \delta\epsilon < 0;$$

therefore condition (2) is not satisfied. Conversely, if condition (2) is not satisfied, there must be a variation in the state of the system for which

$$\delta\epsilon < 0 \quad \text{and} \quad \delta\eta = 0;$$

hence there must also be one for which

$$\delta\epsilon = 0 \quad \text{and} \quad \delta\eta > 0;$$

therefore condition (1) is not satisfied.

The equations which express the condition of equilibrium, as also its statement in words, are to be interpreted in accordance with the general usage in respect to differential equations, that is, infinitesimals of higher orders than the first relatively to those which express the amount of change of the system are to be neglected. But to distinguish the different kinds of equilibrium in respect to stability, we must have regard to the absolute values of the variations. We will use Δ as the sign of variation in those equations which are to be construed *strictly*, i.e., in which infinitesimals of the higher orders are not to be neglected. With this understanding, we may express the necessary and sufficient conditions of the different kinds of equilibrium as follows;—for stable equilibrium

$$(\Delta\eta)_\epsilon < 0, \text{ i.e., } (\Delta\epsilon)_\eta > 0; \qquad (3)$$

for neutral equilibrium there must be some variations in the state of the system for which

$$(\Delta\eta)_\epsilon = 0, \text{ i.e., } (\Delta\epsilon)_\eta = 0; \qquad (4)$$

while in general

$$(\Delta\eta)_\epsilon \leqq 0, \text{ i.e., } (\Delta\epsilon)_\eta \geqq 0; \qquad (5)$$

and for unstable equilibrium there must be some variations for which

$$(\Delta\eta)_\epsilon > 0, \qquad (6)$$

i.e., there must be some for which

$$(\Delta\epsilon)_\eta < 0, \qquad (7)$$

while in general

$$(\delta\eta)_\epsilon \leqq 0, \text{ i.e., } (\delta\epsilon)_\eta \geqq 0. \qquad (8)$$

In these criteria of equilibrium and stability, account is taken only of *possible* variations. It is necessary to explain in what sense this is to be understood. In the first place, all variations in the state of the system which involve the transportation of any matter through any finite distance are of course to be excluded from consideration, although they may be capable of expression by infinitesimal variations of quantities which perfectly determine the state of the system. For example, if the system contains two masses of the same substance, not in contact, nor connected by other masses consisting of or containing the same substance or its components, an infinitesimal increase of the one mass with an equal decrease of the other is not to be considered as a possible variation in the state of the system. In addition to such cases of essential impossibility, if heat can pass by conduction or radiation from every part of the system to every other, only those variations are to be rejected as impossible, which involve changes which are prevented by passive forces or analogous resistances to change. But, if the system consist of parts between which there is supposed to be no thermal communication, it will be necessary to regard as impossible any diminution of the entropy of any of these parts, as such a change can not take place without the passage of heat. This limitation may most conveniently be applied to the second of the above forms of the condition of equilibrium, which will then become

$$(\delta\epsilon)_{\eta', \eta'', \text{etc.}} \geqq 0, \qquad (9)$$

η', η'', etc., denoting the entropies of the various parts between which there is no communication of heat. When the condition of equilibrium is thus expressed, the limitation in respect to the conduction of heat will need no farther consideration.

In order to apply to any system the criteria of equilibrium which have been given, a knowledge is requisite of its passive forces or resistances to change, in so far, at least, as they are capable of *preventing* change. (Those passive forces which only retard change, like viscosity, need not be considered.) Such properties of a system are in general easily recognized upon the most superficial knowledge of its nature. As examples, we may instance the passive force of friction which prevents sliding when two surfaces of solids are pressed together,—that which prevents the different components of a solid, and sometimes of a fluid, from having different motions one from another,—that resistance to change which sometimes prevents either of two forms of the same substance (simple or compound), which are capable of existing, from passing into the other,—that which prevents the changes in solids which imply plasticity, (in other words, changes of the form to which the solid tends to return,) when the deformation does not exceed certain limits.

It is a characteristic of all these passive resistances that they prevent a certain kind of motion or change, however the initial state of the system may be modified, and to whatever external agencies of force and heat it may be subjected, within limits, it may be, but yet within limits which allow finite variations in the values of all the quantities which express the initial state of the system or the mechanical or thermal influences acting on it, without producing the change in question. The equilibrium which is due to such passive properties is thus widely distinguished from that caused by the balance of the active tendencies of the system, where an external influence, or a change in the initial state, infinitesimal in amount, is sufficient to produce change either in the positive or negative direction. Hence the ease with which these passive resistances are recognized. Only in the case that the state of the system lies so near the limit at which the resistances cease to be operative to prevent change, as to create a doubt whether the case falls within or without the limit, will a more accurate knowledge of these resistances be necessary.

To establish the validity of the criterion of equilibrium, we will consider first the sufficiency, and afterwards the necessity, of the condition as expressed in either of the two equivalent forms.

In the first place, if the system is in a state in which its entropy is greater than in any other state of the same energy, it is evidently in equilibrium, as any change of state must involve either a decrease of entropy or an increase of energy, which are alike impossible for an isolated system. We may add that this is a case of *stable* equilibrium, as no infinitely small cause (whether relating to a variation of the initial state or to the action of any external bodies) can produce a finite change of state, as this would involve a finite decrease of entropy or increase of energy.

We will next suppose that the system has the greatest entropy consistent with its energy, and therefore the least energy consistent with its entropy, but that there are other states of the same energy and entropy as its actual state. In this case, it is impossible that any motion of masses should take place; for if any of the energy of the system should come to consist of *vis viva* (of sensible motions), a state of the system identical in other respects but without the motion would have less energy and not less entropy, which would be contrary to the supposition. (But we cannot apply this reasoning to the motion within any mass of its different components in different directions, as in diffusion, when the momenta of the components balance one another.) Nor, in the case supposed, can any conduction of heat take place, for this involves an increase of entropy, as heat is only conducted from bodies of higher to those of lower temperature. It is equally impossible that any changes should be produced by the transfer of heat by radiation. The condition which we have supposed is therefore sufficient for equilibrium, so far as the motion of masses and the transfer of heat are concerned, but to show that the same is true in regard to the motions of diffusion and chemical or molecular changes, when these can occur without being accompanied or followed by the motions of masses or the transfer of heat, we must have recourse to considerations of a more general nature. The following considerations seem to justify the belief that the condition is sufficient for equilibrium in every respect.

Let us suppose, in order to test the tenability of such a hypothesis, that a system may have the greatest entropy consistent with its energy without being in equilibrium. In such a case, changes in the state of the system must take place, but these will necessarily be such that the energy and the entropy will remain unchanged and the system will continue to satisfy the same condition, as initially, of having the greatest entropy consistent with its energy. Let us consider the change which takes place in any time so short that the change may be regarded as uniform in nature throughout that time. This time must be so chosen that the change does not take place in it infinitely slowly, which is always easy, as the change which we suppose to take place cannot be infinitely slow except at particular moments. Now no change whatever in the state of the system, which does not alter the value of the energy, and which commences with the same state in which the system was supposed at the commencement of the short time considered, will cause an increase of entropy. Hence, it will generally be possible by some slight variation in the circumstances of the case to make all changes in the state of the system like or nearly like that which is supposed actually to occur, and not involving a change of energy, to involve a necessary decrease of entropy, which would render any such change impossible. This variation may be in the values of the variables which determine the state of the system, or in the values of the constants which determine the nature of the system, or in the form of the functions which express its laws,—only there must be nothing in the system as modified which is thermodynamically impossible. For example, we might suppose temperature or pressure to be varied, or the composition of the different bodies in the system, or, if no small variations which could be actually realized would produce the required result, we might suppose the properties themselves of the substances to undergo variation, subject to the general laws of matter. If, then, there is any tendency toward change in the system as first supposed, it is a tendency which can be entirely checked by an infinitesimal variation in the circumstances of the case. As this supposition cannot be allowed, we must believe that a system is always in equilibrium when it has the greatest entropy consistent with its energy, or, in other words, when it has the least energy consistent with its entropy.

The same considerations will evidently apply to any case in which a system is in such a state that $\Delta\eta \leqq 0$ for any possible infinitesimal variation of the state for which $\Delta\epsilon = 0$, even if the entropy is not the greatest of which the system is capable with the same energy. (The term *possible* has here the meaning previously defined, and the character Δ is used, as before, to denote that the equations are to be construed strictly, i.e., without neglect of the infinitesimals of the higher orders.)

The only case in which the sufficiency of the condition of equilibrium which has been given remains to be proved is that in which in our notation $\delta\eta \leqq 0$ for all possible variations not affecting the energy, but for some of these variations $\Delta\eta > 0$, that is, when the entropy has in some respects the characteristics of a minimum. In this case the considerations adduced in the last paragraph will not apply without modification, as the change of state may be infinitely slow at first, and it is only in the initial state that the condition $\delta\eta_e \leqq 0$ holds true. But the differential coefficients of all orders of the quantities which determine the state of the system, taken with respect of the time, must be functions of these same quantities. None of these differential coefficients can have any value other than 0, for the state of the system for which $\delta\eta_\epsilon \leqq 0$. For otherwise, as it would generally be possible, as before, by some infinitely small modification of the case, to render impossible any change like or nearly like that which might be supposed to occur, this infinitely small modification of the case would make a finite difference in the value of the differential coefficients which had before the finite values, or in some of lower orders, which is contrary to that continuity which we have reason to expect. Such considerations seem to justify us in regarding such a state as we are discussing as one of theoretical equilibrium; although as the equilibrium is evidently unstable, it cannot be realized.

We have still to prove that the condition enunciated is in every case necessary for equilibrium. It is evidently so in all cases in which the active tendencies of the system are so balanced that changes of every kind, except those excluded in the statement of the condition of equilibrium, can take place *reversibly*, (i.e., both in the positive and the negative direction,) in states of the system differing infinitely little from the state in question. In this case, we may omit the sign of inequality and write as the condition of such a state of equilibrium

$$(\delta\eta)_\epsilon = 0, \text{ i.e., } (\delta\epsilon)_\eta = 0. \tag{10}$$

But to prove that the condition previously enunciated is in every case necessary, it must be shown that whenever an isolated system remains without change, if there is any infinitesimal variation in its state, not involving a finite change of position of any (even an infinitesimal part) of its matter, which would diminish its energy by a quantity which is not infinitely small relatively to the variations of the quantities which determine the state of the system, without altering its entropy,—or, if the system has thermally isolated parts, without altering the entropy of any such part,—this variation involves changes in the system which are prevented by its passive forces or analogous resistances to change. Now, as the described variation in the state of the system diminishes its energy without altering its entropy, it must be regarded as theoretically possible to produce that variation by some process, perhaps a very indirect one, so as to gain a certain amount of work (above all expended on the system). Hence we may conclude that the active forces or tendencies of the system favor the variation in question, and that equilibrium cannot subsist unless the variation is prevented by passive forces.

The preceding considerations will suffice, it is believed, to establish the validity of the criterion of equilibrium which has been given. The criteria of stability may readily be deduced from that of equilibrium. We will now proceed to apply these principles to systems consisting of heterogeneous substances and deduce the special laws which apply to different classes of phenomena. For this purpose we shall use the second form of the criterion of equilibrium, both because it admits more readily the introduction of the condition that there shall be no thermal communication between the different parts of the system, and because it is more convenient, as respects the form of the general equations relating to equilibrium, to make the entropy one of the independent variables which determine the state of the system, than to make the energy one of these variables.

The Conditions of Equilibrium for Heterogeneous Masses in Contact when Uninfluenced by Gravity, Electricity, Distortion of the Solid Masses, or Capillary Tensions.

In order to arrive as directly as possible at the most characteristic and essential laws of chemical equilibrium, we will first give our attention to a case of the simplest kind. We will examine the conditions of equilibrium of a mass of matter of various kinds enclosed in a rigid and fixed envelop, which is impermeable to and unalterable by any of the substances enclosed, and perfectly non-conducting to heat. We will suppose that the case is not complicated by the action of gravity, or by any electrical influences, and that in the solid portions of the mass the pressure is the same in every direction. We will farther simplify the problem by supposing that the variations of the parts of the energy and entropy which depend upon the surfaces separating heterogeneous masses are so small in comparison with the variations of the parts of the energy and entropy which depend upon the quantities of these masses, that the former may be neglected by the side of the latter; in other words, we will exclude the considerations which belong to the theory of capillarity.

It will be observed that the supposition of a rigid and non-conducting envelop enclosing the mass under discussion involves no real loss of generality, for if any mass of matter is in equilibrium, it would also be so, if the whole or any part of it were enclosed in an envelop as supposed; therefore the conditions of equilibrium for a mass thus enclosed are the general conditions which must always be satisfied in case of equilibrium. As for the other suppositions which have been made, all the circumstances and considerations which are here excluded will afterward be made the subject of special discussion.

Conditions relating to the Equilibrium between the initially existing Homogeneous Parts of the given Mass.

Let us first consider the energy of any homogeneous part of the given mass, and its variation for any possible variation in the composition and state of this part. (By *homogeneous* is meant that the part in question is uniform throughout, not only in chemical composition, but also in physical state.) If we consider the amount and kind of matter in this homogeneous mass as fixed, its energy ϵ is a function of its entropy η, and its volume v, and the differentials of these quantities are subject to the relation

$$d\epsilon = t\,d\eta - p\,dv, \tag{11}$$

t denoting the (absolute) temperature of the mass, and p its pressure. For $t\,d\eta$ is the heat received, and $p\,dv$ the work done, by the mass during its change of state. But if we consider the matter in the mass as variable, and write m_1, m_2, ... m_n for the quantities of the various substances S_1, S_2, ... S_n of which the mass is composed, ϵ will evidently be a function of η, v, m_1, m_2, ... m_n, and we shall have for the complete value of the differential of ϵ

$$d\epsilon = t\,d\eta - p\,dv + \mu_1 dm_1 + \mu_2 dm_2 \ldots + \mu_n dm_n, \tag{12}$$

μ_1, μ_2, ... μ_n denoting the differential coefficients of ϵ taken with respect to m_1, m_2, ... m_n.

The substances S_1, S_2, ... S_n, of which we consider the mass composed, must of course be such that the values of the differentials dm_1, dm_2, ... dm_n shall be independent, and shall express every possible variation in the composition of the homogeneous mass considered, including those produced by the absorption of substances different from any initially present. It may therefore be necessary to have terms in the equation relating to component substances which do not initially occur in the homogeneous mass considered, provided, of course, that these substances, or their components, are to be found in some part of the whole given mass.

If the conditions mentioned are satisfied, the choice of the substances which we are to regard as the components of the mass considered, may be determined entirely by convenience, and independently of any theory in regard to the internal constitution of the mass. The number of components will sometimes be greater, and sometimes less, than the number of chemical elements present. For example, in considering the equilibrium in a vessel containing water and free hydrogen and oxygen, we should be obliged to recognize three components in the gaseous part. But in considering the equilibrium of dilute sulphuric acid with the vapor which it yields, we should have only two components to consider in the liquid mass, sulphuric acid (anhydrous, or of any particular degree of concentration) and (additional) water. If, however, we are considering sulphuric acid in a state of maximum concentration in connection with substances which might possibly afford water to the acid, it must be noticed that the condition of the independence of the differentials will require that we consider the acid in the state of maximum concentration as one of the components. The quantity of this component will then be capable of variation both in the positive and in the negative sense, while the quantity of the other component can increase but cannot decrease below the value 0.

For brevity's sake, we may call a substance S_a an *actual component* of any homogeneous mass, to denote that the quantity m_a of that substance in the given mass may be either increased or diminished (although we may have so chosen the other component substances that $m_a = 0$); and we may call a substance S_b a *possible component* to denote that it may be combined with, but cannot be subtracted from the homogeneous mass in question. In this case, as we have seen in the above example, we must so choose the component substances that $m_b = 0$.

The units by which we measure the substances of which we regard the given mass as composed may each be chosen independently. To fix our ideas for the purpose of a general discussion, we may suppose all substances measured by weight or mass. Yet in special cases, it may be more convenient to adopt chemical equivalents as the units of the component substances.

It may be observed that it is not necessary for the validity of equation (12) that the variations of nature and state of the mass to which the equation refers should be such as do not disturb its homogeneity, provided that in all parts of the mass the variations of nature and state are infinitely small. For, if this last condition be not violated, an equation like (12) is certainly valid for all the infinitesimal parts of the (initially) homogeneous mass; i.e., if we write $D\epsilon$, $D\eta$, etc., for the energy, entropy, etc., of any infinitesimal part,

$$dD\epsilon = t\, dD\eta - p\, dDv + \mu_1\, dDm_1 + \mu_2\, dDm_2 \ldots + \mu_n\, dDm_n, \quad (13)$$

whence we may derive equation (12) by integrating for the whole initially homogeneous mass.

We will now suppose that the whole mass is divided into parts so that each part is homogeneous, and consider such variations in the energy of the system as are due to variations in the composition and state of the several parts remaining (at least approximately) homogeneous, and together occupying the whole space within the envelop. We will at first suppose the case to be such that the component substances are the same for each of the parts, each of the substances $S_1, S_2, \ldots S_n$ being an actual component of each part. If we distinguish the letters referring to the different parts by accents, the variation in the energy of the system may be expressed by $\delta\epsilon' + \delta\epsilon'' +$ etc., and the general condition of equilibrium requires that

$$\delta\epsilon' + \delta\epsilon'' + \text{etc.} \geqq 0 \quad (14)$$

for all variations which do not conflict with the *equations of condition*. These equations must express that the entropy of the whole given mass does not vary, nor its volume, nor the total quantities of any of the substances $S_1, S_2, \ldots S_n$. We will suppose that there are no other equations of condition. It will then be necessary for equilibrium that

$$t' \delta\eta' \quad -p' \delta v' \quad +\mu_1' \delta m_1' \quad +\mu_2' \delta m_2' \quad \ldots +\mu_n' \delta m_n'$$
$$+t'' \delta\eta'' -p'' \delta v'' +\mu_1'' \delta m_1'' +\mu_2'' \delta m_2'' \quad \ldots +\mu_n'' \delta m_n''$$
$$+ \text{etc.} \geqq 0 \tag{15}$$

for any values of the variations for which

$$\delta\eta' + \delta\eta'' + \delta\eta''' + \text{etc.} = 0, \tag{16}$$

$$\delta v' + \delta v'' + \delta v''' + \text{etc.} = 0, \tag{17}$$

$$\left.\begin{aligned}
\delta m_1' + \delta m_1'' + \delta m_1''' + \text{etc.} &= 0, \\
\delta m_2' + \delta m_2'' + \delta m_2''' + \text{etc.} &= 0, \\
&\cdots\cdots\cdots\cdots\cdots\cdots\cdots \\
\delta m_n' + \delta m_n'' + \delta m_n''' + \text{etc.} &= 0.
\end{aligned}\right\} \tag{18}$$

For this it is evidently necessary and sufficient that

$$t' = t'' = t''' = \text{etc.} \tag{19}$$

$$p' = p'' = p''' = \text{etc.} \tag{20}$$

$$\left.\begin{aligned}
\mu_1' &= \mu_1'' = \mu_1''' = \text{etc.} \\
\mu_2' &= \mu_2'' = \mu_2''' = \text{etc.} \\
&\cdots\cdots\cdots\cdots\cdots \\
\mu_n' &= \mu_n'' = \mu_n''' = \text{etc.}
\end{aligned}\right\} \tag{21}$$

Equations (19) and (20) express the conditions of thermal and mechanical equilibrium, viz., that the temperature and the pressure must be constant throughout the whole mass. In equations (21) we have the conditions characteristic of chemical equilibrium. If we call a quantity μ_x, as defined by such an equation as (12), the *potential* for the substance S_x in the homogeneous mass considered, these conditions may be expressed as follows:—

The potential for each component substance must be constant throughout the whole mass.

It will be remembered that we have supposed that there is no restriction upon the freedom of motion or combination of the component substances, and that each is an actual component of all parts of the given mass.

The state of the whole mass will be completely determined (if we regard as immaterial the position and form of the various homogeneous parts of which it is composed), when the values are determined of the quantities of which the variations occur in (15). The number of these quantities, which we may call the independent variables, is evidently $(n+2)\nu$, ν denoting the number of homogeneous parts into which the whole mass is divided. All the quantities which occur in (19), (20), (21), are functions of these variables, and may be regarded as known functions, if the energy of each part is known as a function of its entropy, volume, and the quantities of its components. (See eq. (12).) Therefore, equations (19), (20), (21), may be regarded as $(\nu-1)$ $(n+2)$ independent equations between the independent variables. The volume of the whole mass and the total quantities of the various substances being known afford $n+1$ additional equations. If we also know the total energy of the given mass, or its total entropy, we will have as many equations as there are independent variables.

But if any of the substances S_1, S_2, ... S_n are only possible components of some parts of the given mass, the variation δm of the quantity of such a substance in such a part cannot have a negative value, so that the general condition of equilibrium (15) does not require that the potential for that substance in that part should be equal to the potential for the same substance in the parts of which it is an actual component, but only that it shall not be less. In this case instead of (21) we may write

$$\mu_1 = M_1$$

for all parts of which S_1 is an actual component, and

$$\mu_1 \geqq M_1$$

for all parts of which S_1 is a possible (but not actual) component,

$$\mu_2 = M_2$$

for all parts of which S_2 is an actual component, and

$$\mu_2 \geqq M_2$$

for all parts of which S_2 is a possible (but not actual) component,
etc.,

$$\left. \right\} (22)$$

M_1, M_2, etc., denoting constants of which the value is only determined by these equations.

If we now suppose that the components (actual or possible) of the various homogeneous parts of the given mass are not the same, the result will be of the same character as before, provided that all the different components are *independent* (i.e., that no one can be made out of the others), so that the total quantity of each component is fixed. The general condition of equilibrium (15) and the equations of condition (16), (17), (18) will require no change, except that, if any of the substances S_1, S_2, ... S_n is not a component (actual or possible) of any part, the term $\mu \, \delta m$ for that substance and part will be wanting in the former, and the δm in the latter. This will require no change in the form of the particular conditions of equilibrium as expressed by (19), (20), (22); but the number of single conditions contained in (22) is of course less than if all the component substances were components of all the parts. Whenever, therefore, each of the different homogeneous parts of the given mass may be regarded as composed of some or of all of the same set of substances, no one of which can be formed out of the others, the condition which (with equality of temperature and pressure) is necessary and sufficient for equilibrium between the different parts of the given mass may be expressed as follows:—

The potential for each of the component substances must have a constant value in all parts of the given mass of which that substance is an actual component, and have a value not less than this in all parts of which it is a possible component.

The number of *equations* afforded by these conditions, after elimination of $M_1, M_2, \ldots M_n$, will be less than $(n+2)(\nu-1)$ by the number of terms in (15) in which the variation of the form δm is either necessarily nothing or incapable of a negative value. The number of variables to be determined is diminished by the same number, or, if we choose, we may write an equation of the form $m=0$ for each of these terms. But when the substance is a possible component of the part concerned, there will also be a condition (expressed by $\geqq$) to show whether the supposition that the substance is not an actual component is consistent with equilibrium.

We will now suppose that the substances $S_1, S_2, \ldots S_n$ are not all independent of each other, i.e., that some of them can be formed out of others. We will first consider a very simple case. Let S_3 be composed of S_1 and S_2 combined in the ratio of a to b, S_1 and S_2 occurring as actual components in some parts of the given mass, and S_3 in other parts, which do not contain S_1 and S_2 as separately variable components. The general condition of equilibrium will still have the form of (15) with certain of the terms of the form $\mu\delta m$ omitted. It may be written more briefly

$$\Sigma(t\,\delta\eta) - \Sigma(p\,\delta v) + \Sigma(\mu_1\delta m_1) + \Sigma(\mu_2\delta m_2) \ldots + \Sigma(\mu_n\delta m_n) \geqq 0, \quad (23)$$

the sign Σ denoting summation in regard to the different parts of the given mass. But instead of the three equations of condition,

$$\Sigma\,\delta m_1 = 0, \quad \Sigma\,\delta m_2 = 0, \quad \Sigma\,\delta m_3 = 0, \quad (24)$$

we shall have the two,

$$\left.\begin{array}{l} \Sigma\,\delta m_1 + \dfrac{a}{a+b}\Sigma\,\delta m_3 = 0, \\[2mm] \Sigma\,\delta m_2 + \dfrac{b}{a+b}\Sigma\,\delta m_3 = 0. \end{array}\right\} \quad (25)$$

The other equations of condition,

$$\Sigma\,\delta\eta = 0, \quad \Sigma\,\delta v = 0, \quad \Sigma\,\delta m_4 = 0, \quad \text{etc.,} \quad (26)$$

will remain unchanged. Now as all values of the variations which satisfy equations (24) will also satisfy equations (25), it is evident that all the particular conditions of equilibrium which we have already deduced, (19), (20), (22), are necessary in this case also. When these are satisfied, the general condition (23) reduces to

$$M_1 \Sigma\, \delta m_1 + M_2 \Sigma\, \delta m_2 + M_3 \Sigma\, \delta m_3 \geqq 0. \tag{27}$$

For, although it may be that μ_1', for example, is greater than M_1, yet it can only be so when the following $\delta m_1'$ is incapable of a negative value. Hence, if (27) is satisfied, (23) must also be. Again, if (23) is satisfied, (27) must also be satisfied, so long as the variation of the quantity of every substance has the value 0 in all the parts of which it is not an actual component. But as this limitation does not affect the range of the possible values of $\Sigma\, \delta m_1$, $\Sigma\, \delta m_2$, and $\Sigma\, \delta m_3$, it may be disregarded. Therefore the conditions (23) and (27) are entirely equivalent, when (19), (20), (22) are satisfied. Now, by means of the equations of condition (25), we may eliminate $\Sigma\, \delta m_1$ and $\Sigma\, \delta m_2$ from (27), which becomes

$$-a M_1 \Sigma\, \delta m_3 - b M_2 \Sigma\, \delta m_3 + (a+b) M_3 \Sigma\, \delta m_3 \geqq 0, \tag{28}$$

i.e., as the value of $\Sigma\, \delta m_3$ may be either positive or negative,

$$a M_1 + b M_2 = (a+b) M_3, \tag{29}$$

which is the additional condition of equilibrium which is necessary in this case.

The relations between the component substances may be less simple than in this case, but in any case they will only affect the equations of condition, and these may always be found without difficulty, and will enable us to eliminate from the general condition of equilibrium as many variations as there are equations of condition, after which the coefficients of the remaining variations may be set equal to zero, except the coefficients of variations which are incapable of negative values, which coefficients must be equal to or greater than zero. It will be easy to perform these operations in each particular case, but it may be interesting to see the form of the resultant equations in general.

We will suppose that the various homogeneous parts are considered as having in all n components, S_1, S_2, ... S_n, and that there is no restriction upon their freedom of motion and combination. But we will so far limit the generality of the problem as to suppose that each of these components is an actual component of some part of the given mass.* If some of these components can be formed out

---

* When we come to seek the conditions of equilibrium relating to the formation of masses unlike any previously existing, we shall take up *de novo* the whole problem of the equilibrium of heterogeneous masses enclosed in a non-conducting envelop, and give it a more general treatment, which will be free from this limitation.

of others, all such relations can be expressed by equations such as

$$a \, \mathfrak{S}_a + \beta \, \mathfrak{S}_b + \text{etc.} = \kappa \, \mathfrak{S}_k + \lambda \mathfrak{S}_l + \text{etc.} \tag{30}$$

where $\mathfrak{S}_a$, $\mathfrak{S}_b$, $\mathfrak{S}_k$, etc. denote the units of the substances S_a, S_b, S_k, etc., (that is, of certain of the substances S_1, S_2, ... S_n,) and a, β, κ, etc. denote numbers. These are not, it will be observed, equations between abstract quantities, but the sign = denotes qualitative as well as quantitative equivalence. We will suppose that there are r independent equations of this character. The equations of condition relating to the component substances may easily be derived from these equations, but it will not be necessary to consider them particularly. It is evident that they will be satisfied by any values of the variations which satisfy equations (18); hence, the particular conditions of equilibrium (19), (20), (22) must be necessary in this case, and, if these are satisfied, the general equation of equilibrium (15) or (23) will reduce to

$$M_1 \Sigma \, \delta m_1 + M_2 \Sigma \, \delta m_2 \ldots + M_n \Sigma \, \delta m_n \geqq 0. \tag{31}$$

This will appear from the same considerations which were used in regard to equations (23) and (27). Now it is evidently possible to give to $\Sigma \, \delta m_a$, $\Sigma \, \delta m_b$, $\Sigma \, \delta m_k$, etc. values proportional to a, β, $-\kappa$, etc. in equation (30), and also the same values taken negatively, making $\Sigma \, \delta m = 0$ in each of the other terms; therefore

$$a M_a + \beta M_b + \text{etc.} \ldots - \kappa M_k - \lambda M_l - \text{etc.} = 0, \tag{32}$$

or, $$a M_a + \beta M_b + \text{etc.} = \kappa M_k + \lambda M_l + \text{etc.} \tag{33}$$

It will be observed that this equation has the same form and coefficients as equation (30), M taking the place of $\mathfrak{S}$. It is evident that there must be a similar condition of equilibrium for every one of the r equations of which (30) is an example, which may be obtained simply by changing $\mathfrak{S}$ in these equations into M. When these conditions are satisfied, (31) will be satisfied with any possible values of $\Sigma \, \delta m_1$, $\Sigma \, \delta m_2$, ... $\Sigma \, \delta m_n$. For no values of these quantities are possible, except such that the equation

$$(\Sigma \, \delta m_1) \mathfrak{S}_1 + (\Sigma \, \delta m_2) \mathfrak{S}_2 \ldots + (\Sigma \, \delta m_n) \mathfrak{S}_n = 0, \tag{34}$$

after the substitution of these values, can be derived from the r equations like (30), by the ordinary processes of the reduction of linear equations. Therefore, on account of the correspondence between (31) and (34), and between the r equations like (33) and the r equations like (30), the conditions obtained by giving any possible values to the variations in (31) may also be derived from the r equations like (33); that is, the condition (31) is satisfied if the r equations like (33) are satisfied. Therefore the r equations like (33) are with (19), (20), and (22) the equivalent of the general condition (15) or (23).

For determining the state of a given mass when in equilibrium and having a given volume and given energy or entropy, the condition of equilibrium affords an additional equation corresponding to each of the r independent relations between the n component substances. But the equations which express our knowledge of the matter in the given mass will be correspondingly diminished, being $n-r$ in number, like the equations of condition relating to the quantities of the component substances, which may be derived from the former by differentiation.

Conditions relating to the possible Formation of Masses Unlike any Previously Existing.

The variations which we have hitherto considered do not embrace every possible infinitesimal variation in the state of the given mass, so that the particular conditions already formed, although always necessary for equilibrium (when there are no other equations of condition than such as we have supposed), are not always sufficient. For, besides the infinitesimal variations in the state and composition of different parts of the given mass, infinitesimal masses may be formed entirely different in state and composition from any initially existing. Such parts of the whole mass in its varied state as cannot be regarded as parts of the initially existing mass which have been infinitesimally varied in state and composition, we will call *new parts*. These will necessarily be infinitely small. As it is more convenient to regard a vacuum as a limiting case of extreme rarefaction than to give a special consideration to the possible formation of empty spaces within the given mass, the term *new parts* will be used to include any empty spaces which may be formed, when such have not existed initially. We will use $D\epsilon$, $D\eta$, Dv, Dm_1, Dm_2, ... Dm_n to denote the infinitesimal energy, entropy, and volume of any one of these new parts, and the infinitesimal quantities of its components. The component substances S_1, S_2, ... S_n must now be taken to include not only the independently variable components (actual or possible) of all parts of the given mass as initially existing, but also the components of all the new parts, the possible formation of which we have to consider. The character δ will be used as before to express the infinitesimal variations of the quantities relating to those parts which are only infinitesimally varied in state and composition, and which for distinction we will call *original parts*, including under this term the empty spaces, if such exist initially, within the envelop bounding the system. As we may divide the given mass into as many parts as we choose, and as not only the initial boundaries, but also the movements of these boundaries during any variation in the state of the system are arbitrary, we may so define the parts which we have called original, that we may consider them as initially homogeneous and remaining so, and as initially constituting the whole system.

The most general value of the variation of the energy of the whole system is evidently

$$\Sigma\, \delta\epsilon + \Sigma\, D\epsilon, \qquad (35)$$

the first summation relating to all the original parts, and the second to all the new parts. (Throughout the discussion of this problem, the letter δ or D following Σ will sufficiently indicate whether the summation relates to the original or to the new parts.) Therefore the general condition of equilibrium is

$$\Sigma\, \delta\epsilon + \Sigma\, D\epsilon \geqq 0, \qquad (36)$$

or, if we substitute the value of $\delta\epsilon$ taken from equation (12),

$$\Sigma\, D\epsilon + \Sigma(t\, \delta\eta) - \Sigma(p\, \delta v) + \Sigma(\mu_1 \delta m_1) + \Sigma(\mu_2 \delta m_2) \ldots + \Sigma(\mu_n \delta m_n) \geqq 0. \quad (37)$$

If any of the substances $S_1,\ S_2,\ \ldots S_n$ can be formed out of others, we will suppose, as before (see page 69), that such relations are expressed by equations between the units of the different substances. Let these be

$$\left.\begin{aligned} a_1 \mathfrak{S}_1 + a_2 \mathfrak{S}_2 \ldots + a_n \mathfrak{S}_n &= 0 \\ b_1 \mathfrak{S}_1 + b_2 \mathfrak{S}_2 \ldots + b_n \mathfrak{S}_n &= 0 \\ \text{etc.} \end{aligned}\right\} r \text{ equations.} \qquad (38)$$

The equations of condition will be (if there is no restriction upon the freedom of motion and composition of the components)

$$\Sigma\, \delta\eta + \Sigma\, D\eta = 0, \qquad (39)$$

$$\Sigma\, \delta v + \Sigma\, Dv = 0, \qquad (40)$$

and $n - r$ equations of the form

$$\left.\begin{aligned} h_1(\Sigma\, \delta m_1 + \Sigma\, Dm_1) + h_2(\Sigma\, \delta m_2 + \Sigma\, Dm_2) \ldots \\ + h_n(\Sigma\, \delta m_n + \Sigma\, Dm_n) = 0 \\ i_1(\Sigma\, \delta m_1 + \Sigma\, Dm_1) + i_2(\Sigma\, \delta m_2 + \Sigma\, Dm_2) \ldots \\ + i_n(\Sigma\, \delta m_n + \Sigma\, Dm_n) = 0 \end{aligned}\right\} \qquad (41)^*$$

etc.

Now, using Lagrange's "method of multipliers," † we will subtract

---

* In regard to the relation between the coefficients in (41) and those in (38), the reader will easily convince himself that the coefficients of any one of equations (41) are such as would satisfy all the equations (38) if substituted for $S_1,\ S_2,\ \ldots S_n$; and that this is the only condition which these coefficients must satisfy, except that the $n - r$ sets of coefficients shall be independent, i.e., shall be such as to form independent equations; and that this relation between the coefficients of the two sets of equations is a reciprocal one.

† On account of the sign $\geqq$ in (37), and because some of the variations are incapable of negative values, the successive steps in the reasoning will be developed at greater length than would be otherwise necessary.

$T(\Sigma \, \delta\eta + \Sigma \, D\eta) - P(\Sigma \, \delta v + \Sigma \, Dv)$ from the first member of the general condition of equilibrium (37), T and P being constants of which the value is as yet arbitrary. We might proceed in the same way with the remaining equations of condition, but we may obtain the same result more simply in another way. We will first observe that

$$(\Sigma \, \delta m_1 + \Sigma \, Dm_1)\mathfrak{S}_1 + (\Sigma \, \delta m_2 + \Sigma \, Dm_2)\mathfrak{S}_2 \ldots$$
$$+ (\Sigma \, \delta m_n + \Sigma \, Dm_n)\mathfrak{S}_n = 0, \qquad (42)$$

which equation would hold identically for any possible values of the quantities in the parentheses, if for r of the letters $\mathfrak{S}_1, \mathfrak{S}_2, \ldots \mathfrak{S}_n$ were substituted their values in terms of the others as derived from equations (38). (Although $\mathfrak{S}_1, \mathfrak{S}_2, \ldots \mathfrak{S}_n$ do not represent abstract quantities, yet the operations necessary for the reduction of linear equations are evidently applicable to equations (38).) Therefore, equation (42) will hold true if for $\mathfrak{S}_1, \mathfrak{S}_2, \ldots \mathfrak{S}_n$ we substitute n numbers which satisfy equations (38). Let $M_1, M_2, \ldots M_n$ be such numbers, i.e., let

$$\left.\begin{array}{l} a_1 M_1 + a_2 M_2 \ldots + a_n M_n = 0, \\ b_1 M_1 + b_2 M_2 \ldots + b_n M_n = 0, \\ \text{etc.} \end{array}\right\} r \text{ equations}, \qquad (43)$$

then

$$M_1(\Sigma \, \delta m_1 + \Sigma \, Dm_1) + M_2(\Sigma \, \delta m_2 + \Sigma \, Dm_2) \ldots$$
$$+ M_n(\Sigma \, \delta m_n + \Sigma \, Dm_n) = 0. \qquad (44)$$

This expression, in which the values of $n - r$ of the constants M_1, $M_2, \ldots M_n$ are still arbitrary, we will also subtract from the first member of the general condition of equilibrium (37), which will then become

$$\Sigma \, D\epsilon + \Sigma \, (t \, \delta\eta) - \Sigma \, (p \, \delta v) + \Sigma \, (\mu_1 \delta m_1) \ldots + \Sigma \, (\mu_n \delta m_n)$$
$$- T \Sigma \, \delta\eta + P \Sigma \, \delta v - M_1 \Sigma \, \delta m_1 \ldots - M_n \Sigma \, \delta m_n$$
$$- T \Sigma \, D\eta + P \Sigma \, Dv - M_1 \Sigma \, Dm_1 \ldots - M_n \Sigma \, Dm_n \geqq 0. \qquad (45)$$

That is, having assigned to T, P, M_1, $M_2, \ldots M_n$ any values consistent with (43), we may assert that it is necessary and sufficient for equilibrium that (45) shall hold true for any variations in the state of the system consistent with the equations of condition (39), (40), (41). But it will always be possible, in case of equilibrium, to assign such values to T, P, M_1, $M_2, \ldots M_n$, without violating equations (43), that (45) shall hold true for all variations in the state of the system and in the quantities of the various substances composing it, even though these variations are not consistent with the equations of condition (39), (40), (41). For, when it is not possible to do this, it must be possible by applying (45) to variations in the system not necessarily restricted by the equations of condition (39), (40), (41) to obtain conditions in regard to T, P, M_1, M_2 .. M_n, some of which will be inconsistent with others or with equations (43). These conditions we will represent by

$$A \geqq 0, \quad B \geqq 0, \quad \text{etc.}, \tag{46}$$

A, B, etc. being linear functions of T, P, M_1, M_2, ... M_n. Then it will be possible to deduce from these conditions a single condition of the form

$$aA + \beta B + \text{etc.} \geqq 0, \tag{47}$$

a, β, etc. being positive constants, which cannot hold true consistently with equations (43). But it is evident from the form of (47) that, like any of the conditions (46), it could have been obtained directly from (45) by applying this formula to a certain change in the system (perhaps not restricted by the equations of condition (39), (40), (41)). Now as (47) cannot hold true consistently with eqs. (43), it is evident, in the first place, that it cannot contain T or P, therefore in the change in the system just mentioned (for which (45) reduces to (47))

$$\Sigma \, \delta\eta + \Sigma \, D\eta = 0, \quad \text{and} \quad \Sigma \, \delta v + \Sigma \, Dv = 0,$$

so that the equations of condition (39) and (40) are satisfied. Again, for the same reason, the homogeneous function of the first degree of M_1, M_2, ... M_n in (47) must be one of which the value is fixed by eqs. (43). But the value thus fixed can only be zero, as is evident from the form of these equations. Therefore

$$(\Sigma \, \delta m_1 + \Sigma \, Dm_1) \, M_1 + (\Sigma \, \delta m_2 + \Sigma \, Dm_2) \, M_2 \ldots$$
$$+ (\Sigma \, \delta m_n + \Sigma \, Dm_n) \, M_n = 0 \tag{48}$$

for any values of M_1, M_2, ... M_n which satisfy eqs. (43), and therefore

$$(\Sigma \, \delta m_1 + \Sigma \, Dm_1)\mathfrak{S}_1 + (\Sigma \, \delta m_2 + \Sigma \, Dm_2)\mathfrak{S}_2 \ldots$$
$$+ (\Sigma \, \delta m_n + \Sigma \, Dm_n)\mathfrak{S}_n = 0 \tag{49}$$

for any numerical values of $\mathfrak{S}_1$, $\mathfrak{S}_2$, ... $\mathfrak{S}_n$ which satisfy eqs. (38). This equation (49) will therefore hold true, if for r of the letters $\mathfrak{S}_1$, $\mathfrak{S}_2$, ... $\mathfrak{S}_n$ we substitute their values in terms of the others taken from eqs. (38), and therefore it will hold true when we use $\mathfrak{S}_1$, $\mathfrak{S}_2$, ... $\mathfrak{S}_n$, as before, to denote the units of the various components. Thus understood, the equation expresses that the values of the quantities in the parentheses are such as are consistent with the equations of condition (41). The change in the system, therefore, which we are considering, is not one which violates any of the equations of condition, and as (45) does not hold true for this change, and for all values of T, P, M_1, M_2, ... M_n which are consistent with eqs. (43), the state of the system cannot be one of equilibrium. Therefore it is necessary, and it is evidently sufficient for equilibrium, that it shall be possible to assign to T, P, M_1, M_2, ... M_n such values, consistent with eqs. (43), that the condition (45) shall hold true for any change in the system irrespective of the equations of condition (39), (40), (41).

For this it is necessary and sufficient that

$$t = T, \quad p = P, \tag{50}$$

$$\mu_1 \delta m_1 \geqq M_1 \delta m_1, \quad \mu_2 \delta m_2 \geqq M_2 \delta m_2, \ldots \quad \mu_n \delta m_n \geqq M_n \delta m_n \tag{51}$$

for each of the *original parts* as previously defined, and that

$$D\epsilon - T D\eta + P Dv - M_1 Dm_1 - M_2 Dm_2 \ldots - M_n Dm_n \geqq 0, \tag{52}$$

for each of the *new parts* as previously defined. If to these conditions we add equations (43), we may treat T, P, M_1, M_2, ... M_n simply as unknown quantities to be eliminated.

In regard to conditions (51), it will be observed that if a substance S_1, is an actual component of the part of the given mass distinguished by a single accent, $\delta m_1'$ may be either positive or negative, and we shall have $\mu_1' = M_1$; but if S_1 is only a possible component of that part, $\delta m_1'$ will be incapable of a negative value, and we will have $\mu_1' \geqq M_1$.

The formulæ (50), (51), and (43) express the same particular conditions of equilibrium which we have before obtained by a less general process. It remains to discuss (52). This formula must hold true of any infinitesimal mass in the system in its varied state which is not approximately homogeneous with any of the surrounding masses, the expressions $D\epsilon$, $D\eta$, Dv, Dm_1, Dm_2, ... Dm_n denoting the energy, entropy, and volume of this infinitesimal mass, and the quantities of the substances $S_1, S_2, \ldots S_n$ which we regard as composing it (not necessarily as *independently* variable components). If there is more than one way in which this mass may be considered as composed of these substances, we may choose whichever is most convenient. Indeed it follows directly from the relations existing between M_1, M_2, ... and M_n that the result would be the same in any case. Now, if we assume that the values of $D\epsilon$, $D\eta$, Dv, Dm_1, Dm_2, ... Dm_n are proportional to the values of ϵ, η, v, m_1, m_2, ... m_n for any large homogeneous mass of similar composition, and of the same temperature and pressure, the condition is equivalent to this, that

$$\epsilon - T\eta + Pv - M_1 m_1 - M_2 m_2 \ldots - M_n m_n \geqq 0 \tag{53}$$

for any large homogeneous body which can be formed out of the substances $S_1, S_2, \ldots S_n$.

But the validity of this last transformation cannot be admitted without considerable limitation. It is assumed that the relation between the energy, entropy, volume, and the quantities of the different components of a very small mass surrounded by substances of different composition and state is the same as if the mass in question formed a part of a large homogeneous body. We started, indeed, with the assumption that we might neglect the part of the energy, etc., depending upon the surfaces separating heterogeneous masses. Now, in many cases, and for many purposes, as, in general, when the masses are large, such an assumption is quite legitimate, but in the case of these masses which are formed within or among substances of different nature or state, and which at their first formation must be infinitely small, the same assumption is evidently entirely inadmissible, as the surfaces must be regarded as infinitely large in proportion to the masses. We shall see hereafter what modifications are necessary in our formulæ in order to include the parts of the energy, etc., which are due to the surfaces, but this will be on the assumption, which is usual in the theory of capillarity, that the radius of curvature of the surfaces is large in proportion to the radius of sensible molecular action, and also to the thickness of the lamina of matter at the surface which is not (sensibly) homogeneous in all respects with either of the masses which it separates. But although the formulæ thus modified will apply with sensible accuracy to masses (occurring within masses of a different nature) much smaller than if the terms relating to the surfaces were omitted, yet their failure when applied to masses infinitely small in all their dimensions is not less absolute.

Considerations like the foregoing might render doubtful the validity even of (52) as the necessary and sufficient condition of equilibrium in regard to the formation of masses not approximately homogeneous with those previously existing, when the conditions of equilibrium between the latter are satisfied, unless it is shown that in establishing this formula there have been no quantities neglected relating to the mutual action of the new and the original parts, which can affect the result. It will be easy to give such a meaning to the expressions $D\epsilon$, $D\eta$, Dv, Dm_1, Dm_2, ... Dm_n that this shall be evidently the case. It will be observed that the quantities represented by these expressions have not been perfectly defined. In the first place, we have no right to assume the existence of any surface of absolute discontinuity to divide the new parts from the original, so that the position given to the dividing surface is to a certain extent arbitrary. Even if the surface separating the masses were determined, the energy to be attributed to the masses separated would be partly arbitrary, since a part of the total energy depends upon the mutual action of the two masses. We ought perhaps to consider the case the same in regard to the entropy, although the entropy of a system never depends upon the mutual relations of parts at sensible distances from one another. Now the condition (52) will be valid if the quantities $D\epsilon$, $D\eta$, Dv, Dm_1, Dm_2, ... Dm_n are so defined that none of the assumptions which have been made, tacitly or otherwise, relating to the formation of these new parts, shall be violated. These assumptions are the following:—that the relation between the variations of the energy, entropy, volume, etc., of any of the original parts is not affected by the vicinity of the new parts; and that the energy, entropy, volume, etc., of the system in its varied state are correctly represented by the sums of the energies, entropies, volumes, etc., of the various parts (original and new), so far at least as any of these quantities are determined or affected by the formation of the new parts. We will suppose $D\epsilon$, $D\eta$, Dv, Dm_1, Dm_2, ... Dm_n to be so defined that these conditions shall not be violated. This may be done in various ways. We may suppose that the position of the surfaces separating the new and the original parts has been fixed in any suitable way. This will determine the space and the matter belonging to the parts separated. If this does not determine the division of the entropy, we may suppose this determined in any suitable arbitrary way. Thus we may suppose the total energy in and about any new part to be so distributed that equation (12) as applied to the original parts shall not be violated by the formation of the new parts. Or, it may seem more simple to suppose that the imaginary surface which divides any new part from the original is so placed as to include all the matter which is affected by the vicinity of the new formation, so that the part or parts which we regard as original may be left homogeneous in the strictest sense, including uniform *densities of energy and entropy*, up to the very bounding surface. The homogeneity of the new parts is of no consequence, as we have made no assumption in that respect. It may be doubtful whether we can consider the new parts, *as thus bounded*, to be infinitely small even in their earliest stages of development. But if they are not infinitely small, the only way in which this can affect the validity of our formulæ will be that in virtue of the equations of condition, i.e., in virtue of the evident necessities of the case, finite variations of the energy, entropy, volume, etc., of the original parts will be caused, to which it might seem that equation (12) would not apply. But if the nature and state of the mass be not varied, equation (12) will hold true of finite differences. (This appears at once, if we integrate the equation under the above limitation.) Hence, the equation will hold true for finite differences, provided that the nature and state of the mass be infinitely little varied. For the differences may be considered as made up of two parts, of which the first are for a constant nature and state of the mass, and the second are infinitely small. We may therefore regard the new parts to be bounded as supposed without prejudice to the validity of any of our results.

The condition (52) understood in either of these ways (or in others which will suggest themselves to the reader) will have a perfectly definite meaning, and will be valid as the necessary and sufficient condition of equilibrium in regard to the formation of new parts, when the conditions of equilibrium in regard to the original parts, (50), (51), and (43), are satisfied.

In regard to the condition (53), it may be shown that with (50), (51), and (43) it is always sufficient for equilibrium. To prove this, it is only necessary to show that when (50), (51), and (43) are satisfied, and (52) is not, (53) will also not be satisfied.

We will first observe that an expression of the form

$$-\epsilon + T\eta - Pv + M_1 m_1 + M_2 m_2 \ldots + M_n m_n \qquad (54)$$

denotes the work obtainable by the formation (by a reversible process) of a body of which ϵ, η, v, m_1, m_2, $\ldots m_n$ are the energy, entropy, volume, and the quantities of the components, within a medium having the pressure P, the temperature T, and the potentials M_1, M_2, $\ldots M_n$. (The medium is supposed so large that its properties are not sensibly altered in any part by the formation of the body.) For ϵ is the energy of the body formed, and the remaining terms represent (as may be seen by applying equation (12) to the medium) the decrease of the energy of the medium, if, after the formation of the body, the joint entropy of the medium and the body, their joint volumes and joint quantities of matter, were the same as the entropy, etc., of the medium before the formation of the body. This consideration may convince us that for any given finite values of v and of T, P, M_1, etc., this expression cannot be infinite when ϵ, η, m_1, etc., are determined by any real body, whether homogeneous or not (but of the given volume), even when T, P, M_1, etc., do not represent the values of the temperature, pressure, and potentials of any real substance. (If the substances S_1, S_2, $\ldots S_n$ are all actual components of any homogeneous part of the system of which the equilibrium is discussed, that part will afford an example of a body having the temperature, pressure, and potentials of the medium supposed.)

Now by integrating equation (12) on the supposition that the nature and state of the mass considered remain unchanged, we obtain the equation

$$\epsilon = t\eta - pv + \mu_1 m_1 + \mu_2 m_2 \ldots + \mu_n m_n, \qquad (55)$$

which will hold true of any homogeneous mass whatever. Therefore for any one of the original parts, by (50) and (51),

$$\epsilon - T\eta + Pv - M_1 m_1 - M_2 m_2 \ldots - M_n m_n = 0. \qquad (56)$$

If the condition (52) is not satisfied in regard to all possible new parts, let N be a new part occurring in an original part O, for which the condition is not satisfied. It is evident that the value of the expression

$$\epsilon - T\eta + Pv - M_1 m_1 - M_2 m_2 \ldots - M_n m_n \qquad (57)$$

applied to a mass like O including some very small masses like N, will be negative, and will decrease if the number of these masses like N is increased, until there remains within the whole mass no portion of any sensible size without these masses like N, which, it will be remembered, have no sensible size. But it cannot decrease without limit, as the value of (54) cannot become infinite. Now we need not inquire whether the least value of (57) (for constant values of T, P, M_1, M_2, ... M_n) would be obtained by excluding entirely the mass like O, and filling the whole space considered with masses like N, or whether a certain mixture would give a smaller value,—it is certain that the least possible value of (57) per unit of volume, and that a negative value, will be realized by a mass having a certain homogeneity. If the new part N for which the condition (52) is not satisfied occurs between two different original parts O' and O'', the argument need not be essentially varied. We may consider the value of (57) for a body consisting of masses like O' and O'' separated by a lamina N. This value may be decreased by increasing the extent of this lamina, which may be done within a given volume by giving it a convoluted form; and it will be evident, as before, that the least possible value of (57) will be for a homogeneous mass, and that the value will be negative. And such a mass will be not merely an ideal combination, but a body capable of existing, for as the expression (57) has for this mass in the state considered its least possible value per unit of volume, the energy of the mass included in a unit of volume is the least possible for the same matter with the same entropy and volume,—hence, if confined in a non-conducting vessel, it will be in a state of not unstable equilibrium. Therefore when (50), (51), and (43) are satisfied, if the condition (52) is not satisfied in regard to all possible new parts, there will be some homogeneous body which can be formed out of the substances $S_1, S_2, \ldots S_n$ which will not satisfy condition (53).

Therefore, if the initially existing masses satisfy the conditions (50), (51), and (43), and condition (53) is satisfied by every homogeneous body which can be formed out of the given matter, there will be equilibrium.

On the other hand, (53) is not a necessary condition of equilibrium. For we may easily conceive that the condition (52) shall hold true (for any very small formations within or between any of the given masses), while the condition (53) is not satisfied (for all large masses formed of the given matter), and experience shows that this is very often the case. Supersaturated solutions, superheated water, etc., are familiar examples. Such an equilibrium will, however, be *practically* unstable. By this is meant that, although, strictly speaking, an infinitely small disturbance or change may not be sufficient to destroy the equilibrium, yet a very small change in the initial state, perhaps a circumstance which entirely escapes our powers of perception, will be sufficient to do so. The presence of a small portion of the substance for which the condition (53) does not hold true, is sufficient to produce this result, when this substance forms a variable component of the original homogeneous masses. In other cases, when, if the new substances are formed at all, different kinds must be formed simultaneously, the initial presence of the different kinds, and that in immediate proximity, may be necessary.

It will be observed, that from (56) and (53) we can at once obtain (50) and (51), viz., by applying (53) to bodies differing infinitely little from the various homogeneous parts of the given mass. Therefore, the condition (56) (relating to the various homogeneous parts of the given mass) and (53) (relating to any bodies which can be formed of the given matter) with (43) are always sufficient for equilibrium, and always necessary for an equilibrium which shall be practically stable. And, if we choose, we may get rid of limitation in regard to equations (43). For, if we compare these equations with (38), it is easy to see that it is always immaterial, in applying the tests (56) and (53) to any body, how we consider it to be composed. Hence, in applying these tests, we may consider all bodies to be composed of the *ultimate* components of the given mass. Then the terms in (56) and (53) which relate to other components than these will vanish, and we need not regard the equations (43). Such of the constants $M_1, M_2, \dots M_n$ as relate to the ultimate components, may be regarded, like T and P, as unknown quantities subject only to the conditions (56) and (53).

These two conditions, which are sufficient for equilibrium and necessary for a practically stable equilibrium, may be united in one, viz. (if we choose the ultimate components of the given mass for the component substances to which $m_1, m_2, \dots m_n$ relate), that it shall be possible to give such values to the constants $T, P, M_1, M_2, \dots M_n$ in the expression (57) that the value of the expression for each of the homogeneous parts of the mass in question shall be as small as for any body whatever made of the same components.

Effect of Solidity of any Part of the given Mass.

If any of the homogeneous masses of which the equilibrium is in question are solid, it will evidently be proper to treat the proportion of their components as invariable in the application of the criterion of equilibrium, even in the case of *compounds of variable proportions*, i.e., even when bodies can exist which are compounded in proportions infinitesimally varied from those of the solids considered. (Those solids which are capable of absorbing fluids form of course an exception, so far as their fluid components are concerned.) It is true that a solid may be increased by the formation of new solid matter on the surface where it meets a fluid, which is not homogeneous with the previously existing solid, but such a deposit will properly be treated as a distinct part of the system (viz., as one of the parts which we have called *new*). Yet it is worthy of notice that if a homogeneous solid which is a compound of variable proportions is in contact and equilibrium with a fluid, and the actual components of the solid (considered as of variable composition) are also actual components of the fluid, and the condition (53) is satisfied in regard to all bodies which can be formed out of the actual components of the fluid (which will always be the case unless the fluid is practically unstable), all the conditions will hold true of the solid, which would be necessary for equilibrium if it were fluid.

This follows directly from the principles stated on the preceding pages. For in this case the value of (57) will be zero as determined either for the solid or for the fluid considered with reference to their ultimate components, and will not be negative for any body whatever which can be formed of these components; and these conditions are sufficient for equilibrium independently of the solidity of one of the masses. Yet the point is perhaps of sufficient importance to demand a more detailed consideration.

Let $S_a, \ldots S_g$ be the actual components of the solid, and $S_h, \ldots S_k$ its possible components (which occur as actual components in the fluid); then, considering the proportion of the components of the solid as variable, we shall have for this body by equation (12)

$$d\epsilon' = t' d\eta' - p' dv' + \mu_a' dm_a' \ldots + \mu_g' dm_g'$$
$$+ \mu_h' dm_h' \ldots + \mu_k' dm_k'. \tag{58}$$

By this equation the potentials $\mu_a', \ldots \mu_k'$ are perfectly defined. But the differentials $dm_a', \ldots dm_k'$, considered as independent, evidently express variations which are not *possible* in the sense required in the criterion of equilibrium. We might, however, introduce them into the general condition of equilibrium, if we should express the dependence between them by the proper equations of condition. But it will be more in accordance with our method hitherto, if we consider the solid to have only a single independently variable component S_x, of which the nature is represented by the solid itself. We may then write

$$\delta\epsilon' = t' \, \delta\eta' - p' \, \delta v' + \mu' \, \delta m_x'. \tag{59}$$

In regard to the relation of the potential $\mu_x{}'$ to the potentials occurring in equation (58) it will be observed, that as we have by integration of (58) and (59)

$$\epsilon' = t'\eta' - p'v' + \mu_a{}'m_a{}'\ldots + \mu_g{}'m_g{}',\qquad(60)$$

and

$$\epsilon' = t'\eta' - p'v' + \mu_x{}'m_x{}';\qquad(61)$$

therefore

$$\mu_x{}'m_x{}' = \mu_a{}'m_a{}'\ldots + \mu_g{}'m_g{}'.\qquad(62)$$

Now, if the fluid has besides $S_a, \ldots S_g$ and $S_h, \ldots S_k$ the actual components $S_l, \ldots S_n$, we may write for the fluid

$$\delta\epsilon'' = t''\delta\eta'' - p''\delta v'' + \mu_a{}''\delta m_a{}''\ldots + \mu_g{}''\delta m_g{}''$$
$$+ \mu_h{}''\delta m_h{}''\ldots + \mu_k{}''\delta m_k{}'' + \mu_l{}''\delta m_l{}''\ldots + \mu_n{}''\delta m_n{}'',\qquad(63)$$

and as by supposition

$$m_x{}'\mathfrak{S}_x = m_a{}'\mathfrak{S}_a\ldots + m_g{}'\mathfrak{S}_g\qquad(64)$$

equations (43), (50), and (51) will give in this case on elimination of the constants T, P, etc.,

$$t' = t'',\quad p' = p'',\qquad(65)$$

and

$$m_x{}'\mu_x{}' = m_a{}'\mu_a{}''\ldots + m_g{}'\mu_g{}''.\qquad(66)$$

Equations (65) and (66) may be regarded as expressing the conditions of equilibrium between the solid and the fluid. The last condition may also, in virtue of (62), be expressed by the equation

$$m_a{}'\mu_a{}'\ldots + m_g{}'\mu_g{}' = m_a{}'\mu_a{}''\ldots + m_g{}'\mu_g{}''.\qquad(67)$$

But if condition (53) holds true of all bodies which can be formed of $S_a, \ldots S_g$, $S_h, \ldots S_k$, $S_l, \ldots S_n$, we may write for all such bodies

$$\epsilon - t''\eta + p''v - \mu_a{}''m_a\ldots - \mu_g{}''m_g - \mu_h{}''m_h$$
$$\ldots - \mu_k{}''m_k - \mu_l{}''m_l \ldots - \mu_n{}''m_n \geqq 0.\qquad(68)$$

(In applying this formula to various bodies, it is to be observed that only the values of the unaccented letters are to be determined by the different bodies to which it is applied, the values of the accented letters being already determined by the given fluid.) Now, by (60), (65), and (67), the value of the first member of this condition is zero when applied to the solid in its given state. As the condition must hold true of a body differing infinitesimally from the solid, we shall have

$$d\epsilon' - t''d\eta' + p''dv' - \mu_a{}''dm_a{}'\ldots - \mu_g{}''dm_g{}'$$
$$- \mu_h{}''dm_h{}'\ldots - \mu_k{}''dm_k{}' \geqq 0,\qquad(69)$$

or, by equations (58) and (65),

$$(\mu_a{}' - \mu_a{}'')\,dm_a{}'\ldots + (\mu_g{}' - \mu_g{}'')\,dm_g{}'$$
$$+ (\mu_h{}' - \mu_h{}'')\,dm_h{}'\ldots + (\mu_k{}' - \mu_k{}'')\,dm_k{}' \geqq 0.\qquad(70)$$

Therefore, as these differentials are all independent,

$$\mu_a{}' = \mu_a{}'',\ldots \mu_g{}' = \mu_g{}'',\quad \mu_h{}' \geqq \mu_h{}'',\ldots \mu_k{}' \geqq \mu_k{}'';\qquad(71)$$

which with (65) are evidently the same conditions which we would
have obtained if we had neglected the fact of the solidity of one of
the masses.

We have supposed the solid to be homogeneous. But it is evident
that in any case the above conditions must hold for every separate
point where the solid meets the fluid. Hence, the temperature and
pressure and the potentials for all the actual components of the solid
must have a constant value in the solid at the surface where it meets
the fluid. Now, these quantities are determined by the nature and
state of the solid, and exceed in number the independent variations
of which its nature and state are capable. Hence, if we reject as
improbable the supposition that the nature or state of a body can
vary without affecting the value of any of these quantities, we may
conclude that a solid which varies (continuously) in nature or state
at its surface cannot be in equilibrium with a stable fluid which contains, as independently variable components, the variable components
of the solid. (There may be, however, in equilibrium with the same
stable fluid, a *finite* number of different solid bodies, composed of the
variable components of the fluid, and having their nature and state
completely determined by the fluid.)*

Effect of Additional Equations of Condition.

As the equations of condition, of which we have made use, are
such as always apply to matter enclosed in a rigid, impermeable, and
non-conducting envelop, the particular conditions of equilibrium
which we have found will always be sufficient for equilibrium. But
the number of conditions necessary for equilibrium, will be diminished,
in a case otherwise the same, as the number of equations of condition
is increased. Yet the problem of equilibrium which has been treated
will sufficiently indicate the method to be pursued in all cases and the
general nature of the results.

It will be observed that the position of the various homogeneous
parts of the given mass, which is otherwise immaterial, may determine the existence of certain equations of condition. Thus, when
different parts of the system in which a certain substance is a variable
component are entirely separated from one another by parts of which
this substance is not a component, the quantity of this substance will
be invariable for each of the parts of the system which are thus
separated, which will be easily expressed by equations of condition.
Other equations of condition may arise from the passive forces (or
resistances to change) inherent in the given masses. In the problem

---

*The solid has been considered as subject only to isotropic stresses. The effect of
other stresses will be considered hereafter.

which we are next to consider there are equations of condition due to a cause of a different nature.

Effect of a Diaphragm (Equilibrium of Osmotic Forces).

If the given mass, enclosed as before, is divided into two parts, each of which is homogeneous and fluid, by a diaphragm which is capable of supporting an excess of pressure on either side, and is permeable to some of the components and impermeable to others, we shall have the equations of condition

$$\delta\eta' + \delta\eta'' = 0, \tag{72}$$

$$\delta v' = 0, \quad \delta v'' = 0, \tag{73}$$

and for the components which cannot pass the diaphragm

$$\delta m_a' = 0, \quad \delta m_a'' = 0, \quad \delta m_b' = 0, \quad \delta m_b'' = 0, \text{ etc.}, \tag{74}$$

and for those which can

$$\delta m_h' + \delta m_h'' = 0, \quad \delta m_i' + \delta m_i'' = 0, \text{ etc.} \tag{75}$$

With these equations of condition, the general condition of equilibrium (see (15)) will give the following particular conditions:—

$$t' = t'', \tag{76}$$

and for the components which can pass the diaphragm, if actual components of both masses,

$$\mu_h' = \mu_h'', \quad \mu_i' = \mu_i'', \text{ etc.}, \tag{77}$$

but not

$$p' = p'',$$

nor

$$\mu_a' = \mu_a'', \quad \mu_b' = \mu_b'', \text{ etc.}$$

Again, if the diaphragm is permeable to the components in certain proportions only, or in proportions not entirely determined yet subject to certain conditions, these conditions may be expressed by equations of condition, which will be linear equations between $\delta m_1'$, $\delta m_2'$, etc., and if these be known the deduction of the particular conditions of equilibrium will present no difficulties. We will however observe that if the components S_1, S_2, etc. (being actual components on each side) can pass the diaphragm simultaneously in the proportions a_1, a_2, etc. (without other resistances than such as vanish with the velocity of the current), values proportional to a_1, a_2, etc. are possible for $\delta m_1'$, $\delta m_2'$, etc. in the general condition of equilibrium, $\delta m_1''$, $\delta m_2''$, etc., having the same values taken negatively, so that we shall have for one particular condition of equilibrium

$$a_1\mu_1' + a_2\mu_2' + \text{etc.} = a_1\mu_1'' + a_2\mu_2'' + \text{etc.} \tag{78}$$

There will evidently be as many independent equations of this form as there are independent combinations of the elements which can pass the diaphragm.

These conditions of equilibrium do not of course depend in any way upon the supposition that the volume of each fluid mass is kept constant, if the diaphragm is in any case supposed immovable. In fact, we may easily obtain the same conditions of equilibrium, if we suppose the volumes variable. In this case, as the equilibrium must be preserved by forces acting upon the external surfaces of the fluids, the variation of the energy of the sources of these forces must appear in the general condition of equilibrium, which will be

$$\delta\epsilon' + \delta\epsilon'' + P'\,\delta v' + P''\,\delta v'' \geqq 0, \qquad (79)$$

P' and P'' denoting the external forces per unit of area. (Compare (14).) From this condition we may evidently derive the same internal conditions of equilibrium as before, and in addition the external conditions

$$p' = P', \quad p'' = P''. \qquad (80)$$

In the preceding paragraphs it is assumed that the permeability of the diaphragm is perfect, and its impermeability absolute, i.e., that it offers no resistance to the passage of the components of the fluids in certain proportions, except such as vanishes with the velocity, and that in other proportions the components cannot pass at all. How far these conditions are satisfied in any particular case is of course to be determined by experiment.

If the diaphragm is permeable to all the n components without restriction, the temperature and the potentials for all the components must be the same on both sides. Now, as one may easily convince himself, a mass having n components is capable of only $n+1$ independent variations in nature and state. Hence, if the fluid on one side of the diaphragm remains without change, that on the other side cannot (in general) vary in nature or state. Yet the pressure will not necessarily be the same on both sides. For, although the pressure is a function of the temperature and the n potentials, it may be a many-valued function (or any one of several functions) of these variables. But when the pressures are different on the two sides, the fluid which has the less pressure will be *practically unstable*, in the sense in which the term has been used on page 79. For

$$\epsilon'' - t''\eta'' + p''v'' - \mu_1''m_1'' - \mu_2''m_2''\ldots - \mu_n''m_n'' = 0, \qquad (81)$$

as appears from equation (12) if integrated on the supposition that the nature and state of the mass remain unchanged. Therefore, if $p' < p''$ while $t' = t''$, $\mu_1' = \mu_1''$, etc.,

$$\epsilon'' - t'\eta'' + p'v'' - \mu_1'm_1'' - \mu_2'm_2''\ldots - \mu_n'm_n'' < 0. \qquad (82)$$

This relation indicates the instability of the fluid to which the single accents refer. (See page 79.)

But independently of any assumption in regard to the permeability of the diaphragm, the following relation will hold true in any case in which each of the two fluid masses may be regarded as uniform throughout in nature and state. Let the character D be used with the variables which express the nature, state, and quantity of the fluids to denote the increments of the values of these quantities actually occurring in a time either finite or infinitesimal. Then, as the heat received by the two masses cannot exceed t'D$\eta' + t''$Dη'', and as the increase of their energy is equal to the difference of the heat they receive and the work they do,

$$\text{D}\epsilon' + \text{D}\epsilon'' \leqq t'\text{D}\eta' + t''\text{D}\eta'' - p'\text{D}v' - p''\text{D}v'', \qquad (83)$$

i.e., by (12),

$$\mu_1'\text{D}m_1' + \mu_1''\text{D}m_1'' + \mu_2'\text{D}m_2' + \mu_2''\text{D}m_2'' + \text{etc.} \leqq 0, \qquad (84)$$

or

$$(\mu_1'' - \mu_1')\text{D}m_1'' + (\mu_2'' - \mu_2')\text{D}m_2'' + \text{etc.} \leqq 0. \qquad (85)$$

It is evident that the sign = holds true only in the limiting case in which no motion takes place.

Definition and Properties of Fundamental Equations.

The solution of the problems of equilibrium which we have been considering has been made to depend upon the equations which express the relations between the energy, entropy, volume, and the quantities of the various components, for homogeneous combinations of the substances which are found in the given mass. The nature of such equations must be determined by experiment. As, however, it is only *differences* of energy and of entropy that can be measured, or indeed, that have a physical meaning, the values of these quantities are so far arbitrary, that we may choose independently for each simple substance the state in which its energy and its entropy are both zero. The values of the energy and the entropy of any compound body in any particular state will then be fixed. Its energy will be the sum of the work and heat expended in bringing its components from the states in which their energies and their entropies are zero into combination and to the state in question; and its entropy is the value of the integral $\int \frac{dQ}{t}$ for any *reversible* process by which that change is effected (dQ denoting an element of the heat communicated to the matter thus treated, and t the temperature of the matter receiving it). In the determination both of the energy and of the entropy, it is understood that at the close of the process, all bodies which have been used, other than those to which the determinations relate, have been restored to their original state, with the exception of the sources of the work and heat expended, which must be used only as such sources.

We know, however, *a priori*, that if the quantity of any homogeneous mass containing n independently variable components varies and not its nature or state, the quantities ϵ, η, v, m_1, m_2, ... m_n will all vary in the same proportion; therefore it is sufficient if we learn from experiment the relation between all but any one of these quantities for a given constant value of that one. Or, we may consider that we have to learn from experiment the relation subsisting between the $n+2$ ratios of the $n+3$ quantities ϵ, η, v, m_1, m_2, ... m_n. To fix our ideas we may take for these ratios $\dfrac{\epsilon}{v}$, $\dfrac{\eta}{v}$, $\dfrac{m_1}{v}$, $\dfrac{m_2}{v}$, etc., that is, the separate densities of the components, and the ratios $\dfrac{\epsilon}{v}$ and $\dfrac{\eta}{v}$, which may be called the *densities of energy and entropy.* But when there is but one component, it may be more convenient to choose $\dfrac{\epsilon}{m}$, $\dfrac{\eta}{m}$, $\dfrac{v}{m}$ as the three variables. In any case, it is only a function of $n+1$ independent variables, of which the form is to be determined by experiment.

Now if ϵ is a known function of η, v, m_1, m_2, ... m_n, as by equation (12)

$$d\epsilon = t\, d\eta - p\, dv + \mu_1 dm_1 + \mu_2 dm_2 \ldots + \mu_n dm_n, \qquad (86)$$

t, p, μ_1, μ_2, ... μ_n are functions of the same variables, which may be derived from the original function by differentiation, and may therefore be considered as known functions. This will make $n+3$ independent known relations between the $2n+5$ variables, ϵ, η, v, m_1, m_2, ... m_n, t, p, μ_1, μ_2, ... μ_n. These are all that exist, for of these variables, $n+2$ are evidently independent. Now upon these relations depend a very large class of the properties of the compound considered,—we may say in general, all its thermal, mechanical, and chemical properties, so far as *active tendencies* are concerned, in cases in which the form of the mass does not require consideration. A single equation from which all these relations may be deduced we will call a *fundamental equation* for the substance in question. We shall hereafter consider a more general form of the fundamental equation for solids, in which the pressure at any point is not supposed to be the same in all directions. But for masses subject only to isotropic stresses an equation between ϵ, η, v, m_1, m_2, ... m_n is a fundamental equation. There are other equations which possess this same property.[*]

---

[*]M. Massieu (*Comptes Rendus*, T. lxix, 1869, p. 858 and p. 1057) has shown how all the properties of a fluid "which are considered in thermodynamics" may be deduced

Let
$$\psi = \epsilon - t\eta, \tag{87}$$

then by differentiation and comparision with (86) we obtain

$$d\psi = -\eta\, dt - p\, dv + \mu_1 dm_1 + \mu_2 dm_2 \ldots + \mu_n dm_n. \tag{88}$$

If, then, ψ is known as a function of t, v, m_1, m_2, $\ldots m_n$, we can find η, p, μ_1, μ_2, $\ldots \mu_n$ in terms of the same variables. If we then substitute for ψ in our original equation its value taken from eq. (87), we shall have again $n+3$ independent relations between the same $2n+5$ variables as before.

Let
$$\chi = \epsilon + pv, \tag{89}$$

then by (86),

$$d\chi = t\, d\eta + v\, dp + \mu_1 dm_1 + \mu_2 dm_2 \ldots + \mu_n\, dm_n. \tag{90}$$

If, then, χ be known as a function of η, p, m_1, m_2, $\ldots m_n$, we can find t, v, μ_1, μ_2, $\ldots \mu_n$ in terms of the same variables. By eliminating χ, we may obtain again $n+3$ independent relations between the same $2n+5$ variables as at first.

Let
$$\zeta = \epsilon - t\eta + pv, \tag{91}$$

then, by (86),

$$d\zeta = -\eta\, dt + v\, dp + \mu_1 dm_1 + \mu_2 dm_2 \ldots + \mu_n dm_n. \tag{92}$$

If, then, ζ is known as a function of t, p, m_1, m_2, $\ldots m_n$, we can find η, v, μ_1, μ_2, $\ldots \mu_n$ in terms of the same variables. By eliminating ζ, we may obtain again $n+3$ independent relations between the same $2n+5$ variables as at first.

If we integrate (86), supposing the quantity of the compound substance considered to vary from zero to any finite value, its nature and state remaining unchanged, we obtain

$$\epsilon = t\eta - pv + \mu_1 m_1 + \mu_2 m_2 \ldots + \mu_n m_n, \tag{93}$$

and by (87), (89), (91)

$$\psi = -pv + \mu_1 m_1 + \mu_2 m_2 \ldots + \mu_n m_n, \tag{94}$$

$$\chi = t\eta + \mu_1 m_1 + \mu_2 m_2 \ldots + \mu_n m_n, \tag{95}$$

$$\zeta = \mu_1 m_1 + \mu_2 m_2 \ldots + \mu_n m_n. \tag{96}$$

The last three equations may also be obtained directly by integrating (88), (90), and (92).

---

from a single function, which he calls a characteristic function of the fluid considered. In the papers cited, he introduces two different functions of this kind, viz., a function of the temperature and volume, which he denotes by ψ, the value of which in our notation would be $\dfrac{-\epsilon + t\eta}{t}$ or $\dfrac{-\psi}{t}$; and a function of the temperature and pressure, which he denotes by ψ', the value of which in our notation would be $\dfrac{-\epsilon + t\eta - pv}{t}$ or $\dfrac{-\zeta}{t}$. In both cases he considers a constant quantity (one kilogram) of the fluid, which is regarded as invariable in composition.

If we differentiate (93) in the most general manner, and compare the result with (86), we obtain

$$-v\,dp+\eta\,dt+m_1\,d\mu_1+m_2\,d\mu_2\ldots+m_n\,d\mu_n=0, \tag{97}$$

or

$$dp=\frac{\eta}{v}\,dt+\frac{m_1}{v}\,d\mu_1+\frac{m_2}{v}\,d\mu_2\ldots+\frac{m_n}{v}d\mu_n. \tag{98}$$

Hence, there is a relation between the $n+2$ quantities t, p, μ_1, μ_2, $\ldots\mu_n$, which, if known, will enable us to find in terms of these quantities all the ratios of the $n+2$ quantities η, v, m_1, m_2, $\ldots m_n$. With (93), this will make $n+3$ independent relations between the same $2n+5$ variables as at first.

Any equation, therefore, between the quantities

$$\epsilon, \qquad \eta, \qquad v, \qquad m_1, \qquad m_2,\ldots m_n, \tag{99}$$

or

$$\psi, \qquad t, \qquad v, \qquad m_1, \qquad m_2,\ldots m_n, \tag{100}$$

or

$$\chi, \qquad \eta, \qquad p, \qquad m_1, \qquad m_2,\ldots m_n, \tag{101}$$

or

$$\zeta, \qquad t, \qquad p, \qquad m_1, \qquad m_2,\ldots m_n, \tag{102}$$

or

$$t, \qquad p, \qquad \mu_1, \qquad \mu_2,\ldots\mu_n, \tag{103}$$

is a fundamental equation, and any such is entirely equivalent to any other.* For any homogeneous mass whatever, considered (in general) as variable in composition, in quantity, and in thermodynamic state, and having n independently variable components, to which the subscript numerals refer (but not excluding the case in which $n=1$ and the composition of the body is invariable), there is a relation between the quantities enumerated in any one of the above sets, from which, if known, with the aid only of *general* principles and relations, we may deduce all the relations subsisting for such a mass between the quantities ϵ, ψ, χ, ζ, η, v, m_1, m_2, $\ldots m_n$, t, p, μ_1, μ_2, $\ldots\mu_n$. It will be observed that, besides the equations which define ψ, χ, and ζ, there is one finite equation, (93), which subsists between these quantities independently of the form of the fundamental equation.

---

* The distinction between equations which are, and which are not, *fundamental*, in the sense in which the word is here used, may be illustrated by comparing an equation between $\epsilon,\ \eta,\ v,\ m_1,\ m_2,\ \ldots m_n$ with one between $\epsilon,\ t,\ v,\ m_1,\ m_2,\ \ldots m_n$.

As, by (86),

$$t=\left(\frac{d\epsilon}{d\eta}\right)_{vm}$$

the second equation may evidently be derived from the first. But the first equation cannot be derived from the second; for an equation between

$$\epsilon,\ \left(\frac{d\epsilon}{d\eta}\right)_{vm},\ v,\ m_1,\ m_2,\ \ldots m_n$$

is equivalent to one between $\left(\frac{d\eta}{d\epsilon}\right)_{vm},\ \epsilon,\ v,\ m_1,\ m_2,\ \ldots m_n$,

which is evidently not sufficient to determine the value of η in terms of the other variables.

Other sets of quantities might of course be added which possess the same property. The sets (100), (101), (102) are mentioned on account of the important properties of the quanties ψ, χ, ζ, and because the equations (88), (90), (92), like (86), afford convenient definitions of the potentials, viz.,

$$\mu_1 = \left(\frac{d\epsilon}{dm_1}\right)_{\eta,\,v,\,m} = \left(\frac{d\psi}{dm_1}\right)_{t,\,v,\,m} = \left(\frac{d\chi}{dm_1}\right)_{\eta,\,p,\,m} = \left(\frac{d\zeta}{dm_1}\right)_{t,\,p,\,m} \quad (104)$$

etc., where the subscript letters denote the quantities which remain constant in the differentiation, m being written for brevity for all the letters m_1, m_2, ... m_n except the one occurring in the denominator. It will be observed that the quantities in (103) are all independent of the quantity of the mass considered, and are those which must, in general, have the same value in contiguous masses in equilibrium.

On the quantities ψ, χ, ζ.

The quantity ψ has been defined for any homogeneous mass by the equation

$$\psi = \epsilon - t\eta. \quad (105)$$

We may extend this definition to any material system whatever which has a uniform temperature throughout.

If we compare two states of the system of the same temperature, we have

$$\psi' - \psi'' = \epsilon' - \epsilon'' - t(\eta' - \eta''). \quad (106)$$

If we suppose the system brought from the first to the second of these states without change of temperature and by a reversible process in which W is the work done and Q the heat received by the system, then

$$\epsilon' - \epsilon'' = W - Q, \quad (107)$$

and $$t(\eta'' - \eta') = Q. \quad (108)$$

Hence $$\psi' - \psi'' = W; \quad (109)$$

and for an infinitely small reversible change in the state of the system, in which the temperature remains constant, we may write

$$-d\psi = dW. \quad (110)$$

Therefore, $-\psi$ is the force function of the system for constant temperature, just as $-\epsilon$ is the force function for constant entropy. That is, if we consider ψ as a function of the temperature and the variables which express the distribution of the matter in space, for every different value of the temperature $-\psi$ is the different force function required by the system if maintained at that special temperature.

From this we may conclude that when a system has a uniform temperature throughout, the additional conditions which are necessary and sufficient for equilibrium may be expressed by

$$(\delta\psi)_t \geqq 0.^* \tag{111}$$

When it is not possible to bring the system from one to the other of the states to which ψ' and ψ'' relate by a reversible process without altering the temperature, it will be observed that it is not necessary for the validity of (107)–(109) that the temperature of the system should remain constant during the reversible process to which W and Q relate, provided that the only source of heat or cold used has the same temperature as the system in its initial or final state. Any external bodies may be used in the process in any way not affecting the condition of reversibility, if restored to their original condition at the close of the process; nor does the limitation in regard to the use of heat apply to such heat as may be restored to the source from which it has been taken.

It may be interesting to show directly the equivalence of the conditions (111) and (2) when applied to a system of which the temperature in the given state is uniform throughout.

If there are any variations in the state of such a system which do not satisfy (2), then for these variations

$$\delta\epsilon < 0 \quad \text{and} \quad \delta\eta = 0.$$

If the temperature of the system in its varied state is not uniform, we may evidently increase its entropy without altering its energy by supposing heat to pass from the warmer to the cooler parts. And the state having the greatest entropy for the energy $\epsilon + \delta\epsilon$ will necessarily be a state of uniform temperature. For this state (regarded as a variation from the original state)

$$\delta\epsilon < 0 \quad \text{and} \quad \delta\eta > 0.$$

Hence, as we may diminish both the energy and the entropy by

---

* This general condition of equilibrium might be used instead of (2) in such problems of equilibrium as we have considered and others which we shall consider hereafter with evident advantage in respect to the brevity of the formulæ, as the limitation expressed by the subscript t in (111) applies to every part of the system taken separately, and diminishes by one the number of independent variations in the state of these parts which we have to consider. The more cumbersome course adopted in this paper has been chosen, among other reasons, for the sake of deducing *all* the particular conditions of equilibrium from one general condition, and of having the quantities mentioned in this general condition such as are most generally used and most simply defined; and because in the longer formulæ as given, the reader will easily see in each case the form which they would take if we should adopt (111) as the general condition of equilibrium, which would be in effect to take the thermal condition of equilibrium for granted, and to seek only the remaining conditions. For example, in the problem treated on pages 63 ff., we would obtain from (111) by (88) a condition precisely like (15), except that the terms $t\,\delta\eta'$, $t\,\delta\eta''$, etc., would be wanting.

cooling the system, there must be a state of uniform temperature for which (regarded as a variation of the original state)

$$\delta\epsilon < 0 \quad \text{and} \quad \delta\eta = 0.$$

From this we may conclude that for systems of initially uniform temperature condition (2) will not be altered if we limit the variations to such as do not disturb the uniformity of temperature.

Confining our attention, then, to states of uniform temperature, we have by differentiation of (105)

$$\delta\epsilon - t\,\delta\eta = \delta\psi + \eta\,\delta t. \tag{112}$$

Now there are evidently changes in the system (produced by heating or cooling) for which

$$\delta\epsilon - t\,\delta\eta = 0 \quad \text{and therefore} \quad \delta\psi + \eta\,\delta t = 0, \tag{113}$$

neither $\delta\eta$ nor δt having the value zero. This consideration is sufficient to show that the condition (2) is equivalent to

$$\delta\epsilon - t\,\delta\eta \geqq 0, \tag{114}$$

and that the condition (111) is equivalent to

$$\delta\psi + \eta\,\delta t \geqq 0, \tag{115}$$

and by (112) the two last conditions are equivalent.

In such cases as we have considered on pages 62–82, in which the form and position of the masses of which the system is composed are immaterial, uniformity of temperature and pressure are always necessary for equilibrium, and the remaining conditions, when these are satisfied, may be conveniently expressed by means of the function ζ, which has been defined for a homogeneous mass on page 87, and which we will here define for any mass of uniform temperature and pressure by the same equation

$$\zeta = \epsilon - t\eta + pv. \tag{116}$$

For such a mass, the condition of (internal) equilibrium is

$$(\delta\zeta)_{t,p} \geqq 0. \tag{117}$$

That this condition is equivalent to (2) will easily appear from considerations like those used in respect to (111).

Hence, it is necessary for the equilibrium of two contiguous masses identical in composition that the values of ζ as determined for equal quantities of the two masses should be equal. Or, when one of three contiguous masses can be formed out of the other two, it is necessary for equilibrium that the value of ζ for any quantity of the first mass should be equal to the sum of the values of ζ for such quantities of the second and third masses as together contain the same matter. Thus, for the equilibrium of a solution composed of a parts of water and b parts of a salt which is in contact with vapor of water and crystals of the salt, it is necessary that the value of ζ for the quantity $a+b$ of the solution should be equal to the sum of the values of ζ for the quantities a of the vapor and b of the salt. Similar propositions will hold true in more complicated cases. The reader will easily deduce these conditions from the particular conditions of equilibrium given on page 74.

In like manner we may extend the definition of χ to any mass or combination of masses in which the pressure is everywhere the same, using ϵ for the energy and v for the volume of the whole and setting as before

$$\chi = \epsilon + pv. \tag{118}$$

If we denote by Q the heat received by the combined masses from external sources in any process in which the pressure is not varied, and distinguish the initial and final states of the system by accents we have

$$\chi'' - \chi' = \epsilon'' - \epsilon' + p(v'' - v') = Q. \tag{119}$$

This function may therefore be called the *heat function for constant pressure* (just as the energy might be called the heat function for constant volume), the diminution of the function representing in all cases in which the pressure is not varied the heat given out by the system. In all cases of chemical action in which no heat is allowed to escape the value of χ remains unchanged.

Potentials.

In the definition of the potentials μ_1, μ_2, etc., the energy of a homogeneous mass was considered as a function of its entropy, its volume, and the quantities of the various substances composing it. Then the potential for one of these substances was defined as the differential coefficient of the energy taken with respect to the variable expressing the quantity of that substance. Now, as the manner in which we consider the given mass as composed of various substances is in some degree arbitrary, so that the energy may be considered as a function of various different sets of variables expressing quantities of component substances, it might seem that the above definition does not fix the value of the potential of any substance in the given mass, until we have fixed the manner in which the mass is to be considered as composed. For example, if we have a solution obtained by dissolving in water a certain salt containing water of crystallization, we may consider the liquid as composed of m_S weight-units of the hydrate and m_W of water, or as composed of m_s of the anhydrous salt and m_w of water. It will be observed that the values of m_S and

m_s are not the same, nor those of m_W and m_w, and hence it might seem that the potential for water in the given liquid considered as composed of the hydrate and water, viz.,

$$\left(\frac{d\epsilon}{dm_w}\right)_{\eta,\,v,\,m_g},$$

would be different from the potential for water in the same liquid considered as composed of anhydrous salt and water, viz.,

$$\left(\frac{d\epsilon}{dm_w}\right)_{\eta\,v,\,m_s}.$$

The value of the two expressions is, however, the same, for, although m_W is not equal to m_w, we may of course suppose dm_W to be equal to dm_w, and then the numerators in the two fractions will also be equal, as they each denote the increase of energy of the liquid, when the quantity dm_W or dm_w of water is added without altering the entropy and volume of the liquid. Precisely the same considerations will apply to any other case.

In fact, we may give a definition of a potential which shall not presuppose any choice of a particular set of substances as the components of the homogeneous mass considered.

Definition.—If to any homogeneous mass we suppose an infinitesimal quantity of any substance to be added, the mass remaining homogeneous and its entropy and volume remaining unchanged, the increase of the energy of the mass divided by the quantity of the substance added is the *potential* for that substance in the mass considered. (For the purposes of this definition, any chemical element or combination of elements in given proportions may be considered a substance, whether capable or not of existing by itself as a homogeneous body.)

In the above definition we may evidently substitute for entropy, volume, and energy, respectively, either temperature, volume, and the function ψ; or entropy, pressure, and the function χ; or temperature, pressure, and the function ζ. (Compare equation (104).)

In the same homogeneous mass, therefore, we may distinguish the potentials for an indefinite number of substances, each of which has a perfectly determined value.

Between the potentials for different substances in the same homogeneous mass the same equations will subsist as between the units of these substances. That is, if the substances, S_a, S_b, etc., S_k, S_l, etc., are components of any given homogeneous mass, and are such that

$$a\,\mathfrak{S}_a + \beta\,\mathfrak{S}_b + \text{etc.} = \kappa\,\mathfrak{S}_k + \lambda\,\mathfrak{S}_l + \text{etc.,} \tag{120}$$

$\mathfrak{S}_a$, $\mathfrak{S}_b$, etc., $\mathfrak{S}_k$, $\mathfrak{S}_l$, etc., denoting the units of the several substances, and a, β, etc., κ, λ, etc., denoting numbers, then if μ_a, μ_b, etc., μ_k, μ_l, etc., denote the potentials for these substances in the homogeneous mass,

$$a\mu_a + \beta\mu_b + \text{etc.} = \kappa\mu_k + \lambda\mu_l + \text{etc.} \tag{121}$$

To show this, we will suppose the mass considered to be very large. Then, the first member of (121) denotes the increase of the energy of the mass produced by the addition of the matter represented by the first member of (120), and the second member of (121) denotes the increase of energy of the same mass produced by the addition of the matter represented by the second member of (120), the entropy and volume of the mass remaining in each case unchanged. Therefore, as the two members of (120) represent the same matter in kind and quantity, the two members of (121) must be equal.

But it must be understood that equation (120) is intended to denote equivalence of the substances represented *in the mass considered,* and not merely chemical identity; in other words, it is supposed that there are no passive resistances to change in the mass considered which prevent the substances represented by one member of (120) from passing into those represented by the other. For example, in respect to a mixture of vapor of water and free hydrogen and oxygen (at ordinary temperatures), we may not write

$$9 \mathfrak{S}_{Aq} = 1 \mathfrak{S}_H + 8 \mathfrak{S}_O,$$

but water is to be treated as an independent substance, and no necessary relation will subsist between the potential for water and the potentials for hydrogen and oxygen.

The reader will observe that the relations expressed by equations (43) and (51) (which are essentially relations between the potentials for actual components in different parts of a mass in a state of equilibrium) are simply those which by (121) would necessarily subsist between the same potentials in any homogeneous mass containing as variable components all the substances to which the potentials relate.

In the case of a body of invariable composition, the potential for the single component is equal to the value of ζ for one unit of the body, as appears from the equation

$$\zeta = \mu\, m, \tag{122}$$

to which (96) reduces in this case. Therefore, when $n = 1$, the fundamental equation between the quantities in the set (102) (see page 88) and that between the quantities in (103) may be derived either from the other by simple substitution. But, with this single exception, an equation between the quantities in one of the sets (99)–(103) cannot be derived from the equation between the quantities in another of these sets without differentiation.

Also in the case of a body of variable composition, when all the quantities of the components except one vanish, the potential for that one will be equal to the value of ζ for one unit of the body. We may make this occur for any given composition of the body by choosing as one of the components the matter constituting the body itself, so that the value of ζ for one unit of a body may always be considered as a potential. Hence the relations between the values of ζ for contiguous masses given on page 91 may be regarded as relations between potentials.

The two following propositions afford definitions of a potential which may sometimes be convenient.

The potential for any substance in any homogeneous mass is equal to the amount of mechanical work required to bring a unit of the substance by a reversible process from the state in which its energy and entropy are both zero into combination with the homogeneous mass, which at the close of the process must have its original volume, and which is supposed so large as not to be sensibly altered in any part. All other bodies used in the process must by its close be restored to their original state, except those used to supply the work, which must be used only as the source of the work. For, in a reversible process, when the entropies of other bodies are not altered, the entropy of the substance and mass taken together will not be altered. But the original entropy of the substance is zero; therefore the entropy of the mass is not altered by the addition of the substance. Again, the work expended will be equal to the increment of the energy of the mass and substance taken together, and therefore equal, as the original energy of the substance is zero, to the increment of energy of the mass due to the addition of the substance, which by the definition on page 93 is equal to the potential in question.

The potential for any substance in any homogeneous mass is equal to the work required to bring a unit of the substance by a reversible process from a state in which $\psi = 0$ and the temperature is the same as that of the given mass into combination with this mass, which at the close of the process must have the same volume and temperature as at first, and which is supposed so large as not to be sensibly altered in any part. A source of heat or cold of the temperature of the given mass is allowed, with this exception other bodies are to be used only on the same conditions as before. This may be shown by applying equation (109) to the mass and substance taken together.

The last proposition enables us to see very easily how the value of the potential is affected by the arbitrary constants involved in the definition of the energy and the entropy of each elementary substance. For we may imagine the substance brought from the state in which $\psi = 0$ and the temperature is the same as that of the given mass, first to any specified state of the same temperature, and then into combination with the given mass. In the first part of the process the work expended is evidently represented by the value of ψ for the unit of the substance in the state specified. Let this be denoted by ψ', and let μ denote the potential in question, and W the work expended in bringing a unit of the substance from the specified state into combination with the given mass as aforesaid ; then

$$\mu = \psi' + W. \tag{123}$$

Now as the state of the substance for which $\epsilon = 0$ and $\eta = 0$ is arbitrary, we may simultaneously increase the energies of the unit of the substance in all possible states by any constant C, and the entropies of the substance in all possible states by any constant K. The value of ψ, or $\epsilon - t\eta$, for any state would then be increased by $C - tK$, t denoting the temperature of the state. Applying this to ψ' in (123) and observing that the last term in this equation is independent of the values of these constants, we see that the potential would be increased by the same quantity $C - tK$, t being the temperature of the mass in which the potential is to be determined.

On Coexistent Phases of Matter.

In considering the different homogeneous bodies which can be formed out of any set of component substances, it will be convenient to have a term which shall refer solely to the composition and thermodynamic state of any such body without regard to its quantity or form. We may call such bodies as differ in composition or state different *phases* of the matter considered, regarding all bodies which differ only in quantity and form as different examples of the same phase. Phases which can exist together, the dividing surfaces being plane, in an equilibrium which does not depend upon passive resistances to change, we shall call *coexistent*.

If a homogeneous body has n independently variable components, the phase of the body is evidently capable of $n+1$ independent variations. A system of r coexistent phases, each of which has the same n independently variable components is capable of $n+2-r$ variations of phase. For the temperature, the pressure, and the potentials for the actual components have the same values in the different phases, and the variations of these quantities are by (97) subject to as many conditions as there are different phases. Therefore, the number of independent variations in the values of these quantities, i.e., the number of independent variations of phase of the system, will be $n+2-r$.

Or, when the r bodies considered have not the same independently variable components, if we still denote by n the number of independently variable components of the r bodies taken as a whole, the number of independent variations of phase of which the system is capable will still be $n+2-r$. In this case, it will be necessary to consider the potentials for more than n component substances. Let the number of these potentials be $n+h$. We shall have by (97), as before, r relations between the variations of the temperature, of the pressure, and of these $n+h$ potentials, and we shall also have by (43) and (51) h relations between these potentials, of the same form as the relations which subsist between the units of the different component substances.

Hence, if $r=n+2$, no variation in the phases (remaining coexistent) is possible. It does not seem probable that r can ever exceed $n+2$. An example of $n=1$ and $r=3$ is seen in the coexistent solid, liquid, and gaseous forms of any substance of invariable composition. It seems not improbable that in the case of sulphur and some other simple substances there is more than one triad of coexistent phases; but it is entirely improbable that there are four coexistent phases of any simple substance. An example of $n=2$ and $r=4$ is seen in a solution of a salt in water in contact with vapor of water and two different kinds of crystals of the salt.

Concerning $n+1$ Coexistent Phases.

We will now seek the differential equation which expresses the relation between the variations of the temperature and the pressure in a system of $n+1$ coexistent phases (n denoting, as before, the number of independently variable components in the system taken as a whole).

In this case we have $n+1$ equations of the general form of (97) (one for each of the coexistent phases), in which we may distinguish the quantities η, v, m_1, m_2, etc., relating to the different phases by accents. But t and p will each have the same value throughout, and the same is true of μ_1, μ_2, etc., so far as each of these occurs in the different equations. If the total number of these potentials is $n+h$, there will be h independent relations between them, corresponding to the h independent relations between the units of the component substances to which the potentials relate, by means of which we may eliminate the variations of h of the potentials from the equations of the form of (97) in which they occur.

Let one of these equations be

$$v' dp = \eta' dt + m_a' d\mu_a + m_b' d\mu_b + \text{etc.}, \tag{124}$$

and by the proposed elimination let it become

$$v' dp = \eta' dt + A_1' d\mu_1 + A_2' d\mu_2 \dots + A_n' d\mu_n. \tag{125}$$

It will be observed that μ_a, for example, in (124) denotes the potential in the mass considered for a substance S_a which may or may not be identical with any of the substances S_1, S_2, etc., to which the potentials in (125) relate. Now as the equations between the potentials by means of which the elimination is performed are similar to those which subsist between the units of the corresponding substances (compare equations (38), (43), and (51)), if we denote these units by $\mathfrak{S}_a$, $\mathfrak{S}_b$, etc., $\mathfrak{S}_1$, $\mathfrak{S}_2$, etc., we must also have

$$m_a'\mathfrak{S}_a + m_b'\mathfrak{S}_b + \text{etc.} = A_1'\mathfrak{S}_1 + A_2'\mathfrak{S}_2 \ldots + A_n'\mathfrak{S}_n. \qquad (126)$$

But the first member of this equation denotes (in kind and quantity) the matter in the body to which equations (124) and (125) relate. As the same must be true of the second member, we may regard this same body as composed of the quantity A_1' of the substance S_1, with the quantity A_2' of the substance S_2, etc. We will therefore, in accordance with our general usage, write m_1', m_2', etc., for A_1', A_2', etc., in (125), which will then become

$$v'\,dp = \eta'\,dt + m_1'\,d\mu_1 + m_2'\,d\mu_2 \ldots + m_n'\,d\mu_n. \qquad (127)$$

But we must remember that the components to which the m_1', m_2', etc., of this equation relate are not necessarily independently variable, as are the components to which the similar expressions in (97) and (124) relate. The rest of the $n+1$ equations may be reduced to a similar form, viz.,

$$v''\,dp = \eta''\,dt + m_1''\,d\mu_1 + m_2''\,d\mu_2 \ldots + m_n''\,d\mu_n, \qquad (128)$$
etc.

By elimination of $d\mu_1$, $d\mu_2$, ... $d\mu_n$ from these equations we obtain

$$\begin{vmatrix} v' & m_1' & m_2' & \ldots m_n' \\ v'' & m_1'' & m_2'' & \ldots m_n'' \\ v''' & m_1''' & m_2''' & \ldots m_n''' \\ \cdot & \cdot & \cdot & \ldots \cdot \\ \cdot & \cdot & \cdot & \ldots \cdot \end{vmatrix} dp = \begin{vmatrix} \eta' & m_1' & m_2' & \ldots m_n' \\ \eta'' & m_1'' & m_2'' & \ldots m_n'' \\ \eta''' & m_1''' & m_2''' & \ldots m_n''' \\ \cdot & \cdot & \cdot & \ldots \cdot \\ \cdot & \cdot & \cdot & \ldots \cdot \end{vmatrix} dt. \qquad (129)$$

In this equation we may make v', v'', etc., equal to unity. Then m_1', m_2', m_1'', etc., will denote the separate densities of the components in the different phases, and η', η'', etc., the densities of entropy.

When $n = 1$,

$$(m''v' - m'v'')dp = (m''\eta' - m'\eta'')dt, \qquad (130)$$

or, if we make $m' = 1$ and $m'' = 1$, we have the usual formula

$$\frac{dp}{dt} = \frac{\eta' - \eta''}{v' - v''} = \frac{Q}{t(v'' - v')}, \qquad (131)$$

in which Q denotes the heat absorbed by a unit of the substance in passing from one state to the other without change of temperature or pressure.

Concerning Cases in which the Number of Coexistent Phases is less than $n+1$.

When $n > 1$, if the quantities of all the components $S_1, S_2, \ldots S_n$ are proportional in two coexistent phases, the two equations of the form of (127) and (128) relating to these phases will be sufficient for the elimination of the variations of all the potentials. In fact, the condition of the coexistence of the two phases together with the condition of the equality of the $n-1$ ratios of $m_1', m_2', \ldots m_n'$ with the $n-1$ ratios of $m_1'', m_2'', \ldots m_n''$ is sufficient to determine p as a function of t if the fundamental equation is known for each of the phases. The differential equation in this case may be expressed in the form of (130), m' and m'' denoting either the quantities of any one of the components or the total quantities of matter in the bodies to which they relate. Equation (131) will also hold true in this case if the total quantity of matter in each of the bodies is unity. But this case differs from the preceding in that the matter which absorbs the heat Q in passing from one state to another, and to which the other letters in the formula relate, although the same in quantity, is not in general the same in kind at different temperatures and pressures. Yet the case will often occur that one of the phases is essentially invariable in composition, especially when it is a crystalline body, and in this case the matter to which the letters in (131) relate will not vary with the temperature and pressure.

When $n = 2$, two coexistent phases are capable, when the temperature is constant, of a single variation in phase. But as (130) will hold true in this case when $m_1' : m_2' :: m_1'' : m_2''$, it follows that for constant temperature the pressure is in general a maximum or a minimum when the composition of the two phases is identical. In like manner, the temperature of the two coexistent phases is in general a maximum or a minimum, for constant pressure, when the composition of the two phases is identical. Hence, the series of simultaneous values of t and p for which the composition of two coexistent phases is identical separates those simultaneous values of t and p for which no coexistent phases are possible from those for which there are two pair of coexistent phases. This may be applied to a liquid having two independently variable components in connection with the vapor which it yields, or in connection with any solid which may be formed in it.

When $n = 3$, we have for three coexistent phases three equations of the form of (127), from which we may obtain the following,

$$\begin{vmatrix} v' & m_1' & m_2' \\ v'' & m_1'' & m_2'' \\ v''' & m_1''' & m_2''' \end{vmatrix} dp = \begin{vmatrix} \eta' & m_1' & m_2' \\ \eta'' & m_1'' & m_2'' \\ \eta''' & m_1''' & m_2''' \end{vmatrix} dt + \begin{vmatrix} m_1' & m_2' & m_3' \\ m_1'' & m_2'' & m_3'' \\ m_1''' & m_2''' & m_3''' \end{vmatrix} d\mu_3. \quad (132)$$

Now the value of the last of these determinants will be zero, when the composition of one of the three phases is such as can be produced by combining the other two. Hence, the pressure of three coexistent phases will in general be a maximum or minimum for constant temperature, and the temperature a maximum or minimum for constant pressure, when the above condition in regard to the composition of the coexistent phases is satisfied. The series of simultaneous values of t and p for which the condition is satisfied separates those simultaneous values of t and p for which three coexistent phases are not possible, from those for which there are two triads of coexistent phases. These propositions may be extended to higher values of n, and illustrated by the boiling temperatures and pressures of saturated solutions of $n-2$ different solids in solvents having two independently variable components.

Internal Stability of Homogeneous Fluids as indicated by Fundamental Equations.

We will now consider the stability of a fluid enclosed in a rigid envelop which is non-conducting to heat and impermeable to all the components of the fluid. The fluid is supposed initially homogeneous in the sense in which we have before used the word, i.e., uniform in every respect throughout its whole extent. Let S_1, S_2, ... S_n be the *ultimate* components of the fluid; we may then consider every body which can be formed out of the fluid to be composed of $S_1, S_2, ... S_n$, and that in only one way. Let m_1, m_2, ... m_n denote the quantities of these substances in any such body, and let ϵ, η, v, denote its energy, entropy, and volume. The fundamental equation for compounds of $S_1, S_2, ... S_n$, if completely determined, will give us all possible sets of simultaneous values of these variables for homogeneous bodies.

Now, if it is possible to assign such values to the constants T, P, $M_1, M_2, ... M_n$ that the value of the expression

$$\epsilon - T\eta + Pv - M_1 m_1 - M_2 m_2 ... - M_n m_n \tag{133}$$

shall be zero for the given fluid, and shall be positive for every other phase of the same components, i.e., for every homogeneous body* not identical in nature and state with the given fluid (but composed entirely of S_1, S_2, ... S_n), the condition of the given fluid will be stable.

For, in any condition whatever of the given mass, whether or not homogeneous, or fluid, if the value of the expression (133) is not

---

* A vacuum is throughout this discussion to be regarded as a limiting case of an extremely rarified body. We may thus avoid the necessity of the specific mention of a vacuum in propositions of this kind.

negative for any homogeneous part of the mass, its value for the whole mass cannot be negative; and if its value cannot be zero for any homogeneous part which is not identical in phase with the mass in its given condition, its value cannot be zero for the whole except when the whole is in the given condition. Therefore, in the case supposed, the value of this expression for any other than the given condition of the mass is positive. (That this conclusion cannot be invalidated by the fact that it is not entirely correct to regard a composite mass as made up of homogeneous parts having the same properties in respect to energy, entropy, etc., as if they were parts of larger homogeneous masses, will easily appear from considerations similar to those adduced on pages 77–78.) If, then, the value of the expression (133) for the mass considered is less when it is in the given condition than when it is in any other, the energy of the mass in its given condition must be less than in any other condition in which it has the same entropy and volume. The given condition is therefore stable. (See page 57.)

Again, if it is possible to assign such values to the constants in (133) that the value of the expression shall be zero for the given fluid mass, and shall not be negative for any phase of the same components, the given condition will be evidently not unstable. (See page 57.) It will be stable unless it is possible for the given matter in the given volume and with the given entropy to consist of homogeneous parts for all of which the value of the expression (133) is zero, but which are not all identical in phase with the mass in its given condition. (A mass consisting of such parts would be in equilibrium, as we have already seen on pages 78, 79.) In this case, if we disregard the quantities connected with the surfaces which divide the homogeneous parts, we must regard the given condition as one of neutral equilibrium. But in regard to these homogeneous parts, which we may evidently consider to be all different phases, the following conditions must be satisfied. (The accents distinguish the letters referring to the different parts, and the unaccented letters refer to the whole mass.)

$$\left.\begin{array}{l} \eta' + \eta'' + \text{etc.} = \eta, \\ v' + v'' + \text{etc.} = v, \\ m_1' + m_1'' + \text{etc.} = m_1, \\ m_2' + m_2'' + \text{etc.} = m_2, \\ \text{etc.} \end{array}\right\} \qquad (134)$$

Now the values of η, v, m_1, m_2, etc., are determined by the whole fluid mass in its given state, and the values of $\dfrac{\eta'}{v'}$, $\dfrac{\eta''}{v''}$, etc., $\dfrac{m_1'}{v'}$, $\dfrac{m_1''}{v''}$, etc., $\dfrac{m_2'}{v'}$, $\dfrac{m_2''}{v''}$, etc., etc., are determined by the phases of the various parts. But the phases of these parts are evidently determined by the phase of the fluid as given. They form, in fact, the whole set of coexistent phases of which the latter is one. Hence, we may regard (134) as $n+2$ linear equations between v', v'', etc. (The values of v', v'', etc., are also subject to the condition that none of them can be negative.) Now one solution of these equations must give us the given condition of the fluid; and it is not to be expected that they will be capable of any other solution, unless the number of different homogeneous parts, that is, the number of different coexistent phases, is greater than $n+2$. We have already seen (page 97) that it is not probable that this is ever the case.

We may, however, remark that in a certain sense an infinitely large fluid mass will be in neutral equilibrium in regard to the formation of the substances, if such there are, other than the given fluid, for which the value of (133) is zero (when the constants are so determined that the value of the expression is zero for the given fluid, and not negative for any substance); for the tendency of such a formation to be reabsorbed will diminish indefinitely as the mass out of which it is formed increases.

When the substances S_1, S_2, ... S_n are all independently variable components of the given mass, it is evident from (86) that the conditions that the value of (133) shall be zero for the mass as given, and shall not be negative for any phase of the same components, can only be fulfilled when the constants T, P, M_1, M_2, ... M_n are equal to the temperature, the pressure, and the several potentials in the given mass. If we give these values to the constants, the expression (133) will necessarily have the value zero for the given mass, and we shall only have to inquire whether its value is positive for all other phases. But when S_1, S_2, ... S_n are not all independently variable components of the given mass, the values which it will be necessary to give to the constants in (133) cannot be determined entirely from the properties of the given mass; but T and P must be equal to its temperature and pressure, and it will be easy to obtain as many equations connecting M_1, M_2, ... M_n with the potentials in the given mass as it contains independently variable components.

When it is not possible to assign such values to the constants in (133) that the value of the expression shall be zero for the given fluid, and either zero or positive for any phase of the same components, we have already seen (pages 75–79) that if equilibrium subsists without passive resistances to change, it must be in virtue of properties which are peculiar to small masses surrounded by masses of different nature, and which are not indicated by fundamental equations. In this case, the fluid will necessarily be unstable, if we extend this term to embrace all cases in which an initial disturbance confined to a small part of an indefinitely large fluid mass will cause an ultimate change of state not indefinitely small in degree throughout the whole mass. In the discussion of stability as indicated by fundamental equations it will be convenient to use the term in this sense.*

In determining for any given positive values of T and P and any given values whatever of $M_1, M_2, \ldots M_n$ whether the expression (133) is capable of a negative value for any phase of the components $S_1, S_2, \ldots S_n$, and if not, whether it is capable of the value zero for any other phase than that of which the stability is in question, it is only necessary to consider phases having the temperature T and pressure P. For we may assume that a mass of matter represented by any values of $m_1, m_2, \ldots m_n$ is capable of at least one state of not unstable equilibrium (which may or may not be a homogeneous state) at this temperature and pressure. It may easily be shown that for such a state the value of $\epsilon - T\eta + Pv$ must be as small as for any other state of the same matter. The same will therefore be true of the value of (133). Therefore if this expression is capable of a negative value for any mass whatever, it will have a negative value for that mass at the temperature T and pressure P. And if this mass is not homogeneous, the value of (133) must be negative for at least one of its homogeneous parts. So also, if the expression (133) is not capable of a negative value for any phase of the components, any phase for which it has the value zero must have the temperature T and the pressure P.

---

*If we wish to know the stability of the given fluid when exposed to a constant temperature, or to a constant pressure, or to both, we have only to suppose that there is enclosed in the same envelop with the given fluid another body (which cannot combine with the fluid) of which the fundamental equation is $\epsilon = T\eta$, or $\epsilon = - Pv$, or $\epsilon = T\eta - Pv$, as the case may be (T and P denoting the constant temperature and pressure, which of course must be those of the given fluid), and to apply the criteria of page 57 to the whole system. When it is possible to assign such values to the constants in (133) that the value of the expression shall be zero for the given fluid and positive for every other phase of the same components, the value of (133) for the whole system will be less when the system is in its given condition than when it is in any other. (Changes of form and position of the given fluid are of course regarded as immaterial.) Hence the fluid is stable. When it is not possible to assign such values to the constants that the value of (133) shall be zero for the given fluid and zero or positive for any other phase, the fluid is of course unstable. In the remaining case, when it is possible to assign such values to the constants that the value of (133) shall be zero for the given fluid and zero or positive for every other phase, but not without the value zero for some other phase, the state of equilibrium of the fluid as stable or neutral will be determined by the possibility of satisfying, for any other than the given condition of the fluid, equations like (134), in which, however, the first or the second or both are to be stricken out, according as we are considering the stability of the fluid for constant temperature, or for constant pressure, or for both. The number of coexistent phases will sometimes exceed by one or two the number of the remaining equations, and then the equilibrium of the fluid will be neutral in respect to one or two independent changes.

It may easily be shown that the same must be true in the limiting cases in which $T=0$ and $P=0$. For negative values of P, (133) is always capable of negative values, as its value for a vacuum is Pv.

For any body of the temperature T and pressure P, the expression (133) may by (91) be reduced to the form

$$\zeta - M_1 m_1 - M_2 m_2 \ldots - M_n m_n. \tag{135}$$

We have already seen (page 77) that an expression like (133), when T, P, M_1, M_2, $\ldots M_n$ and v have any given finite values, cannot have an infinite negative value as applied to any real body. Hence, in determining whether (133) is capable of a negative value for any phase of the components S_1, S_2, $\ldots S_n$, and if not, whether it is capable of the value zero for any other phase than that of which the stability is in question, we have only to consider the least value of which it is capable for a constant value of v. Any body giving this value must satisfy the condition that for constant volume

$$d\epsilon - T\,d\eta - M_1 dm_1 - M_2 dm_2 \ldots - M_n dm_n \geqq 0, \tag{136}$$

or, if we substitute the value of $d\epsilon$ taken from equation (86), using subscript $a \ldots g$ for the quantities relating to the actual components of the body, and subscript $h \ldots k$ for those relating to the possible,

$$t\,d\eta + \mu_a dm_a \ldots + \mu_g dm_g + \mu_h dm_h \ldots + \mu_k dm_k$$
$$- T\,d\eta - M_1 dm_1 - M_2 dm_2 \ldots - M_n dm_n \geqq 0. \tag{137}$$

That is, the temperature of the body must be equal to T, and the potentials of its components must satisfy the same conditions as if it were in contact and in equilibrium with a body having potentials $M_1, M_2, \ldots M_n$. Therefore the same relations must subsist between $\mu_a \ldots \mu_g$, and $M_1 \ldots M_n$ as between the units of the corresponding substances, so that

$$m_a \mu_a \ldots + m_g \mu_g = m_1 M_1 \ldots + m_n M_n; \tag{138}$$

and as we have by (93)

$$\epsilon = t\eta - pv + \mu_a m_a \ldots + \mu_g m_g, \tag{139}$$

the expression (133) will reduce (for the body or bodies for which it has the least value per unit of volume) to

$$(P-p)v, \tag{140}$$

the value of which will be positive, null, or negative, according as the value of

$$P-p \tag{141}$$

is positive, null, or negative.

Hence, the conditions in regard to the stability of a fluid of which all the ultimate components are independently variable admit a very simple expression. If the pressure of the fluid is greater than that of any other phase of the same components which has the same temperature and the same values of the potentials for its actual components, the fluid is stable without coexistent phases; if its pressure is not as great as that of some other such phase, it will be unstable; if its pressure is as great as that of any other such phase, but not greater than that of every other, the fluid will certainly not be unstable, and in all probability it will be stable (when enclosed in a rigid envelop which is impermeable to heat and to all kinds of matter), but it will be one of a set of coexistent phases of which the others are the phases which have the same pressure.

The considerations of the last two pages, by which the tests relating to the stability of a fluid are simplified, apply to such bodies as actually exist. But if we should form arbitrarily any equation as a fundamental equation, and ask whether a fluid of which the properties were given by that equation would be stable, the tests of stability last given would be insufficient, as some of our assumptions might not be fulfilled by the equation. The test, however, as first given (pages 100–102) would in all cases be sufficient.

Stability in respect to Continuous Changes of Phase.

In considering the changes which may take place in any mass, we have already had occasion to distinguish between infinitesimal changes in existing phases, and the formation of entirely new phases. A phase of a fluid may be stable in regard to the former kind of change, and unstable in regard to the latter. In this case it may be capable of continued existence in virtue of properties which prevent the commencement of discontinuous changes. But a phase which is unstable in regard to continuous changes is evidently incapable of permanent existence on a large scale except in consequence of passive resistances to change. We will now consider the conditions of stability in respect to continuous changes of phase, or, as it may also be called, stability in respect to adjacent phases. We may use the same general test as before, except that the expression (133) is to be applied only to phases which differ infinitely little from the phase of which the stability is in question. In this case the component substances to be considered will be limited to the independently variable components of the fluid, and the constants M_1, M_2, etc., must have the values of the potentials for these components in the given fluid. The constants in (133) are thus entirely determined and the value of the expression for the given phase is necessarily zero. If for any infinitely small variation of the phase the value of (133) can become negative, the fluid will be_ unstable; but if for every infinitely small variation of the phase the value of (133) becomes positive, the fluid will be stable. The only remaining case, in which the phase can be varied without altering the value of (133) can hardly be expected to occur. The phase concerned would in such a case have coexistent adjacent phases. It will be sufficient to discuss the condition of stability (in respect to continuous changes) without coexistent adjacent phases.

This condition, which for brevity's sake we will call the condition of stability, may be written in the form

$$\epsilon'' - t'\eta'' + p'v'' - \mu_1'm_1'' \ldots - \mu_n'm_n'' > 0, \tag{142}$$

in which the quantities relating to the phase of which the stability is in question are distinguished by single accents, and those relating to the other phase by double accents. This condition is by (93) equivalent to

$$\begin{aligned} &\epsilon'' - t'\eta'' + p'v'' - \mu_1'm_1'' \ldots - \mu_n'm_n'' \\ &- \epsilon' + t'\eta' - p'v' + \mu_1'm_1' \ldots + \mu_n'm_n' > 0, \end{aligned} \tag{143}$$

and to

$$\begin{aligned} &- t'\eta'' + p'v'' - \mu_1'm_1'' \ldots - \mu_n'm_n'' \\ &+ t''\eta'' - p''v'' + \mu_1''m_1'' \ldots + \mu_n''m_n'' > 0. \end{aligned} \tag{144}$$

The condition (143) may be expressed more briefly in the form

$$\Delta\epsilon > t\,\Delta\eta - p\,\Delta v + \mu_1\Delta m_1 \ldots + \mu_n\Delta m_n, \tag{145}$$

if we use the character Δ to signify that the condition, although relating to infinitesimal differences, is not to be interpreted in accordance with the usual convention in respect to differential equations with neglect of infinitesimals of higher orders than the first, but is to be interpreted *strictly*, like an equation between finite differences. In fact, when a condition like (145) (interpreted strictly) is satisfied for infinitesimal differences, it must be possible to assign limits within which it shall hold true of finite differences. But it is to be remembered that the condition is not to be applied to any arbitrary values of $\Delta\eta$, Δv, $\Delta m_1, \ldots \Delta m_n$, but only to such as are determined by a change of phase. (If only the quantity of the body which determines the value of the variables should vary and not its phase, the value of the first member of (145) would evidently be zero.) We may free ourselves from this limitation by making v constant, which will cause the term $-p\,\Delta v$ to disappear. If we then divide by the constant v, the condition will become

$$\Delta\frac{\epsilon}{v} > t\,\Delta\frac{\eta}{v} + \mu_1\Delta\frac{m_1}{v} \ldots + \mu_n\Delta\frac{m_n}{v}, \tag{146}$$

in which form it will not be necessary to regard v as constant. As we may obtain from (86)

$$d\frac{\epsilon}{v} = td\frac{\eta}{v} + \mu_1 d\frac{m_1}{v} \ldots + \mu_n d\frac{m_n}{v}, \tag{147}$$

we see that *the stability of any phase in regard to continuous changes depends upon the same conditions in regard to the second and higher differential coefficients of the density of energy regarded as a function of the density of entropy and the densities of the several components, which would make the density of energy a minimum, if the necessary conditions in regard to the first differential coefficients were fulfilled.* When $n=1$, it may be more convenient to regard m as constant in (145) than v. Regarding m a constant, it appears that the stability of a phase depends upon the same conditions in regard to the second and higher differential coefficients of the energy of a unit of mass regarded as a function of its entropy and volume, which would make the energy a minimum, if the necessary conditions in regard to the first differential coefficients were fulfilled.

The formula (144) expresses the condition of stability for the phase to which t', p', etc., relate. But it is evidently the necessary and sufficient condition of the stability of all phases of certain kinds of matter, or of all phases within given limits, that (144) shall hold true of any two infinitesimally differing phases within the same limits, or, as the case may be, in general. For the purpose, therefore, of such *collective* determinations of stability, we may neglect the distinction between the two states compared, and write the condition in the form

$$-\eta \, \Delta t + v \, \Delta p - m_1 \Delta \mu_1 \ldots - m_n \Delta \mu_n > 0, \qquad (148)$$

or
$$\Delta p > \frac{\eta}{v} \Delta t + \frac{m_1}{v} \Delta \mu_1 \ldots + \frac{m_n}{v} \Delta \mu_n. \qquad (149)$$

Comparing (98), we see that it is necessary and sufficient for the stability in regard to continuous changes of all the phases within any given limits, that within those limits the same conditions should be fulfilled in respect to the second and higher differential coefficients of the pressure regarded as a function of the temperature and the several potentials, which would make the pressure a minimum, if the necessary conditions with respect to the first differential coefficients were fulfilled.

By equations (87) and (94), the condition (142) may be brought to the form

$$\begin{aligned} \psi'' + t'' \eta'' + p' v'' - \mu_1' m_1'' \ldots - \mu_n' m_n'' \\ - \psi' - t' \eta'' - p' v' + \mu_1' m_1' \ldots + \mu_n' m_n' > 0. \end{aligned} \qquad (150)$$

For the stability of all phases within any given limits it is necessary and sufficient that within the same limits this condition shall hold true of any two phases which differ infinitely little. This evidently requires that when $v' = v''$, $m_1' = m_1''$, $\ldots m_n' = m_n''$,

$$\psi'' - \psi' + (t'' - t')\eta'' > 0; \qquad (151)$$

and that when $t' = t''$

$$\psi'' + p'v'' - \mu_1'm_1'' \ldots + \mu_n'm_n''$$
$$- \psi' - p'v' + \mu_1'm_1' \ldots + \mu_n'm_n' > 0. \qquad (152)$$

These conditions may be written in the form

$$[\Delta\psi + \eta\,\Delta t]_{v,\,m} < 0, \qquad (153)$$
$$[\Delta\psi + p\,\Delta v - \mu_1\Delta m_1 \ldots - \mu_n\Delta m_n]_t > 0, \qquad (154)$$

in which the subscript letters indicate the quantities which are to be regarded as constant, m standing for all the quantities $m_1 \ldots m_n$. If these conditions hold true within any given limits, (150) will also hold true of any two infinitesimally differing phases within the same limits. To prove this, we will consider a third phase, determined by the equations

$$t''' = t', \qquad (155)$$

and

$$v''' = v'', \; m_1''' = m_1'', \ldots m_n''' = m_n''. \qquad (156)$$

Now by (153),

$$\psi''' - \psi'' + (t''' - t'')\eta'' < 0; \qquad (157)$$

and by (154),

$$\psi''' + p'v''' - \mu_1'm_1''' \ldots - \mu_n'm_n'''$$
$$- \psi' - p'v' + \mu_1'm_1' \ldots + \mu_n'm_n' > 0. \qquad (158)$$

Hence,

$$\psi'' + t''\eta'' + p'v''' - \mu_1'm_1''' \ldots - \mu_n'm_n'''$$
$$- \psi' - t'''\eta'' - p'v' + \mu_1'm_1' \ldots + \mu_n'm_n' > 0, \qquad (159)$$

which by (155) and (156) is equivalent to (150). Therefore, the conditions (153) and (154) in respect to the phases within any given limits are necessary and sufficient for the stability of all the phases within those limits. It will be observed that in (153) we have the condition of thermal stability of a body considered as unchangeable in composition and in volume, and in (154), the condition of mechanical and chemical stability of the body considered as maintained at a constant temperature. Comparing equation (88), we see that the condition (153) will be satisfied, if $\dfrac{d^2\psi}{dt^2} < 0$, i.e., if $\dfrac{d\eta}{dt}$ or $t\dfrac{d\eta}{dt}$ (the specific heat for constant volume) is positive. When $n=1$, i.e., when the composition of the body is invariable, the condition (154) will evidently not be altered, if we regard m as constant, by which the condition will be reduced to

$$[\Delta\psi + p\,\Delta v]_{t,\,m} > 0. \qquad (160)$$

This condition will evidently be satisfied if $\dfrac{d^2\psi}{dv^2} > 0$, i.e., if $-\dfrac{dp}{dv}$ or $-v\dfrac{dp}{dv}$ (the elasticity for constant temperature) is positive. But when $n > 1$, (154) may be abbreviated more symmetrically by making v constant.

Again, by (91) and (96), the condition (142) may be brought to the form

$$\zeta'' + t''\eta'' - p''v'' - \mu_1'm_1'' \ldots - \mu_n'm_n''$$
$$- \zeta' - t'\eta'' + p'v'' + \mu_1'm_1' \ldots + \mu_n'm_n' > 0. \tag{161}$$

Therefore, for the stability of all phases within any given limits it is necessary and sufficient that within the same limits

$$[\Delta\zeta + \eta\,\Delta t - v\,\Delta p]_m < 0, \tag{162}$$

and
$$[\Delta\zeta - \mu_1\Delta m_1 \ldots - \mu_n\Delta m_n]_{t,\,p} > 0, \tag{163}$$

as may easily be proved by the method used with (153) and (154). The first of these formulæ expresses the thermal and mechanical conditions of stability for a body considered as unchangeable in composition, and the second the conditions of chemical stability for a body considered as maintained at a constant temperature and pressure. If $n = 1$, the second condition falls away, and as in this case $\zeta = m\mu$, condition (162) becomes identical with (148).

The foregoing discussion will serve to illustrate the relation of the general condition of stability in regard to continuous changes to some of the principal forms of fundamental equations. It is evident that each of the conditions (146), (149), (154), (162), (163) involves in general several particular conditions of stability. We will now give our attention to the latter. Let

$$\Phi = \epsilon - t'\eta + p'v - \mu_1'm_1 \ldots - \mu_n'm_n, \tag{164}$$

the accented letters referring to one phase and the unaccented to another. It is by (142) the necessary and sufficient condition of the stability of the first phase that, for constant values of the quantities relating to that phase and of v, the value of Φ shall be a minimum when the second phase is identical with the first. Differentiating (164), we have by (86)

$$d\Phi = (t - t')d\eta - (p - p')dv + (\mu_1 - \mu_1')dm_1 \ldots + (\mu_n - \mu_n')dm_n. \tag{165}$$

Therefore, the above condition requires that if we regard v, $m_1, \ldots m_n$ as having the constant values indicated by accenting these letters, t shall be an increasing function of η, when the variable phase differs sufficiently little from the fixed. But as the fixed phase may be any one within the limits of stability, t must be an increasing function of η (within these limits) for any constant values of v, $m_1, \ldots m_n$. This condition may be written

$$\left(\frac{\Delta t}{\Delta \eta}\right)_{v,\,m_1,\,\ldots\,m_n} > 0. \tag{166}$$

When this condition is satisfied, the value of Φ, for any given values of v, $m_1, \ldots m_n$, will be a minimum when $t = t'$. And therefore, in applying the general condition of stability relating to the value of Φ, we need only consider the phases for which $t = t'$.

We see again by (165) that the general condition requires that if we regard t, v, m_2, ... m_n as having the constant values indicated by accenting these letters, μ_1 shall be an increasing function of m_1, when the variable phase differs sufficiently little from the fixed. But as the fixed phase may be any one within the limits of stability, μ_1 must be an increasing function of m_1 (within these limits) for any constant values of t, v, m_2, ... m_n. That is,

$$\left(\frac{\Delta\mu_1}{\Delta m_1}\right)_{t, v, m_2, \ldots m_n} > 0. \tag{167}$$

When this condition is satisfied, as well as (166), Φ will have a minimum value, for any constant values of v, m_2, ... m_n, when $t = t'$ and $\mu_1 = \mu_1'$; so that in applying the general condition of stability we need only consider the phases for which $t = t'$ and $\mu_1 = \mu_1'$.

In this way we may also obtain the following particular conditions of stability :

$$\left(\frac{\Delta\mu_2}{\Delta m_2}\right)_{t, v, \mu_1, m_3, \ldots m_n} > 0, \tag{168}$$

$$\left(\frac{\Delta\mu_n}{\Delta m_n}\right)_{t, v, \mu_1, \ldots \mu_{n-1}} > 0. \tag{169}$$

When the $n+1$ conditions (166)–(169) are all satisfied, the value of Φ, for any constant value of v, will be a minimum when the temperature and the potentials of the variable phase are equal to those of the fixed. The pressures will then also be equal and the phases will be entirely identical. Hence, the general condition of stability will be completely satisfied, when the above particular conditions are satisfied.

From the manner in which these particular conditions have been derived, it is evident that we may interchange in them η, m_1, ... m_n in any way, provided that we also interchange in the same way t, μ_1, ... μ_n. In this way we may obtain different sets of $n+1$ conditions which are necessary and sufficient for stability. The quantity v might be included in the first of these lists, and $-p$ in the second, except in cases when, in some of the phases considered, the entropy or the quantity of one of the components has the value zero. Then the condition that that quantity shall be constant would create a restriction upon the variations of the phase, and cannot be substituted for the condition that the volume shall be constant in the statement of the general condition of stability relative to the minimum value of Φ.

To indicate more distinctly all these particular conditions at once, we observe that the condition (144), and therefore also the condition obtained by interchanging the single and double accents, must hold true of any two infinitesimally differing phases within the limits of stability. Combining these two conditions we have

$$(t''-t')(\eta''-\eta')-(p''-p')(v''-v')$$
$$+(\mu_1''-\mu_1')(m_1''-m_1')\ldots(\mu_n''-\mu_n')(m_n''-m_n')>0, \quad (170)$$

which may be written more briefly

$$\Delta t\,\Delta\eta-\Delta p\,\Delta v+\Delta\mu_1\Delta m_1\ldots+\Delta\mu_n\Delta m_n>0. \quad (171)$$

This must hold true of any two infinitesimally differing phases within the limits of stability. If, then, we give the value zero to one of the differences in every term except one, but not so as to make the phases completely identical, the values of the two differences in the remaining term will have the same sign, except in the case of Δp and Δv, which will have opposite signs. (If both states are stable this will hold true even on the limits of stability.) Therefore, within the limits of stability, either of the two quantities occurring (after the sign Δ) in any term of (171) is an increasing function of the other,— except p and v, of which the opposite is true,—when we regard as constant one of the quantities occurring in each of the other terms, but not such as to make the phases identical.

If we write d for Δ in (166)–(169), we obtain conditions which are always sufficient for stability. If we also substitute $\geqq$ for $>$, we obtain conditions which are necessary for stability. Let us consider the form which these conditions will take when η, v, $m_1,\ldots m_n$ are regarded as independent variables. When $dv=0$, we shall have

$$\left.\begin{aligned} dt&=\frac{dt}{d\eta}\,d\eta+\frac{dt}{dm_1}\,dm_1\ldots+\frac{dt}{dm_n}\,dm_n\\[4pt] d\mu_1&=\frac{d\mu_1}{d\eta}\,d\eta+\frac{d\mu_1}{dm_1}\,dm_1\ldots+\frac{d\mu_n}{dm_n}\,dm_n\\[2pt] &\;\cdot\quad\cdot\quad\cdot\quad\cdot\quad\cdot\quad\cdot\\[2pt] d\mu_n&=\frac{d\mu_n}{d\eta}\,d\eta+\frac{d\mu_n}{dm_1}\,dm_1\ldots+\frac{d\mu_n}{dm_n}\,dm_n \end{aligned}\right\}. \quad (172)$$

Let us write R_{n+1} for the determinant of the order $n+1$:

$$\begin{vmatrix} \dfrac{d^2\epsilon}{d\eta^2} & \dfrac{d^2\epsilon}{dm_1 d\eta} & \cdots & \dfrac{d^2\epsilon}{dm_n d\eta}\\[8pt] \dfrac{d^2\epsilon}{d\eta\,dm_1} & \dfrac{d^2\epsilon}{dm_1{}^2} & \cdots & \dfrac{d^2\epsilon}{dm_n dm_1}\\[6pt] & \cdot & \cdots & \\[4pt] \dfrac{d^2\epsilon}{d\eta\,dm_n} & \dfrac{d^2\epsilon}{dm_1 dm_n} & \cdots & \dfrac{d^2\epsilon}{dm_n{}^2} \end{vmatrix}, \quad (173)$$

of which the constituents are by (86) the same as the coefficients in equations (172), and R_n, R_{n-1}, etc., for the minors obtained by erasing the last column and row in the original determinant and in the minors successively obtained, and R_1 for the last remaining constituent. Then if $dt, d\mu_1, \ldots d\mu_{n-1}$, and dv all have the value zero, we have by (172)

$$R_n d\mu_n = R_{n+1} dm_n, \tag{174}$$

that is,

$$\left(\frac{d\mu_n}{dm_n}\right)_{t, v, \mu_1, \ldots \mu_{n-1}} = \frac{R_{n+1}}{R_n}. \tag{175}$$

In like manner we obtain

$$\left.\begin{array}{c}\left(\dfrac{d\mu_{n-1}}{dm_{n-1}}\right)_{t, v, \mu_1, \ldots \mu_{n-2}, m_n} = \dfrac{R_n}{R_{n-1}}, \\ \text{etc.}\end{array}\right\} \tag{176}$$

Therefore, the conditions obtained by writing d for Δ in (166)–(169) are equivalent to this, that the determinant given above with the n minors obtained from it as above mentioned and the last remaining constituent $\dfrac{d^2\epsilon}{d\eta^2}$ shall all be positive. Any phase for which this condition is satisfied will be stable, and no phase will be stable for which any of these quantities has a negative value. But the conditions (166)–(169) will remain valid, if we interchange in any way $\eta, m_1, \ldots m_n$ (with corresponding interchange of $t, \mu_1, \ldots \mu_n$). Hence the order in which we erase successive columns with the corresponding rows in the determinant is immaterial. Therefore none of the minors of the determinant (173) which are formed by erasing corresponding rows and columns, and none of the constituents of the principal diagonal, can be negative for a stable phase.

We will now consider the conditions which characterize the *limits of stability* (i.e., the limits which divide stable from unstable phases) with respect to continuous changes.* Here, evidently, one of the conditions (166)–(169) must cease to hold true. Therefore, one of the differential coefficients formed by changing Δ into d in the first members of these conditions must have the value zero. (That it is the numerator and not the denominator in the differential coefficient which vanishes at the limit appears from the consideration that the denominator is in each case the differential of a quantity which is necessarily capable of progressive variation, so long at least as the phase is capable of variation at all under the conditions expressed by the subscript letters.) The same will hold true of the set of differential coefficients obtained from these by interchanging in any way $\eta, m_1, \ldots m_n$, and simultaneously interchanging $t, \mu_1, \ldots \mu_n$ in the same way. But we may obtain a more definite result than this.

---

* The limits of stability with respect to discontinuous changes are formed by phases which are coexistent with other phases. Some of the properties of such phases have already been considered. See pages 96–100.

Let us give to η or t, to m_1 or μ_1, ... to m_{n-1} or μ_{n-1}, and to v, the constant values indicated by these letters when accented. Then by (165)

$$d\Phi = (\mu_n - \mu_n')dm_n. \tag{177}$$

Now
$$\mu_n - \mu_n' = \left(\frac{d\mu_n}{dm_n}\right)'(m_n - m_n') \tag{178}$$

approximately, the differential coefficient being interpreted in accordance with the above assignment of constant values to certain variables, and its value being determined for the phase to which the accented letters refer. Therefore,

$$d\Phi = \left(\frac{d\mu_n}{dm_n}\right)'(m_n - m_n')dm_n, \tag{179}$$

and
$$\Phi = \tfrac{1}{2}\left(\frac{d\mu_n}{dm_n}\right)'(m_n - m_n')^2. \tag{180}$$

The quantities neglected in the last equation are evidently of the same order as $(m_n - m_n')^3$. Now this value of Φ will of course be different (the differential coefficient having a different meaning) according as we have made η or t constant, and according as we have made m_1 or μ_1 constant, etc.; but since, within the limits of stability, the value of Φ, for any constant values of m_n and v, will be the least when t, p, μ_1, ... μ_{n-1} have the values indicated by accenting these letters, the value of the differential coefficient will be at least as small when we give these variables these constant values, as when we adopt any other of the suppositions mentioned above in regard to the quantities remaining constant. And in all these relations we may interchange in any way η, m_1, ... m_n if we interchange in the same way t, μ_1, ... μ_n. It follows that, within the limits of stability, when we choose for any one of the differential coefficients

$$\frac{dt}{d\eta}, \quad \frac{d\mu_1}{dm_1}, ... \frac{d\mu_n}{dm_n} \tag{181}$$

the quantities following the sign d in the numerators of the others together with v as those which are to remain constant in differentiation, the value of the differential coefficient as thus determined will be at least as small as when one or more of the constants in differentiation are taken from the denominators, one being still taken from each fraction, and v as before being constant.

Now we have seen that none of these differential coefficients, as determined in any of these ways, can have a negative value within the limit of stability, and that some of them must have the value zero at that limit. Therefore in virtue of the relations just established, one at least of these differential coefficients determined by considering constant the quantities occurring in the numerators of the others together with v, will have the value zero. But if one such has the value zero, all such will in general have the same value. For if

$$\left(\frac{d\mu_n}{dm_n}\right)_{t,\, v,\, \mu_1,\, \ldots\, \mu_{n-1}}, \tag{182}$$

for example, has the value zero, we may change the density of the component S_n without altering (if we disregard infinitesimals of higher orders than the first) the temperature or the potentials, and therefore, by (98), without altering the pressure. That is, we may change the phase without altering any of the quantities t, p, μ_1, $\ldots$ μ_n. (In other words, the phases adjacent to the limits of stability exhibit *approximately* the relations characteristic of neutral equilibrium.) Now this change of phase, which changes the density of one of the components, will in general change the density of the others and the density of entropy. Therefore, all the other differential coefficients formed after the analogy of (182), i.e., formed from the fractions in (181) by taking as constants for each the quantities in the numerators of the others together with v, will in general have the value zero at the limit of stability. And the relation which characterizes the limit of stability may be expressed, in general, by setting any one of these differential coefficients equal to zero. Such an equation, when the fundamental equation is known, may be reduced to the form of an equation between the independent variables of the fundamental equation.

Again, as the determinant (173) is equal to the product of the differential coefficients obtained by writing d for Δ in the first members of (166)–(169), the equation of the limit of stability may be expressed by setting this determinant equal to zero. The form of the differential equation as thus expressed will not be altered by the interchange of the expressions η, m_1, $\ldots$ m_n, but it will be altered by the substitution of v for any one of these expressions, which will be allowable whenever the quantity for which it is substituted has not the value zero in any of the phases to which the formula is to be applied.

The condition formed by setting the expression (182) equal to zero is evidently equivalent to this, that

$$\left[\frac{d\mu_n}{d\dfrac{m_n}{v}}\right]_{t,\, \mu_1,\, \ldots\, \mu_{n-1}} = 0, \tag{183}$$

that is, that

$$\left[\frac{d\dfrac{m_n}{v}}{d\mu_n}\right]_{t,\, \mu_1,\, \ldots\, \mu_{n-1}} = \infty, \tag{184}$$

or by (98), if we regard t, μ_1, ... μ_n as the independent variables,

$$\left(\frac{d^2p}{d\mu_n^2}\right) = \infty .$$ (185)

In like manner we may obtain

$$\frac{d^2p}{dt^2} = \infty , \quad \frac{d^2p}{d\mu_1^2} = \infty , \quad ... \quad \frac{d^2p}{d\mu_{n-1}^2} = \infty .$$ (186)

Any one of these equations, (185), (186), may be regarded, in general, as the equation of the limit of stability. We may be certain that at every phase at that limit one at least of these equations will hold true.

Geometrical Illustrations.

Surfaces in which the Composition of the Body represented is Constant.

In the second paper of this volume (pp. 33-54) a method is described of representing the thermodynamic properties of substances of invariable composition by means of surfaces. The volume, entropy, and energy of a constant quantity of a substance are represented by rectangular co-ordinates. This method corresponds to the first kind of fundamental equation described on pages 85-89. Any other kind of fundamental equation for a substance of invariable composition will suggest an analogous geometrical method. Thus, if we make m constant, the variables in any one of the sets (99)–(103) are reduced to three, which may be represented by rectangular co-ordinates. This will, however, afford but four different methods, for, as has already (page 94) been observed, the two last sets are essentially equivalent when $n = 1$.

The first of the above mentioned methods has certain advantages, especially for the purposes of theoretical discussion, but it may often be more advantageous to select a method in which the properties represented by two of the co-ordinates shall be such as best serve to identify and describe the different states of the substance. This condition is satisfied by temperature and pressure as well, perhaps, as by any other properties. We may represent these by two of the co-ordinates and the potential by the third. (See page 88.) It will not be overlooked that there is the closest analogy between these three quantities in respect to their parts in the general theory of equilibrium. (A similar analogy exists between volume, entropy, and energy.) If we give m the constant value unity, the third co-ordinate will also represent ζ, which then becomes equal to μ.

Comparing the two methods, we observe that in one

$$v = x, \ \eta = y, \ \epsilon = z, \tag{187}$$

$$p = -\frac{dz}{dx}, \ t = \frac{dz}{dy}, \ \mu = \zeta = z - \frac{dz}{dx}x - \frac{dz}{dy}y \ ; \tag{188}$$

and in the other

$$t = x, \ p = y, \ \mu = \zeta = z, \tag{189}$$

$$\eta = -\frac{dz}{dx}, \ v = \frac{dz}{dy}, \ \epsilon = z - \frac{dz}{dx}x - \frac{dz}{dy}y. \tag{190}$$

Now $\frac{dz}{dx}$ and $\frac{dz}{dy}$ are evidently determined by the inclination of the tangent plane, and $z - \frac{dz}{dx}x - \frac{dz}{dy}y$ is the segment which it cuts off on the axis of Z. The two methods, therefore, have this reciprocal relation, that the quantities represented in one by the position of a point in a surface are represented in the other by the position of a tangent plane.

The surfaces defined by equations (187) and (189) may be distinguished as the v-η-ϵ surface, and the t-p-ζ surface, of the substance to which they relate.

In the t-p-ζ surface a line in which one part of the surface cuts another represents a series of pairs of coexistent states. A point through which pass three different parts of the surface represents a triad of coexistent states. Through such a point will evidently pass the three lines formed by the intersection of these sheets taken two by two. The perpendicular projection of these lines upon the p-t plane will give the curves which have recently been discussed by Professor J. Thomson.* These curves divide the space about the projection of the triple point into six parts which may be distinguished as follows: Let $\zeta^{(V)}$, $\zeta^{(L)}$, $\zeta^{(S)}$ denote the three ordinates determined for the same values of p and t by the three sheets passing through the triple point, then in one of the six spaces

$$\zeta^{(V)} < \zeta^{(L)} < \zeta^{(S)}, \tag{191}$$

in the next space, separated from the former by the line for which $\zeta^{(L)} = \zeta^{(S)}$,

$$\zeta^{(V)} < \zeta^{(S)} < \zeta^{(L)}, \tag{192}$$

in the third space, separated from the last by the line for which $\zeta^{(V)} = \zeta^{(S)}$,

$$\zeta^{(S)} < \zeta^{(V)} < \zeta^{(L)}, \tag{193}$$

in the fourth $\zeta^{(S)} < \zeta^{(L)} < \zeta^{(V)}, \tag{194}$

in the fifth $\zeta^{(L)} < \zeta^{(S)} < \zeta^{(V)}, \tag{195}$

in the sixth $\zeta^{(L)} < \zeta^{(V)} < \zeta^{(S)}. \tag{196}$

---

* See the *Reports of the British Association* for 1871 and 1872; and *Philosophical Magazine*, vol. xlvii. (1874), p. 447.

The sheet which gives the least values of ζ is in each case that which represents the stable states of the substance. From this it is evident that in passing around the projection of the triple point we pass through lines representing alternately coexistent stable and coexistent unstable states. But the states represented by the intermediate values of ζ may be called stable *relatively* to the states represented by the highest. The differences $\zeta^{(L)} - \zeta^{(V)}$, etc. represent the amount of work obtained in bringing the substance by a reversible process from one to the other of the states to which these quantities relate, in a medium having the temperature and pressure common to the two states. To illustrate such a process, we may suppose a plane perpendicular to the axis of temperature to pass through the points representing the two states. This will in general cut the double line formed by the two sheets to which the symbols (L) and (V) refer. The intersections of the plane with the two sheets will connect the double point thus determined with the points representing the initial and final states of the process, and thus form a *reversible path* for the body between those states.

The geometrical relations which indicate the stability of any state may be easily obtained by applying the principles stated on pp. 100 ff. to the case in which there is but a single component. The expression (133) as a test of stability will reduce to

$$\epsilon - t'\eta + p'v - \mu'm, \tag{197}$$

the accented letters referring to the state of which the stability is in question, and the unaccented letters to any other state. If we consider the quantity of matter in each state to be unity, this expression may be reduced by equations (91) and (96) to the form

$$\zeta - \zeta' + (t - t')\eta - (p - p')v, \tag{198}$$

which evidently denotes the distance of the point (t', p', ζ') below the tangent plane for the point (t, p, ζ), measured parallel to the axis of ζ. Hence if the tangent plane for every other state passes above the point representing any given state, the latter will be stable. If any of the tangent planes pass below the point representing the given state, that state will be unstable. Yet it is not always necessary to consider these tangent planes. For, as has been observed on page 103, we may assume that (in the case of any real substance) there will be at least one not unstable state for any given temperature and pressure, except when the latter is negative. Therefore the state represented by a point in the surface on the positive side of the plane $p = 0$ will be unstable only when there is a point in the surface for which t and p have the same values and ζ a less value. It follows from what has been stated, that where the surface is doubly convex upwards (in the direction in which ζ is measured) the states represented will be stable in respect to adjacent states. This also appears directly from (162). But where the surface is concave upwards in either of its principal curvatures the states represented will be unstable in respect to adjacent states.

When the number of component substances is greater than unity, it is not possible to represent the fundamental equation by a single surface. We have therefore to consider how it may be represented by an infinite number of surfaces. A natural extension of either of the methods already described will give us a series of surfaces in which every one is the v-η-ϵ surface, or every one the t-p-ζ surface for a body of constant composition, the proportion of the components varying as we pass from one surface to another. But for a simultaneous view of the properties which are exhibited by compounds of two or three components without change of temperature or pressure, we may more advantageously make one or both of the quantities t or p constant in each surface.

Surfaces and Curves in which the Composition of the Body represented is Variable and its Temperature and Pressure are Constant.

When there are three components, the position of a point in the X-Y plane may indicate the composition of a body most simply, perhaps, as follows. The body is supposed to be composed of the quantities m_1, m_2, m_3 of the substances S_1, S_2, S_3, the value of $m_1 + m_2 + m_3$ being unity. Let P_1, P_2, P_3 be any three points in the plane, which are not in the same straight line. If we suppose masses equal to m_1, m_2, m_3 to be placed at these three points, the center of gravity of these masses will determine a point which will indicate the value of these quantities. If the triangle is equiangular and has the height unity, the distances of the point from the three sides will be equal numerically to m_1, m_2, m_3. Now if for every possible phase of the components, of a given temperature and pressure, we lay off from the point in the X-Y plane which represents the composition of the phase a distance measured parallel to the axis of Z and representing the value of ζ (when $m_1 + m_2 + m_3 = 1$), the points thus determined will form a surface, which may be designated us the m_1-m_2-m_3-ζ surface of the substances considered, or simply as their m-ζ surface, for the given temperature and pressure. In like manner, when there are but two component substances, we may obtain a curve, which we will suppose in the X-Z plane. The coordinate y may then represent temperature or pressure. But we will limit ourselves to the consideration of the properties of the m-ζ surface for $n = 3$, or the m-ζ curve for $n = 2$, regarded as a surface, or curve, which varies with the temperature and pressure.

As by (96) and (92)

$$\zeta = \mu_1 m_1 + \mu_2 m_2 + \mu_3 m_3,$$

and (for constant temperature and pressure)

$$d\zeta = \mu_1 dm_1 + \mu_2 dm_2 + \mu_3 dm_3,$$

if we imagine a tangent plane for the point to which these letters relate, and denote by ζ' the ordinate for any point in the plane, and by m_1', m_2', m_3', the distances of the foot of this ordinate from the three sides of the triangle $P_1 P_2 P_3$, we may easily obtain

$$\zeta' = \mu_1 m_1' + \mu_2 m_2' + \mu_3 m_3', \tag{199}$$

which we may regard as the equation of the tangent plane. Therefore the ordinates for this plane at P_1, P_2, and P_3 are equal respectively to the potentials μ_1, μ_2, μ_3. And in general, the ordinate for any point in the tangent plane is equal to the potential (in the phase represented by the point of contact) for a substance of which the composition is indicated by the position of the ordinate. (See page 93.) Among the bodies which may be formed of S_1, S_2, and S_3, there may be some which are incapable of variation in composition, or which are capable only of a single kind of variation. These will be represented by single points and curves in vertical planes. Of the tangent plane to one of these curves only a single line will be fixed, which will determine a series of potentials of which only two will be independent. The phase represented by a separate point will determine only a single potential, viz., the potential for the substance of the body itself, which will be equal to ζ.

The points representing a set of coexistent phases have in general a common tangent plane. But when one of these points is situated on the edge where a sheet of the surface terminates, it is sufficient if the plane is tangent to the edge and passes below the surface. Or, when the point is at the end of a separate line belonging to the surface, or at an angle in the edge of a sheet, it is sufficient if the plane pass through the point and below the line or sheet. If no part of the surface lies below the tangent plane, the points where it meets the plane will represent a stable (or at least not unstable) set of coexistent phases.

The surface which we have considered represents the relation between ζ and m_1, m_2, m_3 for homogeneous bodies when t and p have any constant values and $m_1 + m_2 + m_3 = 1$. It will often be useful to consider the surface which represents the relation between the same variables for bodies which consist of parts in different but coexistent phases. We may suppose that these are stable, at least in regard to adjacent phases, as otherwise the case would be devoid of interest. The point which represents the state of the composite body will evidently be at the center of gravity of masses equal to the parts of the body placed at the points representing the phases of these parts. Hence from the surface representing the properties of homogeneous bodies, which may be called the primitive surface, we may easily construct the surface representing the properties of bodies which are in equilibrium but not homogeneous. This may be called the secondary or derived surface. It will consist, in general, of various portions or sheets. The sheets which represent a combination of two phases may be formed by rolling a double tangent plane upon the primitive surface ; the part of the envelop of its successive positions which lies between the curves traced by the points of contact will belong to the derived surface. When the primitive surface has a triple tangent plane or one of higher order, the triangle in the tangent plane formed by joining the points of contact, or the smallest polygon without re-entrant angles which includes all the points of contact, will belong to the derived surface, and will represent masses consisting in general of three or more phases.

Of the whole thermodynamic surface as thus constructed for any temperature and any positive pressure, that part is especially important which gives the least value of ζ for any given values of m_1, m_2, m_3. The state of a mass represented by a point in this part of the surface is one in which no dissipation of energy would be possible if the mass were enclosed in a rigid envelop impermeable both to matter and to heat; and the state of any mass composed of S_1, S_2, S_3 in any proportions, in which the dissipation of energy has been completed, so far as internal processes are concerned (i.e., under the limitations imposed by such an envelop as above supposed), would be represented by a point in the part which we are considering of the m-ζ surface for the temperature and pressure of the mass. We may therefore briefly distinguish this part of the surface as the *surface of dissipated energy*. It is evident that it forms a continuous sheet, the projection of which upon the X-Y plane coincides with the triangle $P_1P_2P_3$, (except when the pressure for which the m-ζ surface is constructed is negative, in which case there is no surface of dissipated energy), that it nowhere has any convexity upward, and that the states which it represents are in no case unstable.

The general properties of the m-ζ lines for two component substances are so similar as not to require separate consideration. We now proceed to illustrate the use of both the surfaces and the lines by the discussion of several particular cases.

Three coexistent phases of two component substances may be represented by the points A, B, and C, in figure 1, in which ζ is measured toward the top of the page from $P_1 P_2$, m_1 toward the left from $P_2 Q_2$, and m_2 toward the right from $P_1 Q_1$. It is supposed that $P_1 P_2 = 1$. Portions of the curves to which these points belong are seen in the figure, and will be denoted by the symbols (A), (B), (C). We may, for convenience, speak of these as separate curves, without implying anything in regard to their possible continuity in parts of the diagram remote from their common tangent AC. The *line of dissipated energy* includes the straight line AC and portions of the primitive curves (A) and (C). Let us first consider how the diagram will be altered, if the temperature is varied while the pressure remains constant. If the temperature receives the increment dt, an ordinate of which the position is fixed will receive the increment $\left(\dfrac{d\zeta}{dt}\right)_{p,\,m} dt$, or $-\eta\, dt$. (The reader will easily convince himself that this is true of the ordinates for the secondary line AC, as well as of the ordinates for the primitive curves.) Now if we denote by η' the entropy of the phase represented by the point B considered as belonging to the curve (B), and by η'' the entropy of the composite state of the same matter represented by the point B considered as belonging to the tangent to the curves (A) and (C),

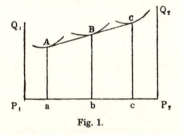

Fig. 1.

$t(\eta' - \eta'')$ will denote the heat yielded by a unit of matter in passing from the first to the second of these states. If this quantity is positive, an elevation of temperature will evidently cause a part of the curve (B) to protrude below the tangent to (A) and (C), which will no longer form a part of the line of dissipated energy. This line will then include portions of the three curves (A), (B), and (C), and of the tangents to (A) and (B) and to (B) and (C). On the other hand, a lowering of the temperature will cause the curve (B) to lie entirely above the tangent to (A) and (C), so that all the phases of the sort represented by (B) will be unstable. If $t(\eta' - \eta'')$ is negative, these effects will be produced by the opposite changes of temperature.

The effect of a change of pressure while the temperature remains constant may be found in a manner entirely analogous. The variation of any ordinate will be $\left(\dfrac{d\zeta}{dp}\right)_{t,\,m} dp$ or $v\, dp$. Therefore, if the volume of the homogeneous phase represented by the point B is greater than the volume of the same matter divided between the phases represented by A and C, an increase of pressure will give a diagram indicating that all phases of the sort represented by curve

(B) are unstable, and a decrease of pressure will give a diagram indicating two stable pairs of coexistent phases, in each of which one of the phases is of the sort represented by the curve (B). When the relation of the volumes is the reverse of that supposed, these results will be produced by the opposite changes of pressure.

When we have four coexistent phases of three component substances, there are two cases which must be distinguished. In the first, one of the points of contact of the primitive surface with the quadruple tangent plane lies within the triangle formed by joining the other three; in the second, the four points may be joined so as to form a quadrilateral without re-entrant angles. Figure 2 represents the projection upon the X-Y plane (in which m_1, m_2, m_3 are measured) of a part of the surface of dissipated energy, when one of the points of contact D falls within the triangle formed by the other three A, B, C. This surface includes the triangle ABC in the quadruple tangent plane, portions of the three sheets of the primitive surface which touch the triangle at its vertices, EAF, GBH, ICK, and portions of the three developable surfaces formed by a tangent plane rolling upon each pair of these sheets. These developable surfaces are represented in the figure by ruled surfaces, the lines indicating the direction of their rectilinear elements. A point within the triangle ABC represents a mass of which the matter is divided, in general, between three or four different phases, in a manner not entirely determined by the position of a point. (The quantities of matter in these phases are such that if placed at the corresponding points, A, B, C, D, their center of gravity would be at the point representing the total mass.) Such a mass, if exposed to constant temperature and pressure, would be in neutral equilibrium. A point in the developable surfaces represents a mass of which the matter is divided between two coexisting phases, which are represented by the extremities of the line in the figure passing through that point. A point in the primitive surface represents of course a homogeneous mass.

To determine the effect of a change of temperature without change of pressure upon the general features of the surface of dissipated energy, we must know whether heat is absorbed or yielded by a mass in passing from the phase represented by the point D *in the primitive surface* to the composite state consisting of the phases A, B, and C which is represented by the same point. If the first is the case, an increase of temperature will cause the sheet (D) (i.e., the sheet of the primitive surface to which the point D belongs) to separate from the plane tangent to the three other sheets, so as to be situated entirely above it, and a decrease of temperature, will cause a part of the sheet (D) to protrude through the plane tangent to the other sheets.   These effects will be produced by the opposite changes of temperature, when heat is yielded by a mass passing from the homogeneous to the composite state above mentioned.

In like manner, to determine the effect of a variation of pressure without change of temperature, we must know whether the volume for the homogeneous phase represented by D is greater or less than the volume of the same matter divided between the phases A, B, and C.   If the homogeneous phase has the greater volume, an increase of pressure will cause the sheet (D) to separate from the plane tangent to the other sheets, and a diminution of pressure will cause a part of the sheet (D) to protrude below that tangent plane.   And these effects will be produced by the opposite changes of pressure, if the homogeneous phase has the less volume.   All this appears from precisely

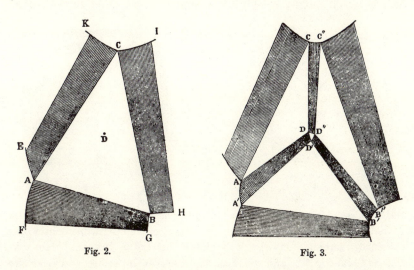

Fig. 2.   Fig. 3.

the same considerations which were used in the analogous case for two component substances.

Now when the sheet (D) rises above the plane tangent to the other sheets, the general features of the surface of dissipated energy are not altered, except by the disappearance of the point D.   But when the sheet (D) protrudes below the plane tangent to the other sheets, the surface of dissipated energy will take the form indicated in figure 3. It will include portions of the four sheets of the primitive surface, portions of the six developable surfaces formed by a double tangent plane rolling upon these sheets taken two by two, and portions of three triple tangent planes for these sheets taken by threes, the sheet (D) being always one of the three.

But when the points of contact with the quadruple tangent plane which represent the four coexistent phases can be joined so as to form a quadrilateral ABCD (fig. 4) without re-entrant angles, the surface of dissipated energy will include this plane quadrilateral, portions of the four sheets of the primitive surface which are tangent to it, and portions of the four developable surfaces formed by double tangent planes rolling upon the four pairs of these sheets which correspond to the four sides of the quadrilateral. To determine the general effect of a variation of temperature upon the surface of dissipated energy, let us consider the composite states represented by the point I at the intersection of the diagonals of the quadrilateral. Among these states (which all relate to the same kind and quantity of matter) there is one which is composed of the phases A and C, and another which is composed of the phases B and D. Now if the entropy of the first of these states is greater than that of the second (i.e., if heat is given out by a body in passing from the first to the second

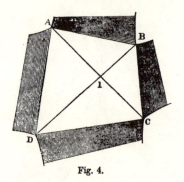

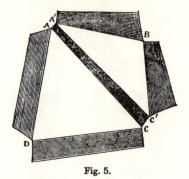

Fig. 4.   Fig. 5.

state at constant temperature and pressure), which we may suppose without loss of generality, an elevation of temperature while the pressure remains constant will cause the triple tangent planes to (B), (D), and (A), and to (B), (D), and (C), to rise above the triple tangent planes to (A), (C), and (B), and to (A), (C), and (D), in the vicinity of the point I. The surface of dissipated energy will therefore take the form indicated in figure 5, in which there are two plane triangles and five developable surfaces besides portions of the four primitive sheets. A diminution of temperature will give a different but entirely analogous form to the surface of dissipated energy. The quadrilateral ABCD will in this case break into two triangles along the diameter BD. The effects produced by variation of the pressure while the temperature remains constant will of course be similar to those described. By considering the difference of volume instead of the difference of entropy of the two states represented by the point I in the quadruple tangent plane, we may distinguish between the effects of increase and diminution of pressure.

It should be observed that the points of contact of the quadruple tangent plane with the primitive surface may be at isolated points or curves belonging to the latter. So also, in the case of two component substances, the points of contact of the triple tangent line may be at isolated points belonging to the primitive curve. Such cases need not be separately treated, as the necessary modifications in the preceding statements, when applied to such cases, are quite evident. And in the remaining discussion of this geometrical method, it will generally be left to the reader to make the necessary limitations or modifications in analogous cases.

The necessary condition in regard to simultaneous variations of temperature and pressure, in order that four coexistent phases of three components, or three coexistent phases of two components, shall remain possible, has already been deduced by purely analytical processes. (See equation (129).)

We will next consider the case of two coexistent phases of identical composition, and first, when the number of components is two. The coexistent phases, if each is variable in composition, will be represented by the point of contact of two curves. One of the curves will in general lie above the other except at the point of contact; therefore, when the temperature and pressure remain constant, one phase cannot be varied in composition without becoming unstable, while the other phase will be stable if the proportion of either component is increased. By varying the temperature or pressure, we may cause the upper curve to protrude below the other, or to rise (relatively) entirely above it. (By comparing the volumes or the entropies of the two coexistent phases, we may easily determine which result would be produced by an increase of temperature or of pressure.) Hence, the temperatures and pressures for which two coexistent phases have the same composition form the limit to the temperatures and pressures for which such coexistent phases are possible. It will be observed that as we pass this limit of temperature and pressure, the pair of coexistent phases does not simply become unstable, like pairs and triads of coexistent phases which we have considered before, but there ceases to be any such pair of coexistent phases. The same result has already been obtained analytically on page 99. But on

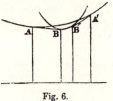

Fig. 6.

that side of the limit on which the coexistent phases are possible, there will be two pairs of coexistent phases for the same values of t and p, as seen in figure 6. If the curve AA′ represents vapor, and the curve BB′ liquid, a liquid (represented by) B may exist in contact with a vapor A, and (at the same temperature and pressure) a liquid B′ in contact with a vapor A′. If we compare these phases in respect to their composition, we see that in one case the vapor is richer than the liquid in a certain component, and in the other case poorer. Therefore, if these liquids are made to boil, the effect on their composition will be opposite. If the boiling is continued under constant pressure, the temperature will rise as the liquids approach each other in composition, and the curve BB′ will rise *relatively* to the curve AA′, until the curves are tangent to each other, when the two liquids become identical in nature, as also the vapors which they yield. In composition, and in the value of ζ per unit of mass, the vapor will then agree with the liquid. But if the curve BB′ (which has the greater curvature) represents vapor, and AA′ represents liquid, the effect of boiling will make the liquids A and A′ differ more in composition. In this case, the relations indicated in the figure will hold for a temperature higher than that for which (with the same pressure) the curves are tangent to one another.

When two coexistent phases of three component substances have the same composition, they are represented by the point of contact of two sheets of the primitive surface. If these sheets do not intersect at the point of contact, the case is very similar to that which we have just considered. The upper sheet except at the point of contact represents unstable phases. If the temperature or pressure are so varied that a part of the upper sheet protrudes through the lower, the points of contact of a double tangent plane rolling upon the two sheets will describe a closed curve on each, and the surface of dissipated energy will include a portion of each sheet of the primitive surface united by a ring-shaped developable surface.

If the sheet having the greater curvatures represents liquid, and the other sheet vapor, the boiling temperature for any given pressure will be a maximum, and the pressure of saturated vapor for any given temperature will be a minimum, when the coexistent liquid and vapor have the same composition.

But if the two sheets, constructed for the temperature and pressure of the coexistent phases which have the same composition, intersect

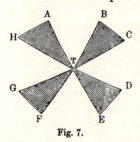

Fig. 7.

at the point of contact, the whole primitive surface as seen from below will in general present four re-entrant furrows, radiating from the point of contact, for each of which a developable surface may be formed by a rolling double tangent plane. The different parts of the surface of dissipated energy in the vicinity of the point of contact are represented in figure 7. ATB, ETF are parts of one sheet of the primitive surface, and CTD, GTH are parts of the other. These are united by the developable surfaces BTC,

DTE, FTG, HTA. Now we may make either sheet of the primitive surface sink relatively to the other by the proper variation of temperature or pressure. If the sheet to which ATB, ETF belong is that which sinks relatively, these parts of the surface of dissipated energy will be merged in one, as well as the developable surfaces BTC, DTE, and also FTG, HTA. (The lines CTD, BTE, ATF, HTG will separate from one another at T, each forming a continuous curve.) But if the sheet of the primitive surface which sinks relatively is that to which CTD and GTH belong, then these parts will be merged in one in the surface of dissipated energy, as will be the developable surfaces BTC, ATH, and also DTE, FTG.

It is evident that this is not a case of maximum or minimum temperature for coexistent phases under constant pressure, or of maximum or minimum pressure for coexistent phases at constant temperature.

Another case of interest is when the composition of one of three coexistent phases is such as can be produced by combining the other two. In this case, the primitive surface must touch the same plane in three points in the same straight line. Let us distinguish the parts of the primitive surface to which these points belong as the sheets (A), (B), and (C), (C) denoting that which is intermediate in position. The sheet (C) is evidently tangent to the developable surface formed upon (A) and (B). It may or it may not intersect it at the point of contact. If it does not, it must lie above the developable surface (unless it represents states which are unstable in regard to continuous changes), and the surface of dissipated energy will include parts of the primitive sheets (A) and (B), the developable surface joining them, and the single point of the sheet (C) in which it meets this developable surface. Now, if the temperature or pressure is varied so as to make the sheet (C) rise above the developable surface formed on the sheets (A) and (B), the surface of dissipated energy will be altered in its general features only by the removal of the single point of the sheet (C). But if the temperature or pressure is altered so as to make a part

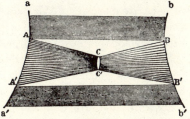

Fig. 8.

of the sheet (C) protrude through the developable surface formed on (A) and (B), the surface of dissipated energy will have the form indicated in figure 8. It will include two plane triangles ABC and A′B′C′, a part of each of the sheets (A) and (B), represented in the figure by the spaces on the left of the line aAA′a′ and on the right of the line bBB′b′, a small part CC′ of the sheet (C), and developable surfaces formed upon these sheets taken by pairs ACC′A′, BCC′B′, aABb, a′A′B′b′, the last two being different portions of the same developable surface.

But if, when the primitive surface is constructed for such a temperature and pressure that it has three points of contact with the same plane in the same straight line, the sheet (C) (which has the middle position) at its point of contact with the triple tangent plane intersects the developable surface formed upon the other sheets (A) and (B), the surface of dissipated energy will not include this developable surface, but will consist of portions of the three primitive sheets with two developable surfaces formed on (A) and (C) and on (B) and (C). These developable surfaces meet one another at the point of contact of (C) with the triple tangent plane, dividing the portion of this sheet which belongs to the surface of dissipated energy into two parts. If now the temperature or pressure are varied so as to make the sheet (C) sink relatively to the developable surface formed on (A) and (B), the only alteration in the general features of the surface of dissipated energy will be that the developable surfaces formed on (A) and (C) and on (B) and (C) will separate from one another, and the two parts of the sheet (C) will be merged in one. But a contrary variation of temperature or pressure will give a surface of dissipated energy such as is represented in figure (9), containing two plane triangles ABC,

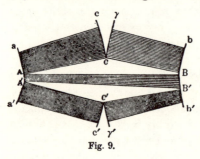

Fig. 9.

A′B′C′ belonging to triple tangent planes, a portion of the sheet (A) on the left of the line aAA′a′, a portion of the sheet (B) on the right of the line bBB′b′, two separate portions cCγ and c′C′γ′ of the sheet (C), two separate portions aACc and a′A′C′c′ of the developable surface formed on (A) and (C), two separate portions bBCγ and b′B′C′γ′ of the developable surface formed on (B) and (C), and the portion A′ABB′ of the developable surface formed on (A) and (B).

From these geometrical relations it appears that (in general) the temperature of three coexistent phases is a maximum or minimum for constant pressure, and the pressure of three coexistent phases a maximum or minimum for constant temperature, when the composition of the three coexistent phases is such that one can be formed by combining the other two. This result has been obtained analytically on page 99.

The preceding examples are amply sufficient to illustrate the use of the m-ζ surfaces and curves. The physical properties indicated by the nature of the surface of dissipated energy have been only occasionally mentioned, as they are often far more distinctly indicated by the diagrams than they could be in words. It will be observed that a knowledge of the lines which divide the various different portions of the surface of dissipated energy and of the direction of the rectilinear elements of the developable surfaces, *as projected upon the X-Y plane*, without a knowledge of the form of the m-ζ surface in space, is sufficient for the determination (in respect to the quantity and composition of the resulting masses) of the combinations and separations of the substances, and of the changes in their states of aggregation, which take place when the substances are exposed to the temperature and pressure to which the projected lines relate, except so far as such transformations are prevented by passive resistances to change.

Critical Phases.

It has been ascertained by experiment that the variations of two coexistent states of the same substance are in some cases limited in one direction by a terminal state at which the distinction of the coexistent states vanishes.* This state has been called the *critical state*. Analogous properties may doubtless be exhibited by compounds of variable composition without change of temperature or pressure. For if, at any given temperature and pressure, two liquids are capable of forming a stable mixture in any ratio $m_1 : m_2$ less than a, and in any greater than b, a and b being the values of that ratio for two coexistent phases, while either can form a stable mixture with a third liquid in all proportions, and any small quantities of the first and second can unite at once with a great quantity of the third to form a stable mixture, it may easily be seen that two coexistent mixtures of the three liquids may be varied in composition, the temperature and pressure remaining the same, from initial phases in each of which the quantity of the third liquid is nothing, to a terminal phase in which the distinction of the two phases vanishes.

In general, we may define a *critical phase* as one at which the distinction between coexistent phases vanishes. We may suppose the coexistent phases to be stable in respect to continuous changes, for although relations in some respects analogous might be imagined to hold true in regard to phases which are unstable in respect to continuous changes, the discussion of such cases would be devoid of interest. But if the coexistent phases and the critical phase are unstable only in respect to the possible formation of phases entirely different from the critical and adjacent phases, the liability to such changes will in no respect affect the relations between the critical and adjacent phases, and need not be considered in a theoretical discussion

---

* See Dr. Andrews "On the continuity of the gaseous and liquid states of matter." *Phil. Trans.*, vol. 159, p. 575.

of these relations, although it may prevent an experimental realization of the phases considered. For the sake of brevity, in the following discussion, phases in the vicinity of the critical phase will generally be called stable, if they are unstable only in respect to the formation of phases entirely different from any in the vicinity of the critical phase.

Let us first consider the number of independent variations of which a critical phase (while remaining such) is capable. If we denote by n the number of independently variable components, a pair of coexistent phases will be capable of n independent variations, which may be expressed by the variations of n of the quantities t, p, μ_1, $\mu_2, \ldots \mu_n$. If we limit these variations by giving to $n-1$ of the quantities the constant values which they have for a certain critical phase, we obtain a linear* series of pairs of coexistent phases terminated by the critical phase. If we now vary infinitesimally the values of these $n-1$ quantities, we shall have for the new set of values considered constant a new linear series of pairs of coexistent phases. Now for every pair of phases in the first series, there must be pairs of phases in the second series differing infinitely little from the pair in the first, and *vice versa*, therefore the second series of coexistent phases must be terminated by a critical phase which differs, but differs infinitely little, from the first. We see, therefore, that if we vary arbitrarily the values of any $n-1$ of the quantities, t, p, μ_1, μ_2, $\ldots \mu_n$, as determined by a critical phase, we obtain one and only one critical phase for each set of varied values; i.e., a critical phase is capable of $n-1$ independent variations.

The quantities t, p, μ_1, μ_2, $\ldots \mu_n$ have the same values in two coexistent phases, but the ratios of the quantities η, v, m_1, m_2, $\ldots m_n$ are in general different in the two phases. Or, if for convenience we compare equal volumes of the two phases (which involves no loss of generality), the quantities η, m_1, m_2, $\ldots m_n$ will in general have different values in two coexistent phases. Applying this to coexistent phases indefinitely near to a critical phase, we see that in the immediate vicinity of a critical phase, if the values of n of the quantities t, p, μ_1, μ_2, $\ldots \mu_n$ are regarded as constant (as well as v), the variations of either of the others will be infinitely small compared with the variations of the quantities η, m_1, m_2, $\ldots m_n$. This condition, which we may write in the form

$$\left(\frac{d\mu_n}{dm_n}\right)_{t,\, v,\, \mu_1,\, \ldots \mu_{n-1}} = 0, \qquad (200)$$

characterizes, as we have seen on page 114, the limits which divide stable from unstable phases in respect to continuous changes.

In fact, if we give to the quantities t, μ_1, μ_2, $\ldots \mu_{n-1}$ constant values

---

* This term is used to characterize a series having a *single* degree of extension.

determined by a pair of coexistent phases, and to $\dfrac{m_n}{v}$ a series of values increasing from the less to the greater of the values which it has in these coexistent phases, we determine a linear series of phases connecting the coexistent phases, in some part of which μ_n—since it has the same value in the two coexistent phases, but not a uniform value throughout the series (for if it had, which is theoretically improbable, all these phases would be coexistent)—must be a decreasing function of $\dfrac{m_n}{v}$, or of m_n, if v also is supposed constant. Therefore, the series must contain phases which are unstable in respect to continuous changes. (See page 111.) And as such a pair of coexistent phases may be taken indefinitely near to any critical phase, the unstable phases (with respect to continuous changes) must approach indefinitely near to this phase.

Critical phases have similar properties with reference to stability as determined with regard to discontinuous changes. For as every stable phase which has a coexistent phase lies upon the limit which separates stable from unstable phases, the same must be true of any stable critical phase. (The same may be said of critical phases which are unstable in regard to discontinuous changes, if we leave out of account the liability to the particular kind of discontinuous change in respect to which the critical phase is unstable.)

The linear series of phases determined by giving to n of the quantities t, p, μ_1, μ_2, ... μ_n the constant values which they have in any pair of coexistent phases consists of unstable phases in the part between the coexistent phases, but in the part beyond these phases in either direction it consists of stable phases. Hence, if a critical phase is varied in such a manner that n of the quantities t, p, μ_1, μ_2, ... μ_n remain constant, it will remain stable in respect both to continuous and to discontinuous changes. Therefore μ_n is an increasing function of m_n when t, v, μ_1, μ_2, ... μ_{n-1} have constant values determined by any critical phase. But as equation (200) holds true at the critical phase, the following conditions must also hold true at that phase:—

$$\left(\frac{d^2\mu_n}{dm_n^2}\right)_{t,\,v,\,\mu_1,\,\ldots\,\mu_{n-1}} = 0, \tag{201}$$

$$\left(\frac{d^3\mu_n}{dm_n^3}\right)_{t,\,v,\,\mu_1,\,\ldots\,\mu_{n-1}} \geqq 0. \tag{202}$$

If the sign of equality holds in the last condition, additional conditions, concerning the differential coefficients of higher orders, must be satisfied.

Equations (200) and (201) may in general be called the equations of critical phases. It is evident that there are only two independent equations of this character, as a critical phase is capable of $n-1$ independent variations.

We are not, however, absolutely certain that equation (200) will always be satisfied by a critical phase. For it is possible that the denominator in the fraction may vanish as well as the numerator for an infinitesimal change of phase in which the quantities indicated are constant. In such a case, we may suppose the subscript n to refer to some different component substance, or use another differential coefficient of the same general form (such as are described on page 114 as characterizing the limits of stability in respect to continuous changes), making the corresponding changes in (201) and (202). We may be certain that some of the formulæ thus formed will not fail. But for a perfectly rigorous method there is an advantage in the use of η, v, m_1, m_2, ... m_n as independent variables. The condition that the phase may be varied without altering any of the quantities t, μ_1, μ_2, ... μ_n will then be expressed by the equation

$$R_{n+1} = 0, \tag{203}$$

in which R_{n+1} denotes the same determinant as on page 111. To obtain the second equation characteristic of critical phases, we observe that as a phase which is critical cannot become unstable when varied so that n of the quantities t, p, μ_1, μ_2, ... μ_n remain constant, the differential of R_{n+1} for constant volume, viz.,

$$\frac{dR_{n+1}}{d\eta}d\eta + \frac{dR_{n+1}}{dm_1}dm_1 \cdots + \frac{dR_{n+1}}{dm_n}dm_n, \tag{204}$$

cannot become negative when n of the equations (172) are satisfied. Neither can it have a positive value, for then its value might become negative by a change of sign of $d\eta$, dm_1, etc. Therefore the expression (204) has the value zero, if n of the equations (172) are satisfied. This may be expressed by an equation

$$S = 0, \tag{205}$$

in which S denotes a determinant in which the constituents are the same as in R_{n+1}, except in a single horizontal line, in which the differential coefficients in (204) are to be substituted. In whatever line this substitution is made, the equation (205), as well as (203), will hold true of every critical phase without exception.

If we choose t, p, m_1, m_2, ... m_n as independent variables, and write U for the determinant

$$\begin{vmatrix} \dfrac{d^2\zeta}{dm_1^2} & \dfrac{d^2\zeta}{dm_2 dm_1} & \cdots & \dfrac{d^2\zeta}{dm_{n-1} dm_1} \\[2ex] \dfrac{d^2\zeta}{dm_1 dm_2} & \dfrac{d^2\zeta}{dm_2^2} & \cdots & \dfrac{d^2\zeta}{dm_{n-1} dm_2} \\[2ex] \cdots & \cdots & \cdots & \cdots \\[2ex] \dfrac{d^2\zeta}{dm_1 dm_{n-1}} & \dfrac{d^2\zeta}{dm_2 dm_{n-1}} & \cdots & \dfrac{d^2\zeta}{dm_{n-1}^2} \end{vmatrix}, \tag{206}$$

and V for the determinant formed from this by substituting for the constituents in any horizontal line the expressions

$$\frac{dU}{dm_1}, \quad \frac{dU}{dm^2}, \quad \cdots \quad \frac{dU}{dm_{n-1}}, \tag{207}$$

the equations of critical phases will be

$$U = 0, \quad V = 0. \tag{208}$$

It results immediately from the definition of a critical phase, that an infinitesimal change in the condition of a mass in such a phase may cause the mass, if it remains in a state of dissipated energy (i.e., in a state in which the dissipation of energy by internal processes is complete), to cease to be homogeneous. In this respect a critical phase resembles any phase which has a coexistent phase, but differs from such phases in that the two parts into which the mass divides when it ceases to be homogeneous differ infinitely little from each other and from the original phase, and that neither of these parts is in general infinitely small. If we consider a change in the mass to be determined by the values of $d\eta$, dv, dm_1, $dm_2, \ldots dm_n$, it is evident that the change in question will cause the mass to cease to be homogeneous whenever the expression

$$\frac{dR_{n+1}}{d\eta}d\eta + \frac{dR_{n+1}}{dv}dv + \frac{dR_{n+1}}{dm_1}dm_1, \ldots + \frac{dR_{n+1}}{dm_n}dm_n \tag{209}$$

has a negative value. For if the mass should remain homogeneous, it would become unstable, as R_{n+1} would become negative. Hence, in general, any change thus determined, or its reverse (determined by giving to $d\eta$, dv, dm_1, $dm_2, \ldots dm_n$ the same values taken negatively), will cause the mass to cease to be homogeneous. The condition which must be satisfied with reference to $d\eta$, dv, dm_1, $dm_2, \ldots dm_n$, in order that neither the change indicated, nor the reverse, shall destroy the homogeneity of the mass, is expressed by equating the above expression to zero.

But if we consider the change in the state of the mass (supposed to remain in a state of dissipated energy) to be determined by arbitrary values of $n+1$ of the differentials dt, dp, $d\mu_1$, $d\mu_2, \ldots d\mu_n$, the case will be entirely different. For, if the mass ceases to be homogeneous, it will consist of two coexistent phases, and as applied to these, only n of the quantities t, p, μ_1, $\mu_2, \ldots \mu_n$ will be independent. Therefore, for arbitrary variations of $n+1$ of these quantities, the mass must in general remain homogeneous.

But if, instead of supposing the mass to remain in a state of dissipated energy, we suppose that it remains homogeneous, it may easily be shown that to certain values of $n+1$ of the above differentials there will correspond three different phases, of which one is stable with respect both to continuous and to discontinuous changes, another is stable with respect to the former and unstable with respect to the latter, and the third is unstable with respect to both.

In general, however, if n of the quantities p, t, μ_1, μ_2, ... μ_n, or n arbitrary functions of these quantities, have the same constant values as at a critical phase, the linear series of phases thus determined will be stable, in the vicinity of the critical phase. But if less than n of these quantities or functions of the same together with certain of the quantities η, v, m_1, m_2, ... m_n, or arbitrary functions of the latter quantities, have the same values as at a critical phase, so as to determine a linear series of phases, the differential of R_{n+1} in such a series of phases will not in general vanish at the critical phase, so that in general a part of the series will be unstable.

We may illustrate these relations by considering separately the cases in which $n = 1$ and $n = 2$. If a mass of invariable composition is in a critical state, we may keep its volume constant, and destroy its homogeneity by changing its entropy (i.e., by adding or subtracting heat— probably the latter), or we may keep its entropy constant and destroy its homogeneity by changing its volume; but if we keep its pressure constant we cannot destroy its homogeneity by any thermal action, nor if we keep its temperature constant can we destroy its homogeneity by any mechanical action.

When a mass having two independently variable components is in a critical phase, and either its volume or its pressure is maintained constant, its homogeneity may be destroyed by a change of entropy or temperature. Or, if either its entropy or its temperature is maintained constant, its homogeneity may be destroyed by a change of volume or pressure. In both these cases it is supposed that the quantities of the components remain unchanged. But if we suppose both the temperature and the pressure to be maintained constant, the mass will remain homogeneous, however the proportion of the components be changed. Or, if a mass consists of two coexistent phases, one of which is a critical phase having two independently variable components, and either the temperature or the pressure of the mass is maintained constant, it will not be possible by mechanical or thermal means, or by changing the quantities of the components, to cause the critical phase to change into a pair of coexistent phases, so as to give three coexistent phases in the whole mass. The statements of this paragraph and of the preceding have reference only to infinitesimal changes.*

---

* A brief abstract (which came to the author's notice after the above was in type) of a memoir by M. Duclaux, "Sur la séparation des liquides mélangés, etc." will be found in *Comptes Rendus*, vol. lxxxi. (1875), p. 815.

On the Values of the Potentials when the Quantity of one of the Components is very small.

If we apply equation (97) to a homogeneous mass having two independently variable components S_1 and S_2, and make t, p, and m_1 constant, we obtain

$$m_1\left(\frac{d\mu_1}{dm_2}\right)_{t,\,p,\,m_1} + m_2\left(\frac{d\mu_2}{dm_2}\right)_{t,\,p,\,m_1} = 0. \tag{210}$$

Therefore, for $m_2 = 0$, either

$$\left(\frac{d\mu_1}{dm_2}\right)_{t,\,p,\,m_1} = 0, \tag{211}$$

or

$$\left(\frac{d\mu_2}{dm_2}\right)_{t,\,p,\,m_1} = \infty. \tag{212}$$

Now, whatever may be the composition of the mass considered, we may always so choose the substance S_1 that the mass shall consist solely of that substance, and in respect to any other variable component S_2, we shall have $m_2 = 0$. But equation (212) cannot hold true *in general* as thus applied. For it may easily be shown (as has been done with regard to the potential on pages 92, 93) that the value of a differential coefficient like that in (212) for any given mass, when the substance S_2 (to which m_2 and μ_2 relate) is determined, is independent of the particular substance which we may regard as the other component of the mass; so that, if equation (212) holds true when the substance denoted by S_1 has been so chosen that $m_2 = 0$, it must hold true without such a restriction, which cannot generally be the case.

In fact, it is easy to prove directly that equation (211) will hold true of any phase which is stable in regard to continuous changes and in which $m_2 = 0$, *if m_2 is capable of negative as well as positive values*. For by (171), in any phase having that kind of stability, μ_1 is an increasing function of m_1 when t, p, and m_2 are regarded as constant. Hence, μ_1 will have its greatest value when the mass consists wholly of S_1, i.e., when $m_2 = 0$. Therefore, if m_2 is capable of negative as well as positive values, equation (211) must hold true for $m_2 = 0$. (This appears also from the geometrical representation of potentials in the m-ζ curve. See page 119.)

But if m_2 is capable only of positive values, we can only conclude from the preceding considerations that the value of the differential coefficient in (211) cannot be positive. Nor, if we consider the physical significance of this case, viz., that an increase of m_2 denotes an addition to the mass in question of a substance not before contained in it, does any reason appear for supposing that this differential coefficient has generally the value zero. To fix our ideas, let us suppose that S_1 denotes water, and S_2 a salt (either anhydrous or any particular hydrate). The addition of the salt to water, previously in a state capable of equilibrium with vapor or with ice, will destroy the possibility of such equilibrium at the same temperature and pressure. The liquid will dissolve the ice, or condense the vapor, which is brought in contact with it under such circumstances, which shows that μ_1 (the potential for water in the liquid mass) is diminished by the addition of the salt, when the temperature and pressure are maintained constant. Now there seems to be no *a priori* reason for supposing that the ratio of this diminution of the potential for water to the quantity of the salt which is added vanishes with this quantity. We should rather expect that, for small quantities of the salt, an effect of this kind would be proportional to its cause, i.e., that the differential coefficient in (211) would have a finite negative value for an infinitesimal value of m_2. That this is the case with respect to numerous watery solutions of salts is distinctly indicated by the experiments of Wüllner * on the tension of the vapor yielded by such solutions, and of Rüdorff † on the temperature at which ice is formed in them; and unless we have experimental evidence that cases are numerous in which the contrary is true, it seems not unreasonable to assume, as a *general* law, that when m_2 has the value zero and is incapable of negative values, the differential coefficient in (211) will have a finite negative value, and that equation (212) will therefore hold true. But this case must be carefully distinguished from that in which m_2 is capable of negative values, which also may be illustrated by a solution of a salt in water. For this purpose let S_1 denote a hydrate of the salt which can be crystallized, and let S_2 denote water, and let us consider a liquid consisting entirely of S_1 and of such temperature and pressure as to be in equilibrium with crystals of S_1. In such a liquid, an increase or a diminution of the quantity of water would alike cause crystals of S_1 to dissolve, which requires that the differential coefficient in (211) shall vanish at the particular phase of the liquid for which $m_2 = 0$.

Let us return to the case in which m_2 is incapable of negative values, and examine, without other restriction in regard to the substances denoted by S_1 and S_2, the relation between μ_2 and $\frac{m_2}{m_1}$ for any constant temperature and pressure and for such small values of $\frac{m_2}{m_1}$ that the differential coefficient in (211) may be regarded as having the same constant value as when $m_2 = 0$, the values of t, p, and m_1 being unchanged. If we denote this value of the differential coefficient by

---

* *Pogg. Ann.*, vol. ciii. (1858), p. 529 ; vol. cv. (1858), p. 85 ; vol. cx. (1860), p. 564.

† *Pogg. Ann.*, vol. cxiv. (1861), p. 63.

$\dfrac{-A}{m_1}$, the value of A will be positive, and will be independent of m_1.
Then for small values of $\dfrac{m_2}{m_1}$ we have by (210), approximately,

$$m_2 \left(\frac{d\mu_2}{dm_2} \right)_{t, \, p, \, m_1} = A, \tag{213}$$

i.e.,

$$\left(\frac{d\mu_2}{d \log m_2} \right)_{t, \, p, \, m_1} = A. \tag{214}$$

If we write the integral of this equation in the form

$$\mu_2 = A \log \frac{Bm_2}{m_1}, \tag{215}$$

B like A will have a positive value depending only upon the temperature and pressure. As this equation is to be applied only to cases in which the value of m_2 is very small compared with m_1, we may regard $\dfrac{m_1}{v}$ as constant, when temperature and pressure are constant, and write

$$\mu_2 = A \log \frac{Cm_2}{v}, \tag{216}$$

C denoting a positive quantity, dependent only upon the temperature and pressure.

We have so far considered the composition of the body as varying only in regard to the proportion of two components. But the argument will be in no respect invalidated, if we suppose the composition of the body to be capable of other variations. In this case, the quantities A and C will be functions not only of the temperature and pressure but also of the quantities which express the composition of the substance of which together with S_2 the body is composed. If the quantities of any of the components besides S_2 are very small (relatively to the quantities of others), it seems reasonable to assume that the value of μ_2, and therefore the values of A and C, will be nearly the same as if these components were absent.

Hence, if the independently variable components of any body are $S_a, \ldots S_g$, and $S_h, \ldots S_k$, the quantities of the latter being very small as compared with the quantities of the former, and are incapable of negative values, we may express approximately the values of the potentials for $S_h, \ldots S_k$ by equations (subject of course to the uncertainties of the assumptions which have been made) of the form

$$\mu_h = A_h \log \frac{C_h m_h}{v}, \tag{217}$$

. . . .

$$\mu_k = A_k \log \frac{C_k m_k}{v}, \tag{218}$$

in which $A_h, C_h, \ldots A_k, C_k$ denote functions of the temperature, the pressure, and the ratios of the quantities $m_a, \ldots m_g$.

We shall see hereafter, when we come to consider the properties of gases, that these equations may be verified experimentally in a very large class of cases, so that we have considerable reason for believing that they express a general law in regard to the limiting values of potentials.*

On Certain Points relating to the Molecular Constitution of Bodies.

It not unfrequently occurs that the number of proximate components which it is necessary to recognize as independently variable in a body exceeds the number of components which would be sufficient to express its ultimate composition. Such is the case, for example, as has been remarked on page 63, in regard to a mixture at ordinary temperatures of vapor of water and free hydrogen and oxygen. This case is explained by the existence of three sorts of molecules in the gaseous mass, viz., molecules of hydrogen, of oxygen, and of hydrogen and oxygen combined. In other cases, which are essentially the same in principle, we suppose a greater number of different sorts of molecules, which differ in composition, and the relations between these may be more complicated. Other cases are explained by molecules which differ in the quantity of matter which they contain, but not in the kind of matter, nor in the proportion of the different kinds. In still other cases, there appear to be different sorts of molecules, which differ neither in the kind nor in the quantity of matter which they contain, but only in the manner in which they are constituted. What is essential in the cases referred to is that a certain number of some sort or sorts of molecules shall be equivalent to a certain number of some other sort or sorts in respect to the kinds and quantities of matter which they collectively contain, and yet the former shall never be transformed into the latter within the body considered, nor the latter into the former, however the proportion of the numbers of the different sorts of molecules may be varied, or the composition of the body in other respects, or its thermodynamic state as represented by temperature and pressure or any other two suitable variables, provided, it may be, that these variations do not exceed certain limits. Thus, in the

---

* The reader will not fail to remark that, if we could assume the universality of this law, the statement of the conditions necessary for equilibrium between different masses in contact would be much simplified. For, as the potential for a substance which is only a *possible* component (see page 64) would always have the value $-\infty$, the case could not occur that the potential for any substance would have a greater value in a mass in which that substance is only a possible component, than in another mass in which it is an actual component; and the conditions (22) and (51) might be expressed with the sign of equality without exception for the case of possible components.

example given above, the temperature must not be raised beyond a certain limit, or molecules of hydrogen and of oxygen may be transformed into molecules of water.

The differences in bodies resulting from such differences in the constitution of their molecules are capable of continuous variation, in bodies containing the same matter and in the same thermodynamic state as determined, for example, by pressure and temperature, as the numbers of the molecules of the different sorts are varied. These differences are thus distinguished from those which depend upon the manner in which the molecules are combined to form sensible masses. The latter do not cause an increase in the number of variables in the fundamental equation; but they may be the cause of different values of which the function is sometimes capable for one set of values of the independent variables, as, for example, when we have several different values of ζ for the same values of t, p, m_1, m_2, ... m_n, one perhaps being for a gaseous body, one for a liquid, one for an amorphous solid, and others for different kinds of crystals, and all being invariable for constant values of the above mentioned independent variables.

But it must be observed that when the differences in the constitution of the molecules are entirely determined by the quantities of the different kinds of matter in a body with the two variables which express its thermodynamic state, these differences will not involve any increase in the number of variables in the fundamental equation. For example, if we should raise the temperature of the mixture of vapor of water and free hydrogen and oxygen, which we have just considered, to a point at which the numbers of the different sorts of molecules are entirely determined by the temperature and pressure and the total quantities of hydrogen and of oxygen which are present, the fundamental equation of such a mass would involve but four independent variables, which might be the four quantities just mentioned. The fact of a certain part of the matter present existing in the form of vapor of water would, of course, be one of the facts which determine the nature of the relation between ζ and the independent variables, which is expressed by the fundamental equation.

But in the case first considered, in which the quantities of the different sorts of molecules are *not* determined by the temperature and pressure and the quantities of the different kinds of matter in the body as determined by its ultimate analysis, the components of which the quantities or the potentials appear in the fundamental equation must be those which are determined by the proximate analysis of the body, so that the variations in their quantities, with two variations relating to the thermodynamic state of the body, shall include all the variations of which the body is capable.* Such cases present no especial difficulty; there is indeed nothing in the physical and chemical properties of such bodies, so far as a certain range of experiments is concerned, which is different from what might be, if the proximate components were incapable of farther reduction or transformation. Yet among the various phases of the kinds of matter concerned, represented by the different sets of values of the variables which satisfy the fundamental equation, there is a certain class which merits especial attention. These are the phases for which the entropy has a maximum value for the same matter, as determined by the ultimate analysis of the body, with the same energy and volume. To fix our ideas let us call the proximate components $S_1, \ldots S_n$, and the ultimate components $S_a, \ldots S_h$; and let $m_1, \ldots m_n$ denote the quantities of the former, and $m_a, \ldots m_h$, the quantities of the latter. It is evident that $m_a, \ldots m_h$ are homogeneous functions of the first degree of $m_1, \ldots m_n$; and that the relations between the substances $S_1, \ldots S_n$ might be expressed by homogeneous equations of the first degree between the units of these substances, equal in number to the difference of the numbers of the proximate and of the ultimate components. The phases in question are those for which η is a maximum for constant values of $\epsilon, v, m_a, \ldots m_h$; or, as they may also be described, those for which ϵ is a minimum for constant values of $\eta, v, m_a, \ldots m_h$; or for which ζ is a minimum for constant values of $t, p, m_a, \ldots m_h$. The phases which satisfy this condition may be readily determined when the fundamental equation (which will contain the quantities $m_1, \ldots m_n$ or $\mu_1, \ldots \mu_n$,) is known. Indeed it is easy to see that we may express the conditions which determine these phases by substituting $\mu_1, \ldots \mu_n$ for the letters denoting the units of the corresponding substances in the equations which express the equivalence in ultimate analysis between these units.

These phases may be called, with reference to the kind of change which we are considering, phases of dissipated energy. That we have used a similar term before, with reference to a different kind of changes, yet in a sense entirely analogous, need not create confusion.

It is characteristic of these phases that we cannot alter the values of $m_1, \ldots m_n$ in any real mass in such a phase, while the volume of the mass as well as its matter remain unchanged, without diminishing the energy or increasing the entropy of some other system. Hence, if the mass is large, its equilibrium can be but slightly disturbed

---

* The terms proximate or ultimate are not necessarily to be understood in an absolute sense. All that is said here and in the following paragraphs will apply to many cases in which components may conveniently be regarded as proximate or ultimate, which are such only in a relative sense.

by the action of any small body, or by a single electric spark, or by any cause which is not in some way proportioned to the effect to be produced. But when the proportion of the proximate components of a mass taken in connection with its temperature and pressure is not such as to constitute a phase of dissipated energy, it may be possible to cause great changes in the mass by the contact of a very small body. Indeed it is possible that the changes produced by such contact may only be limited by the attainment of a phase of dissipated energy. Such a result will probably be produced in a fluid mass by contact with another fluid which contains molecules of all the kinds which occur in the first fluid (or at least all those which contain the same kinds of matter which also occur in other sorts of molecules), but which differs from the first fluid in that the quantities of the various kinds of molecules are entirely determined by the ultimate composition of the fluid and its temperature and pressure. Or, to speak without reference to the molecular state of the fluid, the result considered would doubtless be brought about by contact with another fluid, which absorbs all the proximate components of the first, $S_1, \ldots S_n$ (or all those between which there exist relations of equivalence in respect to their ultimate analysis), independently, and without passive resistances, but for which the phase is completely determined by its temperature and pressure and its ultimate composition (in respect at least to the particular substances just mentioned). By the absorption of the substances $S_1, \ldots S_n$ independently and without passive resistances, it is meant that when the absorbing body is in equilibrium with another containing these substances, it shall be possible by *infinitesimal* changes in these bodies to produce the exchange of all these substances in either direction and independently. An exception to the preceding statement may of course be made for cases in which the result in question is prevented by the occurrence of some other kinds of change; in other words, it is assumed that the two bodies can remain in contact preserving the properties which have been mentioned.

The term *catalysis* has been applied to such action as we are considering. When a body has the property of reducing another, without limitation with respect to the proportion of the two bodies, to a phase of dissipated energy, in regard to a certain kind of molecular change, it may be called a *perfect catalytic agent* with respect to the second body and the kind of molecular change considered.

It seems not improbable that in some cases in which molecular changes take place slowly in homogeneous bodies, a mass of which the temperature and pressure are maintained constant will be finally brought to a state of equilibrium which is entirely determined by its temperature and pressure and the quantities of its ultimate components, while the various transitory states through which the mass passes (which are evidently not completely defined by the quantities just mentioned) may be completely defined by the quantities of certain proximate components with the temperature and pressure, and the matter of the mass may be brought by processes approximately reversible from permanent states to these various transitory states. In such cases, we may form a fundamental equation with reference to all possible phases, whether transitory or permanent; and we may also form a fundamental equation of different import and containing a smaller number of independent variables, which has reference solely to the final phases of equilibrium. The latter are the phases of dissipated energy (with reference to molecular changes), and when the more general form of the fundamental equation is known, it will be easy to derive from it the fundamental equation for these permanent phases alone.

Now, as these relations, theoretically considered, are independent of the rapidity of the molecular changes, the question naturally arises, whether in cases in which we are not able to distinguish such transitory phases, they may not still have a theoretical significance. If so, the consideration of the subject from this point of view, may assist us, in such cases, in discovering the form of the fundamental equation with reference to the ultimate components, which is the only equation required to express all the properties of the bodies which are capable of experimental demonstration. Thus, when the phase of a body is completely determined by the quantities of n independently variable components, with the temperature and pressure, and we have reason to suppose that the body is composed of a greater number n' of proximate components, which are therefore not independently variable (while the temperature and pressure remain constant), it seems quite possible that the fundamental equation of the body may be of the same form as the equation for the phases of dissipated energy of analogous compounds of n' proximate and n ultimate components, in which the proximate components are capable of independent variation (without variation of temperature or pressure). And if such is found to be the case, the fact will be of interest as affording an indication concerning the proximate constitution of the body.

Such considerations seem to be especially applicable to the very common case in which at certain temperatures and pressures, regarded as constant, the quantities of certain proximate components of a mass are capable of independent variations, and all the phases produced by these variations are permanent in their nature, while at other temperatures and pressures, likewise regarded as constant, the quantities of these proximate components are not capable of independent variation, and the phase may be completely defined by the quantities of the ultimate components with the temperature and pressure. There may be, at certain intermediate temperatures and pressures, a condition with respect to the independence of the proximate components intermediate in character, in which the quantities of the proximate components are independently variable when we consider all phases, the essentially transitory as well as the permanent, but in which these quantities are not independently variable when we consider the permanent phases alone. Now we have no reason to believe that the passing of a body in a state of dissipated energy from one to another of the three conditions mentioned has any necessary connection with any discontinuous change of state. Passing the limit which separates one of these states from another will not therefore involve any discontinuous change in the values of any of the quantities enumerated in (99)–(103) on page 88, if m_1, m_2, ... m_n, μ_1, μ_2, ... μ_n are understood as always relating to the ultimate components of the body. Therefore, if we regard masses in the different conditions mentioned above as having different fundamental equations (which we may suppose to be of any one of the five kinds described on page 88), these equations will agree at the limits dividing these conditions not only in the values of all the variables which appear in the equations, but also in all the differential coefficients of the first order involving these variables. We may illustrate these relations by supposing the values of t, p, and ζ for a mass in which the quantities of the ultimate components are constant to be represented by rectilinear coordinates. Where the proximate composition of such a mass is not determined by t and p, the value of ζ will not be determined by these variables, and the points representing connected values of t, p, and ζ will form a solid. This solid will be bounded in the direction opposite to that in which ζ is measured, by a surface which represents the phases of dissipated energy. In a part of the figure, all the phases thus represented may be permanent, in another part only the phases in the bounding surface, and in a third part there may be no such solid figure (for any phases of which the existence is experimentally demonstrable), but only a surface. This surface together with the bounding surfaces representing phases of dissipated energy in the parts of the figure mentioned above forms a continuous sheet, without discontinuity in regard to the direction of its normal at the limits dividing the different parts of the figure which have been mentioned. (There may, indeed, be different sheets representing liquid and gaseous states, etc., but if we limit our consideration to states of one of these sorts, the case will be as has been stated.)

We shall hereafter, in the discussion of the fundamental equations of gases, have an example of the derivation of the fundamental equation for phases of dissipated energy (with respect to the molecular changes on which the proximate composition of the body depends) from the more general form of the fundamental equation.

The Conditions of Equilibrium for Heterogeneous Masses under the Influence of Gravity.

Let us now seek the conditions of equilibrium for a mass of various kinds of matter subject to the influence of gravity. It will be convenient to suppose the mass enclosed in an immovable envelop which is impermeable to matter and to heat, and in other respects, except in regard to gravity, to make the same suppositions as on page 62. The energy of the mass will now consist of two parts, one of which depends upon its intrinsic nature and state, and the other upon its position in space. Let Dm denote an element of the mass, $D\epsilon$ the intrinsic energy of this element, h its height above a fixed horizontal plane, and g the force of gravity; then the total energy of the mass (when without sensible motions) will be expressed by the formula

$$\int D\epsilon + \int gh\, Dm, \tag{219}$$

in which the integrations include all the elements of the mass; and the general condition of equilibrium will be

$$\delta\int D\epsilon + \delta\int gh\, Dm \geqq 0, \tag{220}$$

the variations being subject to certain equations of condition. These must express that the entropy of the whole mass is constant, that the surface bounding the whole mass is fixed, and that the total quantity of each of the component substances is constant. We shall suppose that there are no other equations of condition, and that the independently variable components are the same throughout the whole mass; and we shall at first limit ourselves to the consideration of the conditions of equilibrium with respect to the changes which may be expressed by infinitesimal variations of the quantities which define the initial state of the mass, without regarding the possibility of the formation at any place of infinitesimal masses entirely different from any initially existing in the same vicinity.

Let $D\eta$, Dv, $Dm_1, \ldots Dm_n$ denote the entropy of the element Dm, its volume, and the quantities which it contains of the various components. Then

$$Dm = Dm_1 \ldots + Dm_n, \tag{221}$$

and
$$\delta Dm = \delta Dm_1 \ldots + \delta Dm_n. \tag{222}$$

Also, by equation (12),

$$\delta D\epsilon = t\, \delta D\eta - p\, \delta Dv + \mu_1\, \delta Dm_1 \ldots + \mu_n\, \delta Dm_n. \tag{223}$$

By these equations the general condition of equilibrium may be reduced to the form

$$\int t\, \delta D\eta -\int p\, \delta Dv +\int \mu_1\, \delta Dm_1 \dots +\int \mu_n\, \delta Dm_n$$
$$+\int g\, \delta h\, Dm +\int gh\, \delta Dm_1 \dots +\int gh\, \delta Dm_n \geqq 0. \qquad (224)$$

Now it will be observed that the different equations of condition affect different parts of this condition, so that we must have, separately,

$$\int t\, \delta D\eta \geqq 0, \quad \text{if} \quad \int \delta D\eta =0; \qquad (225)$$

$$-\int p\, \delta Dv +\int g\, \delta h\, Dm \geqq 0, \qquad (226)$$

if the bounding surface is unvaried;

$$\left. \begin{array}{l} \int \mu_1\, \delta Dm_1 +\int gh\, \delta Dm_1 \geqq 0, \quad \text{if} \quad \int \delta Dm_1 =0; \\ \dotfill \\ \int \mu_n\, \delta Dm_n +\int gh\, \delta Dm_n \geqq 0, \quad \text{if} \quad \int \delta Dm_n =0. \end{array} \right\} \qquad (227)$$

From (225) we may derive the condition of thermal equilibrium,

$$t = \text{const.} \qquad (228)$$

Condition (226) is evidently the ordinary mechanical condition of equilibrium, and may be transformed by any of the usual methods. We may, for example, apply the formula to such motions as might take place longitudinally within an infinitely narrow tube, terminated at both ends by the external surface of the mass, but otherwise of indeterminate form. If we denote by m the mass, and by v the volume, included in the part of the tube between one end and a transverse section of variable position, the condition will take the form

$$-\int p\, \delta dv +\int g\, \delta h\, dm \geqq 0, \qquad (229)$$

in which the integrations include the whole contents of the tube. Since no motion is possible at the ends of the tube,

$$\int p\, \delta dv +\int \delta v\, dp =\int d(p\, \delta v) =0. \qquad (230)$$

Again, if we denote by γ the density of the fluid,

$$\int g\, \delta h\, dm =\int g \frac{dh}{dv}\, \delta v\, \gamma\, dv =\int g\gamma\, \delta v\, dh. \qquad (231)$$

By these equations condition (229) may be reduced to the form

$$\int \delta v\, (dp +g\gamma\, dh) \geqq 0. \qquad (232)$$

Therefore, since δv is arbitrary in value,

$$dp = -g\gamma\, dh, \qquad (233)$$

which will hold true at any point in the tube, the differentials being taken with respect to the direction of the tube at that point. Therefore, as the form of the tube is indeterminate, this equation must hold true, without restriction, throughout the whole mass. It evidently requires that the pressure shall be a function of the height alone, and that the density shall be equal to the first derivative of this function, divided by $-g$.

Conditions (227) contain all that is characteristic of chemical equilibrium. To satisfy these conditions it is necessary and sufficient that

$$\left.\begin{aligned} \mu_1 + gh &= \text{const.} \\ \dots\dots\dots\dots\dots & \\ \mu_n + gh &= \text{const.} \end{aligned}\right\} \tag{234}$$

The expressions $\mu_1, \dots \mu_n$ denote quantities which we have called the potentials for the several components, and which are entirely determined at any point in a mass by the nature and state of the mass about that point. We may avoid all confusion between these quantities and the potential of the force of gravity, if we distinguish the former, when necessary, as *intrinsic* potentials. The relations indicated by equations (234) may then be expressed as follows :—

When a fluid mass is in equilibrium under the influence of gravity, and has the same independently variable components throughout, the intrinsic potentials for the several components are constant in any given level, and diminish uniformly as the height increases, the difference of the values of the intrinsic potential for any component at two different levels being equal to the work done by the force of gravity when a unit of matter falls from the higher to the lower level.

The conditions expressed by equations (228), (233), (234) are necessary and sufficient for equilibrium, except with respect to the possible formation of masses which are not approximately identical in phase with any previously existing about the points where they may be formed. The possibility of such formations at any point is evidently independent of the action of gravity, and is determined entirely by the phase or phases of the matter about that point. The conditions of equilibrium in this respect have been discussed on pages 74–79.

But equations (228), (233), and (234) are not entirely independent. For with respect to any mass in which there are no surfaces of discontinuity (i.e., surfaces where adjacent elements of mass have finite differences of phase), one of these equations will be a consequence of the others. Thus by (228) and (234), we may obtain from (97), which will hold true of any continuous variations of phase, the equation

$$v\,dp = -g\,(m_1 \dots + m_n)\,dh\,; \tag{235}$$

or

$$dp = -g\gamma\,dh\,; \tag{236}$$

which will therefore hold true in any mass in which equations (228) and (234) are satisfied, and in which there are no surfaces of discontinuity. But the condition of equilibrium expressed by equation

(233) has no exception with respect to surfaces of discontinuity; therefore in any mass in which such surfaces occur, it will be necessary for equilibrium, in addition to the relations expressed by equations (228) and (234), that there shall be no discontinuous change of pressure at these surfaces.

This superfluity in the particular conditions of equilibrium which we have found, as applied to a mass which is everywhere continuous in phase, is due to the fact that we have made the elements of volume variable in position and size, while the matter initially contained in these elements is not supposed to be confined to them. Now, as the different components may move in different directions when the state of the system varies, it is evidently impossible to define the elements of volume so as always to include the same matter; we must, therefore, suppose the matter contained in the elements of volume to vary; and therefore it would be allowable to make these elements fixed in space. If the given mass has no surfaces of discontinuity, this would be much the simplest plan. But if there are any surfaces of discontinuity, it will be possible for the state of the given mass to vary, not only by infinitesimal changes of phase in the fixed elements of volume, but also by movements of the surfaces of discontinuity. It would therefore be necessary to add to our general condition of equilibrium terms relating to discontinuous changes in the elements of volume about these surfaces,—a necessity which is avoided if we consider these elements movable, as we can then suppose that each element remains always on the same side of the surface of discontinuity.

Method of treating the preceding problem, in which the elements of volume are regarded as fixed.

It may be interesting to see in detail how the particular conditions of equilibrium may be obtained if we regard the elements of volume as fixed in position and size, and consider the possibility of finite as well as infinitesimal changes of phase in each element of volume. If we use the character Δ to denote the differences determined by such finite differences of phase, we may express the variation of the intrinsic energy of the whole mass in the form

$$\int \delta \, D\epsilon + \int \Delta \, D\epsilon, \tag{237}$$

in which the first integral extends over all the elements which are infinitesimally varied, and the second over all those which experience a finite variation. We *may* regard both integrals as extending throughout the whole mass, but their values will be zero except for the parts mentioned.

If we do not wish to limit ourselves to the consideration of masses so small that the force of gravity can be regarded as constant in direction and in intensity, we may use Υ to denote the potential of the force of gravity, and express the variation of the part of the energy which is due to gravity in the form

$$-\int \Upsilon \, \delta Dm - \int \Upsilon \, \Delta Dm. \tag{238}$$

We shall then have, for the general condition of equilibrium,

$$\int \delta De + \int \Delta De - \int \Upsilon \, \delta Dm - \int \Upsilon \, \Delta Dm \geqq 0 ; \tag{239}$$

and the equations of condition will be

$$\int \delta D\eta + \int \Delta D\eta = 0, \tag{240}$$

$$\left. \begin{array}{l} \int \delta Dm_1 + \int \Delta Dm_1 = 0, \\ \cdots\cdots\cdots\cdots\cdots\cdots \\ \int \delta Dm_n + \int \Delta Dm_n = 0. \end{array} \right\} \tag{241}$$

We may obtain a condition of equilibrium independent of these equations of condition, by subtracting these equations, multiplied each by an indeterminate constant, from condition (239). If we denote these indeterminate constants by $T, M_1, \ldots M_n$, we shall obtain after arranging the terms

$$\overline{\int \delta De - \Upsilon \, \delta Dm - T \, \delta D\eta - M_1 \, \delta Dm_1 \ldots - M_n \, \delta Dm_n}$$

$$+ \int \Delta De - \Upsilon \, \Delta Dm - T \, \Delta D\eta - M_1 \, \Delta Dm_1 \ldots - M_n \, \delta Dm_n \geqq 0. \tag{242}$$

The variations, both infinitesimal and finite, in this condition are independent of the equations of condition (240) and (241), and are only subject to the condition that the varied values of De, $D\eta$, $Dm_1, \ldots Dm_n$ for each element are determined by a certain change of phase. But as we do not suppose the same element to experience both a finite and an infinitesimal change of phase, we must have

$$\delta De - \Upsilon \, \delta Dm - T \, \delta D\eta - M_1 \, \delta Dm_1 \ldots - M_n \, \delta Dm_n \geqq 0, \tag{243}$$

and $\quad \Delta De - \Upsilon \, \Delta Dm - T \, \Delta D\eta - M_1 \, \Delta Dm_1 \ldots - M_n \, \Delta Dm_n \geqq 0. \tag{244}$

By equation (12), and in virtue of the necessary relation (222), the first of these conditions reduces to

$$(t - T) \, \delta D\eta + (\mu_1 - \Upsilon - M_1) \, \delta Dm_1 \ldots$$

$$+ (\mu_n - \Upsilon - M_n) \, \delta Dm_n \geqq 0 ; \tag{245}$$

for which it is necessary and sufficient that

$$t = T, \tag{246}$$

$$\left. \begin{array}{l} \mu_1 - \Upsilon = M_1, \\ \cdots\cdots\cdots\cdots \\ \mu_n - \Upsilon = M_n. \end{array} \right\}* \tag{247}$$

---

*The gravitation potential is here supposed to be defined in the usual way. But if it were defined so as to *decrease* when a body falls, we should have the sign + instead of − in these equations; i.e., for each component, the sum of the gravitation and intrinsic potentials would be constant throughout the whole mass.

Condition (244) may be reduced to the form

$$\Delta D\epsilon - T\,\Delta D\eta - (\Upsilon + M_1)\,\Delta Dm_1 \ldots - (\Upsilon + M_n)\,\Delta Dm_n \geqq 0; \quad (248)$$

and by (246) and (247) to

$$\Delta D\epsilon - t\,\Delta D\eta - \mu_1\,\Delta Dm_1 \ldots - \mu_n\,\Delta Dm_n \geqq 0. \quad (249)$$

If values determined subsequently to the change of phase are distinguished by accents, this condition may be written

$$De' - t\,D\eta' - \mu_1 Dm_1' \ldots - \mu_n\,Dm_n'$$
$$- De + t\,D\eta + \mu_1 Dm_1 \ldots + \mu_n Dm_n \geqq 0, \quad (250)$$

which may be reduced by (93) to

$$De' - t\,D\eta' - \mu_1 Dm_1' \ldots - \mu_n Dm_n' + p\,Dv \geqq 0. \quad (251)$$

Now if the element of volume Dv is adjacent to a surface of discontinuity, let us suppose De', $D\eta'$, Dm_1', $\ldots Dm_n'$ to be determined (for the same element of volume) by the phase existing on the other side of the surface of discontinuity. As $t, \mu_1, \ldots \mu_n$ have the same values on both sides of this surface, the condition may be reduced by (93) to

$$-p'\,Dv + p\,Dv \geqq 0. \quad (252)$$

That is, the pressure must not be greater on one side of a surface of discontinuity than on the other.

Applied more generally, (251) expresses the condition of equilibrium with respect to the possibility of discontinuous changes of phases at any point. As $Dv' = Dv$, the condition may also be written

$$De' - t\,D\eta' + p\,Dv' - \mu_1 Dm_1' \ldots - \mu_n\,Dm_n' \geqq 0, \quad (253)$$

which must hold true when $t, p, \mu_1, \ldots \mu_n$ have values determined by any point in the mass, and De', $D\eta'$, Dv', Dm_1', $\ldots Dm_n'$ have values determined by any possible phase of the substances of which the mass is composed. The application of the condition is, however, subject to the limitations considered on pages 74–79. It may easily be shown (see page 104) that for constant values of $t, \mu_1, \ldots \mu_n$, and of Dv', the first member of (253) will have the least possible value when De', $D\eta'$, Dm_1', $\ldots Dm_n'$ are determined by a phase for which the temperature has the value t, and the potentials the values $\mu_1, \ldots \mu_n$. It will be sufficient, therefore, to consider the condition as applied to such phases, in which case it may be reduced by (93) to

$$p - p' \geqq 0. \quad (254)$$

That is, the pressure at any point must be as great as that of any phase of the same components, for which the temperature and the potentials have the same values as at that point. We may also express this condition by saying that the pressure must be as great as is consistent with equations (246), (247). This condition with the equations mentioned will always be sufficient for equilibrium; when the condition is not satisfied, if equilibrium subsists, it will be at least practically unstable.

Hence, the phase at any point of a fluid mass, which is in stable equilibrium under the influence of gravity (whether this force is due to external bodies or to the mass itself), and which has throughout the same independently variable components, is completely determined by the phase at any other point and the difference of the values of the gravitational potential for the two points.

Fundamental Equations of Ideal Gases and Gas-Mixtures.

For a constant quantity of a perfect or ideal gas, the product of the volume and pressure is proportional to the temperature, and the variations of energy are proportional to the variations of temperature. For a unit of such a gas we may write

$$pv = at,$$
$$d\epsilon = c\, dt,$$

a and c denoting constants. By integration, we obtain the equation

$$\epsilon = ct + E,$$

in which E also denotes a constant. If by these equations we eliminate t and p from (11) we obtain

$$d\epsilon = \frac{\epsilon - E}{c} d\eta - \frac{a}{v} \frac{\epsilon - E}{c} dv,$$

or

$$c \frac{d\epsilon}{\epsilon - E} = d\eta - a \frac{dv}{v}.$$

The integral of this equation may be written in the form

$$c \log \frac{\epsilon - E}{c} = \eta - a \log v - H,$$

where H denotes a fourth constant. We may regard E as denoting the energy of a unit of the gas for $t = 0$; H its entropy for $t = 1$ and $v = 1$; a its pressure in the latter state, or its volume for $t = 1$ and $p = 1$; c its specific heat at constant volume. We may extend the application of the equation to any quantity of the gas, without altering the values of the constants, if we substitute $\dfrac{\epsilon}{m}, \dfrac{\eta}{m}, \dfrac{v}{m}$ for ϵ, η, v, respectively. This will give

$$c \log \frac{\epsilon - Em}{cm} = \frac{\eta}{m} - H + a \log \frac{m}{v}. \tag{255}$$

This is a fundamental equation (see pages 85–89) for an ideal gas of invariable composition. It will be observed that if we do not have to consider the properties of the matter which forms the gas as appearing in any other form or combination, but solely as constituting the gas in question (in a state of purity), we may without loss of generality give to E and H the value zero, or any other arbitrary values. But when the scope of our investigations is not thus limited we may have determined the states of the substance of the gas for which $\epsilon = 0$ and $\eta = 0$ with reference to some other form in which the substance appears, or, if the substance is compound, the states of its components for which $\epsilon = 0$ and $\eta = 0$ may be already determined; so that the constants E and H cannot in general be treated as arbitrary.

We obtain from (255) by differentiation

$$\frac{c}{\epsilon - Em}\, d\epsilon = \frac{1}{m}\, d\eta - \frac{a}{v}\, dv + \left(\frac{cE}{\epsilon - Em} + \frac{c+a}{m} - \frac{\eta}{m^2} \right) dm, \qquad (256)$$

whence, in virtue of the general relation expressed by (86),

$$t = \frac{\epsilon - Em}{cm}, \qquad (257)$$

$$p = a\frac{\epsilon - Em}{cv}, \qquad (258)$$

$$\mu = E + \frac{\epsilon - Em}{cm^2}(cm + am - \eta). \qquad (259)$$

We may obtain the fundamental equation between ψ, t, v, and m from equations (87), (255), and (257). Eliminating ϵ we have

$$\psi = Em + cmt - t\eta,$$

and $\qquad\qquad c \log t = \dfrac{\eta}{m} - H + a \log \dfrac{m}{v};$

and eliminating η, we have the fundamental equation

$$\psi = Em + mt\left(c - H - c \log t + a \log \frac{m}{v} \right). \qquad (260)$$

Differentiating this equation, we obtain

$$d\psi = -m\left(H + c \log t + a \log \frac{v}{m} \right) dt - \frac{amt}{v}\, dv$$
$$+ \left(E + t\left(c + a - H - c \log t + a \log \frac{m}{v} \right) \right) dm; \quad (261)$$

whence, by the general equation (88),

$$\eta = m\left(H + c \log t + a \log \frac{v}{m} \right), \qquad (262)$$

$$p = \frac{amt}{v}, \qquad (263)$$

$$\mu = E + t\left(c + a - H - c \log t + a \log \frac{m}{v} \right). \qquad (264)$$

From (260), by (87) and (91), we obtain

$$\zeta = Em + mt\left(c - H - c\log t + a\log\frac{m}{v}\right) + pv,$$

and eliminating v by means of (263), we obtain the fundamental equation

$$\zeta = Em + mt\left(c + a - H - (c+a)\log t + a\log\frac{p}{a}\right). \tag{265}$$

From this, by differentiation and comparison with (92), we may obtain the equations

$$\eta = m\left(H + (c+a)\log t - a\log\frac{p}{a}\right), \tag{266}$$

$$v = \frac{amt}{p}, \tag{267}$$

$$\mu = E + t\left(c + a - H - (c+a)\log t + a\log\frac{p}{a}\right). \tag{268}$$

The last is also a fundamental equation. It may be written in the form

$$\log\frac{p}{a} = \frac{H - c - a}{a} + \frac{c+a}{a}\log t + \frac{\mu - E}{at}, \tag{269}$$

or, if we denote by e the base of the Naperian system of logarithms,

$$p = ae^{\frac{H-c-a}{a}} t^{\frac{c+a}{a}} e^{\frac{\mu - E}{at}}. \tag{270}$$

The fundamental equation between χ, η, p, and m may also be easily obtained; it is

$$(c+a)\log\frac{\chi - Em}{(c+a)m} = \frac{\eta}{m} - H + a\log\frac{p}{a}, \tag{271}$$

which can be solved with respect to χ.

Any one of the fundamental equations (255), (260), (265), (270), and (271), which are entirely equivalent to one another, may be regarded as defining an ideal gas. It will be observed that most of these equations might be abbreviated by the use of different constants. In (270), for example, a single constant might be used for $ae^{\frac{H-c-a}{a}}$, and another for $\frac{c+a}{a}$. The equations have been given in the above form, in order that the relations between the constants occurring in the different equations might be most clearly exhibited. The sum $c+a$ is the specific heat for constant pressure, as appears if we differentiate (266) regarding p and m as constant.[*]

---

[*] We may easily obtain the equation between the temperature and pressure of a saturated vapor, if we know the fundamental equations of the substance both in the gaseous, and in the liquid or solid state. If we suppose that the density and the specific heat at constant pressure of the liquid may be regarded as constant quantities (for such

The preceding fundamental equations all apply to gases *of constant composition*, for which the matter is entirely determined by a single variable (m). We may obtain corresponding fundamental equations for a mixture of gases, in which the proportion of the components shall be variable, from the following considerations.

moderate pressures as the liquid experiences while in contact with the vapor), and denote this specific heat by k, and the volume of a unit of the liquid by V, we shall have for a unit of the liquid

$$t\,d\eta = k\,dt,$$

whence
$$\eta = k\log t + H',$$

where H' denotes a constant. Also, from this equation and (97),

$$d\mu = -(k\log t + H')\,dt + V\,dp,$$

whence
$$\mu = kt - kt\log t - H't + Vp + E', \qquad \text{(A)}$$

where E' denotes another constant. This is a fundamental equation for the substance in the liquid state. If (268) represents the fundamental equation for the same substance in the gaseous state, the two equations will both hold true of coexistent liquid and gas. Eliminating μ we obtain

$$\log\frac{p}{a} = \frac{H-H'+k-c-a}{a} - \frac{k-c-a}{a}\log t - \frac{E-E'}{at} + \frac{V}{a}\frac{p}{t}.$$

If we neglect the last term, which is evidently equal to the density of the vapor divided by the density of the liquid, we may write

$$\log p = A - B\log t - \frac{C}{t},$$

A, B, and C denoting constants. If we make similar suppositions in regard to the substance in the solid state, the equation between the pressure and temperature of coexistent solid and gaseous phases will of course have the same form.

A similar equation will also apply to the phases of an ideal gas which are coexistent with two different kinds of solids, one of which can be formed by the combination of the gas with the other, each being of invariable composition and of constant specific heat and density. In this case we may write for one solid

$$\mu_1 = k't - k't\log t - H't + V'p + E',$$

and for the other
$$\mu_2 = k''t - k''t\log t - H''t + V''p + E'',$$

and for the gas
$$\mu_3 = E + t\left(c + a - H - (c+a)\log t + a\log\frac{p}{a}\right).$$

Now if a unit of the gas unites with the quantity λ of the first solid to form the quantity $1+\lambda$ of the second it will be necessary for equilibrium (see pages 67, 68) that

$$\mu_3 + \lambda\mu_1 = (1+\lambda)\mu_2.$$

Substituting the values of μ_1, μ_2, μ_3, given above, we obtain after arranging the terms and dividing by at

$$\log\frac{p}{a} = A - B\log t - \frac{C}{t} + D\frac{p}{t},$$

when
$$A = \frac{H + \lambda H' - (1+\lambda)H'' - c - a - \lambda k' + (1+\lambda)k''}{a},$$

$$B = \frac{(1+\lambda)k'' - \lambda k' - c - a}{a},$$

$$C = \frac{E + \lambda E' - (1+\lambda)E''}{a}, \qquad D = \frac{(1+\lambda)V'' - \lambda V'}{a}.$$

We may conclude from this that an equation of the same form may be applied to an ideal gas in equilibrium with a liquid of which it forms an independently variable component, when the specific heat and density of the liquid are entirely determined

It is a rule which admits of a very general and in many cases very exact experimental verification, that if several liquid or solid substances which yield different gases or vapors are simultaneously in equilibrium with a mixture of these gases (cases of chemical action between the gases being excluded), the pressure in the gas-mixture is equal to the sum of the pressures of the gases yielded at the same temperature by the various liquid or solid substances taken separately. Now the potential in any of the liquids or solids for the substance which it yields in the form of gas has very nearly the same value when the liquid or solid is in equilibrium with the gas-mixture as when it is in equilibrium with its own gas alone. The difference of the pressure in the two cases will cause a certain difference in the values of the potential, but that this difference will be small, we may infer from the equation

$$\left(\frac{d\mu_1}{dp}\right)_{t,\,m} = \left(\frac{dv}{dm_1}\right)_{t,\,p,\,m}, \tag{272}$$

which may be derived from equation (92). In most cases, there will be a certain absorption by each liquid of the gases yielded by the

---

by its composition, except that the letters A, B, C, and D must in this case be understood to denote quantities which vary with the composition of the liquid. But to consider the case more in detail, we have for the liquid by (A)

$$\frac{\zeta}{m} = \mu = kt - kt \log t - H't + Vp + E',$$

where k, H', V, E' denote quantities which depend only upon the composition of the liquid. Hence, we may write

$$\zeta = \mathbf{k}t - \mathbf{k}t \log t - \mathbf{H}t + \mathbf{V}p + \mathbf{E},$$

where $\mathbf{k}$, $\mathbf{H}$, $\mathbf{V}$, and $\mathbf{E}$ denote functions of m_1, m_2, etc. (the quantities of the several components of the liquid). Hence, by (92),

$$\mu_i = \frac{d\mathbf{k}}{dm_1}t - \frac{d\mathbf{k}}{dm_1}t \log t - \frac{d\mathbf{H}}{dm_1}t + \frac{d\mathbf{V}}{dm_1}p + \frac{d\mathbf{E}}{dm_1}.$$

If the component to which this potential relates is that which also forms the gas, we shall have by (269)

$$\log\frac{p}{a} = \frac{H - c - a}{a} + \frac{c + a}{a}\log t + \frac{\mu_1 - E}{at}.$$

Eliminating μ_1, we obtain the equation

$$\log\frac{p}{a} = A - B \log t - \frac{C}{t} + D\frac{p}{t},$$

in which A, B, C, and D denote quantities which depend only upon the composition of the liquid, viz. :

$$A = \frac{1}{a}\left(H - \frac{d\mathbf{H}}{dm_1} - c - a + \frac{d\mathbf{k}}{dm_1}\right),$$

$$B = \frac{1}{a}\left(\frac{d\mathbf{k}}{dm_1} - c - a\right),$$

$$C = \frac{1}{a}\left(E - \frac{d\mathbf{E}}{dm_1}\right), \qquad D = \frac{1}{a}\frac{d\mathbf{V}}{dm_1}.$$

With respect to some of the equations which have here been deduced, the reader may compare Professor Kirchhoff " Ueber die Spannung des Dampfes von Mischungen aus Wasser und Schwefelsäure," *Pogg. Ann.*, vol. civ. (1858), p. 612; and Dr. Rankine " On Saturated Vapors," *Phil. Mag.*, vol. xxxi. (1866), p. 199.

others, but as it is well known that the above rule does not apply to cases in which such absorption takes place to any great extent, we may conclude that the effect of this circumstance in the cases with which we have to do is of secondary importance. If we neglect the slight differences in the values of the potentials due to these circumstances, the rule may be expressed as follows:—

The pressure in a mixture of different gases is equal to the sum of the pressures of the different gases as existing each by itself at the same temperature and with the same value of its potential.

To form a precise idea of the practical significance of the law as thus stated with reference to the equilibrium of two liquids with a mixture of the gases which they emit, when neither liquid absorbs the gas emitted by the other, we may imagine a long tube closed at each end and bent in the form of a W to contain in each of the descending loops one of the liquids, and above these liquids the gases which they emit, viz., the separate gases at the ends of the tube, and the mixed gases in the middle. We may suppose the whole to be in equilibrium, the difference of the pressures of the gases being balanced by the proper heights of the liquid columns. Now it is evident from the principles established on pages 144-150 that the potential for either gas will have the same value in the mixed and in the separate gas *at the same level,* and therefore according to the rule in the form which we have given, the pressure in the gas-mixture is equal to the sum of the pressures in the separate gases, *all these pressures being measured at the same level.* Now the experiments by which the rule has been established relate rather to the gases in the vicinity of the surfaces of the liquids. Yet, although the differences of level in these surfaces may be considerable, the corresponding differences of pressure in the columns of gas will certainly be very small in all cases which can be regarded as falling under the laws of ideal gases, for which very great pressures are not admitted.

If we apply the above law to a mixture of ideal gases and distinguish by subscript numerals the quantities relating to the different gases, and denote by Σ_1 the sum of all similar terms obtained by changing the subscript numerals, we shall have by (270)

$$p = \Sigma_1 \left(a_1 e^{\frac{H_1 - c_1 - a_1}{a_1}} \, t^{\frac{c_1 + a_1}{a_1}} \, e^{\frac{\mu_1 - E_1}{a_1 t}} \right). \tag{273}$$

It will be legitimate to assume this equation provisionally as the fundamental equation defining an ideal gas-mixture, and afterwards to justify the suitableness of such a definition by the properties which may be deduced from it. In particular, it will be necessary to show that an ideal gas-mixture as thus defined, when the proportion of its components remains constant, has all the properties which have already been assumed for an ideal gas of invariable composition; it will also be desirable to consider more rigorously and more in detail the equilibrium of such a gas-mixture with solids and liquids, with respect to the above rule.

By differentiation and comparison with (98) we obtain

$$\frac{\eta}{v} = \Sigma_1 \left(\left(c_1 + a_1 - \frac{\mu_1 - E_1}{t} \right) e^{\frac{H_1 - c_1 - a_1}{a_1}} t^{\frac{c_1}{a_1}} e^{\frac{\mu_1 - E_1}{a_1 t}} \right), \tag{274}$$

$$\left. \begin{aligned} \frac{m_1}{v} &= e^{\frac{H_1 - c_1 - a_1}{a_1}} t^{\frac{c_1}{a_1}} e^{\frac{\mu_1 - E_1}{a_1 t}}, \\[2mm] \frac{m_2}{v} &= e^{\frac{H_2 - c_2 - a_2}{a_2}} t^{\frac{c_2}{a_2}} e^{\frac{\mu_2 - E_2}{a_2 t}}, \\[2mm] \text{etc.} \end{aligned} \right\} \tag{275}$$

Equations (275) indicate that the relation between the temperature, the density of any component, and the potential for that component, is not affected by the presence of the other components. They may also be written

$$\left. \begin{aligned} \mu_1 = E_1 + t \left(c_1 + a_1 - H_1 - c_1 \log t + a_1 \log \frac{m_1}{v} \right), \\ \text{etc.} \end{aligned} \right\} \tag{276}$$

Eliminating μ_1, μ_2, etc. from (273) and (274) by means of (275) and (276), we obtain

$$p = \Sigma_1 \frac{a_1 m_1 t}{v}, \tag{277}$$

$$\eta = \Sigma_1 \left(m_1 H_1 + m_1 c_1 \log t + m_1 a_1 \log \frac{v}{m_1} \right). \tag{278}$$

Equation (277) expresses the familiar principle that the pressure in a gas-mixture is equal to the sum of the pressures which the component gases would possess if existing separately with the same volume at the same temperature. Equation (278) expresses a similar principle in regard to the entropy of the gas-mixture.

From (276) and (277) we may easily obtain the fundamental equation between ψ, t, v, m_1, m_2, etc. For by substituting in (94) the values of p, μ_1, μ_2, etc. taken from these equations, we obtain

$$\psi = \Sigma_1 \left(E_1 m_1 + m_1 t \left(c_1 - H_1 - c_1 \log t + a_1 \log \frac{m_1}{v} \right) \right). \tag{279}$$

If we regard the proportion of the various components as constant, this equation may be simplified by writing

$$\begin{aligned} m \quad &\text{for} \quad \Sigma_1 m_1, \\ cm \quad &\text{for} \quad \Sigma_1 (c_1 m_1), \\ am \quad &\text{for} \quad \Sigma_1 (a_1 m_1), \\ Em \quad &\text{for} \quad \Sigma_1 (E_1 m_1), \end{aligned}$$

and $\qquad Hm - am \log m \quad$ for $\quad \Sigma_1 (H_1 m_1 - a_1 m_1 \log m_1)$.

The values of c, a, E, and H will then be constant and m will denote the total quantity of gas. As the equation will thus be reduced to the form of (260), it is evident that an ideal gas-mixture, as defined by (273) or (279), when the proportion of its components remains unchanged, will have all the properties which we have assumed for an ideal gas of invariable composition. The relations between the specific heats of the gas mixture at constant volume and at constant pressure and the specific heats of its components are expressed by the equations

$$c = \Sigma_1 \frac{m_1 c_1}{m}, \tag{280}$$

and

$$c + a = \Sigma_1 \frac{m_1 (c_1 + a_1)}{m}. \tag{281}$$

We have already seen that the values of t, v, m_1, μ_1 in a gas-mixture are such as are possible for the component G_1 (to which m_1 and μ_1 relate) existing separately. If we denote by p_1, η_1, ψ_1, ϵ_1, χ_1, ζ_1 the connected values of the several quantities which the letters indicate determined for the gas G_1 as thus existing separately, and extend this notation to the other components, we shall have by (273), (274), and (279)

$$p = \Sigma_1 p_1, \qquad \eta = \Sigma_1 \eta_1, \qquad \psi = \Sigma_1 \psi_1; \tag{282}$$

whence by (87), (89), and (91)

$$\epsilon = \Sigma_1 \epsilon_1, \qquad \chi = \Sigma_1 \chi_1, \qquad \zeta = \Sigma_1 \zeta_1. \tag{283}$$

The quantities p, η, ψ, ϵ, χ, ζ relating to the gas-mixture may therefore be regarded as consisting of parts which may be attributed to the several components in such a manner that between the parts of these quantities which are assigned to any component, the quantity of that component, the potential for that component, the temperature, and the volume, the same relations shall subsist as if that component existed separately. It is in this sense that we should understand the law of Dalton, that every gas is as a vacuum to every other gas.

It is to be remarked that these relations are consistent and possible for a mixture of gases which are not ideal gases, and indeed without any limitation in regard to the thermodynamic properties of the individual gases. They are all consequences of the law that the pressure in a mixture of different gases is equal to the sum of the pressures of the different gases as existing each by itself at the same temperature and with the same value of its potential. For let p_1, η_1, ϵ_1, ψ_1, χ_1, ζ_1; p_2, etc.; etc. be defined as relating to the different gases existing each by itself with the same volume, temperature, and potential as in the gas-mixture; if

$$p = \Sigma_1 p_1,$$

then

$$\left(\frac{dp}{d\mu_1} \right)_{t, \, \mu_2, \, \dots \, \mu_n} = \left(\frac{dp_1}{d\mu_1} \right)_t;$$

and therefore, by (98), the quantity of any component gas G_1 in the gas-mixture, and in the separate gas to which p_1, η_1, etc. relate, is the same and may be denoted by the same symbol m_1. Also

$$\eta = v\left(\frac{dp}{dt}\right)_{\mu_1,\,\ldots\,\mu_n} = v\Sigma_1\left(\frac{dp}{dt}\right)_{\mu_1} = \Sigma_1\eta_1;$$

whence also, by (93)–(96),

$$\epsilon = \Sigma_1\epsilon_1, \qquad \psi = \Sigma_1\psi_1, \qquad \chi = \Sigma_1\chi_1, \qquad \zeta = \Sigma_1\zeta_1.$$

All the same relations will also hold true whenever the value of ψ for the gas-mixture is equal to the sum of the values of this function for the several component gases existing each by itself in the same quantity as in the gas-mixture and with the temperature and volume of the gas-mixture. For if p_1, η_1, ϵ_1, ψ_1, χ_1, ζ_1; p_2, etc.; etc. are defined as relating to the components existing thus by themselves, we shall have

$$\psi = \Sigma_1\psi_1,$$

whence

$$\left(\frac{d\psi}{dm_1}\right)_{t,\,v,\,m} = \left(\frac{d\psi_1}{dm_1}\right)_{t,\,v}.^*$$

Therefore, by (88), the potential μ_1 has the same value in the gas-mixture and in the gas G_1 existing separately as supposed. Moreover,

$$\eta = -\left(\frac{d\psi}{dt}\right)_{v,\,m} = -\Sigma_1\left(\frac{d\psi_1}{dt}\right)_{v,\,m} = \Sigma_1\eta_1,$$

and

$$p = -\left(\frac{d\psi}{dv}\right)_{t,\,m} = -\Sigma_1\left(\frac{d\psi_1}{dv}\right)_{t,\,m} = \Sigma_1 p_1,$$

whence

$$\epsilon = \Sigma_1\epsilon_1, \qquad \chi = \Sigma_1\chi_1, \qquad \zeta = \Sigma_1\zeta_1.$$

Whenever different bodies are combined without communication of work or heat between them and external bodies, the energy of the body formed by the combination is necessarily equal to the sum of the energies of the bodies combined. In the case of ideal gas-mixtures, when the initial temperatures of the gas-masses which are combined are the same (whether these gas-masses are entirely different gases, or gas-mixtures differing only in the proportion of their components), the condition just mentioned can only be satisfied when the temperature of the resultant gas-mixture is also the same. In such combinations, therefore, the final temperature will be the same as the initial.

If we consider a vertical column of an ideal gas-mixture which is

---

* A subscript m after a differential coefficient relating to a body having several independently variable components is used here and elsewhere in this paper to indicate that each of the quantities m_1, m_2, etc., unless its differential occurs in the expression to which the suffix is applied, is to be regarded as constant in the differentiation.

in equilibrium, and denote the densities of one of its components at two different points by γ_1 and γ_1', we shall have by (275) and (234)

$$\frac{\gamma_1}{\gamma_1'} = e^{\frac{\mu_1 - \mu_1'}{a_1 t}} = e^{\frac{g(h'-h)}{a_1 t}}. \tag{284}$$

From this equation, in which we may regard the quantities distinguished by accents as constant, it appears that the relation between the density of any one of the components and the height is not affected by the presence of the other components.

The work obtained or expended in any reversible process of combination or separation of ideal gas-mixtures at constant temperature, or when the temperatures of the initial and final gas-masses and of the only external source of heat or cold which is used are all the same, will be found by taking the difference of the sums of the values of ψ for the initial, and for the final gas-masses. (See pages 89, 90.) It is evident from the form of equation (279) that this work is equal to the sum of the quantities of work which would be obtained or expended in producing in each different component existing separately the same changes of density which that component experiences in the actual process for which the work is sought.*

We will now return to the consideration of the equilibrium of a liquid with the gas which it emits as affected by the presence of different gases, when the gaseous mass in contact with the liquid may be regarded as an ideal gas-mixture.

It may first be observed, that the density of the gas which is emitted by the liquid will not be affected by the presence of other gases which are not absorbed by the liquid, when the liquid is protected in any way from the pressure due to these additional gases. This may be accomplished by separating the liquid and gaseous masses by a diaphragm which is permeable to the liquid. It will then be easy to maintain the liquid at any constant pressure which is not greater than that in the gas. The potential in the liquid for the substance which it yields as gas will then remain constant, and therefore the potential for the same substance in the gas and the density of this substance in the gas and the part of the gaseous pressure due to it will not be affected by the other components of the gas.

But when the gas and liquid meet under ordinary circumstances, i.e., in a free plane surface, the pressure in both is necessarily the same, as also the value of the potential for any common component S_1. Let us suppose the density of an insoluble component of the gas

---

* This result has been given by Lord Rayleigh (*Phil. Mag.*, vol. xlix., 1875, p. 311). It will be observed that equation (279) might be deduced immediately from this principle in connection with equation (260) which expresses the properties ordinarily assumed for perfect gases.

to vary, while the composition of the liquid and the temperature remain unchanged. If we denote the increments of pressure and of the potential for S_1 by dp and $d\mu_1$, we shall have by (272)

$$d\mu_1 = \left(\frac{d\mu_1}{dp}\right)_{t,\,m}^{(\mathrm{L})} dp = \left(\frac{dv}{dm_1}\right)_{t,\,p,\,m}^{(\mathrm{L})} dp,$$

the index (L) denoting that the expressions to which it is affixed refer to the liquid. (Expressions without such an index will refer to the gas alone or to the gas and liquid in common.) Again, since the gas is an ideal gas-mixture, the relation between p_1 and μ_1 is the same as if the component S_1 existed by itself at the same temperature, and therefore by (268)

$$d\mu_1 = a_1 t d \log p_1.$$

Therefore
$$a_1 t d \log p_1 = \left(\frac{dv}{dm_1}\right)_{t,\,p,\,m}^{(\mathrm{L})} dp. \qquad (285)$$

This may be integrated at once if we regard the differential co-efficient in the second member as constant, which will be a very close approximation. We may obtain a result more simple, but not quite so accurate, if we write the equation in the form

$$dp_1 = \gamma_1 \left(\frac{dv}{dm_1}\right)_{t,\,p,\,m}^{(\mathrm{L})} dp, \qquad (286)$$

where γ_1 denotes the density of the component S_1 in the gas, and integrate regarding this quantity also as constant. This will give

$$p_1 - p_1' = \gamma_1 \left(\frac{dv}{dm_1}\right)_{t,\,p,\,m}^{(\mathrm{L})} (p - p'), \qquad (287)$$

where p_1' and p' denote the values of p_1 and p when the insoluble component of the gas is entirely wanting. It will be observed that $p - p'$ is, nearly equal to the pressure of the insoluble component, in the phase of the gas-mixture to which p_1 relates. S_1 is not necessarily the only common component of the gas and liquid. If there are others, we may find the increase of the part of the pressure in the gas-mixture belonging to any one of them by equations differing from the last only in the subscript numerals.

Let us next consider the effect of a gas which is absorbed to some extent, and which must therefore in strictness be regarded as a component of the liquid. We may commence by considering in general the equilibrium of a gas-mixture of two components S_1 and S_2 with a liquid formed of the same components. Using a notation like the previous, we shall have by (98) for constant temperature,

$$dp = \gamma_1 d\mu_1 + \gamma_2 d\mu_2,$$

and
$$dp = \gamma_1^{(\mathrm{L})} d\mu_1 + \gamma_2^{(\mathrm{L})} d\mu_2;$$

whence
$$(\gamma_1^{(\mathrm{L})} - \gamma_1) d\mu_1 = (\gamma_2 - \gamma_2^{(\mathrm{L})}) d\mu_2.$$

Now if the gas is an ideal gas-mixture,

$$d\mu_1 = \frac{a_1 t}{p_1} dp_1 = \frac{dp_1}{\gamma_1}, \quad \text{and} \quad d\mu_2 = \frac{a_2 t}{p_2} dp_2 = \frac{dp_2}{\gamma_2},$$

therefore
$$\left(\frac{\gamma_1^{(L)}}{\gamma_1} - 1\right) dp_1 = \left(1 - \frac{\gamma_2^{(L)}}{\gamma_2}\right) dp_2. \tag{288}$$

We may now suppose that S_1 is the principal component of the liquid, and S_2 is a gas which is absorbed in the liquid to a slight extent. In such cases it is well known that the ratio of the densities of the substance S_2 in the liquid and in the gas is for a given temperature approximately constant. If we denote this constant by A, we shall have

$$\left(\frac{\gamma_1^{(L)}}{\gamma_1} - 1\right) dp_1 = (1 - A) dp_2. \tag{289}$$

It would be easy to integrate this equation regarding γ_1 as variable, but as the variation in the value of p_1 is necessarily very small we shall obtain sufficient accuracy if we regard γ_1 as well as $\gamma_1^{(L)}$ as constant. We shall thus obtain

$$\left(\frac{\gamma_1^{(L)}}{\gamma_1} - 1\right)(p_1 - p_1') = (1 - A) p_2, \tag{290}$$

where p_1' denotes the pressure of the saturated vapor of the pure liquid consisting of S_1. It will be observed that when $A = 1$, the presence of the gas S_2 will not affect the pressure or density of the gas S_1. When $A < 1$, the pressure and density of the gas S_1 are greater than if S_2 were absent, and when $A > 1$, the reverse is true.

The properties of an ideal gas-mixture (according to the definition which we have assumed) when in equilibrium with liquids or solids have been developed at length, because it is only in respect to these properties that there is any variation from the properties usually attributed to perfect gases. As the pressure of a gas saturated with vapor is usually given as a little less than the sum of the pressure of the gas calculated from its density and that of saturated vapor in a space otherwise empty, while our formulæ would make it a little more, when the gas is insoluble, it would appear that in this respect our formulæ are less accurate than the rule which would make the pressure of the gas saturated with vapor equal to the sum of the two pressures mentioned. Yet the reader will observe that the magnitude of the quantities concerned is not such that any stress can be laid upon this circumstance.

It will also be observed that the statement of Dalton's law which we have adopted, while it serves to complete the theory of gas-mixtures (with respect to a certain class of properties), asserts nothing with reference to any solid or liquid bodies. But the common rule that the density of a gas necessary for equilibrium with a solid or liquid is not altered by the presence of a different gas which is not absorbed by the solid or liquid, if construed *strictly*, will involve consequences in regard to solids and liquids which are entirely inadmissible. To show this, we will assume the correctness of the rule mentioned. Let S_1 denote the common component of the gaseous and liquid or solid masses, and S_2 the insoluble gas, and let quantities relating to the gaseous mass be distinguished when necessary by the index (G), and those relating to the liquid or solid by the index (L). Now while the gas is in equilibrium with the liquid or solid, let the quantity which it contains of S_2 receive the increment dm_2, its volume and the quantity which it contains of the other component, as well as the temperature, remaining constant. The potential for S_1 in the gaseous mass will receive the increment

$$\left(\frac{d\mu_1}{dm_2}\right)^{(G)}_{t,\,v,\,m} dm_2$$

and the pressure will receive the increment

$$\left(\frac{dp}{dm_2}\right)^{(G)}_{t,\,v,\,m} dm_2.$$

Now the liquid or solid remaining in equilibrium with the gas must experience the same variations in the values of μ_1 and p. But by (272)

$$\left(\frac{d\mu_1}{dp}\right)^{(L)}_{t,\,m} = \left(\frac{dv}{dm_1}\right)^{(L)}_{t,\,p,\,m}$$

Therefore,
$$\left(\frac{dv}{dm_1}\right)^{(L)}_{t,\,p,\,m} = \frac{\left(\dfrac{d\mu_1}{dm_2}\right)^{(G)}_{t,\,v,\,m,}}{\left(\dfrac{dp}{dm_2}\right)^{(G)}_{t,\,v,\,m}}.$$

It will be observed that the first member of this equation relates solely to the liquid or solid, and the second member solely to the gas. Now we may suppose the same gaseous mass to be capable of equilibrium with several different liquids or solids, and the first member of this equation must therefore have the same value for all such liquids or solids; which is quite inadmissible. In the simplest case, in which the liquid or solid is identical in substance with the vapor which it yields, it is evident that the expression in question denotes the reciprocal of the density of the solid or liquid. Hence, when the gas is in equilibrium with one of its components both in the solid and liquid states (as when a moist gas is in equilibrium with ice and water), it would be necessary that the solid and liquid should have the same density.

The foregoing considerations appear sufficient to justify the definition of an ideal gas-mixture which we have chosen. It is of course immaterial whether we regard the definition as expressed by equation (273), or by (279), or by any other fundamental equation which can be derived from these.

The fundamental equations for an ideal gas-mixture corresponding to (255), (265), and (271) may easily be derived from these equations by using inversely the substitutions given on page 156. They are

$$\Sigma_1(c_1 m_1) \log \frac{\epsilon - \Sigma_1(E_1 m_1)}{\Sigma_1(c_1 m_1)} = \eta + \Sigma_1(a_1 m_1 \log \frac{m_1}{v} - H_1 m_1), \tag{291}$$

$$\Sigma_1(c_1 m_1 + a_1 m_1) \log \frac{\chi - \Sigma_1(E_1 m_1)}{\Sigma_1(c_1 m_1 + a_1 m_1)}$$

$$= \eta + \Sigma_1 \left(a_1 m_1 \log \frac{p m_1}{\Sigma_1(a_1 m_1)} - H_1 m_1 \right), \tag{292}$$

$$\zeta = \Sigma_1 \left(E_1 m_1 + m_1 t (c_1 + a_1 - H_1) \right)$$

$$- \Sigma_1(c_1 m_1 + a_1 m_1) t \log t + \Sigma_1 \left(a_1 m_1 t \log \frac{p m_1}{\Sigma_1(a_1 m_1)} \right). \tag{293}$$

The components to which the fundamental equations (273), (279), (291), (292), (293) refer, may themselves be gas-mixtures. We may for example apply the fundamental equations of a binary gas-mixture to a mixture of hydrogen and air, or to any ternary gas-mixture in which the proportion of two of the components is fixed. In fact, the form of equation (279) which applies to a gas-mixture of any particular number of components may easily be reduced, when the proportions of some of these components are fixed, to the form which applies to a gas-mixture of a smaller number of components. The necessary substitutions will be analogous to those given on page 156. But the components must be entirely different from one another with respect to the gases of which they are formed by mixture. We cannot, for example, apply equation (279) to a gas-mixture in which the components are oxygen and air. It would indeed be easy to form a fundamental equation for such a gas-mixture with reference to the designated gases as components. Such an equation might be derived from (279) by the proper substitutions. But the result would be an equation of more complexity than (279). A *chemical* compound, however, with respect to Dalton's law, and with respect to all the equations which have been given, is to be regarded as entirely different from its components. Thus, a mixture of hydrogen, oxygen, and vapor of water is to be regarded as a ternary gas-mixture, having the three components mentioned. This is certainly true when the quantities of the compound gas and of its components are all independently variable in the gas-mixture, without change of temperature or pressure. Cases in which these quantities are not thus independently variable will be considered hereafter.

Inferences in regard to Potentials in Liquids and Solids.

Such equations as (264), (268), (276), by which the values of potentials in pure or mixed gases may be derived from quantities capable of direct measurement, have an interest which is not confined to the theory of gases. For as the potentials of the independently variable components which are common to coexistent liquid and gaseous masses have the same values in each, these expressions will generally afford the means of determining for liquids, at least approximately, the potential for any independently variable component which is capable of existing in the gaseous state. For although every state of a liquid is not such as can exist in contact with a gaseous mass, it will always be possible, when any of the components of the liquid are volatile, to bring it by a change of pressure alone, its temperature and composition remaining unchanged, to a state for which there is a coexistent phase of vapor, in which the values of the potentials of the volatile components of the liquid may be estimated from the density of these substances in the vapor. The variations of the potentials in the liquid due to the change of pressure will in general be quite trifling as compared with the variations which are connected with changes of temperature or of composition, and may moreover be readily estimated by means of equation (272). The same considerations will apply to volatile solids with respect to the determination of the potential for the substance of the solid.

As an application of this method of determining the potentials in liquids, let us make use of the law of Henry in regard to the absorption of gases by liquids to determine the relation between the quantity of the gas contained in any liquid mass and its potential. Let us consider the liquid as in equilibrium with the gas, and let $m_1^{(G)}$ denote the quantity of the gas existing as such, $m_1^{(L)}$ the quantity of the same substance contained in the liquid mass, μ_1 the potential for this substance common to the gas and liquid, $v^{(G)}$ and $v^{(L)}$ the volumes of the gas and liquid. When the absorbed gas forms but a very small part of the liquid mass, we have by Henry's law

$$\frac{m_1^{(L)}}{v^{(L)}} = A \frac{m_1^{(G)}}{v^{(G)}}, \tag{294}$$

where A is a function of the temperature; and by (276)

$$\mu_1 = B + C \log \frac{m_1^{(G)}}{v^{(G)}}, \tag{295}$$

B and C also denoting functions of the temperature. Therefore

$$\mu_1 = B + C \log \frac{m_1^{(L)}}{A v^{(L)}}. \tag{296}$$

It will be seen (if we disregard the difference of notation) that this equation is equivalent in form to (216), which was deduced from *a priori* considerations as a probable relation between the quantity and the potential of a small component. When a liquid absorbs several gases at once, there will be several equations of the form of (296), which will hold true simultaneously, and which we may regard as equivalent to equations (217), (218). The quantities A and C in (216), with the corresponding quantities in (217), (218), were regarded as functions of the temperature and pressure, but since the potentials in liquids are but little affected by the pressure, we might anticipate that these quantities in the case of liquids might be regarded as functions of the temperature alone.

In regard to equations (216), (217), (218), we may now observe that by (264) and (276) they are shown to hold true in ideal gases or gas-mixtures, not only for components which form only a small part of the whole gas-mixture, but without any such limitation, and not only approximately but absolutely. It is noticeable that in this case quantities A and C are functions of the temperature alone, and do not even depend upon the nature of the gaseous mass, except upon the particular component to which they relate. As all gaseous bodies are generally supposed to approximate to the laws of ideal gases when sufficiently rarefied, we may regard these equations as approximately valid for gaseous bodies in general when the density is sufficiently small. When the density of the gaseous mass is very great, but the separate density of the component in question is small, the equations will probably hold true, but the values of A and C may not be entirely independent of the pressure, or of the composition of the mass in respect to its principal components. These equations will also apply, as we have just seen, to the potentials in liquid bodies for components of which the density in the liquid is very small, whenever these components exist also in the gaseous state, and conform to the law of Henry. This seems to indicate that the law expressed by these equations has a very general application.

Considerations relating to the Increase of Entropy due to the Mixture of Gases by Diffusion.

From equations (278) we may easily calculate the increase of entropy which takes place when two different gases are mixed by diffusion, at a constant temperature and pressure. Let us suppose that the quantities of the gases are such that each occupies initially one half of the total volume. If we denote this volume by V, the increase of entropy will be

$$m_1a_1 \log V + m_2a_2 \log V - m_1a_1 \log \frac{V}{2} - m_2a_2 \log \frac{V}{2},$$

or

$$(m_1a_1 + m_2a_2) \log 2.$$

Now

$$m_1a_1 = \frac{pV}{2t}, \quad \text{and} \quad m_2a_2 = \frac{pV}{2t}.$$

Therefore the increase of entropy may be represented by the expression

$$\frac{pV}{t} \log 2. \tag{297}$$

It is noticeable that the value of this expression does not depend upon the kinds of gas which are concerned, if the quantities are such as has been supposed, except that the gases which are mixed must be of different kinds. If we should bring into contact two masses of the same kind of gas, they would also mix, but there would be no increase of entropy. But in regard to the relation which this case bears to the preceding, we must bear in mind the following considerations. When we say that when two different gases mix by diffusion, as we have supposed, the energy of the whole remains constant, and the entropy receives a certain increase, we mean that the gases could be separated and brought to the same volume and temperature which they had at first by means of certain changes in external bodies, for example, by the passage of a certain amount of heat from a warmer to a colder body. But when we say that when two gas-masses of the same kind are mixed under similar circumstances there is no change of energy or entropy, we do not mean that the gases which have been mixed can be separated without change to external bodies. On the contrary, the separation of the gases is entirely impossible. We call the energy and entropy of the gas-masses when mixed the same as when they were unmixed, because we do not recognize any difference in the substance of the two masses. So when gases of different kinds are mixed, if we ask what changes in external bodies are necessary to bring the system to its original state, we do not mean a state in which each particle shall occupy more or less exactly the same position as at some previous epoch, but only a state which shall be undistinguishable from the previous one in its sensible properties. It is to states of systems thus incompletely defined that the problems of thermodynamics relate.

But if such considerations explain why the mixture of gas-masses of the same kind stands on a different footing from the mixture of gas-masses of different kinds, the fact is not less significant that the increase of entropy due to the mixture of gases of different kinds, in such a case as we have supposed, is independent of the nature of the gases.

Now we may without violence to the general laws of gases which are embodied in our equations suppose other gases to exist than such as actually do exist, and there does not appear to be any limit to the resemblance which there might be between two such kinds of gas. But the increase of entropy due to the mixing of given volumes of the gases at a given temperature and pressure would be independent of the degree of similarity or dissimilarity between them. We might also imagine the case of two gases which should be absolutely identical in all the properties (sensible and molecular) which come into play while they exist as gases either pure or mixed with each other, but which should differ in respect to the attractions between their atoms and the atoms of some other substances, and therefore in their tendency to combine with such substances. In the mixture of such gases by diffusion an increase of entropy would take place, although the process of mixture, dynamically considered, might be absolutely identical in its minutest details (even with respect to the precise path of each atom) with processes which might take place without any increase of entropy. In such respects, entropy stands strongly contrasted with energy. Again, when such gases have been mixed, there is no more impossibility of the separation of the two kinds of molecules in virtue of their ordinary motions in the gaseous mass without any especial external influence, than there is of the separation of a homogeneous gas into the same two parts into which it has once been divided, after these have once been mixed. In other words, the impossibility of an uncompensated decrease of entropy seems to be reduced to improbability.

There is perhaps no fact in the molecular theory of gases so well established as that the number of molecules in a given volume at a given temperature and pressure is the same for every kind of gas when in a state to which the laws of ideal gases apply. Hence the quantity $\dfrac{pV}{t}$ in (297) must be entirely determined by the number of molecules which are mixed. And the increase of entropy is therefore determined by the number of these molecules and is independent of their dynamical condition and of the degree of difference between them.

The result is of the same nature when the volumes of the gases which are mixed are not equal, and when more than two kinds of gas are mixed. If we denote by v_1, v_2, etc., the initial volumes of the different kinds of gas, and by V as before the total volume, the increase of entropy may be written in the form

$$\Sigma_1(m_1 a_1) \log V - \Sigma_1(m_1 a_1 \log v_1).$$

And if we denote by r_1, r_2, etc., the numbers of the molecules of the several different kinds of gas, we shall have

$$r_1 = C m_1 a_1, \quad r_2 = C m_2 a_2, \quad \text{etc.},$$

where C denotes a constant. Hence

$$v_1 : V :: m_1 a_1 : \Sigma_1(m_1 a_1) :: r_1 : \Sigma_1 r_1 ;$$

and the increase of entropy may be written

$$\frac{\Sigma_1 r_1 \log \Sigma_1 r_1 - \Sigma_1(r_1 \log r_1)}{C}. \tag{298}$$

The Phases of Dissipated Energy of an Ideal Gas-mixture with Components which are Chemically Related.

We will now pass to the consideration of the phases of dissipated energy (see page 140) of an ideal gas-mixture, in which the number of the proximate components exceeds that of the ultimate.

Let us first suppose that an ideal gas-mixture has for proximate components the gases G_1, G_2, and G_3, the units of which are denoted by $\mathfrak{G}_1$, $\mathfrak{G}_2$, $\mathfrak{G}_3$, and that in ultimate analysis

$$\mathfrak{G}_3 = \lambda_1 \mathfrak{G}_1 + \lambda_2 \mathfrak{G}_2, \tag{299}$$

λ_1 and λ_2 denoting positive constants, such that $\lambda_1 + \lambda_2 = 1$. The phases which we are to consider are those for which the energy of the gas-mixture is a minimum for constant entropy and volume and constant quantities of G_1 and G_2, as determined in ultimate analysis. For such phases, by (86),

$$\mu_1 \delta m_1 + \mu_2 \delta m_2 + \mu_3 \delta m_3 \geqq 0 \tag{300}$$

for such values of the variations as do not affect the quantities of G_1 and G_2 as determined in ultimate analysis. Values of δm_1, δm_2, δm_3 proportional to λ_1, λ_2, -1, and only such, are evidently consistent with this restriction: therefore

$$\lambda_1 \mu_1 + \lambda_2 \mu_2 = \mu_3. \tag{301}$$

If we substitute in this equation values of μ_1, μ_2, μ_3 taken from (276), we obtain, after arranging the terms and dividing by t,

$$\lambda_1 a_1 \log \frac{m_1}{v} + \lambda_2 a_2 \log \frac{m_2}{v} - a_3 \log \frac{m_3}{v} = A + B \log t - \frac{C}{t}, \tag{302}$$

where

$$A = \lambda_1 H_1 + \lambda_2 H_2 - H_3 - \lambda_1 c_1 - \lambda_2 c_2 + c_3 - \lambda_1 a_1 - \lambda_2 a_2 + a_3, \tag{303}$$

$$B = \lambda_1 c_1 + \lambda_2 c_2 - c_3, \tag{304}$$

$$C = \lambda_1 E_1 + \lambda_2 E_2 - E_3. \tag{305}$$

If we denote by β_1 and β_2 the volumes (determined under standard conditions of temperature and pressure) of the quantities of the gases G_1 and G_2 which are contained in a unit of volume of the gas G_3, we shall have

$$\beta_1 = \frac{\lambda_1 a_1}{a_3}, \quad \text{and} \quad \beta_2 = \frac{\lambda_2 a_2}{a_3}, \tag{306}$$

and (302) will reduce to the form

$$\log \frac{m_1{}^{\beta_1} m_2{}^{\beta_2}}{m_3 v^{\beta_1 + \beta_2 - 1}} = \frac{A}{a_3} + \frac{B}{a_3} \log t - \frac{C}{a_3 t}. \tag{307}$$

Moreover, as by (277)

$$pv = (a_1 m_1 + a_2 m_2 + a_3 m_3)t, \tag{308}$$

we have on eliminating v

$$\log \frac{m_1{}^{\beta_1} m_2{}^{\beta_2} p^{\beta_1 + \beta_2 - 1}}{m_3 (a_1 m_1 + a_2 m_2 + a_3 m_3)^{\beta_1 + \beta_2 - 1}} = \frac{A}{a_3} + \frac{B'}{a_3} \log t - \frac{C}{a_3 t}, \tag{309}$$

where

$$B' = \lambda_1 c_1 + \lambda_2 c_2 - c_3 + \lambda_1 a_1 + \lambda_2 a_2 - a_3. \tag{310}$$

It will be observed that the quantities β_1, β_2 will always be positive and have a simple relation to unity, and that the value of $\beta_1 + \beta_2 - 1$ will be positive or zero, according as gas G_3 is formed of G_1 and G_2 with or without condensation. If we should assume, according to the rule often given for the specific heat of compound gases, that the thermal capacity at constant volume of any quantity of the gas G_3 is equal to the sum of the thermal capacities of the quantities which it contains of the gases G_1 and G_2, the value of B would be zero. The heat evolved in the formation of a unit of the gas G_3 out of the gases G_1 and G_2, without mechanical action, is by (283) and (257)

$$\lambda_1(c_1 t + E_1) + \lambda_2(c_2 t + E_2) - (c_3 t + E_3),$$

or

$$Bt + C,$$

which will reduce to C when the above relation in regard to the specific heats is satisfied. In any case the quantity of heat thus evolved divided by $a_3 t^2$ will be equal to the differential coefficient of the second member of equation (307) with respect to t. Moreover, the heat evolved in the formation of a unit of the gas G_3 out of the gases G_1 and G_2 under constant pressure is

$$Bt + C + \lambda_1 a_1 t + \lambda_2 a_2 t - a_3 t = B't + C,$$

which is equal to the differential coefficient of the second member of (309) with respect to t, multiplied by $a_3 t^2$.

It appears by (307) that, except in the case when $\beta_1 + \beta_2 = 1$, for any given finite values of m_1, m_2, m_3, and t (infinitesimal values being excluded as well as infinite), it will always be possible to assign such a finite value to v that the mixture shall be in a state

of dissipated energy.   Thus, if we regard a mixture of hydrogen
oxygen, and vapor of water as an ideal gas-mixture, for a mixture
containing any given quantities of these three gases at any given
temperature there will be a certain volume at which the mixture will be
in a state of dissipated energy.   In such a state no such phenomenon
as explosion will be possible, and no formation of water by the action
of platinum.   (If the mass should be expanded beyond this volume,
the only possible action of a catalytic agent would be to resolve the
water into its components.)   It may indeed be true that at ordinary
temperatures, except when the quantity either of hydrogen or of
oxygen is very small compared with the quantity of water, the state
of dissipated energy is one of such extreme rarefaction as to lie
entirely beyond our power of experimental verification.   It is also to
be noticed that a state of great rarefaction is so unfavorable to any
condensation of the gases, that it is quite probable that the catalytic
action of platinum may cease entirely at a degree of rarefaction far
short of what is necessary for a state of dissipated energy.   But with
respect to the theoretical demonstration, such states of great rarefaction are precisely those to which we should suppose that the laws of
ideal gas-mixtures would apply most perfectly.

But when the compound gas G_3 is formed of G_1 and G_2 without
condensation (i.e., when $\beta_1 + \beta_2 = 1$), it appears from equation (307)
that the relation between m_1, m_2, and m_3 which is necessary for a
phase of dissipated energy is determined by the temperature alone.

In any case, if we regard the total quantities of the gases G_1 and
G_2 (as determined by the ultimate analysis of the gas-mixture), and
also the volume, as constant, the quantities of these gases which
appear uncombined in a phase of dissipated energy will increase with
the temperature, if the formation of the compound G_3 *without
change of volume* is attended with evolution of heat.   Also, if we
regard the total quantities of the gases G_1 and G_2, and also the
pressure, as constant, the quantities of these gases which appear uncombined in a phase of dissipated energy, will increase with the
temperature, if the formation of the compound G_3 *under constant
pressure* is attended with evolution of heat.   If $B = 0$ (a case, as
has been seen, of especial importance), the heat obtained by the
formation of a unit of G_3 out of G_1 and G_2 without change of volume
or of temperature will be equal to C.   If this quantity is positive,
and the total quantities of the gases G_1 and G_2 and also the volume
have given finite values, for an infinitesimal value of t we shall have
(for a phase of dissipated energy) an infinitesimal value either of m_1
or of m_2, and for an infinite value of t we shall have finite (neither infinitesimal nor infinite) values of m_1, m_2, and m_3.   But if we suppose
the pressure instead of the volume to have a given finite value (with

suppositions otherwise the same), we shall have for infinitesimal values of t an infinitesimal value either of m_1 or m_2, and for infinite values of t finite or infinitesimal values of m_3 according as $\beta_1 + \beta_2$ is equal to or greater than unity.

The case which we have considered is that of a ternary gas-mixture, but our results may easily be generalized in this respect. In fact, whatever the number of component gases in a gas-mixture, if there are relations of equivalence in ultimate analysis between these components, such relations may be expressed by one or more equations of the form

$$\lambda_1 \mathfrak{G}_1 + \lambda_2 \mathfrak{G}_2 + \lambda_3 \mathfrak{G}_3 + \text{etc.} = 0, \tag{311}$$

where $\mathfrak{G}_1$, $\mathfrak{G}_2$, etc. denote the units of the various component gases, and λ_1, λ_2, etc. denote positive or negative constants such that $\Sigma_1 \lambda_1 = 0$. From (311) with (86) we may derive for phases of dissipated energy,

$$\lambda_1 \mu_1 + \lambda_2 \mu_2 + \lambda_3 \mu_3 + \text{etc.} = 0,$$

or

$$\Sigma_1 (\lambda_1 \mu_1) = 0. \tag{312}$$

Hence, by (276),

$$\Sigma_1 \left(\lambda_1 a_1 \log \frac{m_1}{v} \right) = A + B \log t - \frac{C}{t}, \tag{313}$$

where A, B and C are constants determined by the equations

$$A = \Sigma_1 (\lambda_1 H_1 - \lambda_1 c_1 - \lambda_1 a_1), \tag{314}$$

$$B = \Sigma_1 (\lambda_1 c_1), \tag{315}$$

$$C = \Sigma_1 (\lambda_1 E_1). \tag{316}$$

Also, since $\quad pv = \Sigma_1 (a_1 m_1) t,$

$$\Sigma_1 (\lambda_1 a_1 \log m_1) - \Sigma (\lambda_1 a_1) \log \Sigma_1 (a_1 m_1)$$
$$+ \Sigma (\lambda_1 a_1) \log p = A + B' \log t - \frac{C}{t}, \tag{317}$$

where

$$B' = \Sigma_1 (\lambda_1 c_1 + \lambda_1 a_1). \tag{318}$$

If there is more than one equation of the form (311), we shall have more than one of each of the forms (313) and (317), which will hold true simultaneously for phases of dissipated energy.

It will be observed that the relations necessary for a phase of dissipated energy between the volume and temperature of an ideal gas-mixture, and the quantities of the components which take part in the chemical processes, and the pressure due to these components, are not affected by the presence of neutral gases in the gas-mixture.

From equations (312) and (234) it follows that if there is a phase of dissipated energy at any point in an ideal gas-mixture in equilibrium under the influence of gravity, the whole gas-mixture must consist of such phases.

The equations of the phases of dissipated energy of a binary gas-mixture, the components of which are identical in substance, are comparatively simple in form. In this case the two components have the same potential, and if we write β for $\dfrac{a_1}{a_2}$ (the ratio of the volumes of equal quantities of the two components under the same conditions of temperature and pressure), we shall have

$$\log \frac{m_1{}^\beta}{m_2 v^{\beta-1}} = \frac{A}{a_2} + \frac{B}{a_2} \log t - \frac{C}{a_2 t}, \tag{319}$$

$$\log \frac{m_1{}^\beta p^{\beta-1}}{m_2 (a_1 m_1 + a_2 m_2)^{\beta-1}} = \frac{A}{a_2} + \frac{B'}{a_2} \log t - \frac{C}{a_2 t}; \tag{320}$$

where

$$A = H_1 - H_2 - c_1 + c_2 - a_1 + a_2, \tag{321}$$

$$B = c_1 - c_2, \qquad B' = c_1 - c_2 + a_1 - a_2, \tag{322}$$

$$C = E_1 - E_2. \tag{323}$$

Gas-mixtures with Convertible Components.

The equations of the phases of dissipated energy of ideal gas-mixtures which have components of which some are identical in ultimate analysis to others have an especial interest in relation to the theory of gas-mixtures in which the components are not only thus equivalent, but are actually transformed into each other within the gas-mixture on variations of temperature and pressure, so that quantities of these (proximate) components are entirely determined, at least in any permanent phase of the gas-mixture, by the quantities of a smaller number of ultimate components, with the temperature and pressure. Such gas-mixtures may be distinguished as having *convertible components*. The very general considerations adduced on pages 138–144, which are not limited in their application to gaseous bodies, suggest the hypothesis that the equations of the phases of dissipated energy of ideal gas-mixtures may apply to such gas-mixtures as have been described. It will, however, be desirable to consider the matter more in detail.

In the first place, if we consider the case of a gas-mixture which only differs from an ordinary ideal gas-mixture for which some of the components are equivalent in that there is perfect freedom in regard to the transformation of these components, it follows at once from the general formula of equilibrium (1) or (2) that equilibrium is only possible for such phases as we have called phases of dissipated energy, for which some of the characteristic equations have been deduced in the preceding pages.

If it should be urged, that regarding a gas-mixture which has convertible components as an ideal gas-mixture of which, for some reason, only a part of the phases are actually capable of existing, we might still suppose the particular phases which alone can exist to be determined by some other principle than that of the free convertibility of the components (as if, perhaps, the case were analogous to one of *constraint* in mechanics), it may easily be shown that such a hypothesis is entirely untenable, when the quantities of the proximate components may be varied independently by suitable variations of the temperature and pressure, and of the quantities of the ultimate components, and it is admitted that the relations between the energy, entropy, volume, temperature, pressure, and the quantities of the several proximate components in the gas-mixture are the same as for an ordinary ideal gas-mixture, in which the components are not convertible. Let us denote the quantities of the n' proximate components of a gas-mixture A by m_1, m_2, etc., and the quantities of its n ultimate components by $\mathbf{m}_1$, $\mathbf{m}_2$, etc. (n denoting a number less than n'), and let us suppose that for this gas-mixture the quantities ϵ, η, v, t, p, m_1, m_2, etc. satisfy the relations characteristic of an ideal gas-mixture, while the phase of the gas-mixture is entirely determined by the values of $\mathbf{m}_1$, $\mathbf{m}_2$, etc., with two of the quantities ϵ, η, v, t, p. We may evidently imagine such an ideal gas-mixture B having n' components (not convertible), that every phase of A shall correspond with one of B in the values of ϵ, η, v, t, p, m_1, m_2, etc. Now let us give to the quantities $\mathbf{m}_1$, $\mathbf{m}_2$, etc. in the gas-mixture A any fixed values, and for the body thus defined let us imagine the v-η-ϵ surface (see page 116) constructed; likewise for the ideal gas-mixture B let us imagine the v-η-ϵ surface constructed for every set of values of m_1, m_2, etc. which is consistent with the given values of $\mathbf{m}_1$, $\mathbf{m}_2$, etc., i.e., for every body of which the *ultimate* composition would be expressed by the given values of $\mathbf{m}_1$, $\mathbf{m}_2$, etc. It follows immediately from our supposition, that every point in the v-η-ϵ surface relating to A must coincide with some point of one of the v-η-ϵ surfaces relating to B not only in respect to position but also in respect to its tangent plane (which represents temperature and pressure); therefore the v-η-ϵ surface relating to A must be tangent to the various v-η-ϵ surfaces relating to B, and therefore must be an envelop of these surfaces. From this it follows that the points which represent phases common to both gas-mixtures must represent the phases of dissipated energy of the gas-mixture B.

The properties of an ideal gas-mixture which are assumed in regard to the gas-mixture of convertible components in the above demonstration are expressed by equations (277) and (278) with the equation

$$\epsilon = \Sigma_1(c_1 m_1 t + m_1 E_1). \tag{324}$$

It is usual to assume in regard to gas-mixtures having convertible components that the convertibility of the components does not affect the relations (277) and (324). The same cannot be said of the equation (278). But in a very important class of cases it will be sufficient if the applicability of (277) and (324) is admitted. The cases referred to are those in which in certain phases of a gas-mixture the components are convertible, and in other phases of the same proximate composition the components are not convertible, and the equations of an ideal gas-mixture hold true.

If there is only a single degree of convertibility between the components (i.e., if only a single kind of conversion, with its reverse, can take place among the components), it will be sufficient to assume, in regard to the phases in which conversion takes place, the validity of equation (277) and of the following, which can be derived from (324) by differentiation, and comparison with equation (11), which expresses a necessary relation,

$$[t \, d\eta - p \, dv - \Sigma_1(c_1 m_1) \, dt]_m = 0.^* \qquad (325)$$

We shall confine our demonstration to this case. It will be observed that the physical signification of (325) is that if the gas-mixture is subjected to such changes of volume and temperature as do not alter its proximate composition, the heat absorbed or yielded may be calculated by the same formula as if the components were not convertible.

Let us suppose the thermodynamic state of a gaseous mass M, of such a kind as has just been described, to be varied while within the limits within which the components are not convertible. (The quantities of the proximate components, therefore, as well as of the ultimate, are supposed constant.) If we use the same method of geometrical representation as before, the point representing the volume, entropy, and energy of the mass will describe a line in the v-η-ϵ surface of an ideal gas-mixture of inconvertible components, the form and position of this surface being determined by the proximate composition of M. Let us now suppose the same mass to be carried beyond the limit of inconvertibility, the variations of state after passing the limit being such as not to alter its proximate composition. It is evident that this will in general be possible. Exceptions can only occur when the limit is formed by phases in which the proximate composition is uniform. The line traced in the region of convertibility must belong to the same v-η-ϵ surface of an ideal gas-mixture of inconvertible components as before, continued beyond the limit of inconvertibility for the components of M, since the variations of volume, entropy, and energy are the same as would be possible if the components were not convertible. But it must also belong to the v-η-ϵ surface of the body M, which is here a gas-mixture of con-

---

* This notation is intended to indicate that m_1, m_2, etc. are regarded as constant.

vertible components. Moreover, as the inclination of each of these surfaces must indicate the temperature and pressure of the phases through which the body passes, these two surfaces must be tangent to each other along the line which has been traced. As the v-η-ϵ surface of the body M in the region of convertibility must thus be tangent to all the surfaces representing ideal gas-mixtures of every possible proximate composition consistent with the ultimate composition of M, continued beyond the region of inconvertibility, in which alone their form and position may be capable of experimental demonstration, the former surface must be an envelop of the latter surfaces, and therefore a continuation of the surface of the phases of dissipated energy in the region of inconvertibility.

The foregoing considerations may give a measure of *a priori* probability to the results which are obtained by applying the ordinary laws of ideal gas-mixtures to cases in which the components are convertible. It is only by experiments upon gases in phases in which their components are convertible that the validity of any of these results can be established.

The very accurate determinations of density which have been made for the peroxide of nitrogen enable us to subject some of our equations to a very critical test. That this substance in the gaseous state is properly regarded as a mixture of different gases can hardly be doubted, as the proportion of the components derived from its density on the supposition that one component has the molecular formula NO_2 and the other the formula N_2O_4 is the same as that derived from the depth of the color on the supposition that the absorption of light is due to one of the components alone, and is proportioned to the separate density of that component.*

MM. Sainte-Claire Deville and Troost † have given a series of determinations of what we shall call the *relative densities* of peroxide of nitrogen at various temperatures under atmospheric pressure. We use the term *relative density* to denote what it is usual in treatises on chemistry to denote by the term *density*, viz., the actual density of a gas divided by the density of a standard perfect gas at the same pressure and temperature, the standard gas being air, or more strictly, an ideal gas which has the same density as air at the zero of the centigrade scale and the pressure of one atmosphere. In order to test our equations by these determinations, it will be convenient to transform equation (320), so as to give directly the relation between the relative density, the pressure, and the temperature.

As the density of the standard gas at any given temperature and

---

* Salet, "Sur la coloration du peroxyde d'azote," *Comptes Rendus*, vol. lxvii. p. 488.

† *Comptes Rendus*, vol. lxiv. p. 237.

pressure may by (263) be expressed by the formula $\dfrac{p}{a_s t}$, the relative density of a binary gas-mixture may be expressed by

$$D = (m_1 + m_2)\frac{a_s t}{pv}. \tag{326}$$

Now by (263)
$$a_1 m_1 + a_2 m_2 = \frac{pv}{t}. \tag{327}$$

By giving to m_2 and m_1 successively the value zero in these equations, we obtain

$$D_1 = \frac{a_s}{a_1}, \qquad D_2 = \frac{a_s}{a_2}, \tag{328}$$

where D_1 and D_2 denote the values of D when the gas consists wholly of one or of the other component. If we assume that

$$D_2 = 2D_1, \tag{329}$$

we shall have
$$a_1 = 2a_2. \tag{330}$$

From (326) we have
$$m_1 + m_2 = D\frac{pv}{a_s t},$$

and from (327), by (328) and (330),

$$2m_1 + m_2 = D_2\frac{pv}{a_s t} = 2D_1\frac{pv}{a_s t},$$

whence
$$m_1 = (D_2 - D)\frac{pv}{a_s t}, \tag{331}$$

$$m_2 = 2(D - D_1)\frac{pv}{a_s t}. \tag{332}$$

By (327), (331), and (332) we obtain from (320)

$$\log\frac{(D_2 - D)^2 p}{2(D - D_1)a_s} = \frac{A}{a_2} + \frac{B'}{a_2}\log t - \frac{C}{a_2 t}. \tag{333}$$

This formula will be more convenient for purposes of calculation if we introduce common logarithms (denoted by $\log_{10}$) instead of hyperbolic, the temperature of the ordinary centigrade scale t_c instead of the absolute temperature t, and the pressure in atmospheres p_{at} instead of p the pressure in a rational system of units. If we also add the logarithm of a_s to both sides of the equation, we obtain

$$\log_{10}\frac{(D_2 - D)^2 p_{at}}{2(D - D_1)} = \mathbf{A} + \frac{B'}{a_2}\log_{10}(t_c + 273) - \frac{\mathbf{C}}{t_c + 273}, \tag{334}$$

where $\mathbf{A}$ and $\mathbf{C}$ denote constants, the values of which are closely connected with those of A and C.

From the molecular formulæ of peroxide of nitrogen NO_2 and N_2O_4, we may calculate the relative densities

$$D_1 = \frac{14 + 32}{2}\cdot 0691 = 1\cdot589, \quad \text{and} \quad D_2 = \frac{28 + 64}{2}\cdot 0691 = 3\cdot178. \tag{335}$$

The determinations of MM. Deville and Troost are satisfactorily represented by the equation

$$\log_{10}\frac{(3\cdot178-D)^2 p_{at}}{2(D-1\cdot589)}=9\cdot47056-\frac{3118\cdot6}{t_c+273},\qquad(336)$$

which gives $\qquad D=3\cdot178+\Theta-\sqrt{\Theta(3\cdot178+\Theta)}$,

where $\qquad \log_{10}\Theta=9\cdot47056-\dfrac{3118\cdot6}{t_c+273}-\log_{10}p_{at}$.

In the first part of the following table are given in successive columns the temperature and pressure of the gas in the several experiments of MM. Deville and Troost, the relative densities calculated from these numbers by equation (336), the relative densities as observed, and the difference of the observed and calculated relative densities. It will be observed that these differences are quite small, in no case reaching ·03, and on the average scarcely exceeding ·01. The significance of such correspondence in favour of the hypothesis by means of which equation (336) has been established is of course diminished by the fact that two constants in the equation have been determined from these experiments. If the same equation can be shown to give correctly the relative densities at other pressures than that for which the constants have been determined, such correspondence will be much more decisive.

t_c	p_{at}	D calculated by eq. (336).	D observed.	diff.	Observers.
26·7	1	2·676	2·65	− ·026	D. & T.
35·4	1	2·524	2·53	+ ·006	D. & T.
39·8	1	2·443	2·46	+ ·017	D. & T.
49·6	1	2·256	2·27	+ ·014	D. & T.
60·2	1	2·067	2·08	+ ·013	D. & T.
70·0	1	1·920	1·92	·000	D. & T.
80·6	1	1·801	1·80	− ·001	D. & T.
90·0	1	1·728	1·72	− ·008	D. & T.
100·1	1	1·676	1·68	+ ·004	D. & T.
111·3	1	1·641	1·65	+ ·009	D. & T.
121·5	1	1·622	1·62	− ·002	D. & T.
135·0	1	1·607	1·60	− ·007	D. & T.
154·0	1	1·597	1·58	− ·017	D. & T.
183·2	1	1·592	1·57	− ·022	D. & T.
97·5	1	1·687			
97·5	$\frac{19459}{26367}$	1·631	1·783	+ ·152	P. & W.
24·5	1	2·711			
24·5	$\frac{18090}{42529}$	2·524	2·52	− ·004	P. & W.
11·3	1	2·891			
11·3	$\frac{9265}{44205}$	2·620	2·645	+ ·025	P. & W.
4·2	1	2·964			
4·2	$\frac{6023}{35438}$	2·708	2·588	− ·120	P. & W.

Messrs. Playfair and Wanklyn have published[*] four determinations of the relative density of peroxide of nitrogen at various temperatures

---

[*] *Transactions of the Royal Society of Edinburgh*, vol. xxii. p. 441.

when diluted with nitrogen.   Since the relations expressed by equations (319) and (320) are not affected by the presence of a third gas
which is different from the gases G_1 and G_2 (to which m_1 and m_2
relate) and neutral to them (see the remark at the foot of page 171),
—provided that we take p to denote the pressure which we attribute
to the gases G_1 and G_2, i.e., the total pressure diminished by the
pressure which the third gas would exert if occupying alone the
same space at the same temperature,—it follows that the relations
expressed for peroxide of nitrogen by (333), (334), and (336) will
not be affected by the presence of free nitrogen, if the pressure
expressed by p or p_{at} and contained implicitly in the symbol D (see
equation (326) by which D is defined) is understood to denote the
total pressure diminished by the pressure due to the free nitrogen.
The determinations of Playfair and Wanklyn are given in the latter
part of the above table.   The pressures given are those obtained by
subtracting the pressure due to the free nitrogen from the total
pressure.   We may suppose such reduced pressures to have been
used in the reduction of the observations by which the numbers
in the column of observed relative densities were obtained.   Besides
the relative densities calculated by equation (336) for the temperatures
and (reduced) pressures of the observations, the table contains the
relative densities calculated for the same temperatures and the pressure
of one atmosphere.

The reader will observe that in the second and third experiments
of Playfair and Wanklyn there is a very close accordance between
the calculated and observed values of D, while in the second and
fourth experiments there is a considerable difference.   Now the weight
to be attributed to the several determinations is very different.   The
quantities of peroxide of nitrogen which were used in the several
experiments were respectively ·2410, ·5893, ·3166, and ·2016 grammes.
For a rough approximation, we may assume that the probable errors
of the relative densities are inversely proportional to these numbers.
This would make the probable error of the first and fourth observations
two or three times as great as that of the second and considerably
greater than that of the third.   We must also observe that in the
first of these experiments, the observed relative density 1·783 is
greater than 1·687, the relative density calculated by equation (336)
for the temperature of the experiment and the pressure of one
atmosphere.   Now the number 1·687 we may regard as established
directly by the experiments of Deville and Troost.   For in seven
successive experiments in this part of the series the calculated relative
densities differ from the observed by less than ·01.   If then we accept
the numbers given by experiment, the effect of diluting the gas with
nitrogen is to increase its relative density.   As this result is entirely

at variance with the facts observed in the case of other gases, and in the case of this gas at lower temperatures, as appears from the three other determinations of Playfair and Wanklyn, it cannot possibly be admitted on the strength of a single observation. The first experiment of this series cannot therefore properly be used as a test of our equations. Similar considerations apply with somewhat less force to the last experiment. By comparing the temperatures and pressures of the three last experiments with the observed relative densities, the reader may easily convince himself that if we admit the substantial accuracy of the determinations in the two first of these experiments (the second and third of the series, which have the greatest weight) the last determination of relative density 2·588 must be too small. In fact, it should evidently be greater than the number in the preceding experiment 2·645.

If we confine our attention to the second and third experiments of the series, the agreement is as good as could be desired. Nor will the admission of errors of ·152 and ·120 (certainly not large in determinations of this kind) in the first and fourth experiments involve any serious doubt of the substantial accuracy of the second and third, when the difference of weight of the determinations is considered. Yet it is much to be desired that the relation expressed by (336), or with more generality by (334), should be tested by more numerous experiments.

It should be stated that the numbers in the column of pressures are not quite accurate. In the experiments of Deville and Troost the gas was subject to the actual atmospheric pressure at the time of the experiment. This varied from 747 to 764 millimeters of mercury. The precise pressure for each experiment is not given. In the experiments of Playfair and Wanklyn the mixture of nitrogen and peroxide of nitrogen was subject to the actual atmospheric pressure at the time of the experiment. The numbers in the column of pressures express the fraction of the whole pressure which remains after subtracting the part due to the free nitrogen. But no indication is given in the published account of the experiments in regard to the height of the barometer. Now it may easily be shown that a variation of $\frac{13}{760}$ in the value of p can in no case cause a variation of more than ·005 in the value of D as calculated by equation (336). In any of the experiments of Playfair and Wanklyn a variation of more than 30^{mm} in the height of the barometer would be necessary to produce a variation of ·01 in the value of D. The errors due to this source cannot therefore be very serious. They might have been avoided altogether in the discussion of the experiments of Deville and Troost by using instead of (336) a formula expressing the relation between the relative density, the temperature, and the actual density, as the reciprocal of the latter quantity is given for each experiment of this series. It seemed best, however, to make a trifling sacrifice of accuracy for the sake of simplicity.

It might be thought that the experiments under discussion would be better represented by a formula in which the term containing $\log t$ (see equation (333)) was retained. But an examination of the figures in the table will show that nothing important can be gained in this respect, and there is hardly sufficient motive for adding another term to the formula of calculation. Any attempt to determine the *real* values of A, B' and C in equation (333) (assuming the absolute validity of such an equation for peroxide of nitrogen), from the experiments under discussion would be entirely misleading, as the reader may easily convince himself.

From equation (336), however, the following conclusions may be deduced. By comparison with (334) we obtain

$$\mathbf{A} + \frac{B'}{a_2} \log_{10} t - \frac{\mathbf{C}}{t} = 9\text{·}47056 - \frac{3118\text{·}6}{t},$$

which must hold true approximately between the temperatures 11^C and 90^C. (At higher temperatures the relative densities vary too slowly with the temperatures to afford a critical test of the accuracy of this relation.) By differentiation we obtain

$$\frac{MB'}{a_2 t} + \frac{\mathbf{C}}{t^2} = \frac{3118\text{·}6}{t^2},$$

where M denotes the modulus of the common system of logarithms. Now by comparing equations (333) and (334) we see that

$$\mathbf{C} = \frac{MC}{a_2} = \text{·}43429\, \frac{C}{a_2}.$$

Hence $B't + C = 7181\, a_2 = 3590 a_1,$

which may be regarded as a close approximation at 40^C or 50^C, and a tolerable approximation between the limits of temperature above mentioned. Now $B't + C$ represents the heat evolved by the conversion of a unit of NO_2 into N_2O_4 under constant pressure. Such conversion cannot take place at constant pressure without change of temperature, which renders the experimental verification of the last equation less simple. But since by equations (322)

$$B' = B + a_1 - a_2 = B + \tfrac{1}{2} a_1,$$

we shall have for the temperature of 40^C

$$Bt + C = 3434 a_1.$$

Now $Bt + C$ represents the decrease of energy when a unit of NO_2 is transformed into N_2O_4 without change of temperature. It therefore represents the excess of the heat evolved over the work done by external forces when a mass of the gas is compressed at constant temperature until a unit of NO_2 has been converted into N_2O_4. This quantity will be constant if $B=0$, i.e., if the specific heats at constant volume of NO_2 and N_2O_4 are the same. This assumption would be more simple from a theoretical stand-point and perhaps safer than the assumption that $B'=0$. If $B=0$, $B'=a_2$. If we wish to embody this assumption in the equation between D, p, and t, we may substitute

$$6{\cdot}5228 + \log_{10}(t_c+273) - \frac{2977{\cdot}4}{t_c+273}$$

for the second member of equation (336). The relative densities calculated by the equation thus modified from the temperatures and pressures of the experiments under discussion will not differ from those calculated from the unmodified equation by more than ·002 in any case, or by more than ·001 in the first series of experiments.

It is to be noticed that if we admit the validity of the volumetrical relation expressed by equation (333), which is evidently equivalent to an equation between p, t, v, and m (this letter denoting the quantity of the gas without reference to its molecular condition), or if we admit the validity of the equation only between certain limits of temperature and for densities less than a certain limit of density, and also admit that between the given limits of temperature the specific heat of the gas at constant volume may be regarded as a constant quantity when the gas is sufficiently rarefied to be regarded as consisting wholly of NO_2,—or, to speak without reference to the molecular state of the gas, when it is rarefied until its relative density D approximates to its limiting value D_1,—we must also admit the validity (within the same limits of temperature and density) of all the calorimetrical relations which belong to ideal gas-mixtures with convertible components. The premises are evidently equivalent to this,—that we may imagine an ideal gas with convertible components such that between certain limits of temperature and above a certain limit of density the relation between p, t, and v shall be the same for a unit of this ideal gas as for a unit of peroxide of nitrogen, and for a very great value of v (within the given limits of temperature) the thermal capacity at constant volume of the ideal and actual gases shall be the same. Let us regard t and v as independent variables; we may let these letters and p refer alike to the ideal and real gases, but we must distinguish the entropy η' of the ideal gas from the entropy η of the real gas. Now by (88)

$$\frac{d\eta}{dv} = \frac{dp}{dt}, \tag{337}$$

therefore

$$\frac{d}{dv}\frac{d\eta}{dt} = \frac{d}{dt}\frac{d\eta}{dv} = \frac{d}{dt}\frac{dp}{dt} = \frac{d^2p}{dt^2}. \tag{338}$$

Since a similar relation will hold true for η', we obtain

$$\frac{d}{dv}\frac{d\eta}{dt}=\frac{d}{dv}\frac{d\eta'}{dt},\tag{339}$$

which must hold true within the given limits of temperature and density. Now it is granted that

$$\frac{d\eta}{dt}=\frac{d\eta'}{dt}\tag{340}$$

for very great values of v at any temperature within the given limits (for the two members of the equation represent the thermal capacities at constant volume of the real and ideal gases divided by t), hence, in virtue of (339), this equation must hold true in general within the given limits of temperature and density. Again, as an equation like (337) will hold true of η', we shall have

$$\frac{d\eta}{dv}=\frac{d\eta'}{dv}.\tag{341}$$

From the two last equations it is evident that in all calorimetrical relations the ideal and real gases are identical. Moreover the energy and entropy of the ideal gas are evidently so far arbitrary that we may suppose them to have the same values as in the real gas for any given values of t and v. Hence the entropies of the two gases are the same within the given limits; and on account of the necessary relation

$$d\epsilon = t\,d\eta - p\,dv,$$

the energies of the two gases are in like manner identical. Hence the fundamental equation between the energy, entropy, volume, and quantity of matter must be the same for the ideal gas as for the actual.

We may easily form a fundamental equation for an ideal gas-mixture with convertible components, which shall relate only to the phases of equilibrium. For this purpose, we may use the equations of the form (312) to eliminate from the equation of the form (273), which expresses the relation between the pressure, the temperature, and the potentials for the proximate components, as many of the potentials as there are equations of the former kind, leaving the potentials for those components which it is convenient to regard as the ultimate components of the gas-mixture.

In the case of a binary gas-mixture with convertible components, the components will have the same potential, which may be denoted by μ, and the fundamental equation will be

$$p = a_1 L_1 t^{\frac{c_1+a_1}{a_1}} e^{\frac{\mu-E_1}{a_1 t}} + a_2 L_2 t^{\frac{c_2+a_2}{a_2}} e^{\frac{\mu-E_2}{a_2 t}},\tag{342}$$

where
$$L_1 = e^{\frac{H_1-c_1-a_1}{a_1}},\quad L_2 = e^{\frac{H_2-c_2-a_2}{a_2}}.\tag{343}$$

From this equation, by differentiation and comparison with (98), we obtain

$$\frac{\eta}{v}=L_1\Big(c_1+a_1-\frac{\mu-E_1}{t}\Big)t^{\frac{c_1}{a_1}}e^{\frac{\mu-E_1}{a_1t}}$$

$$+L_2\Big(c_2+a_2-\frac{\mu-E_2}{t}\Big)t^{\frac{c_2}{a_2}}e^{\frac{\mu-E_2}{a_2t}},\tag{344}$$

$$\frac{m}{v}=L_1t^{\frac{c_1}{a_1}}e^{\frac{\mu-E_1}{a_1t}}+L_2t^{\frac{c_2}{a_2}}e^{\frac{\mu-E_2}{a_2t}}.\tag{345}$$

From the general equation (93) with the preceding equations the following is easily obtained,—

$$\frac{\epsilon}{v}=L_1(c_1t+E_1)t^{\frac{c_1}{a_1}}e^{\frac{\mu-E_1}{a_1t}}+L_2(c_2t+E_2)t^{\frac{c_2}{a_2}}e^{\frac{\mu-E_2}{a_2t}}.\tag{346}$$

We may obtain the relation between p, t, v, and m by eliminating μ from (342) and (345). For this purpose we may proceed as follows. From (342) and (345) we obtain

$$p-a_2t\frac{m}{v}=(a_1-a_2)L_1t^{\frac{c_1+a_1}{a_1}}e^{\frac{\mu-E_1}{a_1t}},\tag{347}$$

$$a_1t\frac{m}{v}-p=(a_1-a_2)L_2t^{\frac{c_2+a_2}{a_2}}e^{\frac{\mu-E_2}{a_2t}};\tag{348}$$

and from these equations we obtain

$$a_1\log\Big(p-a_2t\frac{m}{v}\Big)-a_2\log\Big(a_1t\frac{m}{v}-p\Big)=(a_1-a_2)\log(a_1-a_2)$$

$$+a_1\log L_1-a_2\log L_2+(c_1-c_2+a_1-a_2)\log t-\frac{E_1-E_2}{t}.\tag{349}$$

(In the particular case when $a_1=2a_2$ this equation will be equivalent to (333).) By (347) and (348) we may easily eliminate μ from (346).

The reader will observe that the relations thus deduced from the fundamental equation (342) without any reference to the different components of the gaseous mass are equivalent to those which relate to the phases of dissipated energy of a binary gas-mixture with components which are equivalent in substance but not convertible, except that the equations derived from (342) do not give the quantities of the proximate components, but relate solely to those properties which are capable of direct experimental verification without the aid of any theory of the constitution of the gaseous mass.

The practical application of these equations is rendered more simple by the fact that the ratio $a_1:a_2$ will always bear a simple relation to unity. When a_1 and a_2 are equal, if we write a for their common value, we shall have by (342) and (345)

$$pv=amt,\tag{350}$$

and by (345) and (346)

$$\frac{\epsilon}{m} = \frac{L_1(c_1 t + E_1) + L_2(c_2 t + E_2) t^{\frac{c_2 - c_1}{a}} e^{\frac{E_1 - E_2}{at}}}{L_1 + L_2 t^{\frac{c_2 - c_1}{a}} e^{\frac{E_1 - E_2}{at}}}. \tag{351}$$

By this equation we may calculate directly the amount of heat required to raise a given quantity of the gas from one given temperature to another at constant volume. The equation shows that the amount of heat will be independent of the volume of the gas. The heat necessary to produce a given change of temperature in the gas at constant pressure, may be found by taking the difference of the values of χ, as defined by equation (89), for the initial and final states of the gas. From (89), (350), and (351) we obtain

$$\frac{\chi}{m} = \frac{L_1(c_1 t + at + E_1) + L_2(c_2 t + at + E_2) t^{\frac{c_2 - c_1}{a}} e^{\frac{E_1 - E_2}{at}}}{L_1 + L_2 t^{\frac{c_2 - c_1}{a}} e^{\frac{E_1 - E_2}{at}}}. \tag{352}$$

By differentiation of the two last equations we may obtain directly the specific heats of the gas at constant volume and at constant pressure.

The fundamental equation of an ideal ternary gas-mixture with a single relation of convertibility between its components is

$$\begin{aligned}
p = &\, a_1 e^{\frac{H_1 - c_1 - a_1}{a_1}} t^{\frac{c_1 + a_1}{a_1}} e^{\frac{\mu_1 - E_1}{a_1 t}} \\
&+ a_2 e^{\frac{H_2 - c_2 - a_2}{a_2}} t^{\frac{c_2 + a_2}{a_2}} e^{\frac{\mu_2 - E_2}{a_2 t}} \\
&+ a_3 e^{\frac{H_3 - c_3 - a_3}{a_3}} t^{\frac{c_3 + a_3}{a_3}} e^{\frac{\lambda_1 \mu_1 + \lambda_2 \mu_2 - E_3}{a_3 t}},
\end{aligned} \tag{353}$$

where λ_1 and λ_2 have the same meaning as on page 168.

*The Conditions of Internal and External Equilibrium for Solids in contact with Fluids with regard to all possible States of Strain of the Solids.

In treating of the physical properties of a solid, it is necessary to consider its *state of strain.* A body is said to be *strained* when the relative position of its parts is altered, and by its *state of strain* is meant its state in respect to the relative position of its parts. We have hitherto considered the equilibrium of solids only in the case in which their state of strain is determined by pressures having the same values in all directions about any point. Let us now consider the subject without this limitation.

If x', y', z' are the rectangular co-ordinates of a point of a solid body in any completely determined state of strain, which we shall call

---

*[This paper was originally printed in two parts, divided at this point. For dates see heading, p. 55.]

the *state of reference*, and x, y, z, the rectangular co-ordinates of the same point of the body in the state in which its properties are the subject of discussion, we may regard x, y, z as functions of x', y', z', the form of the functions determining the second state of strain. For brevity, we may sometimes distinguish the variable state, to which x, y, z relate, and the constant state (state of reference) to which x', y', z' relate, as the *strained* and *unstrained* states; but it must be remembered that these terms have reference merely to the change of form or *strain* determined by the functions which express the relations of x, y, z and x', y', z', and do not imply any particular physical properties in either of the two states, nor prevent their possible coincidence. The axes to which the co-ordinates x, y, z and x', y', z' relate will be distinguished as the axes of X, Y, Z and X', Y', Z'. It is not necessary, nor always convenient, to regard these systems of axes as identical, but they should be similar, i.e., capable of superposition.

The state of strain of any element of the body is determined by the values of the differential coefficients of x, y, and z with respect to x', y', and z'; for changes in the values of x, y, z, when the differential coefficients remain the same, only cause motions of translation of the body. When the differential coefficients of the first order do not vary sensibly except for distances greater than the radius of sensible molecular action, we may regard them as completely determining the state of strain of any element. There are nine of these differential coefficients, viz.,

$$\left. \begin{array}{ccc} \dfrac{dx}{dx'}, & \dfrac{dx}{dy'}, & \dfrac{dx}{dz'}, \\[2mm] \dfrac{dy}{dx'}, & \dfrac{dy}{dy'}, & \dfrac{dy}{dz'}, \\[2mm] \dfrac{dz}{dx'}, & \dfrac{dz}{dy'}, & \dfrac{dz}{dz'}. \end{array} \right\} \tag{354}$$

It will be observed that these quantities determine the orientation of the element as well as its strain, and both these particulars must be given in order to determine the nine differential coefficients. Therefore, since the orientation is capable of three independent variations, which do not affect the strain, the strain of the element, considered without regard to directions in space, must be capable of six independent variations.

The physical state of any given element of a solid in any unvarying state of strain is capable of one variation, which is produced by addition or subtraction of heat. If we write $\epsilon_{V'}$ and $\eta_{V'}$ for the energy and entropy of the element divided by its volume in the state of reference, we shall have for any constant state of strain

$$\delta \epsilon_{V'} = t \, \delta \eta_{V'}.$$

But if the strain varies, we may consider $\epsilon_{V'}$ as a function of $\eta_{V'}$ and the nine quantities in (354), and may write

$$
\begin{aligned}
\delta\epsilon_{V'} = {}& t\,\delta\eta_{V'} + X_{X'}\delta\frac{dx}{dx'} + X_{Y'}\delta\frac{dx}{dy'} + X_{Z'}\delta\frac{dx}{dz'} \\
& + Y_{X'}\delta\frac{dy}{dx'} + Y_{Y'}\delta\frac{dy}{dy'} + Y_{Z'}\delta\frac{dy}{dz'} \quad\quad (355) \\
& + Z_{X'}\delta\frac{dz}{dx'} + Z_{Y'}\delta\frac{dz}{dy'} + Z_{Z'}\delta\frac{dz}{dz'},
\end{aligned}
$$

where $X_{X'}, \ldots Z_{Z'}$ denote the differential coefficients of ϵ_V taken with respect to $\dfrac{dx}{dx'}, \ldots \dfrac{dz}{dz'}$. The physical signification of these quantities will be apparent, if we apply the formula to an element which in the state of reference is a right parallelopiped having the edges dx', dy', dz', and suppose that in the strained state the face in which x' has the smaller constant value remains fixed, while the opposite face is moved parallel to the axis of X. If we also suppose no heat to be imparted to the element, we shall have, on multiplying by $dx'\,dy'\,dz'$,

$$
\delta\epsilon_{V'}dx'\,dy'\,dz' = X_{X'}\delta\frac{dx}{dx'}dx'\,dy'\,dz'.
$$

Now the first member of this equation evidently represents the work done upon the element by the surrounding elements; the second member must therefore have the same value. Since we must regard the forces acting on opposite faces of the elementary parallelopiped as equal and opposite, the whole work done will be zero except for the face which moves parallel to X. And since $\delta\dfrac{dx}{dx'}dx'$ represents the distance moved by this face, $X_{X'}dy'\,dz'$ must be equal to the component parallel to X of the force acting upon this face. In general, therefore, if by the positive side of a surface for which x' is constant we understand the side on which x' has the greater value, we may say that $X_{X'}$ denotes the component parallel to X of the force exerted by the matter on the positive side of a surface for which x' is constant upon the matter on the negative side of that surface per unit of the surface measured in the state of reference. The same may be said, *mutatis mutandis*, of the other symbols of the same type.

It will be convenient to use Σ and Σ' to denote summation with respect to quantities relating to the axes X, Y, Z, and to the axes X', Y', Z', respectively. With this understanding we may write

$$
\delta\epsilon_{V'} = t\,\delta\eta_{V'} + \Sigma\Sigma'\left(X_{X'}\delta\frac{dx}{dx'}\right). \quad\quad (356)
$$

This is the complete value of the variation of $\epsilon_{V'}$ for a given element of the solid. If we multiply by $dx'\,dy'\,dz'$, and take the integral for the whole body, we shall obtain the value of the variation of the total energy of the body, when this is supposed invariable in substance. But if we suppose the body to be increased or diminished in substance at its surface (the increment being continuous in nature and state with the part of the body to which it is joined), to obtain the complete value of the variation of the energy of the body, we must add the integral

$$\int \epsilon_{V'} \delta N' Ds'$$

in which Ds' denotes an element of the surface measured in the state of reference, and $\delta N'$ the change in position of this surface (due to the substance added or taken away) measured normally and outward in the state of reference. The complete value of the variation of the intrinsic energy of the solid is therefore

$$\iiint t \, \delta \eta_{V'} dx' dy' dz' + \iiint \Sigma\Sigma \left(X_x \delta \frac{dx}{dx'} \right) dx' dy' dz' + \int \epsilon_{V'} \delta N' Ds'. \quad (357)$$

This is entirely independent of any supposition in regard to the homogeneity of the solid.

To obtain the conditions of equilibrium for solid and fluid masses in contact, we should make the variation of the energy of the whole equal to or greater than zero. But since we have already examined the conditions of equilibrium for fluids, we need here only seek the conditions of equilibrium for the interior of a solid mass and for the surfaces where it comes in contact with fluids. For this it will be necessary to consider the variations of the energy of the fluids only so far as they are immediately connected with the changes in the solid. We may suppose the solid with so much of the fluid as is in close proximity to it to be enclosed in a fixed envelop, which is impermeable to matter and to heat, and to which the solid is firmly attached wherever they meet. We may also suppose that in the narrow space or spaces between the solid and the envelop, which are filled with fluid, there is no motion of matter or transmission of heat across any surfaces which can be generated by moving normals to the surface of the solid, since the terms in the condition of equilibrium relating to such processes may be cancelled on account of the internal equilibrium of the fluids. It will be observed that this method is perfectly applicable to the case in which a fluid mass is entirely enclosed in a solid. A detached portion of the envelop will then be necessary to separate the great mass of the fluid from the small portion adjacent to the solid, which alone we have to consider. Now the variation of the energy of the fluid mass will be, by equation (13),

$$\int^F t \, \delta D\eta - \int^F p \, \delta Dv + \Sigma_1 \int^F \mu_1 \, \delta Dm_1, \quad (358)$$

where $\int^F$ denotes an integration extending over all the elements of the fluid (within the envelop), and Σ_1 denotes a summation with regard to those independently variable components of the fluid of which the solid is composed. Where the solid does not consist of substances which are components, actual or possible (see page 64), of the fluid, this term is of course to be cancelled.

If we wish to take account of gravity, we may suppose that it acts in the negative direction of the axis of Z. It is evident that the variation of the energy due to gravity for the whole mass considered is simply

$$\iiint g \, \Gamma' \, \delta z \, dx' \, dy' \, dz', \tag{359}$$

where g denotes the force of gravity, and Γ' the density of the element in the state of reference, and the triple integration, as before, extends throughout the solid.

We have, then, for the general condition of equilibrium,

$$\iiint t \, \delta \eta_{V'} dx' \, dy' \, dz' + \iiint \Sigma\Sigma' \left(X_X \cdot \delta \frac{dx}{dx'} \right) dx' \, dy' \, dz'$$

$$+ \iiint g \Gamma' \delta z \, dx' \, dy' \, dz' + \int \epsilon_{V'} \delta N' \, Ds'$$

$$+ \int^F t \, \delta D\eta - \int^F p \, \delta Dv + \Sigma_1 \int^F \mu_1 \, \delta Dm_1 \geqq 0. \tag{360}$$

The equations of condition to which these variations are subject are: (1) that which expresses the constancy of the total entropy,

$$\iiint \delta \eta_{V'} dx' \, dy' \, dz' + \int \eta_{V'} \, \delta N' \, Ds' + \int^F \delta D\eta = 0 ; \tag{361}$$

(2) that which expresses how the value of δDv for any element of the fluid is determined by changes in the solid,

$$\delta Dv = -(a\delta x + \beta \delta y + \gamma \delta z) Ds - v_{V'} \delta N' Ds', \tag{362}$$

where a, β, γ denote the direction cosines of the normal to the surface of the body in the state to which x, y, z relate, Ds the element of the surface in this state corresponding to Ds' in the state of reference, and $v_{V'}$ the volume of an element of the solid divided by its volume in the state of reference;

(3) those which express how the values of δDm_1, δDm_2, etc. for any element in the fluid are determined by the changes in the solid,

$$\left. \begin{aligned} \delta Dm_1 &= -\Gamma_1' \delta N' Ds', \\ \delta Dm_2 &= -\Gamma_2' \delta N' Ds', \\ \text{etc.,} & \end{aligned} \right\} \tag{363}$$

where Γ_1', Γ_2', etc. denote the separate densities of the several components in the solid in the state of reference.

Now, since the variations of entropy are independent of all the other variations, the condition of equilibrium (360), considered with regard to the equation of condition (361), evidently requires that throughout the whole system

$$t = \text{const.} \tag{364}$$

We may therefore use (361) to eliminate the fourth and fifth integrals from (360). If we multiply (362) by p, and take the integrals for the whole surface of the solid and for the fluid in contact with it, we obtain the equation

$$\int^\text{F} p\,\delta Dv = -\int p(a\delta x + \beta\delta y + \gamma\delta z)Ds - \int pv_{\text{v}'}\,\delta N'Ds', \qquad (365)$$

by means of which we may eliminate the sixth integral from (360). If we add equations (363) multiplied respectively by μ_1, μ_2, etc., and take the integrals, we obtain the equation

$$\Sigma_1 \int^\text{F} \mu_1\,\delta Dm_1 = -\int \Sigma_1(\mu_1\Gamma_1')\,\delta N'Ds', \qquad (366)$$

by means of which we may eliminate the last integral from (360).

The condition of equilibrium is thus reduced to the form

$$\iiint \Sigma\Sigma'\left(X_{\text{x}'}\delta\frac{dx}{dx'}\right)dx'\,dy'\,dz' + \iiint g\,\Gamma'\delta z\,dx'\,dy'\,dz'$$

$$+\int\epsilon_{\text{v}'}\delta N'Ds' - \int t\eta_{\text{v}'}\delta N'Ds' + \int p(a\delta x + \beta\delta y + \gamma\delta z)Ds$$

$$+\int pv_{\text{v}'}\delta N'Ds' - \int\Sigma_1(\mu_1\Gamma_1')\delta N'Ds' \geqq 0, \qquad (367)$$

in which the variations are independent of the equations of condition, and in which the only quantities relating to the fluids are p and μ_1, μ_2, etc.

Now by the ordinary method of the calculus of variations, if we write a', β', γ' for the direction-cosines of the normal to the surface of the solid in the state of reference, we have

$$\iiint X_{\text{x}'}\delta\frac{dx}{dx'}\,dx'\,dy'\,dz'$$

$$=\int a'X_{\text{x}'}\,\delta x\,Ds' - \iiint \frac{dX_{\text{x}'}}{dx'}\,\delta x\,dx'\,dy'\,dz', \qquad (368)$$

with similar expressions for the other parts into which the first integral in (367) may be divided. The condition of equilibrium is thus reduced to the form

$$-\iiint \Sigma\Sigma'\left(\frac{dX_{\text{x}'}}{dx'}\,\delta x\right)dx'\,dy'\,dz' + \iiint g\,\Gamma'\delta z\,dx'\,dy'\,dz'$$

$$+\int\Sigma\Sigma'(a'X_{\text{x}'}\,\delta x)Ds' + \int p\,\Sigma(a\delta x)Ds$$

$$+\int[\epsilon_{\text{v}'} - t\eta_{\text{v}'} + pv_{\text{v}'} - \Sigma_1(\mu_1\Gamma_1')]\delta N'Ds' \geqq 0. \qquad (369)$$

It must be observed that if the solid mass is not continuous throughout in nature and state, the surface-integral in (368), and therefore the first surface-integral in (369), must be taken to apply not only to the external surface of the solid, but also to every surface of discontinuity within it, and that with reference to each of the two masses separated by the surface. To satisfy the condition of equilibrium, as thus understood, it is necessary and sufficient that throughout the solid mass

$$\Sigma\Sigma'\left(\frac{dX_{X'}}{dx'}\delta x\right) - g\Gamma'\delta z = 0 ; \qquad (370)$$

that throughout the surfaces where the solid meets the fluid

$$Ds'\Sigma\Sigma'(a'X_{X'}\delta x) + Dsp\Sigma(a\delta x) = 0, \qquad (371)$$

and

$$[\epsilon_{V'} - t\eta_{V'} + pv_{V'} - \Sigma_1(\mu_1\Gamma_1')]\,\delta N' \geqq 0 ; \qquad (372)$$

and that throughout the internal surfaces of discontinuity

$$\Sigma\Sigma'(a'X_{X'}\delta x)_1 + \Sigma\,\Sigma'(a'X_{X'}\delta x)_2 = 0, \qquad (373)$$

where the suffixed numerals distinguish the expressions relating to the masses on opposite sides of a surface of discontinuity.

Equation (370) expresses the mechanical conditions of internal equilibrium for a continuous solid under the influence of gravity. If we expand the first term, and set the coefficients of δx, δy, and δz separately equal to zero, we obtain

$$\left.\begin{aligned}
\frac{dX_{X'}}{dx'} + \frac{dX_{Y'}}{dy'} + \frac{dX_{Z'}}{dz'} &= 0, \\[4pt]
\frac{dY_{X'}}{dx'} + \frac{dY_{Y'}}{dy'} + \frac{dY_{Z'}}{dz'} &= 0, \\[4pt]
\frac{dZ_{X'}}{dx'} + \frac{dZ_{Y'}}{dy'} + \frac{dZ_{Z'}}{dz'} &= g\Gamma'.
\end{aligned}\right\} \qquad (374)$$

The first member of any one of these equations multiplied by $dx'\,dy'\,dz'$ evidently represents the sum of the components parallel to one of the axes X, Y, Z of the forces exerted on the six faces of the element $dx'\,dy'\,dz'$ by the neighboring elements.

As the state which we have called the state of reference is arbitrary, it may be convenient for some purposes to make it coincide with the state to which x, y, z relate, and the axes X', Y', Z' with the axes X, Y, Z. The values of $X_{X'}, \ldots Z_{Z'}$ on this particular supposition may be represented by the symbols $X_X, \ldots Z_Z$. Since

$$X_{Y'} = \frac{d\epsilon_{V'}}{d\dfrac{dx}{dy'}}, \quad \text{and} \quad Y_{X'} = \frac{d\epsilon_{V'}}{d\dfrac{dy}{dx'}},$$

and since, when the states, x, y, z and x' y' z' coincide, and the axes X, Y, Z, and X', Y', Z', $d\dfrac{dx}{dy'}$ and $d\dfrac{dy}{dx'}$ represent displacements which differ only by a rotation, we must have

$$X_Y = Y_X, \qquad (375)$$

and for similar reasons,

$$Y_Z = Z_Y, \qquad Z_X = X_Z. \qquad (376)$$

The six quantities X_X, Y_Y, Z_Z, X_Y or Y_X, Y_Z or Z_Y, and Z_X or X_Z are called the *rectangular components of stress*, the three first being the *longitudinal stresses* and the three last the *shearing stresses*. The mechanical conditions of internal equilibrium for a solid under the influence of gravity may therefore be expressed by the equations

$$
\left.
\begin{aligned}
\frac{dX_X}{dx} + \frac{dX_Y}{dy} + \frac{dX_Z}{dz} &= 0, \\
\frac{dY_X}{dx} + \frac{dY_Y}{dy} + \frac{dY_Z}{dz} &= 0, \\
\frac{dZ_X}{dx} + \frac{dZ_Y}{dy} + \frac{dZ_Z}{dz} &= g\Gamma,
\end{aligned}
\right\}
\tag{377}
$$

where Γ denotes the density of the element to which the other symbols relate. Equations (375), (376) are rather to be regarded as expressing necessary relations (when $X_X, \ldots Z_Z$ are regarded as internal forces determined by the state of strain of the solid) than as expressing conditions of equilibrium. They will hold true of a solid which is not in equilibrium,—of one, for example, through which vibrations are propagated,—which is not the case with equations (377).

Equation (373) expresses the mechanical conditions of equilibrium for a surface of discontinuity within the solid. If we set the coefficients of δx, δy, δz, separately equal to zero we obtain

$$
\left.
\begin{aligned}
(a'X_{X'} + \beta'X_{Y'} + \gamma'X_{Z'})_1 + (a'X_{X'} + \beta'X_{Y'} + \gamma'X_{Z'})_2 &= 0, \\
(a'Y_{X'} + \beta'Y_{Y'} + \gamma'Y_{Z'})_1 + (a'Y_{X'} + \beta'Y_{Y'} + \gamma'Y_{Z'})_2 &= 0, \\
(a'Z_{X'} + \beta'Z_{Y'} + \gamma'Z_{Z'})_1 + (a'Z_{X'} + \beta'Z_{Y'} + \gamma'Z_{Z'})_2 &= 0.
\end{aligned}
\right\}
\tag{378}
$$

Now when the a', β', γ' represent the direction-cosines of the normal in the state of reference on the positive side of any surface within the solid, an expression of the form

$$
a'X_{X'} + \beta'X_{Y'} + \gamma'X_{Z'}
\tag{379}
$$

represents the component parallel to X of the force exerted upon the surface in the strained state by the matter on the positive side per unit of area measured in the state of reference. This is evident from the consideration that in estimating the force upon any surface we may substitute for the given surface a broken one consisting of elements for each of which either x' or y' or z' is constant. Applied to a surface bounding a solid, or any portion of a solid which may not be continuous with the rest, when the normal is drawn outward as usual, the same expression taken negatively represents the component parallel to X of the force exerted upon the surface (per unit of surface measured in the state of reference) by the interior of the solid, or of the portion considered. Equations (378) therefore express the condition that the force exerted upon the surface of discontinuity by the matter on one side and determined by its state of strain shall be equal and opposite to that exerted by the matter on the other side. Since

$$(a')_1 = -(a')_2, \qquad (\beta')_1 = -(\beta')_2, \qquad (\gamma')_1 = -(\gamma')_2,$$

we may also write

$$a'(X_{X'})_1 + \beta'(X_{Y'})_1 + \gamma'(X_{Z'})_1 = a'(X_{X'})_2 + \beta'(X_{Y'})_2 + \gamma'(X_{Z'})_2, \Big\} \quad (380)$$
etc.,

where the signs of a', β', γ' may be determined by the normal on either side of the surface of discontinuity.

Equation (371) expresses the mechanical condition of equilibrium for a surface where the solid meets a fluid. It involves the separate equations

$$\left. \begin{aligned} a'X_{X'} + \beta'X_{Y'} + \gamma'X_{Z'} &= -a\,p\,\frac{Ds}{Ds'}, \\[2mm] a'Y_{X'} + \beta'Y_{Y'} + \gamma'Y_{Z'} &= -\beta\,p\,\frac{Ds}{Ds'}, \\[2mm] a'Z_{X'} + \beta'Z_{Y'} + \gamma'Z_{Z'} &= -\gamma\,p\,\frac{Ds}{Ds'}, \end{aligned} \right\} \quad (381)$$

the fraction $\dfrac{Ds}{Ds'}$ denoting the ratio of the areas of the same element of the surface in the strained and unstrained states of the solid. These equations evidently express that the force exerted by the interior of the solid upon an element of its surface, and determined by the strain of the solid, must be normal to the surface and equal (but acting in the opposite direction) to the pressure exerted by the fluid upon the same element of surface.

If we wish to replace a and Ds by a', β', γ', and the quantities which express the strain of the element, we may make use of the following considerations. The product $a\,Ds$ is the projection of the element Ds on the Y-Z plane. Now since the ratio $\dfrac{Ds}{Ds'}$ is independent of the form of the element, we may suppose that it has any convenient form. Let it be bounded by the three surfaces $x' = \text{const.}$, $y' = \text{const.}$, $z' = \text{const.}$, and let the parts of each of these surfaces included by the two others with the surface of the body be denoted by L, M, and N, or by L', M', and N', according as we have reference to the strained or unstrained state of the body. The areas of L', M', and N' are evidently $a'Ds'$, $\beta'Ds'$, and $\gamma'Ds'$; and the sum of the projections of L, M, and N upon any plane is equal to the projection of Ds upon that plane, since L, M, and N with Ds include a solid figure. (In propositions of this kind the *sides* of surfaces must be distinguished. If the normal to Ds falls outward from the small solid figure, the normals to L, M, and N must fall inward, and *vice versa*.) Now L' is a right-angled triangle of which the perpendicular sides may be called dy' and dz'.

The projection of L on the Y-Z plane will be a triangle, the angular points of which are determined by the co-ordinates

$$y,\ z;\quad y+\frac{dy}{dy'}dy',\ \ z+\frac{dz}{dy'}dy';\quad y+\frac{dy}{dz'}dz',\ \ z+\frac{dz}{dz'}dz';$$

the area of such a triangle is

$$\tfrac{1}{2}\Big(\frac{dy}{dy'}\frac{dz}{dz'}-\frac{dz}{dy'}\frac{dy}{dz'}\Big)dy'\,dz',$$

or, since $\tfrac{1}{2}\,dy'\,dz'$ represents the area of L',

$$\Big(\frac{dy}{dy'}\frac{dz}{dz'}-\frac{dz}{dy'}\frac{dy}{dz'}\Big)a'\,Ds'.$$

(That this expression has the proper sign will appear if we suppose for the moment that the strain vanishes.) The areas of the projections of M and N upon the same plane will be obtained by changing y', z' and a' in this expression into z', x', and β', and into x', y', and γ'. The sum of the three expressions may be substituted for $a\,Ds$ in (381).

We shall hereafter use Σ' to denote the sum of the three terms obtained by rotary substitutions of quantities relating to the axes X', Y', Z' (i.e., by changing x', y', z' into y', z', x', and into z', x', y', with similar changes in regard to a', β', γ', and other quantities relating to these axes), and Σ to denote the sum of the three terms obtained by similar rotary changes of quantities relating to the axes X, Y, Z. This is only an extension of our previous use of these symbols.

With this understanding, equations (381) may be reduced to the form

$$\left.\begin{aligned}\Sigma'(a'X_{x'})+p\Sigma'\Big\{a'\Big(\frac{dy}{dy'}\frac{dz}{dz'}-\frac{dz}{dy'}\frac{dy}{dz'}\Big)\Big\}=0,\\ \text{etc.}\end{aligned}\right\} \qquad (382)$$

The formula (372) expresses the additional condition of equilibrium which relates to the dissolving of the solid, or its growth without discontinuity. If the solid consists entirely of substances which are actual components of the fluid, and there are no passive resistances which impede the formation or dissolving of the solid, $\delta N'$ may have either positive or negative values, and we must have

$$\epsilon_{V'}-t\eta_{V'}-pv_{V'}=\Sigma_1(\mu_1\Gamma_1'). \qquad (383)$$

But if some of the components of the solid are only possible components (see page 64) of the fluid, $\delta N'$ is incapable of positive values, as the quantity of the solid cannot be increased, and it is sufficient for equilibrium that

$$\epsilon_{V'}-t\eta_{V'}+pv_{V'}\leqq\Sigma_1(\mu_1\Gamma_1'). \qquad (384)$$

To express condition (383) in a form independent of the state of reference, we may use ϵ_V, η_V, Γ_1, etc., to denote the densities of energy, of entropy, and of the several component substances in the *variable* state of the solid. We shall obtain, on dividing the equation by $v_{\mathrm{V}'}$,

$$\epsilon_{\mathrm{V}} - t\eta_{\mathrm{V}} + p = \Sigma_1(\mu_1 \Gamma_1). \tag{385}$$

It will be remembered that the summation relates to the several components of the solid. If the solid is of uniform composition throughout, or if we only care to consider the contact of the solid and the fluid at a single point, we may treat the solid as composed of a single substance. If we use μ_1 to denote the potential for this substance in the fluid, and Γ to denote the density of the solid in the variable state (Γ', as before denoting its density in the state of reference), we shall have

$$\epsilon_{\mathrm{V}'} - t\eta_{\mathrm{V}'} + pv_{\mathrm{V}'} = \mu_1 \Gamma', \tag{386}$$

and

$$\epsilon_{\mathrm{V}} - t\eta_{\mathrm{V}} + p = \mu_1 \Gamma. \tag{387}$$

To fix our ideas in discussing this condition, let us apply it to the case of a solid body which is homogeneous in nature and in state of strain. If we denote by ϵ, η, v, and m, its energy, entropy, volume, and mass, we have

$$\epsilon - t\eta + pv = \mu_1 m. \tag{388}$$

Now the mechanical conditions of equilibrium for the surface where a solid meets a fluid require that the traction upon the surface determined by the state of strain of the solid shall be normal to the surface. This condition is always satisfied with respect to three surfaces at right angles to one another. In proving this well-known proposition, we shall lose nothing in generality, if we make the state of reference, which is arbitrary, coincident with the state under discussion, the axes to which these states are referred being also coincident. We shall then have, for the normal component of the traction per unit of surface across any surface for which the direction-cosines of the normal are a, β, γ (compare (379), and for the notation X_{X}, etc., page 190),

$$S = a(aX_{\mathrm{X}} + \beta X_{\mathrm{Y}} + \gamma X_{\mathrm{Z}})$$
$$+ \beta(aY_{\mathrm{X}} + \beta Y_{\mathrm{Y}} + \gamma Y_{\mathrm{Z}})$$
$$+ \gamma(aZ_{\mathrm{X}} + \beta Z_{\mathrm{Y}} + \gamma Z_{\mathrm{Z}}),$$

or, by (375), (376),

$$S = a^2 X_{\mathrm{X}} + \beta^2 Y_{\mathrm{Y}} + \gamma^2 Z_{\mathrm{Z}}$$
$$+ 2a\beta X_{\mathrm{Y}} + 2\beta\gamma Y_{\mathrm{Z}} + 2\gamma a Z_{\mathrm{X}}. \tag{389}$$

We may also choose any convenient directions for the co-ordinate axes. Let us suppose that the direction of the axis of X is so chosen that the value of S for the surface perpendicular to this axis is as great as for any other surface, and that the direction of the axis of Y (supposed at right angles to X) is such that the value of S for the surface perpendicular to it is as great as for any other surface passing through the axis of X. Then, if we write $\dfrac{dS}{da}, \dfrac{dS}{d\beta}, \dfrac{dS}{d\gamma}$ for the differential coefficients derived from the last equation by treating a, β, and γ as *independent* variables,

$$\frac{dS}{da}\,da+\frac{dS}{d\beta}\,d\beta+\frac{dS}{d\gamma}\,d\gamma=0,$$

when

$$a\,da+\beta\,d\beta+\gamma\,d\gamma=0,$$

and

$$a=1, \quad \beta=0, \quad \gamma=0.$$

That is,

$$\frac{dS}{d\beta}=0, \quad \text{and} \quad \frac{dS}{d\gamma}=0,$$

when

$$a=1, \quad \beta=0, \quad \gamma=0.$$

Hence

$$X_Y=0, \quad \text{and} \quad Z_X=0. \tag{390}$$

Moreover,

$$\frac{dS}{d\beta}\,d\beta+\frac{dS}{d\gamma}\,d\gamma=0,$$

when

$$a=0, \quad da=0,$$

$$\beta\,d\beta+\gamma\,d\gamma=0,$$

and

$$\beta=1, \quad \gamma=0.$$

Hence

$$Y_Z=0. \tag{391}$$

Therefore, when the co-ordinate axes have the supposed directions, which are called the *principal axes of stress*, the rectangular components of the traction across any surface (a, β, γ) are by (379)

$$a\,X_X, \quad \beta\,Y_Y, \quad \gamma\,Z_Z. \tag{392}$$

Hence, the traction across any surface will be normal to that surface,—

(1), when the surface is perpendicular to a principal axis of stress;

(2), if two of the *principal tractions* X_X, Y_Y, Z_Z are equal, when the surface is perpendicular to the plane containing the two corresponding axes (in this case the traction across any such surface is equal to the common value of the two principal tractions);

(3), if the principal tractions are all equal, the traction is normal and constant for all surfaces.

It will be observed that in the second and third cases the positions of the principal axes of stress are partially or wholly indeterminate (so that these cases may be regarded as included in the first), but the values of the principal tractions are always determinate, although not always different.

If, therefore, a solid which is homogeneous in nature and in state of strain is bounded by six surfaces perpendicular to the principal axes of stress, the mechanical conditions of equilibrium for these surfaces may be satisfied by the contact of fluids having the proper pressures (see (381)), which will in general be different for the different pairs of opposite sides, and may be denoted by p', p'', p'''. (These pressures are equal to the principal tractions of the solid taken negatively.) It will then be necessary for equilibrium with respect to the tendency of the solid to dissolve that the potential for the substance of the solid in the fluids shall have values μ_1', μ_1'', μ_1''', determined by the equations

$$\epsilon - t\eta + p'v = \mu_1'm, \tag{393}$$

$$\epsilon - t\eta + p''v = \mu_1''m, \tag{394}$$

$$\epsilon - t\eta + p'''v = \mu_1'''m. \tag{395}$$

These values, it will be observed, are entirely determined by the nature and state of the solid, and their differences are equal to the differences of the corresponding pressures divided by the density of the solid.

It may be interesting to compare one of these potentials, as μ_1', with the potential (for the same substance) in a fluid of the same temperature t and pressure p' which would be in equilibrium with the same solid subjected on all sides to the uniform pressure p'. If we write $[\epsilon]_{p'}$, $[\eta]_{p'}$, $[v]_{p'}$, and $[\mu_1]_{p'}$ for the values which ϵ, η, v, and μ_1 would receive on this supposition, we shall have

$$[\epsilon]_{p'} - t[\eta]_{p'} + p'[v]_{p'} = [\mu_1]_{p'}m. \tag{396}$$

Subtracting this from (393), we obtain

$$\epsilon - [\epsilon]_{p'} - t\eta + t[\eta]_{p'} + p'v - p'[v]_{p'} = \mu_1m - [\mu_1]_{p'}m. \tag{397}$$

Now it follows immediately from the definitions of energy and entropy that the first four terms of this equation represent the work spent upon the solid in bringing it from the state of hydrostatic stress to the other state without change of temperature, and $p'v - p'[v]_{p'}$ evidently denotes the work done in displacing a fluid of pressure p' surrounding the solid during the operation. Therefore, the first number of the equation represents the total work done in bringing the solid *when surrounded by a fluid of pressure p'* from the state of hydrostatic stress p' to the state of stress p', p'', p'''. This quantity is necessarily positive, except of course in the limiting case when $p' = p'' = p'''$. If the quantity of matter of the solid body be unity, the increase of the potential in the fluid on the side of the solid on which the pressure remains constant, which will be necessary to maintain equilibrium, is equal to the work done as above described. Hence, μ_1' is greater than $[\mu_1]_{p'}$, and for similar reasons μ_1'' is greater than the value of the potential which would be necessary for equilibrium if the solid were subjected to the uniform pressure p'', and μ_1''' greater than that which would be necessary for equilibrium if the solid were subjected to the uniform pressure p'''. That is (if we adapt our language to what we may regard as the most general case, viz., that in which the fluids contain the substance of the solid but are not wholly composed of that substance), the fluids in equilibrium with the solid are all supersaturated with respect to the substance of the solid, except when the solid is in a state of hydrostatic stress; so that if there were present in any one of these fluids any small fragment of the same kind of solid subject to the hydrostatic pressure of the fluid, such a fragment would tend to increase. Even when no such fragment is present, although there must be perfect equilibrium so far as concerns the tendency of the solid to dissolve or to increase by the accretion of similarly strained matter, yet the presence of the solid which is subject to the distorting stresses, will doubtless facilitate the commencement of the formation of a solid of hydrostatic stress upon its surface, to the same extent, perhaps, in the case of an amorphous body, as if it were itself subject only to hydrostatic stress. This may sometimes, or perhaps generally, make it a necessary condition of equilibrium in cases of contact between a fluid and an amorphous solid which can be formed out of it, that the solid at the surface where it meets the fluid shall be sensibly in a state of hydrostatic stress.

But in the case of a solid of continuous crystalline structure, subjected to distorting stresses and in contact with solutions satisfying the conditions deduced above, although crystals of hydrostatic stress would doubtless commence to form upon its surface (if the distorting stresses and consequent supersaturation of the fluid should be carried too far), before they would commence to be formed within the fluid or on the surface of most other bodies, yet within certain limits the relations expressed by equations (393)–(395) must admit of realization, especially when the solutions are such as can be easily supersaturated.*

It may be interesting to compare the variations of p, the pressure in the fluid which determines in part the stresses and the state of strain of the solid, with other variations of the stresses or strains in the solid, with respect to the relation expressed by equation (388). To examine this point with complete generality, we may proceed in the following manner.

Let us consider so much of the solid as has in the state of reference the form of a cube, the edges of which are equal to unity, and parallel to the co-ordinate axes. We may suppose this body to be homogeneous in nature and in state of strain both in its state of

---

* The effect of distorting stresses in a solid on the phenomena of crystallization and liquefaction, as well as the effect of change of hydrostatic pressure common to the solid and liquid, was first described by Professor James Thomson. See *Trans. R. S. Edin.*, vol. xvi, p. 575; and *Proc. Roy. Soc.*, vol. xi, p. 473, or *Phil. Mag.*, ser. 4, vol. xxiv, p. 395.

reference and in its variable state. (This involves no loss of generality, since we may make the unit of length as small as we choose.) Let the fluid meet the solid on one or both of the surfaces for which Z' is constant. We may suppose these surfaces to remain perpendicular to the axis of Z in the variable state of the solid, and the edges in which y' and z' are both constant to remain parallel to the axis of X. It will be observed that these suppositions only fix the position of the strained body relatively to the co-ordinate axes, and do not in any way limit its state of strain.

It follows from the suppositions which we have made that

$$\frac{dz}{dx'} = \text{const.} = 0, \quad \frac{dz}{dy'} = \text{const.} = 0, \quad \frac{dy}{dx'} = \text{const.} = 0 \; ; \qquad (398)$$

and

$$X_{Z'} = 0, \quad Y_{Z'} = 0, \quad Z_{Z'} = -p \frac{dx}{dx'} \frac{dy}{dy'}. \qquad (399)$$

Hence, by (355),

$$d\epsilon_{V'} = t \, d\eta_{V'} + X_{X'} d\frac{dx}{dx'} + X_{Y'} d\frac{dx}{dy'} + Y_{Y'} d\frac{dy}{dy'} - p \frac{dx}{dx'} \frac{dy}{dy'} d\frac{dz}{dz'}. \quad (400)$$

Again, by (388),

$$d\epsilon = t \, d\eta + \eta \, dt - p \, dv - v \, dp + m \, d\mu_1. \qquad (401)$$

Now the suppositions which have been made require that

$$v = \frac{dx}{dx'} \frac{dy}{dy'} \frac{dz}{dz'}, \qquad (402)$$

and

$$dv = \frac{dy}{dy'} \frac{dz}{dz'} d\frac{dx}{dx'} + \frac{dz}{dz'} \frac{dx}{dx'} d\frac{dy}{dy'} + \frac{dx}{dx'} \frac{dy}{dy'} d\frac{dz}{dz'}. \qquad (403)$$

Combining equations (400), (401), and (403), and observing that $\epsilon_{V'}$ and $\eta_{V'}$ are equivalent to ϵ and η, we obtain

$$\eta \, dt - v \, dp + m \, d\mu_1$$
$$= \left(X_{X'} + p \frac{dy}{dy'} \frac{dz}{dz'} \right) d\frac{dx}{dx'} + X_{Y'} d\frac{dx}{dy'} + \left(Y_{Y'} + p \frac{dz}{dz'} \frac{dx}{dx'} \right) d\frac{dy}{dy'}. \qquad (404)$$

The reader will observe that when the solid is subjected on all sides to the uniform normal pressure p, the coefficients of the differentials in the second member of this equation will vanish. For the expression $\frac{dy}{dy'} \frac{dz}{dz'}$ represents the projection on the Y-Z plane of a side of the parallelopiped for which x' is constant, and multiplied by p it will be equal to the component parallel to the axis of X of the total pressure across this side, i.e., it will be equal to $X_{X'}$ taken negatively. The case is similar with respect to the coefficient of $d\frac{dy}{dy'}$; and $X_{Y'}$ evidently denotes a force tangential to the surface on which it acts.

It will also be observed, that if we regard the forces acting upon the sides of the solid parallelopiped as composed of the hydrostatic pressure p together with additional forces, the work done in any infinitesimal variation of the state of strain of the solid by these additional forces will be represented by the second member of the equation.

We will first consider the case in which the fluid is identical in substance with the solid. We have then, by equation (97), for a mass of the fluid equal to that of the solid,

$$\eta_F \, dt - v_F \, dp + m \, d\mu_1 = 0, \tag{405}$$

η_F and v_F denoting the entropy and volume of the fluid. By subtraction we obtain

$$-(\eta_F - \eta)dt + (v_F - v)dp$$
$$= \left(X_X + p\frac{dy}{dy'}\frac{dz}{dz'}\right)d\frac{dx}{dx'} + X_Y d\frac{dx}{dy'} + \left(Y_Y + p\frac{dz}{dz'}\frac{dx}{dx'}\right)d\frac{dy}{dy'}. \tag{406}$$

Now if the quantities $\dfrac{dx}{dx'}$, $\dfrac{dx}{dy'}$, $\dfrac{dy}{dy'}$ remain constant, we shall have for the relation between the variations of temperature and pressure which is necessary for the preservation of equilibrium

$$\frac{dt}{dp} = \frac{v_F - v}{\eta_F - \eta} = t\frac{v_F - v}{Q}, \tag{407}$$

where Q denotes the heat which would be absorbed if the solid body should pass into the fluid state without change of temperature or pressure. This equation is similar to (131), which applies to bodies subject to hydrostatic pressure. But the value of $\dfrac{dt}{dp}$ will not generally be the same as if the solid were subject on all sides to the uniform normal pressure p; for the quantities v and η (and therefore Q) will in general have different values. But when the pressures on all sides are normal and equal, the value of $\dfrac{dt}{dp}$ will be the same, whether we consider the pressure when varied as still normal and equal on all sides, or consider the quantities $\dfrac{dx}{dx'}$, $\dfrac{dx}{dy'}$, $\dfrac{dy}{dy'}$ as constant.

But if we wish to know how the temperature is affected if the pressure between the solid and fluid remains constant, but the strain of the solid is varied in any way consistent with this supposition, the differential coefficients of t with respect to the quantities which express the strain are indicated by equation (406). These differential coefficients all vanish, when the pressures on all sides are normal and equal, but the differential coefficient $\dfrac{dt}{dp}$, when $\dfrac{dx}{dx'}$, $\dfrac{dx}{dy'}$, $\dfrac{dy}{dy'}$ are constant, or when the pressures on all sides are normal and equal, vanishes only when the density of the fluid is equal to that of the solid.

The case is nearly the same when the fluid is not identical in substance with the solid, if we suppose the composition of the fluid to remain unchanged. We have necessarily with respect to the fluid

$$d\mu_1 = \left(\frac{d\mu_1}{dt}\right)_{p,\,m}^{(F)} dt + \left(\frac{d\mu_1}{dp}\right)_{t,\,m}^{(F)} dp,^* \tag{408}$$

where the index (F) is used to indicate that the expression to which it is affixed relates to the fluid. But by equation (92)

$$\left(\frac{d\mu_1}{dt}\right)_{p,\,m}^{(F)} = -\left(\frac{d\eta}{dm_1}\right)_{t,\,p,\,m}^{(F)}, \quad \text{and} \quad \left(\frac{d\mu_1}{dp}\right)_{t,\,m}^{(F)} = \left(\frac{dv}{dm_1}\right)_{t,\,p,\,m}^{(F)}. \tag{409}$$

Substituting these values in the preceding equation, transposing terms, and multiplying by m, we obtain

$$m\left(\frac{d\eta}{dm_1}\right)_{t,\,p,\,m}^{(F)} dt - m\left(\frac{dv}{dm_1}\right)_{t,\,p,\,m}^{(F)} dp + m\,d\mu_1 = 0. \tag{410}$$

By subtracting this equation from (404) we may obtain an equation similar to (406), except that in place of η_F and v_F we shall have the expressions

$$m\left(\frac{d\eta}{dm_1}\right)_{t,\,p,\,m}^{(F)} \quad \text{and} \quad m\left(\frac{dv}{dm_1}\right)_{t,\,p,\,m}^{(F)}$$

The discussion of equation (406) will therefore apply *mutatis mutandis* to this case.

We may also wish to find the variations in the composition of the fluid which will be necessary for equilibrium when the pressure p or the quantities $\frac{dx}{dx'}, \frac{dx}{dy'}, \frac{dy}{dy'}$ are varied, the temperature remaining constant. If we know the value for the fluid of the quantity represented by ζ on page 87 in terms of t, p, and the quantities of the several components m_1, m_2, m_3, etc., the first of which relates to the substance of which the solid is formed, we can easily find the value of μ_1 in terms of the same variables. Now in considering variations in the composition of the fluid, it will be sufficient if we make all but one of the components variable. We may therefore give to m_1 a constant value, and making t also constant, we shall have

$$d\mu_1 = \left(\frac{d\mu_1}{dp}\right)_{t,\,m}^{(F)} dp + \left(\frac{d\mu_1}{dm_2}\right)_{t,\,p,\,m}^{(F)} dm_2 + \left(\frac{d\mu_1}{dm_3}\right)_{t,\,p,\,m}^{(F)} dm_3 + \text{etc.}$$

---

* A suffixed m stands here, as elsewhere in this paper, for all the symbols m_1, m_2, etc., except such as may occur in the differential coefficient.

Substituting this value in equation (404), and cancelling the term containing dt, we obtain

$$\left\{ m\left(\frac{d\mu_1}{dp}\right)^{(F)}_{t,\,m} - v\right\}dp + m\left(\frac{d\mu_1}{dm_2}\right)^{(F)}_{t,\,p,\,m} dm_2$$

$$+ m\left(\frac{d\mu_1}{dm_3}\right)^{(F)}_{t,\,p,\,m} dm_3 + \text{etc.} = \left(X_{\mathrm{X}'} + p\frac{dy}{dy'}\frac{dz}{dz'}\right)d\frac{dx}{dx'}$$

$$+ X_{\mathrm{Y}'}d\frac{dx}{dy'} + \left(Y_{\mathrm{Y}'} + p\frac{dz}{dz'}\frac{dx}{dx'}\right)d\frac{dy}{dy'}. \tag{411}$$

This equation shows the variation in the quantity of any one of the components of the fluid (other than the substance which forms the solid) which will balance a variation of p, or of $\frac{dx}{dx'}$, $\frac{dx}{dy'}$, $\frac{dy}{dy'}$, with respect to the tendency of the solid to dissolve.

Fundamental Equations for Solids.

The principles developed in the preceding pages show that the solution of problems relating to the equilibrium of a solid, or at least their reduction to purely analytical processes, may be made to depend upon our knowledge of the composition and density of the solid at every point in some particular state, which we have called the state of reference, and of the relation existing between the quantities which have been represented by $\epsilon_{\mathrm{V}'}$, $\eta_{\mathrm{V}'}$, $\frac{dx}{dx'}$, $\frac{dx}{dy'}$, ... $\frac{dz}{dz'}$, x', y', and z'. When the solid is in contact with fluids, a certain knowledge of the properties of the fluids is also requisite, but only such as is necessary for the solution of problems relating to the equilibrium of fluids among themselves.

If in any state of which a solid is capable, it is homogeneous in its nature and in its state of strain, we may choose this state as the state of reference, and the relation between $\epsilon_{\mathrm{V}'}$, $\eta_{\mathrm{V}'}$, $\frac{dx}{dx'}$, . . . $\frac{dz}{dz'}$, will be independent of x', y', z'. But it is not always possible, even in the case of bodies which are homogeneous in nature, to bring all the elements simultaneously into the same state of strain. It would not be possible, for example, in the case of a Prince Rupert's drop.

If, however, we know the relation between $\epsilon_{\mathrm{V}'}$, $\eta_{\mathrm{V}'}$, $\frac{dx}{dx'}$, . . . $\frac{dz}{dz'}$, for any kind of homogeneous solid, with respect to any given state of reference, we may derive from it a similar relation with respect to any other state as a state of reference. For if x', y', z' denote the co-ordinates of points of the solid in the first state of reference, and

x'', y'', z'' the co-ordinates of the same points in the second state of reference, we shall have necessarily

$$\frac{dx}{dx'} = \frac{dx}{dx''}\frac{dx''}{dx'} + \frac{dx}{dy''}\frac{dy''}{dx'} + \frac{dx}{dz''}\frac{dz''}{dx'}, \text{ etc. (nine equations),} \quad (412)$$

and if we write R for the volume of an element in the state (x'', y'', z'') divided by its volume in the state (x', y', z'), we shall have

$$R = \begin{vmatrix} \dfrac{dx''}{dx'} & \dfrac{dx''}{dy'} & \dfrac{dx''}{dz'} \\[2mm] \dfrac{dy''}{dx'} & \dfrac{dy''}{dy'} & \dfrac{dy''}{dz'} \\[2mm] \dfrac{dz''}{dx'} & \dfrac{dz''}{dy'} & \dfrac{dz''}{dz'} \end{vmatrix}, \quad (413)$$

$$\epsilon_{\mathrm{V}'} = R\epsilon_{\mathrm{V}''}, \quad \eta_{\mathrm{V}'} = R\eta_{\mathrm{V}''}. \quad (414)$$

If, then, we have ascertained by experiment the value of $\epsilon_{\mathrm{V}'}$ in terms of $\eta_{\mathrm{V}'}$, $\dfrac{dx}{dx'}$, $\ldots \dfrac{dz}{dz'}$, and the quantities which express the composition of the body, by the substitution of the values given in (412)–(414), we shall obtain $\epsilon_{\mathrm{V}''}$ in terms of $\eta_{\mathrm{V}''}$, $\dfrac{dx}{dx''}$, $\ldots \dfrac{dz}{dz''}$, $\dfrac{dx''}{dx'}$, $\ldots \dfrac{dz''}{dz'}$, and the quantities which express the composition of the body.

We may apply this to the elements of a body which may be variable from point to point in composition and state of strain in a given state of reference (x'', y'', z''), and if the body is fully described in that state of reference, both in respect to its composition and to the displacement which it would be necessary to give to a homogeneous solid of the same composition, for which $\epsilon_{\mathrm{V}'}$ is known in terms of $\eta_{\mathrm{V}'}$, $\dfrac{dx}{dx'}$, $\ldots \dfrac{dz}{dz'}$, and the quantities which express its composition, to bring it from the state of reference (x', y', z') into a similar and similarly situated state of strain with that of the element of the non-homogeneous body, we may evidently regard $\dfrac{dx''}{dx'}$, $\ldots \dfrac{dz''}{dz'}$ as known for each element of the body, that is, as known in terms of x'', y'', z''. We shall then have $\epsilon_{\mathrm{V}''}$ in terms of $\eta_{\mathrm{V}''}$, $\dfrac{dx}{dx''}$, $\ldots \dfrac{dz}{dz''}$, x'', y'', z'' ; and since the composition of the body is known in terms of x'', y'', z'', and the density, if not given directly, can be determined from the density of the homogeneous body in its state of reference (x', y', z'), this is sufficient for determining the equilibrium of any given state of the non-homogeneous solid.

An equation, therefore, which expresses for any kind of solid, and with reference to any determined state of reference, the relation between the quantities denoted by $\epsilon_{V'}$, $\eta_{V'}$, $\dfrac{dx}{dx'}$, ... $\dfrac{dz}{dz'}$, involving also the quantities which express the composition of the body, when that is capable of continuous variation, or any other equation from which the same relations may be deduced, may be called a *fundamental equation* for that kind of solid. It will be observed that the sense in which this term is here used, is entirely analogous to that in which we have already applied the term to fluids and solids which are subject only to hydrostatic pressure.

When the fundamental equation between $\epsilon_{V'}$, $\eta_{V'}$, $\dfrac{dx}{dx'}$, ... $\dfrac{dz}{dz'}$ is known, we may obtain by differentiation the values of t, $X_{X'}$, ... $Z_{Z'}$ in terms of the former quantities, which will give eleven independent relations between the twenty-one quantities

$$\epsilon_{V'}, \eta_{V'}, \frac{dx}{dx'}, \cdots \frac{dz}{dz'}, t, X_{X'}, \ldots Z_{Z'}, \qquad (415)$$

which are all that exist, since ten of these quantities are independent. All these equations may also involve variables which express the composition of the body, when that is capable of continuous variation.

If we use the symbol $\psi_{V'}$ to denote the value of ψ (as defined on page 89) for any element of a solid divided by the volume of the element in the state of reference, we shall have

$$\psi_{V'} = \epsilon_{V'} - t\eta_{V'}. \qquad (416)$$

The equation (356) may be reduced to the form

$$\delta\psi_{V'} = -\eta_{V'}\,\delta t + \Sigma\Sigma'\left(X_{X'}\,\delta\frac{dx}{dx'}\right). \qquad (417)$$

Therefore, if we know the value of $\psi_{V'}$ in terms of the variables t, $\dfrac{dx}{dx'}$, ... $\dfrac{dz}{dz'}$, together with those which express the composition of the body, we may obtain by differentiation the values of $\eta_{V'}$, $X_{X'}$, ... $Z_{Z'}$ in terms of the same variables. This will make eleven independent relations between the same quantities as before, except that we shall have $\psi_{V'}$ instead of $\epsilon_{V'}$. Or if we eliminate $\psi_{V'}$ by means of equation (416), we shall obtain eleven independent equations between the quantities in (415) and those which express the composition of the body. An equation, therefore, which determines the value of $\psi_{V'}$ as a function of the quantities t, $\dfrac{dx}{dx'}$, ... $\dfrac{dz}{dz'}$, and the quantities which express the composition of the body when it is capable of continuous variation, is a fundamental equation for the kind of solid to which it relates.

In the discussion of the conditions of equilibrium of a solid, we might have started with the principle that it is necessary and sufficient for equilibrium that the temperature shall be uniform throughout the whole mass in question, and that the variation of the force-function $(-\psi)$ of the same mass shall be null or negative for any variation in the state of the mass not affecting its temperature. We might have assumed that the value of ψ for any same element of the solid is a function of the temperature and the state of strain, so that for constant temperature we might write

$$\delta\psi_{\mathrm{V}'} = \Sigma\Sigma'\left(X_{\mathrm{X}'}\cdot\delta\frac{dx}{dx'}\right),$$

the quantities $X_{\mathrm{X}}, \ldots Z_{\mathrm{Z}'}$, being defined by this equation. This would be only a formal change in the definition of $X_{\mathrm{X}'}, \ldots Z_{\mathrm{Z}'}$ and would not affect their values, for this equation holds true of $X_{\mathrm{X}'}, \ldots Z_{\mathrm{Z}'}$ as defined by equation (355). With such data, by transformations similar to those which we have employed, we might obtain similar results.* It is evident that the only difference in the equations would be that $\psi_{\mathrm{V}'}$ would take the place of $\epsilon_{\mathrm{V}'}$, and that the terms relating to entropy would be wanting. Such a method is evidently preferable with respect to the directness with which the results are obtained. The method of this paper shows more distinctly the *rôle* of *energy* and *entropy* in the theory of equilibrium, and can be extended more naturally to those dynamical problems in which motions take place under the condition of constancy of entropy of the elements of a solid (as when vibrations are propagated through a solid), just as the other method can be more naturally extended to dynamical problems in which the temperature is constant. (See note on page 90.)

We have already had occasion to remark that the state of strain of any element considered without reference to directions in space is capable of only six independent variations. Hence, it must be possible to express the state of strain of an element by six functions of $\frac{dx}{dx'}, \ldots \frac{dz}{dz'}$, which are independent of the position of the element. For these quantities we may choose the squares of the ratios of elongation of lines parallel to the three co-ordinate axes in the state of reference, and the products of the ratios of elongation for each pair of these lines multiplied by the cosine of the angle which they include in the variable state of the solid. If we denote these quantities by A, B, C, a, b, c we shall have

---

* For an example of this method, see Thomson and Tait's *Natural Philosophy*, vol. i, p. 705. With regard to the general theory of elastic solids, compare also Thomson's Memoir "On the Thermo-elastic and Thermo-magnetic Properties of Matter" in the *Quarterly Journal of Mathematics*, vol. i, p. 57 (1855), and Green's memoirs on the propagation, reflection, and refraction of light in the *Transactions of the Cambridge Philosophical Society*, vol. vii.

$$A = \Sigma\left(\frac{dx}{dx'}\right)^2, \qquad B = \Sigma\left(\frac{dx}{dy'}\right)^2, \qquad C = \Sigma\left(\frac{dx}{dz'}\right)^2, \tag{418}$$

$$a = \Sigma\left(\frac{dx}{dy'}\frac{dx}{dz'}\right), \qquad b = \Sigma\left(\frac{dx}{dz'}\frac{dx}{dx'}\right), \qquad c = \Sigma\left(\frac{dx}{dx'}\frac{dx}{dy'}\right). \tag{419}$$

The determination of the fundamental equation for a solid is thus reduced to the determination of the relation between $\epsilon_{V'}$, $\eta_{V'}$, A, B, C, a, b, c, or of the relation between $\psi_{V'}$, t, A, B, C, a, b, c.

In the case of isotropic solids, the state of strain of an element, so far as it can affect the relation of $\epsilon_{V'}$ and $\eta_{V'}$, or of $\psi_{V'}$ and t, is capable of only three independent variations. This appears most distinctly as a consequence of the proposition that for any given strain of an element there are three lines in the element which are at right angles to one another both in its unstrained and in its strained state. If the unstrained element is isotropic, the ratios of elongation for these three lines must with $\eta_{V'}$ determine the value $\epsilon_{V'}$, or with t determine the value of $\psi_{V'}$.

To demonstrate the existence of such lines, which are called the *principal axes of strain*, and to find the relations of the elongations of such lines to the quantities $\frac{dx}{dx'}, \dots \frac{dz}{dz'}$, we may proceed as follows. The ratio of elongation r of any line of which α', β', γ' are the direction-cosines in the state of reference is evidently given by the equation

$$r^2 = \left(\frac{dx}{dx'}\alpha' + \frac{dx}{dy'}\beta' + \frac{dx}{dz'}\gamma'\right)^2$$
$$+ \left(\frac{dy}{dx'}\alpha' + \frac{dy}{dy'}\beta' + \frac{dy}{dz'}\gamma'\right)^2$$
$$+ \left(\frac{dz}{dx'}\alpha' + \frac{dz}{dy'}\beta' + \frac{dz}{dz'}\gamma'\right)^2. \tag{420}$$

Now the proposition to be established is evidently equivalent to this —that it is always possible to give such directions to the two systems of rectangular axes X', Y', Z', and X, Y, Z, that

$$\left.\begin{array}{ccc} \dfrac{dx}{dy'} = 0, & \dfrac{dx}{dz'} = 0, & \dfrac{dy}{dz'} = 0, \\[2mm] \dfrac{dy}{dx'} = 0, & \dfrac{dz}{dx'} = 0, & \dfrac{dz}{dy'} = 0. \end{array}\right\} \tag{421}$$

We may choose a line in the element for which the value of r is at least as great as for any other, and make the axes of X and X' parallel to this line in the strained and unstrained states respectively.

Then
$$\frac{dy}{dx'} = 0, \qquad \frac{dz}{dx'} = 0. \tag{422}$$

Moreover, if we write $\dfrac{d(r^2)}{da'}, \dfrac{d(r^2)}{d\beta'}, \dfrac{d(r^2)}{d\gamma'}$ for the differential coefficients obtained from (420) by treating a', β', γ' as *independent* variables,

$$\frac{d(r^2)}{da'} da' + \frac{d(r^2)}{d\beta'} d\beta' + \frac{d(r^2)}{d\gamma'} d\gamma' = 0,$$

when

$$a' da' + \beta' d\beta' + \gamma' d\gamma' = 0,$$

and

$$a' = 1, \quad \beta' = 0, \quad \gamma' = 0.$$

That is,

$$\frac{d(r^2)}{d\beta'} = 0, \text{ and } \frac{d(r^2)}{d\gamma'} = 0,$$

when

$$a' = 1, \quad \beta' = 0, \quad \gamma' = 0.$$

Hence,

$$\frac{dx}{dy'} = 0, \quad \frac{dx}{dz'} = 0. \tag{423}$$

Therefore a line of the element which in the unstrained state is perpendicular to X' is perpendicular to X in the strained state. Of all such lines we may choose one for which the value of r is at least as great as for any other, and make the axes of Y' and Y parallel to this line in the unstrained and in the strained state respectively. Then

$$\frac{dz}{dy'} = 0 ; \tag{424}$$

and it may easily be shown by reasoning similar to that which has just been employed that

$$\frac{dy}{dz'} = 0. \tag{425}$$

Lines parallel to the axes of X', Y', and Z' in the unstrained body will therefore be parallel to X, Y, and Z in the strained body, and the ratios of elongation for such lines will be

$$\frac{dx}{dx'}, \quad \frac{dy}{dy'}, \quad \frac{dz}{dz'}.$$

These lines have the common property of a stationary value of the ratio of elongation for varying directions of the line. This appears from the form to which the general value of r^2 is reduced by the positions of the co-ordinate axes, viz.,

$$r^2 = \left(\frac{dx}{dx'}\right)^2 a'^2 + \left(\frac{dy}{dy'}\right)^2 \beta'^2 + \left(\frac{dz}{dz'}\right)^2 \gamma'^2.$$

Having thus proved the existence of lines, with reference to any particular strain, which have the properties mentioned, let us proceed to find the relations between the ratios of elongation for these lines (the *principal axes of strain*) and the quantities

$\dfrac{dx}{dx'}, \ldots \dfrac{dz}{dz'}$ under the most general supposition with respect to the position of the co-ordinate axes.

For any principal axis of strain we have

$$\frac{d(r^2)}{da'}da' + \frac{d(r^2)}{d\beta'}d\beta' + \frac{d(r^2)}{d\gamma'}d\gamma' = 0,$$

when

$$\alpha' \, da' + \beta' \, d\beta' + \gamma' \, d\gamma' = 0,$$

the differential coefficients in the first of these equations being determined from (420) as before. Therefore,

$$\frac{1}{\alpha'}\frac{d(r^2)}{da'} = \frac{1}{\beta'}\frac{d(r^2)}{d\beta'} = \frac{1}{\gamma'}\frac{d(r^2)}{d\gamma'}. \tag{426}$$

From (420) we obtain directly

$$\frac{\alpha'}{2}\frac{d(r^2)}{da'} + \frac{\beta'}{2}\frac{d(r^2)}{d\beta'} + \frac{\gamma'}{2}\frac{d(r^2)}{d\gamma'} = r^2. \tag{427}$$

From the two last equations, in virtue of the necessary relation $\alpha'^2 + \beta'^2 + \gamma'^2 = 1$, we obtain

$$\tfrac{1}{2}\frac{d(r^2)}{da'} = \alpha' \, r^2, \quad \tfrac{1}{2}\frac{d(r^2)}{d\beta'} = \beta' \, r^2, \quad \tfrac{1}{2}\frac{d(r^2)}{d\gamma'} = \gamma' \, r^2, \tag{428}$$

or, if we substitute the values of the differential coefficients taken from (420),

$$\left.\begin{aligned}
\alpha' \, \Sigma\left(\frac{dx}{dx'}\right)^2 + \beta' \, \Sigma\left(\frac{dx}{dx'}\frac{dx}{dy'}\right) + \gamma' \, \Sigma\left(\frac{dx}{dx'}\frac{dx}{dz'}\right) &= \alpha' \, r^2, \\
\alpha' \, \Sigma\left(\frac{dx}{dy'}\frac{dx}{dx'}\right) + \beta' \, \Sigma\left(\frac{dx}{dy'}\right)^2 + \gamma' \, \Sigma\left(\frac{dx}{dy'}\frac{dx}{dz'}\right) &= \beta' \, r^2, \\
\alpha' \, \Sigma\left(\frac{dx}{dz'}\frac{dx}{dx'}\right) + \beta' \, \Sigma\left(\frac{dx}{dz'}\frac{dx}{dy'}\right) + \gamma' \, \Sigma\left(\frac{dx}{dz'}\right)^2 &= \gamma' r^2.
\end{aligned}\right\} \tag{429}$$

If we eliminate α', β', γ' from these equations, we may write the result in the form,

$$\begin{vmatrix}
\Sigma\left(\dfrac{dx}{dx'}\right)^2 - r^2 & \Sigma\left(\dfrac{dx}{dx'}\dfrac{dx}{dy'}\right) & \Sigma\left(\dfrac{dx}{dx'}\dfrac{dx}{dz'}\right) \\[2mm]
\Sigma\left(\dfrac{dx}{dy'}\dfrac{dx}{dx'}\right) & \Sigma\left(\dfrac{dx}{dy'}\right)^2 - r^2 & \Sigma\left(\dfrac{dx}{dy'}\dfrac{dx}{dz'}\right) \\[2mm]
\Sigma\left(\dfrac{dx}{dz'}\dfrac{dx}{dx'}\right) & \Sigma\left(\dfrac{dx}{dz'}\dfrac{dx}{dy'}\right) & \Sigma\left(\dfrac{dx}{dz'}\right)^2 - r^2
\end{vmatrix} = 0. \tag{430}$$

We may write

$$-r^6 + Er^4 - Fr^2 + G = 0. \tag{431}$$

Then

$$E = \Sigma'\Sigma\left(\frac{dx}{dx'}\right)^2. \tag{432}$$

Also*

$$F = \Sigma' \left\{ \Sigma \left(\frac{dx}{dx'}\right)^2 \Sigma \left(\frac{dx}{dy'}\right)^2 - \Sigma \left(\frac{dx}{dx'}\frac{dx}{dy'}\right) \Sigma \left(\frac{dx}{dx'}\frac{dx}{dy'}\right) \right\}$$

$$= \Sigma' \Sigma \left\{ \left(\frac{dx}{dx'}\right)^2 \Sigma \left(\frac{dx}{dy'}\right)^2 - \frac{dx}{dx'}\frac{dx}{dy'} \Sigma \left(\frac{dx}{dx'}\frac{dx}{dy'}\right) \right\}$$

$$= \Sigma' \Sigma \left\{ \left(\frac{dx}{dx'}\right)^2 \left(\frac{dy}{dy'}\right)^2 + \left(\frac{dx}{dx'}\right)^2 \left(\frac{dz}{dy'}\right)^2 - \frac{dx}{dx'}\frac{dx}{dy'}\frac{dy}{dx'}\frac{dy}{dy'} - \frac{dx}{dx'}\frac{dx}{dy'}\frac{dz}{dx'}\frac{dz}{dy'} \right\}$$

$$= \Sigma' \Sigma \left\{ \left(\frac{dx}{dx'}\right)^2 \left(\frac{dy}{dy'}\right)^2 + \left(\frac{dy}{dx'}\right)^2 \left(\frac{dx}{dy'}\right)^2 - 2\frac{dx}{dx'}\frac{dx}{dy'}\frac{dy}{dx'}\frac{dy}{dy'} \right\}$$

$$= \Sigma' \Sigma \left(\frac{dx}{dx'}\frac{dy}{dy'} - \frac{dy}{dx'}\frac{dx}{dy'}\right)^2. \tag{433}$$

This may also be written

$$F = \Sigma' \Sigma \begin{vmatrix} \dfrac{dx}{dx'} & \dfrac{dx}{dy'} \\ \dfrac{dy}{dx'} & \dfrac{dy}{dy'} \end{vmatrix}^2. \tag{434}$$

In the reduction of the value of G, it will be convenient to use the symbol $\underset{3+3}{\Sigma}$ to denote the sum of the *six* terms formed by changing x, y, z, into y, z, x; z, x, y; x, z, y; y, x, z; and z, y, x; and the symbol $\underset{3-3}{\Sigma}$ in the same sense except that the last three terms are to be taken negatively; also to use Σ' in a similar sense with respect to x', y', z'; and to use $\underset{3-3}{x', y', z'}$ as equivalent to x', y', z', except that they are not to be affected by the sign of summation. With this understanding we may write

$$G = \underset{3-3}{\Sigma'} \left\{ \Sigma \left(\frac{dx}{dx'}\frac{dx}{dx'}\right) \Sigma \left(\frac{dx}{dy'}\frac{dx}{dy'}\right) \Sigma \left(\frac{dx}{dz'}\frac{dx}{dz'}\right) \right\}. \tag{435}$$

In expanding the product of the three sums, we may cancel on account of the sign Σ' the terms which do not contain all the three expressions dx, dy, and dz. Hence we may write

$$G = \underset{3-3}{\Sigma'} \underset{3+3}{\Sigma} \left(\frac{dx}{dx'}\frac{dx}{dx'}\frac{dy}{dy'}\frac{dy}{dy'}\frac{dz}{dz'}\frac{dz}{dz'}\right)$$

$$= \underset{3+3}{\Sigma} \left\{ \frac{dx}{dx'}\frac{dy}{dy'}\frac{dz}{dz'} \underset{3-3}{\Sigma'} \left(\frac{dx}{dx'}\frac{dy}{dy'}\frac{dz}{dz'}\right) \right\}$$

$$= \underset{3-3}{\Sigma} \left(\frac{dx}{dx'}\frac{dy}{dy'}\frac{dz}{dz'}\right) \underset{3-3}{\Sigma'} \left(\frac{dx}{dx'}\frac{dy}{dy'}\frac{dz}{dz'}\right). \tag{436}$$

---

* The values of F and G given in equations (434) and (438), which are here deduced at length, may be derived from inspection of equation (430) by means of the usual theorems relating to the multiplication of determinants. See Salmon's *Lessons Introductory to the Modern Higher Algebra*, 2d ed., Lesson III; or Baltzer's *Theorie und Anwendung der Determinanten*, § 5.

Or, if we set

$$H = \begin{vmatrix} \dfrac{dx}{dx'} & \dfrac{dx}{dy'} & \dfrac{dx}{dz'} \\[2mm] \dfrac{dy}{dx'} & \dfrac{dy}{dy'} & \dfrac{dy}{dz'} \\[2mm] \dfrac{dz}{dx'} & \dfrac{dz}{dy'} & \dfrac{dz}{dz'} \end{vmatrix}, \tag{437}$$

we shall have

$$G = H^2. \tag{438}$$

It will be observed that F represents the sum of the squares of the nine minors which can be formed from the determinant in (437), and that E represents the sum of the squares of the nine constituents of the same determinant.

Now we know by the theory of equations that equation (431) will be satisfied in general by three different values of r^2, which we may denote by r_1^2, r_2^2, r_3^2, and which must represent the squares of the ratios of elongation for the three principal axes of strain; also that E, F, G are symmetrical functions of r_1^2, r_2^2, r_3^2, viz.,

$$\left. \begin{array}{c} E = r_1^2 + r_2^2 + r_3^2, \quad F = r_1^2 r_2^2 + r_2^2 r_3^2 + r_3^2 r_1^2, \\[2mm] G = r_1^2 r_2^2 r_3^2. \end{array} \right\} \tag{439}$$

Hence, although it is possible to solve equation (431) by the use of trigonometrical functions, it will be more simple to regard $\epsilon_{V'}$ as a function of $\eta_{V'}$ and the quantities E, F, G (or H), which we have expressed in terms of $\dfrac{dx}{dx'}, \dots \dfrac{dz}{dz'}$. Since $\epsilon_{V'}$ is a single-valued function of $\eta_{V'}$ and r_1^2, r_2^2, r_3^2 (with respect to all the changes of which the body is capable), and a symmetrical function with respect to r_1^2, r_2^2, r_3^2, and since r_1^2, r_2^2, r_3^2 are *collectively* determined without ambiguity by the values of E, F, and H, the quantity $\epsilon_{V'}$ must be a single-valued function of $\eta_{V'}$, E, F, and H. The determination of the fundamental equation for isotropic bodies is therefore reduced to the determination of this function, or (as appears from similar considerations) the determination of $\psi_{V'}$ as a function of t, E, F, and H.

It appears from equations (439) that E represents the sum of the squares of the ratios of elongation for the principal axes of strain, that F represents the sum of the squares of the ratios of enlargement for the three surfaces determined by these axes, and that G represents the square of the ratio of enlargement of volume. Again, equation (432) shows that E represents the sum of the squares of the ratios of elongation for lines parallel to X', Y', and Z'; equation (434) shows that F represents the sum of the squares of the ratios of enlargement for surfaces parallel to the planes X'-Y', Y'-Z', Z'-X'; and equation (438), like (439), shows that G represents the square of the ratio of enlargement of volume. Since the position of the co-ordinate axes is arbitrary, it follows that the sum of the squares of the ratios of elongation or enlargement of three lines or surfaces which in the unstrained state are at right angles to one another, is otherwise independent of the direction of the lines or surfaces. Hence, $\frac{1}{3}E$ and $\frac{1}{3}F$ are the mean squares of the ratios of linear elongation and of superficial enlargement, for all possible directions in the unstrained solid.

There is not only a practical advantage in regarding the strain as determined by E, F, and H, instead of E, F, and G, because H is more simply expressed in terms of $\frac{dx}{dx'}$, ... $\frac{dz}{dz'}$, but there is also a certain theoretical advantage on the side of E, F, H. If the systems of co-ordinate axes X, Y, Z, and X', Y', Z', are either identical or such as are capable of superposition, which it will always be convenient to suppose, the determinant H will always have a positive value for any strain of which a body can be capable. But it is possible to give to x, y, z such values as functions of x', y', z' that H shall have a negative value. For example, we may make

$$x = x', \quad y = y', \quad z = -z'. \tag{440}$$

This will give $H = -1$, while

$$x = x', \quad y = y', \quad z = z' \tag{441}$$

will give $H = 1$. Both (440) and (441) give $G = 1$. Now although such a change in the position of the particles of a body as is represented by (440) cannot take place while the body remains solid, yet a method of representing strains may be considered incomplete, which confuses the cases represented by (440) and (441).

We may avoid all such confusion by using E, F, and H to represent a strain. Let us consider an element of the body strained which in the state (x', y', z') is a cube with its edges parallel to the axes of X', Y', Z', and call the edges dx', dy', dz' according to the axes to which they are parallel, and consider the ends of the edges as positive for which the values of x', y', or z' are the greater. Whatever may be the nature of the parallelopiped in the state (x, y, z) which corresponds to the cube dx', dy', dz' and is determined by the quantities $\frac{dx}{dx'}$, ... $\frac{dz}{dz'}$, it may always be brought by continuous changes to the form of a cube and to a position in which the edges dx', dy' shall be parallel to the axes of X and Y, the positive ends of the edges toward the positive directions of the axes, and this may be done without giving the volume of the parallelopiped the value zero, and therefore without changing the sign of H. Now two cases are possible;—the positive end of the edge dz' may be turned toward the positive or toward the negative direction of the axis of Z. In the first case, H is evidently positive; in the second, negative. The determinant H will therefore be positive or negative,—we may say, if we choose, that the volume will be positive or negative,—according as the element can or cannot be brought from the state (x, y, z) to the state (x', y', z') by continuous changes without giving its volume the value zero.

If we now recur to the consideration of the principal axes of strain and the principal ratios of elongation r_1, r_2, r_3, and denote by U_1, U_2, U_3 and U_1', U_2', U_3' the principal axes of strain in the strained and unstrained element respectively, it is evident that the sign of r_1, for example, depends upon the direction in U_1 which we regard as corresponding to a given direction in U_1'. If we choose to associate directions in these axes so that r_1, r_2, r_3 shall all be positive, the positive or negative value of H will determine whether the system of axes U_1, U_2, U_3 is or is not capable of superposition upon the system U_1', U_2', U_3' so that corresponding directions in the axes shall coincide. Or, if we prefer to associate directions in the two systems of axes so that they shall be capable of superposition, corresponding directions coinciding, the positive or negative value of H will determine whether an even or an odd number of the quantities r_1, r_2, r_3 are negative. In this case we may write

$$r_1 r_2 r_3 = H = \begin{vmatrix} \dfrac{dx}{dx'} & \dfrac{dx}{dy'} & \dfrac{dx}{dz'} \\ \dfrac{dy}{dx'} & \dfrac{dy}{dy'} & \dfrac{dy}{dz'} \\ \dfrac{dz}{dx'} & \dfrac{dz}{dy'} & \dfrac{dz}{dz'} \end{vmatrix}. \tag{442}$$

It will be observed that to change the signs of two of the quantities r_1, r_2, r_3 is simply to give a certain rotation to the body without changing its state of strain.

Whichever supposition we make with respect to the axes U_1, U_2, U_3, it is evident that the state of strain is completely determined by the values E, F, and H, not only when we limit ourselves to the consideration of such strains as are consistent with the idea of solidity, but also when we regard any values of $\dfrac{dx}{dx'}, \dots \dfrac{dz}{dz'}$ as possible.

Approximative Formulæ.—For many purposes the value of $\epsilon_{V'}$ for an isotropic solid may be represented with sufficient accuracy by the formula

$$\epsilon_{V'} = i' + e'E + f'F + h'H, \tag{443}$$

where i', e', f', and h' denote functions of $\eta_{V'}$; or the value of $\psi_{V'}$ by the formula

$$\psi_{V'} = i + eE + fF + hH, \tag{444}$$

where $i, e, f,$ and h denote functions of t. Let us first consider the second of these formulæ. Since $E, F,$ and H are symmetrical functions of r_1, r_2, r_3, if $\psi_{V'}$ is any function of t, E, F, H, we must have

$$\left.\begin{array}{l} \dfrac{d\psi_{V'}}{dr_1} = \dfrac{d\psi_{V'}}{dr_2} = \dfrac{d\psi_{V'}}{dr_3}, \\[2mm] \dfrac{d^2\psi_{V'}}{dr_1{}^2} = \dfrac{d^2\psi_{V'}}{dr_2{}^2} = \dfrac{d^2\psi_{V'}}{dr_3{}^2}, \\[2mm] \dfrac{d^2\psi_{V'}}{dr_1 dr_2} = \dfrac{d^2\psi_{V'}}{dr_2 dr_3} = \dfrac{d^2\psi_{V'}}{dr_3 dr_1}, \end{array}\right\} \qquad (445)$$

whenever $r_1 = r_2 = r_3$. Now $i, e, f,$ and h may be determined (as functions of t) so as to give to

$$\psi_{V'}, \quad \dfrac{d\psi_{V'}}{dr_1}, \quad \dfrac{d^2\psi_{V'}}{dr_1{}^2}, \quad \dfrac{d^2\psi_{V'}}{dr_1 dr_2}$$

their proper values at every temperature for some isotropic state of strain, which may be determined by any desired condition. We shall suppose that they are determined so as to give the proper values to $\psi_{V'}$, etc., when the stresses in the solid vanish. If we denote by r_0 the common value of r_1, r_2, r_3 which will make the stresses vanish at any given temperature, and imagine the true value of $\psi_{V'}$, and also the value given by equation (444) to be expressed in terms of the ascending powers of

$$r_1 - r_0, \quad r_2 - r_0, \quad r_3 - r_0, \qquad (446)$$

it is evident that the expressions will coincide as far as the terms of the second degree *inclusive*. That is, the errors of the values of $\psi_{V'}$ given by equation (444) are of the same order of magnitude as the cubes of the above differences. The errors of the values of

$$\dfrac{d\psi_{V'}}{dr_1}, \quad \dfrac{d\psi_{V'}}{dr_2}, \quad \dfrac{d\psi_{V'}}{dr_3}$$

will be of the same order of magnitude as the squares of the same differences. Therefore, since

$$\dfrac{d\psi_{V'}}{d\dfrac{dx}{dx'}} = \dfrac{d\psi_{V'}}{dr_1}\dfrac{dr_1}{d\dfrac{dx}{dx'}} + \dfrac{d\psi_{V'}}{dr_2}\dfrac{dr_2}{d\dfrac{dx}{dx'}} + \dfrac{d\psi_{V'}}{dr_3}\dfrac{dr_3}{d\dfrac{dx}{dx'}} \qquad (447)$$

whether we regard the true value of $\psi_{V'}$ or the value given by equation (444), and since the error in (444) does not affect the values of

$$\dfrac{dr_1}{d\dfrac{dx}{dx'}}, \quad \dfrac{dr_2}{d\dfrac{dx}{dx'}}, \quad \dfrac{dr_3}{d\dfrac{dx}{dx'}},$$

which we may regard as determined by equations (431), (432), (434), (437) and (438), the errors in the values of $X_{x'}$ derived from (444) will be of the same order of magnitude as the squares of the differences in (446). The same will be true with respect to $X_{Y'}$, $X_{Z'}$, $Y_{X'}$, etc., etc.

It will be interesting to see how the quantities e, f, and h are related to those which most simply represent the elastic properties of isotropic solids. If we denote by V and R the *elasticity of volume* and the *rigidity**(both determined under the condition of constant temperature and for states of vanishing stress), we shall have as definitions

$$V = -v\left(\frac{dp}{dv}\right)_t, \text{ when } v = r_0{}^3 v', \tag{448}$$

where p denotes a uniform pressure to which the solid is subjected, v its volume, and v' its volume in the state of reference; and

$$\left. \begin{aligned} r_0 R &= \frac{dX_{Y'}}{d\frac{dx}{dy'}} = \frac{d^2 \psi_{V'}}{\left(d\frac{dx}{dy'}\right)^2}, \\[2mm] \text{when} \qquad \frac{dx}{dx'} &= \frac{dy}{dy'} = \frac{dz}{dz'} = r_0, \\[2mm] \text{and} \qquad \frac{dx}{dy'} &= \frac{dx}{dz'} = \frac{dy}{dz'} = \frac{dy}{dx'} = \frac{dz}{dx'} = \frac{dz}{dy'} = 0. \end{aligned} \right\} \tag{449}$$

Now when the solid is subject to uniform pressure on all sides, if we consider so much of it as has the volume unity in the state of reference, we shall have

$$r_1 = r_2 = r_3 = v^{\frac{1}{3}}, \tag{450}$$

and by (444) and (439),

$$\psi_{V'} = i + 3ev^{\frac{2}{3}} + 3fv^{\frac{4}{3}} + hv. \tag{451}$$

Hence, by equation (88), since $\psi_{V'}$ is equivalent to ψ,

$$-p = \left(\frac{d\psi}{dv}\right)_t = 2ev^{-\frac{1}{3}} + 4fv^{\frac{1}{3}} + h, \tag{452}$$

$$-v\left(\frac{dp}{dv}\right)_t = -\tfrac{2}{3}ev^{-\frac{1}{3}} + \tfrac{4}{3}fv^{\frac{1}{3}}; \tag{453}$$

and by (448),

$$V = -\tfrac{2}{3}\frac{e}{r_0} + \tfrac{4}{3}fr_0. \tag{454}$$

To obtain the value of R in accordance with the definition (449), we may suppose the values of E, F, and H given by equations (432), (434), and (437) to be substituted in equation (444). This will give for the value of R

$$R = \frac{2e}{r_0} + 2fr_0. \tag{455}$$

---

* See Thomson and Tait's *Natural Philosophy*, vol. i, p. 711.

Moreover, since p must vanish in (452) when $v=r_0{}^3$, we have

$$2e+4fr_0{}^2+hr_0=0. \tag{456}$$

From the three last equations may be obtained the values of e, f, h, in terms of r_0, V, and R; viz.,

$$e=\tfrac{1}{3}r_0R-\tfrac{1}{2}r_0V, \quad f=\frac{R+3V}{6r_0}, \quad h=-\tfrac{4}{3}R-V. \tag{457}$$

The quantity r_0, like R and V, is a function of the temperature, the differential coefficient $\dfrac{d\log r_0}{dt}$ representing the rate of linear expansion of the solid when without stress.

It will not be necessary to discuss equation (443) at length, as the case is entirely analogous to that which has just been treated. (It must be remembered that $\eta_{V'}$, in the discussion of (443), will take the place everywhere of the temperature in the discussion of (444).) If we denote by V' and R' the *elasticity of volume* and the *rigidity*, both determined under the condition of *constant entropy*, (i.e., of *no transmission of heat*,) and for states of vanishing stress, we shall have the equations:—

$$V'=-\frac{2e'}{3r_0}+\tfrac{4}{3}f'r_0, \tag{458}$$

$$R'=\frac{2e'}{r_0}+2f'r_0, \tag{459}$$

$$2e'+4f'r_0{}^2+h'r_0=0. \tag{460}$$

Whence

$$e'=\tfrac{1}{3}r_0R'-\tfrac{1}{2}r_0V', \quad f'=\frac{R'+3V'}{6r_0}, \quad h'=-\tfrac{4}{3}R'-V'. \tag{461}$$

In these equations r_0, R', and V' are to be regarded as functions of the quantity $\eta_{V'}$.

If we wish to change from one state of reference to another (also isotropic), the changes required in the fundamental equation are easily made. If a denotes the length of any line of the solid in the second state of reference divided by its length in the first, it is evident that when we change from the first state of reference to the second the values of the symbols $\epsilon_{V'}$, $\eta_{V'}$, $\psi_{V'}$, H are divided by a^3, that of E by a^2, and that of F by a^4. In making the change of the state of reference, we must therefore substitute in the fundamental equation of the form (444) $a^3\psi_{V'}$, a^2E, a^4F, a^3H for $\psi_{V'}$, E, F, and H, respectively. In the fundamental equation of the form (443), we must make the analogous substitutions, and also substitute $a^3\eta_{V'}$ for $\eta_{V'}$. (It will be remembered that i', e', f', and h' represent functions of $\eta_{V'}$, and that it is only when their values in terms of $\eta_{V'}$ are substituted, that equation (443) becomes a fundamental equation.)

Concerning Solids which absorb Fluids.

There are certain bodies which are solid with respect to some of their components, while they have other components which are fluid. In the following discussion, we shall suppose both the solidity and the fluidity to be perfect, so far as any properties are concerned which can affect the conditions of equilibrium,—i.e., we shall suppose that the solid matter of the body is entirely free from plasticity and that there are no passive resistances to the motion of the fluid components except such as vanish with the velocity of the motion,— leaving it to be determined by experiment how far and in what cases these suppositions are realized.

It is evident that equation (356) must hold true with regard to such a body, when the quantities of the fluid components contained in a given element of the solid remain constant. Let Γ_a', Γ_b', etc., denote the quantities of the several fluid components contained in an element of the body divided by the volume of the element in the state of reference, or, in other words, let these symbols denote the densities which the several fluid components would have, if the body should be brought to the state of reference while the matter contained in each element remained unchanged. We may then say that equation (356) will hold true, when Γ_a', Γ_b', etc., are constant. The complete value of the differential of $\epsilon_{V'}$ will therefore be given by an equation of the form

$$d\epsilon_{V'} = t\, d\eta_{V'} + \Sigma\Sigma'\left(X_{X'} d\frac{dx}{dx'}\right) + L_a d\Gamma_a' + L_b d\Gamma_b' + \text{etc.} \qquad (462)$$

Now when the body is in a state of hydrostatic stress, the term in this equation containing the signs of summation will reduce to $-p\, dv_{V'}$ ($v_{V'}$ denoting, as elsewhere, the volume of the element divided by its volume in the state of reference). For in this case

$$X_{X'} = -p\left(\frac{dy}{dy'}\frac{dz}{dz'} - \frac{dz}{dy'}\frac{dy}{dz'}\right), \qquad (463)$$

$$\Sigma\Sigma'\left(X_{X'} d\frac{dx}{dx'}\right) = -p\Sigma\Sigma'\left\{\left(\frac{dy}{dy'}\frac{dz}{dz'} - \frac{dz}{dy'}\frac{dy}{dz'}\right)d\frac{dx}{dx'}\right\}$$

$$= -p\, d\begin{vmatrix} \dfrac{dx}{dx'} & \dfrac{dx}{dy'} & \dfrac{dx}{dz'} \\[2mm] \dfrac{dy}{dx'} & \dfrac{dy}{dy'} & \dfrac{dy}{dz'} \\[2mm] \dfrac{dz}{dx'} & \dfrac{dz}{dy'} & \dfrac{dz}{dz'} \end{vmatrix}$$

$$= -p\, dv_{V'}. \qquad (464)$$

We have, therefore, for a state of hydrostatic stress,

$$d\epsilon_{V'} = t\,d\eta_{V'} - p\,dv_{V'} + L_a d\Gamma_a' + L_b d\Gamma_b' + \text{etc.}, \tag{465}$$

and multiplying by the volume of the element in the state of reference, which we may regard as constant,

$$d\epsilon = t\,d\eta - p\,dv + L_a dm_a + L_b dm_b + \text{etc.}, \tag{466}$$

where ϵ, η, v, m_a, m_b, etc., denote the energy, entropy, and volume of the element, and the quantities of its several fluid components. It is evident that the equation will also hold true, if these symbols are understood as relating to a homogeneous body of finite size. The only limitation with respect to the variations is that the element or body to which the symbols relate shall always contain the same solid matter. The varied state may be one of hydrostatic stress or otherwise.

But when the body is in a state of hydrostatic stress, and the solid matter is considered invariable, we have by equation (12)

$$d\epsilon = t\,d\eta - p\,dv + \mu_a dm_a + \mu_b dm_b + \text{etc.} \tag{467}$$

It should be remembered that the equation cited occurs in a discussion which relates only to bodies of hydrostatic stress, so that the varied state as well as the initial is there regarded as one of hydrostatic stress. But a comparison of the two last equations shows that the last will hold true without any such limitation, and moreover, that the quantities L_a, L_b, etc., when determined for a state of hydrostatic stress, are equal to the potentials μ_a, μ_b, etc.

Since we have hitherto used the term *potential* solely with reference to bodies of hydrostatic stress, we may apply this term as we choose with regard to other bodies. We may therefore call the quantities L_a, L_b, etc., the *potentials* for the several fluid components in the body considered, whether the state of the body is one of hydrostatic stress or not, since this use of the term involves only an extension of its former definition. It will also be convenient to use our ordinary symbol for a potential to represent these quantities. Equation (462) may then be written

$$d\epsilon_{V'} = t\,d\eta_{V'} + \Sigma\Sigma'\left(X_{X'}\,d\frac{dx}{dx'}\right) + \mu_a d\Gamma_a' + \mu_b d\Gamma_b' + \text{etc.} \tag{468}$$

This equation holds true of solids having fluid components without any limitation with respect to the initial state or to the variations, except that the solid matter to which the symbols relate shall remain the same.

In regard to the conditions of equilibrium for a body of this kind, it is evident in the first place that if we make Γ_a', Γ_b', etc., constant, we shall obtain from the general criterion of equilibrium all the conditions which we have obtained for ordinary solids, and which are expressed by the formulæ (364), (374), (380), (382)–(384). The quantities Γ_1', Γ_2', etc., in the last two formulæ include of course those which have just been represented by Γ_a', Γ_b', etc., and which relate to the fluid components of the body, as well as the corresponding quantities relating to its solid components. Again, if we suppose the solid matter of the body to remain without variation in quantity or position, it will easily appear that the potentials for the substances which form the fluid components of the solid body must satisfy the same conditions in the solid body and in the fluids in contact with it, as in the case of entirely fluid masses. See eqs. (22).

The above conditions must however be slightly modified in order to make them sufficient for equilibrium. It is evident that if the solid is dissolved at its surface, the fluid components which are set free may be absorbed by the solid as well as by the fluid mass, and in like manner if the quantity of the solid is increased, the fluid components of the new portion may be taken from the previously existing solid mass. Hence, whenever the *solid* components of the solid body are actual components of the fluid mass, (whether the case is the same with the *fluid* components of the solid body or not,) an equation of the form (383) must be satisfied, in which the potentials μ_a, μ_b, etc., contained implicitly in the second member of the equation are determined from the solid body. Also if the *solid* components of the solid body are all possible but not all actual components of the fluid mass, a condition of the form (384) must be satisfied, the values of the potentials in the second member being determined as in the preceding case.

The quantities

$$t, \quad X_{X'}, \ldots Z_{Z'}, \quad \mu_a, \quad \mu_b, \quad \text{etc.,} \tag{469}$$

being differential coefficients of $\epsilon_{V'}$ with respect to the variables

$$\eta_{V'}, \quad \frac{dx}{dx'}, \ldots \frac{dz}{dz'}, \quad \Gamma_a', \quad \Gamma_b', \quad \text{etc.,} \tag{470}$$

will of course satisfy the necessary relations

$$\frac{dt}{d\frac{dx}{dx'}} = \frac{dX_{X'}}{d\eta_{V'}}, \quad \text{etc.} \tag{471}$$

This result may be generalized as follows. Not only is the second member of equation (468) a complete differential in its present form, but it will remain such if we transfer the sign of differentiation (d) from one factor to the other of any term (the sum indicated by the symbol $\Sigma\Sigma'$ is here supposed to be expanded into nine terms), and at the same time change the sign of the term from $+$ to $-$. For to substitute $-\eta_{V'}dt$ for $td\eta_{V'}$, for example, is equivalent to subtracting the complete differential $d(t\eta_{V'})$. Therefore, if we consider the quantities in (469) and (470) which occur in any same term in equation (468) as forming a pair, we may choose as independent variables either quantity of each pair, and the differential coefficient of the remaining quantity of any pair with respect to the independent variable of another pair will be equal to the differential coefficient of the remaining quantity of the second pair with respect to the independent variable of the first, taken positively, if the independent variables of these pairs are both affected by the sign d in equation (468), or are neither thus affected, but otherwise taken negatively. Thus

$$\left(\frac{dX_{X'}}{d\Gamma_a'}\right)_{\frac{dx}{dx'}} = \left(\frac{d\mu_a}{d\frac{dx}{dx'}}\right)_{\Gamma_a'}, \quad \left(\frac{dX_{X'}}{d\mu_a}\right)_{\frac{dx}{dx'}} = -\left(\frac{d\Gamma_a}{d\frac{dx}{dx'}}\right)_{\mu_a}, \qquad (472)$$

$$\left(\frac{d\frac{dx}{dx'}}{d\mu_a}\right)_{X_{X'}} = \left(\frac{d\Gamma_a'}{dX_{X'}}\right)_{\mu_a}, \quad \left(\frac{d\frac{dx}{dx'}}{d\Gamma_a'}\right)_{X_{X'}} = -\left(\frac{d\mu_a}{dX_{X'}}\right)_{\Gamma_a'}, \qquad (473)$$

where in addition to the quantities indicated by the suffixes, the following are to be considered as constant:—either t or $\eta_{V'}$, either $X_{Y'}$ or $\frac{dx}{dy'}$, ... either $Z_{Z'}$ or $\frac{dz}{dz'}$, either μ_b or Γ_b', etc.

It will be observed that when the temperature is constant the conditions $\mu_a = \text{const.}$, $\mu_b = \text{const.}$, represent the physical condition of a body in contact with a fluid of which the phase does not vary, and which contains the components to which the potentials relate. Also that when Γ_a', Γ_b', etc., are constant, the heat absorbed by the body in any infinitesimal change of condition per unit of volume measured in the state of reference is represented by $td\eta_{V'}$. If we denote this quantity by $dQ_{V'}$, and use the suffix $_Q$ to denote the condition of no transmission of heat, we may write

$$\left(\frac{d\log t}{d\frac{dx}{dx'}}\right)_Q = \left(\frac{dX_{X'}}{dQ_{V'}}\right)_{\frac{dx}{dx'}}, \quad \left(\frac{d\log t}{dX_{X'}}\right)_Q = -\left(\frac{d\frac{dx}{dx'}}{dQ_{V'}}\right)_{X_{X'}}, \qquad (474)$$

$$\left(\frac{dQ_{V'}}{dX_{X'}}\right)_t = \left(\frac{d\frac{dx}{dx'}}{d\log t}\right)_{X_{X'}}, \quad \left(\frac{dQ_{V'}}{d\frac{dx}{dx'}}\right)_t = -\left(\frac{dX_{X'}}{d\log t}\right)_{\frac{dx}{dx'}}, \qquad (475)$$

where Γ_a', Γ_b', etc., must be regarded as constant in all the equations, and either $X_{Y'}$ or $\frac{dx}{dy'}$, ... either $Z_{Z'}$ or $\frac{dz}{dz'}$, in each equation.

Influence of Surfaces of Discontinuity upon the Equilibrium of Heterogeneous Masses.—Theory of Capillarity.

We have hitherto supposed, in treating of heterogeneous masses in contact, that they might be considered as separated by mathematical surfaces, each mass being unaffected by the vicinity of the others, so that it might be homogeneous quite up to the separating surfaces both with respect to the density of each of its various components and also with respect to the densities of energy and entropy. That such is not rigorously the case is evident from the consideration that if it were so with respect to the densities of the components it could not be so in general with respect to the density of energy, as the sphere of molecular action is not infinitely small. But we know from observation that it is only within very small distances of such a surface that any mass is sensibly affected by its vicinity,—a natural consequence of the exceedingly small sphere of sensible molecular action,—and this fact renders possible a simple method of taking account of the variations in the densities of the component substances and of energy and entropy, which occur in the vicinity of surfaces of discontinuity. We may use this term, for the sake of brevity, without implying that the discontinuity is absolute, or that the term distinguishes any surface with mathematical precision. It may be taken to denote the non-homogeneous film which separates homogeneous or nearly homogeneous masses.

Let us consider such a surface of discontinuity in a fluid mass which is in equilibrium and uninfluenced by gravity. For the precise measurement of the quantities with which we have to do, it will be convenient to be able to refer to a geometrical surface, which shall be sensibly coincident with the physical surface of discontinuity, but shall have a precisely determined position. For this end, let us take some point in or very near to the physical surface of discontinuity, and imagine a geometrical surface to pass through this point and all other points which are similarly situated with respect to the condition of the adjacent matter. Let this geometrical surface be called the *dividing surface*, and designated by the symbol S. It will be observed that the position of this surface is as yet to a certain extent arbitrary, but that the directions of its normals are already everywhere determined, since all the surfaces which can be formed in the manner described are evidently parallel to one another. Let us also imagine a closed surface cutting the surface S and including a part of the homogeneous mass on each side. We will so far limit the form of this closed surface as to suppose that on each side of S, as far as there is any want of perfect homogeneity in the fluid masses, the closed surface is such as may be generated by a moving normal to S.

Let the portion of S which is included by the closed surface be denoted by **s**, and the area of this portion by s. Moreover, let the mass contained within the closed surface be divided into three parts by two surfaces, one on each side of S, and very near to that surface, although at such distance as to lie entirely beyond the influence of the discontinuity in its vicinity. Let us call the part which contains the surface **s** (with the physical surface of discontinuity) M, and the homogeneous parts M' and M'', and distinguish by ϵ, ϵ', ϵ'', η, η', η'', m_1, m_1', m_1'', m_2, m_2', m_2'', etc., the energies and entropies of these masses, and the quantities which they contain of their various components.

It is necessary, however, to define more precisely what is to be understood in cases like the present by the energy of masses which are only separated from other masses by imaginary surfaces. A part of the total energy which belongs to the matter in the vicinity of the separating surface, relates to pairs of particles which are on different sides of the surface, and such energy is not in the nature of things referable to either mass by itself. Yet, to avoid the necessity of taking separate account of such energy, it will often be convenient to include it in the energies which we refer to the separate masses. When there is no break in the homogeneity at the surface, it is natural to treat the energy as distributed with a uniform density. This is essentially the case with the initial state of the system which we are considering, for it has been divided by surfaces passing in general through homogeneous masses. The only exception—that of the surface which cuts at right angles the non-homogeneous film—(apart from the consideration that without any important loss of generality we may regard the part of this surface within the film as very small compared with the other surfaces) is rather apparent than real, as there is no change in the state of the matter *in the direction perpendicular to this surface*. But in the variations to be considered in the state of the system, it will not be convenient to limit ourselves to such as do not create any discontinuity at the surfaces bounding the masses M, M', M''; we must therefore determine how we will estimate the energies of the masses in case of such infinitesimal discontinuities as may be supposed to arise. Now the energy of each mass will be most easily estimated by neglecting the discontinuity, i.e., if we estimate the energy on the supposition that beyond the bounding surface the phase is identical with that within the surface. This will evidently be allowable, if it does not affect the total amount of energy. To show that it does not affect this quantity, we have only to observe that, if the energy of the mass on one side of a surface where there is an infinitesimal discontinuity of phase is greater as determined by this rule than if determined by any other (suitable) rule, the energy of the mass on the other side must be less by the same amount when determined by the first rule than when determined by the second, since the discontinuity relative to the second mass is equal but opposite in character to the discontinuity relative to the first.

If the entropy of the mass which occupies any one of the spaces considered is not in the nature of things determined without reference to the surrounding masses, we may suppose a similar method to be applied to the estimation of entropy.

With this understanding, let us return to the consideration of the equilibrium of the three masses M, M', and M''. We shall suppose that there are no limitations to the possible variations of the system due to any want of perfect mobility of the components by means of which we express the composition of the masses, and that these components are independent, i.e., that no one of them can be formed out of the others.

With regard to the mass M, which includes the surface of discontinuity, it is necessary for its internal equilibrium that when its boundaries are considered constant, and when we consider only *reversible* variations (i.e., those of which the opposite are also possible), the variation of its energy should vanish with the variations of its entropy and of the quantities of its various components. For changes within this mass will not affect the energy or the entropy of the surrounding masses (when these quantities are estimated on the principle which we have adopted), and it may therefore be treated as an isolated system. For fixed boundaries of the mass M, and for reversible variations, we may therefore write

$$\delta \epsilon = A_0 \delta \eta + A_1 \delta m_1 + A_2 \delta m_2 + \text{etc.}, \qquad (476)$$

where A_0, A_1, A_2, etc., are quantities determined by the initial (unvaried) condition of the system. It is evident that A_0 is the temperature of the lamelliform mass to which the equation relates, or the *temperature at the surface of discontinuity*. By comparison of this equation with (12) it will be seen that the definition of A_1, A_2, etc., is entirely analogous to that of the potentials in homogeneous masses, although the mass to which the former quantities relate is not homogeneous, while in our previous definition of potentials, only homogeneous masses were considered. By a natural extension of the term *potential*, we may call the quantities A_1, A_2, etc., the *potentials at the surface of discontinuity*. This designation will be farther justified by the fact, which will appear hereafter, that the value of these quantities is independent of the thickness of the lamina (M) to which they relate. If we employ our ordinary symbols for temperature and potentials, we may write

$$\delta \epsilon = t \, \delta \eta + \mu_1 \delta m_1 + \mu_2 \delta m_2 + \text{etc.} \qquad (477)$$

If we substitute $\geqq$ for $=$ in this equation, the formula will hold true of all variations whether reversible or not;[*] for if the variation of energy could have a value less than that of the second member of the equation, there must be variation in the condition of M in which its energy is diminished without change of its entropy or of the quantities of its various components.

It is important, however, to observe that for any given values of $\delta\eta$, δm_1, δm_2, etc., while there *may* be possible variations of the nature and state of M for which the value of $\delta\epsilon$ is greater than that of the second member of (477), there *must* always be possible variations for which the value of $\delta\epsilon$ is equal to that of the second member. It will be convenient to have a notation which will enable us to express this by an equation. Let $\mathfrak{d}\epsilon$ denote the smallest value (i.e., the value nearest to $-\infty$) of $\delta\epsilon$ consistent with given values of the other variations, then

$$\mathfrak{d}\epsilon = t\,\delta\eta + \mu_1\delta m_1 + \mu_2\delta m_2 + \text{etc.} \tag{478}$$

For the internal equilibrium of the whole mass which consists of the parts M, M', M'', it is necessary that

$$\delta\epsilon + \delta\epsilon' + \delta\epsilon'' \geqq 0 \tag{479}$$

for all variations which do not affect the enclosing surface or the total entropy or the total quantity of any of the various components. If we also regard the surfaces separating M, M', and M'' as invariable, we may derive from this condition, by equations (478) and (12), the following as a *necessary* condition of equilibrium :—

$$\begin{aligned} &t\,\delta\eta + \mu_1\delta m_1 + \mu_2\delta m_2 + \text{etc.}\\ &+ t'\,\delta\eta' + \mu_1{}'\delta m_1{}' + \mu_2{}'\delta m_2{}' + \text{etc.}\\ &+ t''\,\delta\eta'' + \mu_1{}''\delta m_1{}'' + \mu_2{}''\delta m_2{}'' + \text{etc.} \geqq 0, \end{aligned} \tag{480}$$

---

[*] To illustrate the difference between variations which are reversible, and those which are not, we may conceive of two entirely different substances meeting in equilibrium at a mathematical surface without being at all mixed. We may also conceive of them as mixed in a thin film about the surface where they meet, and then the amount of mixture is capable of variation both by increase and by diminution. But when they are absolutely unmixed, the amount of mixture can be increased, but is incapable of diminution, and it is then consistent with equilibrium that the value of $\delta\epsilon$ (for a variation of the system in which the substances commence to mix) should be greater than the second member of (477). It is not necessary to determine whether precisely such cases actually occur; but it would not be legitimate to overlook the possible occurrence of cases in which variations may be possible while the opposite variations are not.

It will be observed that the sense in which the term *reversible* is here used is entirely different from that in which it is frequently used in treatises on thermodynamics, where a process by which a system is brought from a state A to a state B is called reversible, to signify that the system may also be brought from the state B to the state A through the same series of intermediate states taken in the reverse order by means of external agencies of the opposite character. The variation of a system from a state A to a state B (supposed to differ infinitely little from the first) is here called reversible when the system is capable of another state B' which bears the same relation to the state A that A bears to B.

the variations being subject to the equations of condition

$$\left.\begin{aligned}
\delta\eta + \delta\eta' + \delta\eta'' &= 0, \\
\delta m_1 + \delta m_1' + \delta m_1'' &= 0, \\
\delta m_2 + \delta m_2' + \delta m_2'' &= 0, \\
\text{etc.} &
\end{aligned}\right\} \tag{481}$$

It may also be the case that some of the quantities $\delta m_1'$, $\delta m_1''$, $\delta m_2'$, $\delta m_2''$, etc., are incapable of negative values or can only have the value zero. This will be the case when the substances to which these quantities relate are not actual or possible components of M′ or M″. (See page 64.) To satisfy the above condition it is necessary and sufficient that

$$t = t' = t'', \tag{482}$$

$$\mu_1' \delta m_1' \gtreqless \mu_1 \delta m_1', \qquad \mu_2' \delta m_2' \gtreqless \mu_2 \delta m_2', \quad \text{etc.}, \tag{483}$$

$$\mu_1'' \delta m_1'' \gtreqless \mu_1 \delta m_1'', \qquad \mu_2'' \delta m_2'' \gtreqless \mu_2 \delta m_2'', \quad \text{etc.} \tag{484}$$

It will be observed that, if the substance to which μ_1, for instance, relates is an actual component of each of the homogeneous masses, we shall have $\mu_1 = \mu_1' = \mu_1''$. If it is an actual component of the first only of these masses, we shall have $\mu_1 = \mu_1'$. If it is also a possible component of the second homogeneous mass, we shall also have $\mu_1 \geqq \mu_1''$. If this substance occurs only at the surface of discontinuity, the value of the potential μ_1 will not be determined by any equation, but cannot be greater than the potential for the same substance in either of the homogeneous masses in which it may be a possible component.

It appears, therefore, that the particular conditions of equilibrium *relating to temperature and the potentials* which we have before obtained by neglecting the influence of the surfaces of discontinuity (pp. 65, 66, 74) are not invalidated by the influence of such discontinuity in their application to homogeneous parts of the system bounded like M′ and M″ by imaginary surfaces lying within the limits of homogeneity,—a condition which may be fulfilled by surfaces very near to the surfaces of discontinuity. It appears also that similar conditions will apply to the non-homogeneous films like M, which separate such homogeneous masses. The properties of such films, which are of course different from those of homogeneous masses, require our farther attention.

The volume occupied by the mass M is divided by the surface **s** into two parts which we will call v''' and v'''', v''' lying next to M′, and v'''' to M″. Let us imagine these volumes filled by masses having throughout the same temperature, pressure and potentials, and the same densities of energy and entropy, and of the various components, as the masses M′ and M″ respectively. We shall then have, by equation (12), if we regard the volumes as constant,

$$\delta\epsilon''' = t'\,\delta\eta''' + \mu_1'\,\delta m_1''' + \mu_2'\,\delta m_2''' + \text{etc.,} \tag{485}$$

$$\delta\epsilon'''' = t''\,\delta\eta'''' + \mu_1''\,\delta m_1'''' + \mu_2''\,\delta m_2'''' + \text{etc.;} \tag{486}$$

whence, by (482)–(484), we have for reversible variations

$$\delta\epsilon''' = t\,\delta\eta''' + \mu_1\,\delta m_1''' + \mu_2\,\delta m_2''' + \text{etc.,} \tag{487}$$

$$\delta\epsilon'''' = t\,\delta\eta'''' + \mu_1\,\delta m_1'''' + \mu_2\,\delta m_2'''' + \text{etc.} \tag{488}$$

From these equations and (477), we have for reversible variations

$$\delta(\epsilon - \epsilon''' - \epsilon'''') = t\,\delta(\eta - \eta''' - \eta'''')$$
$$+ \mu_1\delta(m_1 - m_1''' - m_1'''') + \mu_2\delta(m_2 - m_2''' - m_2'''') + \text{etc.} \tag{489}$$

Or, if we set*

$$\epsilon^{\text{s}} = \epsilon - \epsilon''' - \epsilon'''', \qquad \eta^{\text{s}} = \eta - \eta''' - \eta'''', \tag{490}$$

$$m_1^{\text{s}} = m_1 - m_1''' - m_1'''', \qquad m_2^{\text{s}} = m_2 - m_2''' - m_2'''', \text{ etc.,} \tag{491}$$

we may write

$$\delta\epsilon^{\text{s}} = t\,\delta\eta^{\text{s}} + \mu_1\delta m_1^{\text{s}} + \mu_2\delta m_2^{\text{s}} + \text{etc.} \tag{492}$$

This is true of reversible variations in which the surfaces which have been considered are fixed. It will be observed that ϵ^{s} denotes the excess of the energy of the actual mass which occupies the total volume which we have considered over that energy which it would have, if on each side of the surface S the density of energy had the same uniform value quite up to that surface which it has at a sensible distance from it; and that η^{s}, m_1^{s}, m_2^{s}, etc., have analogous significations. It will be convenient, and need not be a source of any misconception, to call ϵ^{s} and η^{s} the energy and entropy *of the surface* (or the *superficial* energy and entropy), $\dfrac{\epsilon^{\text{s}}}{s}$ and $\dfrac{\eta^{\text{s}}}{s}$ the *superficial densities* of energy and entropy, $\dfrac{m_1^{\text{s}}}{s}$, $\dfrac{m_2^{\text{s}}}{s}$, etc., the *superficial densities* of the several components.

Now these quantities (ϵ^{s}, η^{s}, m_1^{s}, etc.) are determined partly by the state of the physical system which we are considering, and partly by the various imaginary surfaces by means of which these quantities have been defined. The position of these surfaces, it will be remembered, has been regarded as fixed in the variation of the system. It is evident, however, that the form of that portion of these surfaces which lies in the region of homogeneity on either side of the surface of discontinuity cannot affect the values of these quantities. To obtain the complete value of $\delta\epsilon^{\text{s}}$ for reversible variations, we have

---

*It will be understood that the ˢ here used is not an algebraic exponent, but is only intended as a distinguishing mark. The Roman letter S has not been used to denote any *quantity*.

therefore only to regard variations in the position and form of the limited surface s, as this determines all of the surfaces in question lying within the region of non-homogeneity. Let us first suppose the form of s to remain unvaried and only its position in space to vary, either by translation or rotation. No change in (492) will be necessary to make it valid in this case. For the equation is valid if s remains fixed and the material system is varied in position; also, if the material system and s are both varied in position, while their relative position remains unchanged. Therefore, it will be valid if the surface alone varies its position.

But if the form of s be varied, we must add to the second member of (492) terms which shall represent the value of

$$\delta\epsilon^s - t\,\delta\eta^s - \mu_1\delta m_1^s - \mu_2\delta m_2^s - \text{etc.}$$

due to such variation in the form of s. If we suppose s to be sufficiently small to be considered uniform throughout in its curvatures and in respect to the state of the surrounding matter, the value of the above expression will be determined by the variation of its area δs and the variations of its principal curvatures δc_1 and δc_2, and we may write

$$\delta\epsilon^s = t\,\delta\eta^s + \mu_1\delta m_1^s + \mu_2\delta m_2^s + \text{etc.}$$
$$+\sigma\,\delta s + C_1\delta c_1 + C_2\,\delta c_2, \tag{493}$$

or

$$\delta\epsilon^s = t\,\delta\eta^s + \mu_1\delta m_1^s + \mu_2\delta m_2^s + \text{etc.}$$
$$+\sigma\,\delta s + \tfrac{1}{2}(C_1+C_2)\,\delta(c_1+c_2) + \tfrac{1}{2}(C_1-C_2)\,\delta(c_1-c_2), \tag{494}$$

σ, C_1, and C_2 denoting quantities which are determined by the initial state of the system and the position and form of s. The above is the complete value of the variation of ϵ^s for reversible variations of the system. But it is always possible to give such a position to the surface s that C_1+C_2 shall vanish.

To show this, it will be convenient to write the equation in the longer form {see (490), (491)}

$$\delta\epsilon - t\,\delta\eta - \mu_1\delta m_1 - \mu_2\delta m_2 - \text{etc.}$$
$$-\,\delta\epsilon''' + t\,\delta\eta''' + \mu_1\delta m_1''' + \mu_2\delta m_2''' + \text{etc.}$$
$$-\,\delta\epsilon'''' + t\,\delta\eta'''' + \mu_1\delta m_1'''' + \mu_2\delta m_2'''' + \text{etc.}$$
$$=\sigma\,\delta s + \tfrac{1}{2}(C_1+C_2)\,\delta(c_1+c_2) + \tfrac{1}{2}(C_1-C_2)\,\delta(c_1-c_2), \tag{495}$$

i.e., by (482)–(484) and (12),

$$\delta\epsilon - t\,\delta\eta - \mu_1\delta m_1 - \mu_2\delta m_2 - \text{etc.} + p'\,\delta v''' + p''\,\delta v''''$$
$$=\sigma\,\delta s + \tfrac{1}{2}(C_1+C_2)\,\delta(c_1+c_2) + \tfrac{1}{2}(C_1-C_2)\,\delta(c_1-c_2). \tag{496}$$

From this equation it appears in the first place that the pressure is the same in the two homogeneous masses separated by a plane surface of discontinuity. For let us imagine the material system to remain unchanged, while the plane surface **s** without change of area or of form moves in the direction of its normal. As this does not affect the boundaries of the mass M,

$$\delta\epsilon - t\,\delta\eta - \mu_1\delta m_1 - \mu_2\delta m_2 - \text{etc.} = 0.$$

Also $\delta s = 0$, $\delta(c_1 + c_2) = 0$, $\delta(c_1 - c_2) = 0$, and $\delta v''' = -\delta v''''$. Hence $p' = p''$, when the surface of discontinuity is plane.

Let us now examine the effect of different positions of the surface **s** in the same material system upon the value of $C_1 + C_2$, supposing at first that in the initial state of the system the surface of discontinuity is plane. Let us give the surface **s** some particular position. In the initial state of the system this surface will of course be plane like the physical surface of discontinuity, to which it is parallel. In the varied state of the system, let it become a portion of a spherical surface having positive curvature; and at sensible distances from this surface let the matter be homogeneous and with the same phases as in the initial state of the system; also at and about the surface let the state of the matter so far as possible be the same as at and about the plane surface in the initial state of the system. (Such a variation in the system may evidently take place negatively as well as positively, as the surface may be curved toward either side. But whether such a variation is consistent with the maintenance of equilibrium is of no consequence, since in the preceding equations only the initial state is supposed to be one of equilibrium.) Let the surface **s**, placed as supposed, whether in the initial or the varied state of the surface, be distinguished by the symbol **s'**. Without changing either the initial or the varied state of the material system, let us make another supposition with respect to the imaginary surface **s**. In the unvaried system let it be parallel to its former position but removed from it a distance λ on the side on which lie the centers of positive curvature. In the varied state of the system, let it be spherical and concentric with **s'**, and separated from it by the same distance λ. It will of course lie on the same side of **s'** as in the unvaried system. Let the surface **s**, placed in accordance with this second supposition, be distinguished by the symbol **s''**. Both in the initial and the varied state, let the perimeters of **s'** and **s''** be traced by a common normal. Now the value of

$$\delta\epsilon - t\,\delta\eta - \mu_1\delta m_1 - \mu_2\delta m_2 - \text{etc.}$$

in equation (496) is not affected by the position of **s**, being determined simply by the body M. The same is true of $p'\delta v''' + p''\delta v''''$ or $p'\delta(v''' + v'''')$, $v''' + v''''$ being the volume of M. Therefore the second member of (496) will have the same value whether the expressions relate to **s'** or **s''**. Moreover, $\delta(c_1 - c_2) = 0$ both for **s'** and **s''**. If we distinguish the quantities determined for s' and for s'' by the marks $'$ and $''$, we may therefore write

$$\sigma' \delta s' + \tfrac{1}{2}(C_1' + C_2') \,\delta(c_1' + c_2') = \sigma'' \delta s'' + \tfrac{1}{2}(C_1'' + C_2'') \,\delta(c_1'' + c_2'').$$

Now if we make
$$\delta s'' = 0,$$

we shall have by geometrical necessity
$$\delta s' = s\lambda \, \delta(c_1'' + c_2'').$$

Hence
$$\sigma' s\lambda \, \delta(c_1'' + c_2'') + \tfrac{1}{2}(C_1' + C_2') \,\delta(c_1' + c_2') = \tfrac{1}{2}(C_1'' + C_2'') \,\delta(c_1'' + c_2'').$$

But
$$\delta(c_1' + c_2') = \delta(c_1'' + c_2'').$$

Therefore,
$$C_1' + C_2' + 2\sigma' s\lambda = C_1'' + C_2''.$$

This equation shows that we may give a positive or negative value to $C_1'' + C_2''$ by placing s'' a sufficient distance on one or on the other side of s'. Since this is true when the (unvaried) surface is plane, it must also be true when the surface is nearly plane. And for this purpose a surface may be regarded as nearly plane, when the radii of curvature are very large in proportion to the thickness of the non-homogeneous film. This is the case when the radii of curvature have any sensible size. In general, therefore, whether the surface of discontinuity is plane or curved it is possible to place the surface s so that $C_1 + C_2$ in equation (494) shall vanish.

Now we may easily convince ourselves by equation (493) that if s is placed within the non-homogeneous film, and $s=1$, the quantity σ is of the same order of magnitude as the values of ϵ^s, η^s, m_1^s, m_2^s, etc., while the values of C_1 and C_2 are of the same order of magnitude as the changes in the values of the former quantities caused by increasing the curvature of s by unity. Hence, on account of the thinness of the non-homogeneous film, since it can be very little affected by such a change of curvature in s, the values of C_1 and C_2 must in general be very small relatively to σ. And hence, if s' be placed within the non-homogeneous film, the value of λ which will make $C_1'' + C_2''$ vanish must be very small (of the same order of magnitude as the thickness of the non-homogeneous film). The position of s, therefore, which will make $C_1 + C_2$ in (494) vanish, will in general be sensibly coincident with the physical surface of discontinuity.

We shall hereafter suppose, when the contrary is not distinctly indicated, that the surface s, in the unvaried state of the system, has such a position as to make $C_1 + C_2 = 0$. It will be remembered that the surface s is a part of a larger surface S, which we have called the *dividing surface*, and which is coextensive with the physical surface of discontinuity. We may suppose that the position of the dividing surface is everywhere determined by similar considerations. This is evidently consistent with the suppositions made on page 219 with regard to this surface.

We may therefore cancel the term

$$\tfrac{1}{2}(C_1+C_2)\,\delta(c_1+c_2)$$

in (494). In regard to the following term, it will be observed that C_1 must necessarily be equal to C_2, when $c_1=c_2$, which is the case when the surface of discontinuity is plane. Now on account of the thinness of the non-homogeneous film, we may always regard it as composed of parts which are approximately plane. Therefore, without danger of sensible error, we may also cancel the term

$$\tfrac{1}{2}(C_1-C_2)\,\delta(c_1-c_2).$$

Equation (494) is thus reduced to the form

$$\delta\epsilon^s=t\,\delta\eta^s+\sigma\,\delta s+\mu_1\delta m_1^s+\mu_2\delta m_2^s+\text{etc.} \qquad (497)$$

We may regard this as the complete value of $\delta\epsilon^s$, for all reversible variations in the state of the system supposed initially in equilibrium, when the dividing surface has its initial position determined in the manner described.

The above equation is of fundamental importance in the theory of capillarity. It expresses a relation with regard to surfaces of discontinuity analogous to that expressed by equation (12) with regard to homogeneous masses. From the two equations may be directly deduced the conditions of equilibrium of heterogeneous masses in contact, subject or not to the action of gravity, without disregard of the influence of the surfaces of discontinuity. The general problem, including the action of gravity, we shall take up hereafter; at present we shall only consider, as hitherto, a small part of a surface of discontinuity with a part of the homogeneous mass on either side, in order to deduce the additional condition which may be found when we take account of the motion of the dividing surface.

We suppose as before that the mass especially considered is bounded by a surface of which all that lies in the region of non-homogeneity is such as may be traced by a moving normal to the dividing surface. But instead of dividing the mass as before into four parts, it will be sufficient to regard it as divided into two parts by the dividing surface. The energy, entropy, etc., of these parts, estimated on the supposition that its nature (including density of energy, etc.) is uniform quite up to the dividing surface, will be denoted by ϵ', η', etc., ϵ'', η'', etc. Then the total energy will be $\epsilon^s+\epsilon'+\epsilon''$, and the general condition of internal equilibrium will be that

$$\delta\epsilon^s+\delta\epsilon'+\delta\epsilon''\geqq 0, \qquad (498)$$

when the bounding surface is fixed, and the total entropy and total quantities of the various components are constant. We may suppose η^s, η', η'', m_1^s, m_1', m_1'', m_2^s, m_2', m_2'', etc., to be all constant. Then by (497) and (12) the condition reduces to

$$\sigma\,\delta s - p'\,\delta v' - p''\,\delta v'' = 0. \tag{499}$$

(We may set $=$ for $\geqq$, since changes in the position of the dividing surface can evidently take place in either of two opposite directions.) This equation has evidently the same form as if a membrane without rigidity and having a tension σ, uniform in all directions, existed at the dividing surface.   Hence the particular position which we have chosen for this surface may be called the surface of tension, and σ the superficial tension.   If all parts of the dividing surface move a uniform normal distance δN, we shall have

$$\delta s = (c_1 + c_2)s\,\delta N, \quad \delta v' = s\,\delta N, \quad \delta v'' = -s\,\delta N\,;$$

whence $$\sigma(c_1 + c_2) = p' - p'', \tag{500}$$

the curvatures being positive when their centers lie on the side to which p' relates.   This is the condition which takes the place of that of equality of pressure (see pp. 65, 74) for heterogeneous fluid masses in contact, when we take account of the influence of the surfaces of discontinuity.   We have already seen that the conditions relating to temperature and the potentials are not affected by these surfaces.

Fundamental Equations for Surfaces of Discontinuity between Fluid Masses.

In equation (497) the initial state of the system is supposed to be one of equilibrium.   The only limitation with respect to the varied state is that the variation shall be reversible, i.e., that an opposite variation shall be possible.   Let us now confine our attention to variations in which the system remains in equilibrium.   To distinguish this case, we may use the character d instead of δ, and write

$$d\epsilon^s = t\,d\eta^s + \sigma\,ds + \mu_1 dm_1^s + \mu_2 dm_2^s + \text{etc.} \tag{501}$$

Both the states considered being states of equilibrium, the limitation with respect to the reversibility of the variations may be neglected, since the variations will always be reversible in at least one of the states considered.

If we integrate this equation, supposing the area s to increase from zero to any finite value s, while the material system to a part of which the equation relates remains without change, we obtain

$$\epsilon^s = t\eta^s + \sigma s + \mu_1 m_1^s + \mu_2 m_2^s + \text{etc.}, \tag{502}$$

which may be applied to any portion of any surface of discontinuity (in equilibrium) which is of the same nature throughout, or through-out which the values of t, σ, μ_1, μ_2, etc., are constant.

If we differentiate this equation, regarding all the quantities as variable, and compare the result with (501), we obtain

$$\eta^s dt + s\, d\sigma + m_1^s d\mu_1 + m_2^s d\mu_2 + \text{etc.} = 0. \tag{503}$$

If we denote the *superficial densities* of energy, of entropy, and of the several component substances (see page 224) by ϵ_s, η_s, Γ_1, Γ_2, etc., we have

$$\epsilon_s = \frac{\epsilon^s}{s}, \qquad \eta_s = \frac{\eta^s}{s}, \tag{504}$$

$$\Gamma_1 = \frac{m_1^s}{s}, \quad \Gamma_2 = \frac{m_2^s}{s}, \quad \text{etc.,} \tag{505}$$

and the preceding equations may be reduced to the form

$$d\epsilon_s = t\, d\eta_s + \mu_1 d\Gamma_1 + \mu_2 d\Gamma_2 + \text{etc.,} \tag{506}$$

$$\epsilon_s = t\eta_s + \sigma + \mu_1 \Gamma_1 + \mu_2 \Gamma_2 + \text{etc.,} \tag{507}$$

$$d\sigma = -\eta_s dt - \Gamma_1 d\mu_1 - \Gamma_2 d\mu_2 - \text{etc.} \tag{508}$$

Now the contact of the two homogeneous masses does not impose any restriction upon the variations of phase of either, except that the temperature and the potentials for actual components shall have the same value in both. {See (482)–(484) and (500).} For however the values of the pressures in the homogeneous masses may vary (on account of arbitrary variations of the temperature and potentials), and however the superficial tension may vary, equation (500) may always be satisfied by giving the proper curvature to the surface of tension, so long, at least, as the difference of pressures is not great. Moreover, if any of the potentials μ_1, μ_2, etc., relate to substances which are found only at the surface of discontinuity, their values may be varied by varying the superficial densities of those sub-stances. The values of t, μ_1, μ_2, etc., are therefore independently variable, and it appears from equation (508) that σ is a function of these quantities. If the form of this function is known, we may derive from it by differentiation $n+1$ equations (n denoting the total number of component substances) giving the values of η_s, Γ_1, Γ_2, etc., in terms of the variables just mentioned. This will give us, with (507), $n+3$ independent equations between the $2n+4$ quantities which occur in that equation. These are all that exist, since $n+1$ of these quantities are independently variable. Or, we may consider that we have $n+3$ independent equations between the $2n+5$ quantities occurring in equation (502), of which $n+2$ are independently variable.

An equation, therefore, between

$$\sigma, \ t, \ \mu_1, \ \mu_2, \ \text{etc.,} \tag{509}$$

may be called a fundamental equation for the surface of discontinuity. An equation between

$$\epsilon^s, \ \eta^s, \ s, \ m_1^s, \ m_2^s, \ \text{etc.,} \tag{510}$$

or between $\qquad \epsilon_s, \ \eta_s, \ \Gamma_1, \ \Gamma_2, \ \text{etc.} \tag{511}$

may also be called a fundamental equation in the same sense. For it is evident from (501) that an equation may be regarded as subsisting between the variables (510), and if this equation be known, since $n+2$ of the variables may be regarded as independent (viz., $n+1$ for the $n+1$ variations in the nature of the surface of discontinuity, and one for the area of the surface considered), we may obtain by differentiation and comparison with (501), $n+2$ additional equations between the $2n+5$ quantities occurring in (502). Equation (506) shows that equivalent relations can be deduced from an equation between the variables (511). It is moreover quite evident that an equation between the variables (510) must be reducible to the form of an equation between the ratios of these variables, and therefore to an equation between the variables (511).

The same designation may be applied to any equation from which, by differentiation and the aid only of general principles and relations, $n+3$ independent relations between the same $2n+5$ quantities may be obtained.

If we set $\qquad \psi^s = \epsilon^s - t\eta^s, \tag{512}$

we obtain by differentiation and comparison with (501)

$$d\psi^s = -\eta^s dt + \sigma \, ds + \mu_1 dm_1^s + \mu_2 dm_2^s + \text{etc.} \tag{513}$$

An equation, therefore, between $\psi^s, \ t, \ s, \ m_1^s, \ m_2^s$, etc., is a fundamental equation, and is to be regarded as entirely equivalent to either of the other fundamental equations which have been mentioned.

The reader will not fail to notice the analogy between these fundamental equations, which relate to surfaces of discontinuity, and those relating to homogeneous masses, which have been described on pages 85–89.

On the Experimental Determination of Fundamental Equations for Surfaces of Discontinuity between Fluid Masses.

When all the substances which are found at a surface of discontinuity are components of one or the other of the homogeneous masses, the potentials μ_1, μ_2, etc., as well as the temperature, may be determined from these homogeneous masses.* The tension σ may

---

* It is here supposed that the thermodynamic properties of the homogeneous masses have already been investigated, and that the fundamental equations of these masses may be regarded as known.

be determined by means of the relation (500).   But our measurements are practically confined to cases in which the difference of the pressures in the homogeneous masses is small; for with increasing differences of pressure the radii of curvature soon become too small for measurement.   Therefore, although the equation $p'=p''$ (which is equivalent to an equation between t, μ_1, μ_2, etc., since p' and p'' are both functions of these variables) may not be exactly satisfied in cases in which it is convenient to measure the tension, yet this equation is so nearly satisfied in all the measurements of tension which we can make, that we must regard such measurements as simply establishing the values of σ for values of t, μ_1, μ_2, etc., which satisfy the equation $p'=p''$, but not as sufficient to establish the rate of change in the value of σ for variations of t, μ_1, μ_2, etc., which are inconsistent with the equation $p'=p''$.

To show this more distinctly, let t, μ_2, m_3, etc., remain constant, then by (508) and (98)

$$d\sigma = -\Gamma_1 d\mu_1,$$

$$dp' = \gamma_1{}' d\mu_1,$$

$$dp'' = \gamma_1{}'' d\mu_1,$$

$\gamma_1{}'$ and $\gamma_1{}''$ denoting the densities $\dfrac{m_1{}'}{v'}$ and $\dfrac{m_1{}''}{v''}$.   Hence,

$$dp' - dp'' = (\gamma_1{}' - \gamma_1{}'')d\mu_1,$$

and
$$\Gamma_1 d(p' - p'') = (\gamma_1{}'' - \gamma_1{}')\, d\sigma.$$

But by (500)
$$(c_1 + c_2)\, d\sigma + \sigma\, d(c_1 + c_2) = d(p' - p'').$$

Therefore,
$$\Gamma_1(c_1 + c_2)\, d\sigma + \Gamma_1 \sigma\, d(c_1 + c_2) = (\gamma_1{}'' - \gamma_1{}')\, d\sigma,$$

or
$$\{\gamma_1{}'' - \gamma_1{}' - \Gamma_1(c_1 + c_2)\} d\sigma = \Gamma_1 \sigma\, d(c_1 + c_2).$$

Now $\Gamma_1(c_1 + c_2)$ will generally be very small compared with $\gamma_1{}'' - \gamma_1{}'$. Neglecting the former term, we have

$$\frac{d\sigma}{\sigma} = \frac{\Gamma_1}{\gamma_1{}'' - \gamma_1{}'} d(c_1 + c_2).$$

To integrate this equation, we may regard Γ_1, $\gamma_1{}'$, $\gamma_1{}''$ as constant. This will give, as an approximate value,

$$\log \frac{\sigma}{\sigma'} = \frac{\Gamma_1}{\gamma_1{}'' - \gamma_1{}'}(c_1 + c_2),$$

σ' denoting the value of σ when the surface is plane. From this it appears that when the radii of curvature have any sensible magnitude, the value of σ will be sensibly the same as when the surface is plane and the temperature and all the potentials except one have the same values, unless the component for which the potential has not the same value has very nearly the same density in the two homogeneous masses, in which case, the condition under which the variations take place is nearly equivalent to the condition that the pressures shall remain equal.

Accordingly, we cannot in general expect to determine the superficial density Γ_1 from its value $-\left(\dfrac{d\sigma}{d\mu_1}\right)^{*}_{t,\,\mu}$ by measurements of superficial tensions. The case will be the same with Γ_2, Γ_3, etc., and also with η_S, the superficial density of entropy.

The quantities ϵ_S, η_S, Γ_1, Γ_2, etc., are evidently too small in general to admit of direct measurement. When one of the components, however, is found only at the surface of discontinuity, it may be more easy to measure its superficial density than its potential. But except in this case, which is of secondary interest, it will generally be easy to determine σ in terms of t, μ_1, μ_2, etc., with considerable accuracy for plane surfaces, and extremely difficult or impossible to determine the fundamental equation more completely.

Fundamental Equations for Plane Surfaces of Discontinuity between Fluid Masses.

An equation giving σ in terms of t, μ_1, μ_2, etc., which will hold true only so long as the surface of discontinuity is plane, may be called a fundamental equation for a plane surface of discontinuity. It will be interesting to see precisely what results can be obtained from such an equation, especially with respect to the energy and entropy and the quantities of the component substances in the vicinity of the surface of discontinuity.

These results can be exhibited in a more simple form, if we deviate to a certain extent from the method which we have been following. The particular position adopted for the dividing surface (which determines the superficial densities) was chosen in order to make the term $\frac{1}{2}(C_1 + C_2)\,\delta(c_1 + c_2)$ in (494) vanish. But when the curvature of the surface is not supposed to vary, such a position of the dividing surface is not necessary for the simplification of the formula. It is evident that equation (501) will hold true for plane surfaces (supposed to remain such) without reference to the position of the dividing surface, except that it shall be parallel to the surface of discontinuity. We are therefore at liberty to choose such a position for the dividing surface as may for any purpose be convenient.

None of the equations (502)–(513), which are either derived from (501), or serve to define new symbols, will be affected by such a

---

* The suffixed μ is used to denote that all the potentials except that occurring in the denominator of the differential coefficient are to be regarded as constant.

change in the position of the dividing surface. But the expressions ϵ^s, η^s, m_1^s, m_2^s, etc., as also ϵ_s, η_s, Γ_1, Γ_2, etc., and ψ^s, will of course have different values when the position of that surface is changed. The quantity σ, however, which we may regard as defined by equations (501), or, if we choose, by (502) or (507), will not be affected in value by such a change. For if the dividing surface be moved a distance λ measured normally and toward the side to which v'' relates, the quantities
$$\epsilon_s, \quad \eta_s, \quad \Gamma_1, \quad \Gamma_2, \quad \text{etc.,}$$
will evidently receive the respective increments
$$\lambda(\epsilon_V'' - \epsilon_V'), \quad \lambda(\eta_V'' - \eta_V'), \quad \lambda(\gamma_1'' - \gamma_1'), \quad \lambda(\gamma_2'' - \gamma_2'), \quad \text{etc.,}$$
ϵ_V', ϵ_V'', η_V', η_V'' denoting the densities of energy and entropy in the two homogeneous masses. Hence, by equation (507), σ will receive the increment
$$\lambda(\epsilon_V'' - \epsilon_V') - t\lambda(\eta_V'' - \eta_V') - \mu_1\lambda(\gamma_1'' - \gamma_1') - \mu_2\lambda(\gamma_2'' - \gamma_2') - \text{etc.}$$
But by (93)
$$-p'' = \epsilon_V'' - t\eta_V'' - \mu_1\gamma_1'' - \mu_2\gamma_2'' - \text{etc.,}$$
$$-p' = \epsilon_V' - t\eta_V' - \mu_1\gamma_1' - \mu_2\gamma_2' - \text{etc.}$$
Therefore, since $p' = p''$, the increment in the value of σ is zero. The value of σ is therefore independent of the position of the dividing surface, when this surface is plane. But when we call this quantity the superficial tension, we must remember that it will not have its characteristic properties as a tension with reference to any arbitrary surface. Considered as a tension, its position is in the surface which we have called the surface of tension, and, strictly speaking, nowhere else. The positions of the dividing surface, however, which we shall consider, will not vary from the surface of tension sufficiently to make this distinction of any practical importance.

It is generally possible to place the dividing surface so that the total quantity of any desired component in the vicinity of the surface of discontinuity shall be the same as if the density of that component were uniform on each side quite up to the dividing surface. In other words, we may place the dividing surface so as to make any one of the quantities Γ_1, Γ_2, etc., vanish. The only exception is with regard to a component which has the same density in the two homogeneous masses. With regard to a component which has very nearly the same density in the two masses such a location of the dividing surface might be objectionable, as the dividing surface might fail to coincide sensibly with the physical surface of discontinuity. Let us suppose that γ_1' is not equal (nor very nearly equal) to γ_1'', and that the dividing surface is so placed as to make $\Gamma_1 = 0$. Then equation (508) reduces to
$$d\sigma = -\eta_{s(1)}dt - \Gamma_{2(1)}d\mu_2 - \Gamma_{3(1)}d\mu_3 - \text{etc.,} \tag{514}$$

where the symbols $\eta_{8(1)}$, $\Gamma_{2(1)}$, etc., are used for greater distinctness to denote the values of η_8, Γ_2, etc., as determined by a dividing surface placed so that $\Gamma_1 = 0$. Now we may consider all the differentials in the second member of this equation as independent, without violating the condition that the surface shall remain plane, i.e., that $dp' = dp''$. This appears at once from the values of dp' and dp'' given by equation (98). Moreover, as has already been observed, when the fundamental equations of the two homogeneous masses are known, the equation $p' = p''$ affords a relation between the quantities t, μ_1, μ_2, etc. Hence, when the value of σ is also known for plane surfaces in terms of t, μ_1, μ_2, etc., we can eliminate μ_1 from this expression by means of the relation derived from the equality of pressures, and obtain the value of σ for plane surfaces in terms of t, μ_2, μ_3, etc. From this, by differentiation, we may obtain directly the values of $\eta_{8(1)}$, $\Gamma_{2(1)}$, $\Gamma_{3(1)}$, etc., in terms of t, μ_2, μ_3, etc. This would be a convenient form of the fundamental equation. But, if the elimination of p', p'', and μ_1 from the finite equations presents algebraic difficulties, we can in all cases easily eliminate dp', dp'', $d\mu_1$ from the corresponding differential equations and thus obtain a differential equation from which the values of $\eta_{8(1)}$, $\Gamma_{2(1)}$, $\Gamma_{3(1)}$, etc., in terms of t, μ_1, μ_2, etc., may be at once obtained by comparison with (514).*

---

* If liquid mercury meets the mixed vapors of water and mercury in a plane surface, and we use μ_1 and μ_2 to denote the potentials of mercury and water respectively, and place the dividing surface so that $\Gamma_1 = 0$, i.e., so that the total quantity of mercury is the same as if the liquid mercury reached this surface on one side and the mercury vapor on the other without change of density on either side, then $\Gamma_{2(1)}$ will represent the amount of water in the vicinity of this surface, per unit of surface, above that which there would be, if the water-vapor just reached the surface without change of density, and this quantity (which we may call the quantity of water condensed upon the surface of the mercury) will be determined by the equation

$$\Gamma_{2(1)} = -\frac{d\sigma}{d\mu_2}.$$

(In this differential coefficient as well as the following, the temperature is supposed to remain constant and the surface of discontinuity plane. Practically, the latter condition may be regarded as fulfilled in the case of any ordinary curvatures.)

If the pressure in the mixed vapors conforms to the law of Dalton (see pp. 155, 157), we shall have for constant temperature

$$dp_2 = \gamma_2 \, d\mu_2,$$

where p_2 denotes the part of the pressure in the vapor due to the water-vapor, and γ_2 the density of the water-vapor. Hence we obtain

$$\Gamma_{2(1)} = -\gamma_2 \frac{d\sigma}{dp_2}.$$

For temperatures below 100° centigrade, this will certainly be accurate, since the pressure due to the vapor of mercury may be neglected.

The value of σ for $p_2 = 0$ and the temperature of 20° centigrade must be nearly the same as the superficial tension of mercury in contact with air, or 55·03 grammes per linear meter according to Quincke (*Pogg. Ann.*, Bd. 139, p. 27). The value of σ at the same temperature, when the condensed water begins to have the properties of water

The same physical relations may of course be deduced without giving up the use of the surface of tension as a dividing surface, but the formulæ which express them will be less simple. If we make t, μ_3, μ_4, etc., constant, we have by (98) and (508)

$$dp' = \gamma_1' d\mu_1 + \gamma_2' d\mu_2,$$
$$dp'' = \gamma_1'' d\mu_1 + \gamma_2'' d\mu_2,$$
$$d\sigma = -\Gamma_1 d\mu_1 - \Gamma_2 d\mu_2,$$

where we may suppose Γ_1 and Γ_2 to be determined with reference to the surface of tension. Then, if $dp' = dp''$,

$$(\gamma_1' - \gamma_1'')d\mu_1 + (\gamma_2' - \gamma_2'')d\mu_2 = 0,$$

and

$$d\sigma = \Gamma_1 \frac{\gamma_2' - \gamma_2''}{\gamma_1' - \gamma_1''} d\mu_2 - \Gamma_2 d\mu_2.$$

That is,

$$\left(\frac{d\sigma}{d\mu_2}\right)_{p'-p'', \, t, \, \mu_3, \, \mu_4, \, \text{etc.}} = -\Gamma_2 + \Gamma_1 \frac{\gamma_2' - \gamma_2''}{\gamma_1' - \gamma_1''}. \tag{515}$$

The reader will observe that $\dfrac{\Gamma_1}{\gamma_1' - \gamma_1''}$ represents the distance between the surface of tension and that dividing surface which would make $\Gamma_1 = 0$; the second number of the last equation is therefore equivalent to $-\Gamma_{2(1)}$.

If any component substance has the same density in the two homogeneous masses separated by a plane surface of discontinuity, the value of the superficial density for that component is independent of the position of the dividing surface. In this case alone we may derive the value of the superficial density of a component with reference to the surface of tension from the fundamental equation for plane surfaces alone. Thus in the last equation, when $\gamma_2' = \gamma_2''$, the second member will reduce to $-\Gamma_2$. It will be observed that to

---

in mass, will be equal to the sum of the superficial tensions of mercury in contact with water and of water in contact with its own vapor. This will be, according to the same authority, 42·58 + 8·25, or 50·83 grammes per meter, if we neglect the difference of the tensions of water with its vapor and water with air. As p_2, therefore, increases from zero to 236400 grammes per square meter (when water begins to be condensed *in mass*), σ diminishes from about 55·03 to about 50·83 grammes per linear meter. If the general course of the values of σ for intermediate values of p_2 were determined by experiment, we could easily form an approximate estimate of the values of the superficial density $\Gamma_{2(1)}$ for different pressures less than that of saturated vapor. It will be observed that the determination of the superficial density does not by any means depend upon inappreciable differences of superficial tension. The greatest difficulty in the determination would doubtless be that of distinguishing between the diminution of superficial tension due to the water and that due to other substances which might accidentally be present. Such determinations are of considerable practical importance on account of the use of mercury in measurements of the specific gravity of vapors.

make $p' - p''$, t, μ_3, μ_4, etc. constant is in this case equivalent to making t, μ_1, μ_3, μ_4, etc. constant.

Substantially the same is true of the superficial density of entropy or of energy, when either of these has the same density in the two homogeneous masses.*

Concerning the Stability of Surfaces of Discontinuity between Fluid Masses.

We shall first consider the stability of a film separating homogeneous masses with respect to changes in its nature, while its position and the nature of the homogeneous masses are not altered. For this purpose, it will be convenient to suppose that the homogeneous masses are very large, and thoroughly stable with respect to the possible formation of any different homogeneous masses out of their components, and that the surface of discontinuity is plane and uniform.

Let us distinguish the quantities which relate to the actual components of one or both of the homogeneous masses by the suffixes $_{a, b}$, etc., and those which relate to components which are found only at the surface of discontinuity by the suffixes $_{g, h}$, etc., and consider the variation of the energy of the whole system in consequence of a given change in the nature of a small part of the surface of discontinuity, while the entropy of the whole system and the total quantities of the several components remain constant, as well as the volume of each of the homogeneous masses, as determined by the surface of tension. This small part of the surface of discontinuity in its changed state is supposed to be still uniform in nature, and such as may subsist in equilibrium between the given homogeneous masses, which will evidently not be sensibly altered in nature or thermodynamic state. The remainder of the surface of discontinuity is also supposed to

---

* With respect to questions which concern only the *form* of surfaces of discontinuity, such precision as we have employed in regard to the position of the dividing surface is evidently quite unnecessary. This precision has not been used for the sake of the mechanical part of the problem, which does not require the surface to be defined with greater nicety than we can employ in our observations, but in order to give determinate values to the superficial densities of energy, entropy, and the component substances, which quantities, as has been seen, play an important part in the relations between the tension of a surface of discontinuity, and the composition of the masses which it separates.

The product σs of the superficial tension and the area of the surface, may be regarded as the *available energy* due to the surface in a system in which the temperature and the potentials μ_1, μ_2, etc.—or the differences of these potentials and the gravitational potential (see page 148) when the system is subject to gravity—are maintained sensibly constant. The value of σ, as well as that of s, is sensibly independent of the precise position which we may assign to the dividing surface (so long as this is sensibly coincident with the surface of discontinuity), but ϵ_s, the *superficial density of energy*, as the term is used in this paper, like the superficial densities of entropy and of the component substances, requires a more precise localization of the dividing surface.

remain uniform, and on account of its infinitely greater size to be infinitely less altered in its nature than the first part. Let $\Delta\epsilon^s$ denote the increment of the superficial energy of this first part, $\Delta\eta^s$, Δm_a^s, Δm_b^s, etc., Δm_g^s, Δm_h^s, etc., the increments of its superficial entropy and of the quantities of the components which we regard as belonging to the surface. The increments of entropy and of the various components which the rest of the system receive will be expressed by

$$-\Delta\eta^s, \quad -\Delta m_a^s, \quad -\Delta m_b^s, \quad \text{etc.}, \quad -\Delta m_g^s, \quad -\Delta m_h^s, \quad \text{etc.},$$

and the consequent increment of energy will be by (12) and (501)

$$-t\,\Delta\eta^s - \mu_a\Delta m_a^s - \mu_b\Delta m_b^s - \text{etc.} - \mu_g\Delta m_g^s - \mu_h\Delta m_h^s - \text{etc.}$$

Hence the total increment of energy in the whole system will be

$$\left.\begin{aligned}\Delta\epsilon^s - t\,\Delta\eta^s - \mu_a\Delta m_a^s - \mu_b\Delta m_b^s - \text{etc.}\\ -\mu_g\Delta m_g^s - \mu_h\Delta m_h^s - \text{etc.}\end{aligned}\right\} \tag{516}$$

If the value of this expression is necessarily positive, for finite changes as well as infinitesimal in the nature of the part of the film to which $\Delta\epsilon^s$, etc. relate,* the increment of energy of the whole system will be positive for any possible changes in the nature of the film, and the film will be stable, at least with respect to changes in its nature, as distinguished from its position. For, if we write

$$D\epsilon^s, \quad D\eta^s, \quad Dm_a^s, \quad Dm_b^s, \quad \text{etc.}, \quad Dm_g^s, \quad Dm_h^s, \quad \text{etc.},$$

for the energy, etc. of any element of the surface of discontinuity, we have from the supposition just made

$$\Delta\,D\epsilon^s - t\,\Delta\,D\eta^s - \mu_a\Delta\,Dm_a^s, - \mu_b\Delta\,Dm_b^s - \text{etc.}$$
$$- \mu_g\Delta Dm_g^s - \mu_h\Delta Dm_h^s - \text{etc.} > 0\,; \tag{517}$$

and integrating for the whole surface, since

$$\Delta\!\int\! Dm_g^s = 0, \quad \Delta\!\int\! Dm_h^s = 0, \quad \text{etc.},$$

we have

$$\Delta\!\int\! D\epsilon^s - t\,\Delta\!\int\! D\eta^s - \mu_a\,\Delta\!\int\! Dm_a^s - \mu_b\,\Delta\!\int\! Dm_b^s - \text{etc.} > 0. \tag{518}$$

Now $\Delta\!\int\! D\eta^s$ is the increment of the entropy of the whole surface, and $-\Delta\!\int\! D\eta^s$ is therefore the increment of the entropy of the two homogeneous masses. In like manner, $-\Delta\!\int\! Dm_a^s$, $-\Delta\!\int\! Dm_b^s$, etc., are the increments of the quantities of the components in these masses. The expression

$$-t\,\Delta\!\int\! D\eta^s - \mu_a\,\Delta\!\int\! Dm_a^s - \mu_b\,\Delta\!\int\! Dm_b^s - \text{etc.}$$

denotes therefore, according to equation (12), the increment of energy of the two homogeneous masses, and since $\Delta\!\int\! D\epsilon^s$ denotes the

---

*In the case of infinitesimal changes in the nature of the film, the sign Δ must be interpreted, as elsewhere in this paper, without neglect of infinitesimals of the higher orders. Otherwise, by equation (501), the above expression would have the value zero.

increment of energy of the surface, the above condition expresses that the increment of the total energy of the system is positive. That we have only considered the possible formation of such films as are capable of existing in equilibrium between the given homogeneous masses can not invalidate the conclusion in regard to the stability of the film, for in considering whether any state of the system will have less energy than the given state, we need only consider the state of least energy, which is necessarily one of equilibrium.

If the expression (516) is capable of a negative value for an infinitesimal change in the nature of the part of the film to which the symbols relate, the film is obviously unstable.

If the expression is capable of a negative value, but only for finite and not for infinitesimal changes in the nature of this part of the film, the film is *practically unstable*,* i.e., if such a change were made in a small part of the film, the disturbance would tend to increase. But it might be necessary that the initial disturbance should also have a finite magnitude in respect to the extent of surface in which it occurs; for we cannot suppose that the thermodynamic relations of an infinitesimal part of a surface of discontinuity are independent of the adjacent parts. On the other hand, the changes which we have been considering are such that every part of the film remains in equilibrium with the homogeneous masses on each side; and if the energy of the system can be diminished by a finite change satisfying this condition, it may perhaps be capable of diminution by an infinitesimal change which does not satisfy the same condition. We must therefore leave it undetermined whether the film, which in this case is practically unstable, is or is not unstable in the strict mathematical sense of the term.

Let us consider more particularly the condition of practical stability, in which we need not distinguish between finite and infinitesimal changes. To determine whether the expression (516) is capable of a negative value, we need only consider the least value of which it is capable. Let us write it in the fuller form

$$\epsilon^{s''} - \epsilon^{s'} - t(\eta^{s''} - \eta^{s'}) - \mu_a(m_a^{s''} - m_a^{s'}) - \mu_b(m_b^{s''} - m_b^{s'}) - \text{etc.} \atop - \mu_g'(m_g^{s''} - m_g^{s'}) - \mu_h'(m_h^{s''} - m_h^{s'}) - \text{etc.,}} \right\} \quad (519)$$

where the single and double accents distinguish the quantities which relate to the first and second states of the film, the letters without accents denoting those quantities which have the same value in both states. The differential of this expression when the quantities distinguished by double accents are alone considered variable, and the area of the surface is constant, will reduce by (501) to the form

$$(\mu_g'' - \mu_g')dm_g^{s''} + (\mu_h'' - \mu_h')dm_h^{s''} + \text{etc.}$$

---

* With respect to the sense in which this term is used, compare page 79.

To make this incapable of a negative value, we must have

$$\mu_g'' = \mu_g', \quad \text{unless} \quad m_g^{s''} = 0,$$
$$\mu_h'' = \mu_h', \quad \text{unless} \quad m_h^{s''} = 0.$$

In virtue of these relations and by equation (502), the expression (519), i.e., (516), will reduce to

$$\sigma'' s - \sigma' s,$$

which will be positive or negative according as

$$\sigma'' - \sigma' \tag{520}$$

is positive or negative.

That is, if the tension of the film is less than that of any other film of the same components which can exist between the same homogeneous masses (which has therefore the same values of t, μ_a, μ_b, etc.), and which moreover has the same values of the potentials μ_g, μ_h, etc., so far as it contains the substances to which these relate, then the first film will be stable. But the film will be practically unstable, if any other such film has a less tension. (Compare the expression (141), by which the practical stability of homogeneous masses is tested.)

It is, however, evidently necessary for the stability of the surface of discontinuity with respect to *deformation*, that the value of the superficial tension should be positive. Moreover, since we have by (502) for the surface of discontinuity

$$\epsilon^s - t\eta^s - \mu_a m_a^s - \mu_b m_b^s - \text{etc.} - \mu_g m_g^s - \mu_h m_h^s - \text{etc.} = \sigma s,$$

and by (93) for the two homogeneous masses

$$\epsilon' - t\eta' + pv' - \mu_a m_a' - \mu_b m_b' - \text{etc.} = 0,$$
$$\epsilon'' - t\eta'' + pv'' - \mu_a m_a'' - \mu_b m_b'' - \text{etc.} = 0,$$

if we denote by

$$\epsilon, \quad \eta, \quad v, \quad m_a, \quad m_b, \quad \text{etc.}, \quad m_g, \quad m_h, \quad \text{etc.},$$

the total energy, etc. of a composite mass consisting of two such homogeneous masses divided by such a surface of discontinuity, we shall have by addition of these equations

$$\epsilon - t\eta + pv - \mu_a m_a - \mu_b m_b - \text{etc.} - \mu_g m_g - \mu_h m_h - \text{etc.} = \sigma s.$$

Now if the value of σ is negative, the value of the first member of this equation will decrease as s increases, and may therefore be decreased by making the mass to consist of thin alternate strata of the two kinds of homogeneous masses which we are considering. There will be no limit to the decrease which is thus possible with a given value of v, so long as the equation is applicable, i.e., so long as the strata have the properties of similar bodies in mass. But it may easily be shown (as in a similar case on pages 77, 78) that when the values of

$$t, \quad p, \quad \mu_a, \quad \mu_b, \quad \text{etc.}, \quad \mu_g, \quad \mu_h, \quad \text{etc.},$$

are regarded as fixed, being determined by the surface of discontinuity in question, and the values of

$$\epsilon, \quad \eta, \quad m_a, \quad m_b, \quad \text{etc.}, \quad m_g, \quad m_h, \quad \text{etc.},$$

are variable and may be determined by any body having the given volume v, the first member of this equation cannot have an infinite negative value, and must therefore have a least possible value, which will be negative, if any value is negative, that is, if σ is negative.

The body determining ϵ, η, etc. which will give this least value to this expression will evidently be sensibly homogeneous. With respect to the formation of such a body, the system consisting of the two homogeneous masses and the surface of discontinuity with the negative tension is by (53) (see also page 79) at least practically unstable, if the surface of discontinuity is very large, so that it can afford the requisite material without sensible alteration of the values of the potentials. (This limitation disappears, if all the component substances are found in the homogeneous masses.) Therefore, in a system satisfying the conditions of practical stability with respect to the possible formation of all kinds of homogeneous masses, negative tensions of the surfaces of discontinuity are necessarily excluded.

Let us now consider the condition which we obtain by applying (516) to infinitesimal changes. The expression may be expanded as before to the form (519), and then reduced by equation (502) to the form

$$s(\sigma'' - \sigma') + m_g^{s''}(\mu_g'' - \mu_g') + m_h^{s''}(\mu_h'' - \mu_h') + \text{etc.}$$

That the value of this expression shall be positive when the quantities are determined by two films which differ infinitely little is a necessary condition of the stability of the film to which the single accents relate. But if one film is stable, the other will in general be so too, and the distinction between the films with respect to stability is of importance only at the limits of stability. If all films for all values of μ_g, μ_h, etc. are stable, or all within certain limits, it is evident that the value of the expression must be positive when the quantities are determined by any two infinitesimally different films within the same limits. For such collective determinations of stability the condition may be written

$$-s\Delta\sigma - m_g^s\Delta\mu_g - m_h^s\Delta\mu_h - \text{etc.} > 0,$$

or

$$\Delta\sigma < -\Gamma_g\Delta\mu_g - \Gamma_h\Delta\mu_h - \text{etc.} \tag{521}$$

On comparison of this formula with (508), it appears that within the limits of stability the second and higher differential coefficients of the tension considered as a function of the potentials for the substances which are found only at the surface of discontinuity (the potentials for the substances found in the homogeneous masses and the temperature being regarded as constant) satisfy the conditions which would make the tension a maximum if the necessary conditions relative to the first differential coefficients were fulfilled.

In the foregoing discussion of stability, the surface of discontinuity is supposed plane. In this case, as the tension is supposed positive, there can be no tendency to a change of form of the surface. We now pass to the consideration of changes consisting in or connected with motion and change of form of the surface of tension, which we shall at first suppose to be and to remain spherical and uniform throughout.

In order that the equilibrium of a spherical mass entirely surrounded by an indefinitely large mass of different nature shall be neutral with respect to changes in the value of r, the radius of the sphere, it is evidently necessary that equation (500), which in this case may be written

$$2\sigma = r(p'-p''),\tag{522}$$

as well as the other conditions of equilibrium, shall continue to hold true for varying values of r. Hence, for a state of equilibrium which is on the limit between stability and instability, it is necessary that the equation

$$2d\sigma = (p'-p'')\,dr + r\,dp'$$

shall be satisfied, when the relations between $d\sigma$, dp', and dr are determined from the fundamental equations on the supposition that the conditions of equilibrium relating to temperature and the potentials remain satisfied. (The differential coefficients in the equations which follow are to be determined on this supposition.) Moreover, if

$$r\frac{dp'}{dr} < 2\frac{d\sigma}{dr} - p' + p'',\tag{523}$$

i.e., if the pressure of the interior mass increases less rapidly (or decreases more rapidly) with increasing radius than is necessary to preserve neutral equilibrium, the equilibrium is stable. But if

$$r\frac{dp'}{dr} > 2\frac{d\sigma}{dr} - p' + p'',\tag{524}$$

the equilibrium is unstable. In the remaining case, when

$$r\frac{dp'}{dr} = 2\frac{d\sigma}{dr} - p' + p'',\tag{525}$$

farther conditions are of course necessary to determine absolutely whether the equilibrium is stable or unstable, but in general the equilibrium will be stable in respect to change in one direction and unstable in respect to change in the opposite direction, and is therefore to be considered unstable. In general, therefore, we may call (523) the condition of stability.

When the interior mass and the surface of discontinuity are formed entirely of substances which are components of the external mass, p' and σ cannot vary, and condition (524) being satisfied the equilibrium is unstable.

But if either the interior homogeneous mass or the surface of discontinuity contains substances which are not components of the enveloping mass, the equilibrium may be stable. If there is but one such substance, and we denote its densities and potential by γ'_1, Γ_1, and μ_1, the condition of stability (523) will reduce to the form

$$\left(r\frac{dp'}{d\mu_1}-2\frac{d\sigma}{d\mu_1}\right)\frac{d\mu_1}{dr}<p''-p',$$

or, by (98) and (508),

$$(r\gamma_1'+2\Gamma_1)\frac{d\mu_1}{dr}<p''-p'. \tag{526}$$

In these equations and in all which follow in the discussion of this case, the temperature and the potentials μ_2, μ_3, etc. are to be regarded as constant. But

$$\gamma_1'v'+\Gamma_1 s,$$

which represents the total quantity of the component specified by the suffix, must be constant. It is evidently equal to

$$\tfrac{4}{3}\pi r^3\gamma_1'+4\pi r^2\Gamma_1.$$

Dividing by 4π and differentiating, we obtain

$$(r^2\gamma_1'+2r\Gamma_1)dr+\tfrac{1}{3}r^3\,d\gamma_1'+r^2\,d\Gamma_1=0,$$

or, since γ_1' and Γ_1 are functions of μ_1,

$$(r\gamma_1'+2\Gamma_1)dr+\left(\frac{r^2}{3}\frac{d\gamma_1'}{d\mu_1}+r\frac{d\Gamma_1}{d\mu_1}\right)d\mu_1=0. \tag{527}$$

By means of this equation, the condition of stability is brought to the form

$$\frac{(r\gamma_1'+2\Gamma_1)^2}{\dfrac{r^2}{3}\dfrac{d\gamma_1'}{d\mu_1}+r\dfrac{d\Gamma_1}{d\mu_1}}>p'-p''. \tag{528}$$

If we eliminate r by equation (522), we have

$$\frac{\left(\dfrac{\gamma_1'}{p'-p''}+\dfrac{\Gamma_1}{\sigma}\right)^2}{\dfrac{1}{3(p'-p'')}\dfrac{d\gamma_1'}{d\mu_1}+\dfrac{1}{2\sigma}\dfrac{d\Gamma_1}{d\mu_1}}>1. \tag{529}$$

If p' and σ are known in terms of t, μ_1, μ_2, etc., we may express the first member of this condition in terms of the same variables and p''.

This will enable us to determine, for any given state of the external mass, the values of μ_1 which will make the equilibrium stable or unstable.

If the component to which γ_1' and Γ_1 relate is found only at the surface of discontinuity, the condition of stability reduces to

$$\frac{\Gamma_1^2}{\sigma}\frac{d\mu_1}{d\Gamma_1}>\frac{1}{2}. \tag{530}$$

Since

$$\Gamma_1=-\frac{d\sigma}{d\mu_1},$$

we may also write

$$\frac{\Gamma_1}{\sigma}\frac{d\sigma}{d\Gamma_1}<-\frac{1}{2}, \text{ or } \frac{d\log\sigma}{d\log\Gamma_1}<-\frac{1}{2}. \tag{531}$$

Again, if $\Gamma_1=0$ and $\dfrac{d\Gamma_1}{d\mu_1}=0$, the condition of stability reduces to

$$\frac{3\gamma_1'^2}{p'-p''}\frac{d\mu_1}{d\gamma_1'}>1. \tag{532}$$

Since

$$\gamma_1'=\frac{dp'}{d\mu_1},$$

we may also write

$$\frac{\gamma_1'}{p'-p''}\frac{dp'}{d\gamma_1'}>\frac{1}{3}, \text{ or } \frac{d\log(p'-p'')}{d\log\gamma_1'}>\frac{1}{3}. \tag{533}$$

When r is large, this will be a close approximation for any values of Γ_1, unless γ_1' is very small. The two special conditions (531) and (533) might be derived from very elementary considerations.

Similar conditions of stability may be found when there are more substances than one in the inner mass or the surface of discontinuity, which are not components of the enveloping mass. In this case, we have instead of (526) a condition of the form

$$(r\gamma_1'+2\Gamma_1)\frac{d\mu_1}{dr}+(r\gamma_2'+2\Gamma_2)\frac{d\mu_2}{dr}+\text{etc.}<p''-p', \tag{534}$$

from which $\dfrac{d\mu_1}{dr}, \dfrac{d\mu_2}{dr}$, etc. may be eliminated by means of equations derived from the conditions that

$$\gamma_1'v'+\Gamma_1s, \quad \gamma_2'v'+\Gamma_2s, \text{ etc.}$$

must be constant.

Nearly the same method may be applied to the following problem. Two different homogeneous fluids are separated by a diaphragm having a circular orifice, their volumes being invariable except by the motion of the surface of discontinuity, which adheres to the edge of the orifice;—to determine the stability or instability of this surface when in equilibrium.

The condition of stability derived from (522) may in this case be written

$$r\frac{d(p'-p'')}{dv'} < 2\frac{d\sigma}{dv'} - (p'-p'')\frac{dr}{dv'}, \tag{535}$$

where the quantities relating to the concave side of the surface of tension are distinguished by a single accent.

If both the masses are infinitely large, or if one which contains all the components of the system is infinitely large, $p'-p''$ and σ will be constant, and the condition reduces to

$$\frac{dr}{dv'} < 0.$$

The equilibrium will therefore be stable or unstable according as the surface of tension is less or greater than a hemisphere.

To return to the general problem:—if we denote by x the part of the axis of the circular orifice intercepted between the center of the orifice and the surface of tension, by R the radius of the orifice, and by V' the value of v' when the surface of tension is plane, we shall have the geometrical relations

$$R^2 = 2rx - x^2,$$

and
$$v' = V' + \tfrac{2}{3}\pi r^2 x - \tfrac{1}{3}\pi R^2(r-x)$$
$$= V' + \pi r x^2 - \tfrac{1}{3}\pi x^3.$$

By differentiation we obtain

$$(r-x)dx + x\,dr = 0,$$

and
$$dv' = \pi x^2\,dr + (2\pi r x - \pi x^2)dx;$$

whence
$$(r-x)dv' = -\pi r x^2\,dr. \tag{536}$$

By means of this relation, the condition of stability may be reduced to the form

$$\frac{dp'}{dv'} - \frac{dp''}{dv'} - \frac{2}{r}\frac{d\sigma}{dv'} < (p'-p'')\frac{r-x}{\pi r^2 x^2}. \tag{537}$$

Let us now suppose that the temperature and all the potentials except one, μ_1, are to be regarded as constant. This will be the case when one of the homogeneous masses is very large and contains all the components of the system except one, or when both these masses are very large and there is a single substance at the surface of discontinuity which is not a component of either; also when the whole system contains but a single component, and is exposed to a constant temperature at its surface. Condition (537) will reduce by (98) and (508) to the form

$$\left(\gamma_1' - \gamma_1'' + \frac{2\Gamma_1}{r}\right)\frac{d\mu_1}{dv'} < (p'-p'')\frac{r-x}{\pi r^2 x^2}. \tag{538}$$

But $$\gamma_1'v' + \gamma_1''v'' + \Gamma_1 s$$

(the total quantity of the component specified by the suffix) must be constant; therefore, since

$$dv'' = -dv', \text{ and } ds = \frac{2}{r}dv',$$

$$\left(v'\frac{d\gamma_1'}{d\mu_1} + v''\frac{d\gamma_1''}{d\mu_1} + s\frac{d\Gamma_1}{d\mu_1}\right)d\mu_1 + \left(\gamma_1' - \gamma_1'' + \frac{2\Gamma_1}{r}\right)dv' = 0. \quad (539)$$

By this equation, the condition of stability is brought to the form

$$\frac{\left(\gamma_1' - \gamma_1'' + \frac{2\Gamma_1}{r}\right)^2}{v'\dfrac{d\gamma_1'}{d\mu_1} + v''\dfrac{d\gamma_1''}{d\mu_1} + s\dfrac{d\Gamma_1}{d\mu_1}} > (p' - p'')\frac{x - r}{\pi x^2 r^2}. \quad (540)$$

When the substance specified by the suffix is a component of either of the homogeneous masses, the terms $\dfrac{2\Gamma_1}{r}$ and $s\dfrac{d\Gamma_1}{d\mu_1}$ may generally be neglected. When it is not a component of either, the terms γ_1', γ_1'', $v'\dfrac{d\gamma_1'}{d\mu_1}$, $v''\dfrac{d\gamma_1''}{d\mu_1}$ may of course be cancelled, but we must not apply the formula to cases in which the substance spreads over the diaphragm separating the homogeneous masses.

In the cases just discussed, the problem of the stability of certain surfaces of tension has been solved by considering the case of neutral equilibrium,—a condition of neutral equilibrium affording the equation of the limit of stability. This method probably leads as directly as any to the result, when that consists in the determination of the value of a certain quantity at the limit of stability, or of the relation which exists at that limit between certain quantities specifying the state of the system. But problems of a more general character may require a more general treatment.

Let it be required to ascertain the stability or instability of a fluid system in a given state of equilibrium with respect to motion of the surfaces of tension and accompanying changes. It is supposed that the conditions of internal stability for the separate homogeneous masses are satisfied, as well as those conditions of stability for the surfaces of discontinuity which relate to small portions of these surfaces with the adjacent masses. (The conditions of stability which are here supposed to be satisfied have been already discussed in part and will be farther discussed hereafter.) The fundamental equations for all the masses and surfaces occurring in the system are supposed to be known. In applying the general criteria of stability which are given on page 57, we encounter the following difficulty.

The question of the stability of the system is to be determined by the consideration of states of the system which are slightly varied from that of which the stability is in question. These varied states of the system are not in general states of equilibrium, and the relations expressed by the fundamental equations may not hold true of them. More than this,—if we attempt to describe a varied state of the system by varied values of the quantities which describe the initial state, if these varied values are such as are inconsistent with equilibrium, they may fail to determine with precision any state of the system. Thus, when the phases of two contiguous homogeneous masses are specified, if these phases are such as satisfy all the conditions of equilibrium, the nature of the surface of discontinuity (if without additional components) is entirely determined; but if the phases do not satisfy all the conditions of equilibrium, the nature of the surface of discontinuity is not only undetermined, but incapable of determination by specified values of such quantities as we have employed to express the nature of surfaces of discontinuity in equilibrium. For example, if the temperatures in contiguous homogeneous masses are different, we cannot specify the thermal state of the surface of discontinuity by assigning to it any particular temperature. It would be necessary to give the law by which the temperature passes over from one value to the other. And if this were given, we could make no use of it in the determination of other quantities, unless the rate of change of the temperature were so gradual that at every point we could regard the thermodynamic state as unaffected by the change of temperature in its vicinity. It is true that we are also ignorant in respect to surfaces of discontinuity *in equilibrium* of the law of change of those quantities which are different in the two phases in contact, such as the densities of the components, but this, although unknown to us, is entirely determined by the nature of the phases in contact, so that no vagueness is occasioned in the definition of any of the quantities which we have occasion to use with reference to such surfaces of discontinuity.

It may be observed that we have established certain differential equations, especially (497), in which only the initial state is necessarily one of equilibrium. Such equations may be regarded as establishing certain properties of states bordering upon those of equilibrium. But these are properties which hold true only when we disregard quantities proportional to the square of those which express the degree of variation of the system from equilibrium. Such equations are therefore sufficient for the determination of the conditions of equilibrium, but not sufficient for the determination of the conditions of stability.

We may, however, use the following method to decide the question of stability in such a case as has been described.

Beside the real system of which the stability is in question, it will be convenient to conceive of another system, to which we shall attribute in its initial state the same homogeneous masses and surfaces of discontinuity which belong to the real system. We shall also suppose that the homogeneous masses and surfaces of discontinuity of this system, which we may call the imaginary system, have the same fundamental equations as those of the real system. But the imaginary system is to differ from the real in that the variations of its state are limited to such as do not violate the conditions of equilibrium relating to temperature and the potentials, and that the fundamental equations of the surfaces of discontinuity hold true for these varied states, although the condition of equilibrium expressed by equation (500) may not be satisfied.

Before proceeding farther, we must decide whether we are to examine the question of stability under the condition of a constant external temperature, or under the condition of no transmission of heat to or from external bodies, and in general, to what external influences we are to regard the system as subject. It will be convenient to suppose that the exterior of the system is fixed, and that neither matter nor heat can be transmitted through it. Other cases may easily be reduced to this, or treated in a manner entirely analogous.

Now if the real system in the given state is unstable, there must be some slightly varied state in which the energy is less, but the entropy and the quantities of the components the same as in the given state, and the exterior of the system unvaried. But it may easily be shown that the given state of the system may be made stable by constraining the surfaces of discontinuity to pass through certain fixed lines situated in the unvaried surfaces. Hence, if the surfaces of discontinuity are constrained to pass through corresponding fixed lines in the surfaces of discontinuity belonging to the varied state just mentioned, there must be a state of stable equilibrium for the system thus constrained which will .differ infinitely little from the given state of the system, the stability of which is in question, and will have the same entropy, quantities of components, and exterior, but less energy. The imaginary system will have a similar state, since the real and imaginary systems do not differ in respect to those states which satisfy all the conditions of equilibrium for each surface of discontinuity. That is, the imaginary system has a state, differing infinitely little from the given state, and with the same entropy, quantities of components, and exterior, but with less energy.

Conversely, if the imaginary system has such a state as that just described, the real system will also have such a state. This may be shown by fixing certain lines in the surfaces of discontinuity of the imaginary system in its state of less energy and then making the energy a minimum under the conditions. The state thus determined will satisfy all the conditions of equilibrium for each surface of discontinuity, and the real system will therefore have a corresponding state, in which the entropy, quantities of components, and exterior will be the same as in the given state, but the energy less.

We may therefore determine whether the given system is or is not unstable, by applying the general criterion of instability (7) to the imaginary system.

If the system is not unstable, the equilibrium is either neutral or stable. Of course we can determine which of these is the case by reference to the imaginary system, since the determination depends upon states of equilibrium, in regard to which the real and imaginary systems do not differ. We may therefore determine whether the equilibrium of the given system is stable, neutral, or unstable, by applying the criteria (3)–(7) to the imaginary system.

The result which we have obtained may be expressed as follows:— In applying to a fluid system which is in equilibrium, and of which all the small parts taken separately are stable, the criteria of stable, neutral, and unstable equilibrium, we may regard the system as under constraint to satisfy the conditions of equilibrium relating to temperature and the potentials, and as satisfying the relations expressed by the fundamental equations for masses and surfaces, even when the condition of equilibrium relating to pressure {equation (500)} is not satisfied.

It follows immediately from this principle, in connection with equations (501) and (86), that in a stable system each surface of tension must be a surface of minimum area for constant values of the volumes which it divides, when the other surfaces bounding these volumes and the perimeter of the surface of tension are regarded as fixed; that in a system in neutral equilibrium each surface of tension will have as small an area as it can receive by any slight variations under the same limitations; and that in seeking the remaining conditions of stable or neutral equilibrium, when these are satisfied, it is only necessary to consider such varied surfaces of tension as have similar properties with reference to the varied volumes and perimeters.

We may illustrate the method which has been described by applying it to a problem but slightly different from one already (pp. 244, 245) discussed by a different method. It is required to determine the conditions of stability for a system in equilibrium, consisting of two different homogeneous masses meeting at a surface of discontinuity, the perimeter of which is invariable, as well as the exterior of the whole system, which is also impermeable to heat.

To determine what is necessary for stability in addition to the condition of minimum area for the surface of tension, we need only consider those varied surfaces of tension which satisfy the same condition. We may therefore regard the surface of tension as determined by v', the volume of one of the homogeneous masses. But the state of the system would evidently be completely determined by the position of the surface of tension and the temperature and potentials, if the entropy and the quantities of the components were variable; and therefore, since the entropy and the quantities of the components are constant, the state of the system must be completely determined by the position of the surface of tension. We may therefore regard all the quantities relating to the system as functions of v', and the condition of stability may be written

$$\frac{d\epsilon}{dv'}dv' + \frac{1}{2}\frac{d^2\epsilon}{dv'^2}dv'^2 + \text{etc.} > 0,$$

where ϵ denotes the total energy of the system. Now the conditions of equilibrium require that

$$\frac{d\epsilon}{dv'} = 0.$$

Hence, the general condition of stability is that

$$\frac{d^2\epsilon}{dv'^2} > 0. \tag{541}$$

Now if we write ϵ', ϵ'', ϵ^s for the energies of the two masses and of the surface, we have by (86) and (501), since the total entropy and the total quantities of the several components are constant,

$$d\epsilon = d\epsilon' + d\epsilon'' + d\epsilon^s = -p'dv' - p''dv'' + \sigma ds,$$

or, since $dv'' = -dv'$,

$$\frac{d\epsilon}{dv'} = -p' + p'' + \sigma\frac{ds}{dv'}. \tag{542}$$

Hence,

$$\frac{d^2\epsilon}{dv'^2} = -\frac{dp'}{dv'} + \frac{dp''}{dv'} + \frac{d\sigma}{dv'}\frac{ds}{dv'} + \sigma\frac{d^2s}{dv'^2}, \tag{543}$$

and the condition of stability may be written

$$\sigma\frac{d^2s}{dv'^2} > \frac{dp'}{dv'} - \frac{dp''}{dv'} - \frac{d\sigma}{dv'}\frac{ds}{dv'}. \tag{544}$$

If we now simplify the problem by supposing, as in the similar case on page 245, that we may disregard the variations of the temperature and of all the potentials except one, the condition will reduce to

$$\sigma\frac{d^2s}{dv'^2} > \left(\gamma_1' - \gamma_1'' + \Gamma_1\frac{ds}{dv'}\right)\frac{d\mu_1}{dv'}. \tag{545}$$

The total quantity of the substance indicated by the suffix $_1$ is

$$\gamma_1' \, v' + \gamma_1'' \, v'' + \Gamma_1 \, s.$$

Making this constant, we have

$$\left(\gamma_1' - \gamma_1'' + \Gamma_1 \frac{ds}{dv'}\right) dv' + \left(v' \frac{d\gamma_1'}{d\mu_1} + v'' \frac{d\gamma_1''}{d\mu_1} + s \frac{d\Gamma_1}{d\mu_1}\right) d\mu_1 = 0. \qquad (546)$$

The condition of equilibrium is thus reduced to the form

$$\sigma \frac{d^2 s}{dv'^2} > - \frac{\left(\gamma_1' - \gamma_1'' + \Gamma_1 \frac{ds}{dv'}\right)^2}{v' \frac{d\gamma_1'}{d\mu_1} + v'' \frac{d\gamma_1''}{d\mu_1} + s \frac{d\Gamma_1}{d\mu_1}}, \qquad (547)$$

where $\dfrac{ds}{dv'}$ and $\dfrac{d^2 s}{dv'^2}$ are to be determined from the form of the surface
of tension by purely geometrical considerations, and the other differential coefficients are to be determined from the fundamental equations of the homogeneous masses and the surface of discontinuity. Condition (540) may be easily deduced from this as a particular case.

The condition of stability with reference to motion of surfaces of discontinuity admits of a very simple expression when we can treat the temperature and potentials as constant. This will be the case when one or more of the homogeneous masses, containing together all the component substances, may be considered as indefinitely large, the surfaces of discontinuity being finite. For if we write $\Sigma \Delta \epsilon$ for the sum of the variations of the energies of the several homogeneous masses, and $\Sigma \Delta \epsilon^s$ for the sum of the variations of the energies of the several surfaces of discontinuity, the condition of stability may be written

$$\Sigma \, \Delta \epsilon + \Sigma \, \Delta \epsilon^s > 0, \qquad (548)$$

the total entropy and the total quantities of the several components being constant. The variations to be considered are infinitesimal, but the character Δ signifies, as elsewhere in this paper, that the expression is to be interpreted without neglect of infinitesimals of the higher orders. Since the temperature and potentials are sensibly constant, the same will be true of the pressures and surface-tensions, and by integration of (86) and (501) we may obtain for any homogeneous mass

$$\Delta \epsilon = t \, \Delta \eta - p \, \Delta v + \mu_1 \Delta m_1 + \mu_2 \Delta m_2 + \text{etc.},$$

and for any surface of discontinuity

$$\Delta \epsilon^s = t \, \Delta \eta^s + \sigma \, \Delta s + \mu_1 \Delta m_1^s + \mu_2^s \Delta m_2 + \text{etc.}$$

These equations will hold true of finite differences, when t, p, σ, μ_1, μ_2, etc. are constant, and will therefore hold true of infinitesimal differences, under the same limitations, without neglect of the infinitesimals of the higher orders. By substitution of these values, the condition of stability will reduce to the form

$$-\Sigma(p\,\Delta v)+\Sigma(\sigma\,\Delta s) > 0,$$

or $$\Sigma(p\,\Delta v)-\Sigma(\sigma\,\Delta s) < 0. \qquad (549)$$

That is, the sum of the products of the volumes of the masses by their pressures, diminished by the sum of the products of the areas of the surfaces of discontinuity by their tensions, must be a maximum. This is a purely geometrical condition, since the pressures and tensions are constant. This condition is of interest, because it is always *sufficient* for stability with reference to motion of surfaces of discontinuity. For any system may be reduced to the kind described by putting certain parts of the system in communication (by means of fine tubes if necessary) with large masses of the proper temperatures and potentials. This may be done without introducing any new movable surfaces of discontinuity. The condition (549) when applied to the altered system is therefore the same as when applied to the original system. But it is sufficient for the stability of the altered system, and therefore sufficient for its stability if we diminish its freedom by breaking the connection between the original system and the additional parts, and therefore sufficient for the stability of the original system.

On the Possibility of the Formation of a Fluid of different Phase within any Homogeneous Fluid.

The study of surfaces of discontinuity throws considerable light upon the subject of the stability of such homogeneous fluid masses as have a less pressure than others formed of the same components (or some of them) and having the same temperature and the same potentials for their actual components.[*]

In considering this subject, we must first of all inquire how far our method of treating surfaces of discontinuity is applicable to cases in which the radii of curvature of the surfaces are of insensible magnitude. That it should not be applied to such cases without limitation is evident from the consideration that we have neglected the term $\frac{1}{2}(C_1 - C_2)\,\delta(c_1 - c_2)$ in equation (494) on account of the magnitude of the radii of curvature compared with the thickness of the non-homogeneous film. (See page 228.) When, however, only spherical masses are considered, this term will always disappear, since C_1 and C_2 will necessarily be equal.

---

[*] See page 104, where the term stable is used (as indicated on page 103) in a less strict sense than in the discussion which here follows.

Again, the surfaces of discontinuity have been regarded as separating homogeneous masses. But we may easily conceive that a globular mass (surrounded by a large homogeneous mass of different nature) may be so small that no part of it will be homogeneous, and that even at its center the matter cannot be regarded as having any phase of matter *in mass*. This, however, will cause no difficulty, if we regard the phase of the interior mass as determined by the same relations to the exterior mass as in other cases. Beside the phase of the exterior mass, there will always be another phase having the same temperature and potentials, but of the general nature of the small globule which is surrounded by that mass and in equilibrium with it. This phase is completely determined by the system considered, and in general entirely stable and perfectly capable of realization in mass, although not such that the exterior mass could exist in contact with it at a plane surface. This is the phase which we are to attribute to the mass which we conceive as existing within the dividing surface.*

With this understanding with regard to the phase of the fictitious interior mass, there will be no ambiguity in the meaning of any of the symbols which we have employed, when applied to cases in which the surface of discontinuity is spherical, however small the radius may be. Nor will the demonstration of the general theorems require any material modification. The dividing surface which determines the value of ϵ^s, η^s, m_1^s, m_2^s, etc. is as in other cases to be placed so as to make the term $\frac{1}{2}(C_1+C_2)\,\delta(c_1+c_2)$ in equation (494) vanish, i.e., so as to make equation (497) valid. It has been shown on pages 225–227 that when thus placed it will sensibly coincide with the physical surface of discontinuity, when this consists of a non-homogeneous film separating homogeneous masses, and having radii of curvature which are large compared with its thickness. But in regard to globular masses too small for this theorem to have any application, it will be worth while to examine how far we may be certain that the radius of the dividing surface will have a real and positive value, since it is only then that our method will have any natural application.

The value of the radius of the dividing surface, supposed spherical, of any globule in equilibrium with a surrounding homogeneous fluid may be most easily obtained by eliminating σ from equations (500) and (502), which have been derived from (497), and contain the radius implicitly. If we write r for this radius, equation (500) may be written

$$2\sigma = (p'-p'')r, \tag{550}$$

---

* For example, in applying our formulæ to a microscopic globule of water in steam, by the density or pressure of the interior mass we should understand, not the actual density or pressure at the center of the globule, but the density of liquid water (in large quantities) which has the temperature and potential of the steam.

the single and double accents referring respectively to the interior and exterior masses. If we write $[\epsilon]$, $[\eta]$, $[m_1]$, $[m_2]$, etc., for the excess of the total energy, entropy, etc., in and about the globular mass above what would be in the same space if it were uniformly filled with matter of the phase of the exterior mass, we shall have necessarily with reference to the whole dividing surface

$$\epsilon^s = [\epsilon] - v'(\epsilon_V' - \epsilon_V''), \qquad \eta^s = [\eta] - v'(\eta_V' - \eta_V''),$$

$$m_1^s = [m_1] - v'(\gamma_1' - \gamma_1''), \quad m_2^s = [m_2] - v'(\gamma_2' - \gamma_2''), \text{ etc.},$$

where ϵ_V', ϵ_V'', η_V', η_V'', γ_1', γ_1'', etc. denote, in accordance with our usage elsewhere, the volume-densities of energy, of entropy, and of the various components, in the two homogeneous masses. We may thus obtain from equation (502)

$$\sigma s = [\epsilon] - v'(\epsilon_V' - \epsilon_V'') - t[\eta] + tv'(\eta_V' - \eta_V'')$$
$$- \mu_1[m_1] + \mu_1 v'(\gamma_1' - \gamma_1'') - \mu_2[m_2] + \mu_2 v'(\gamma_2' - \gamma_2'') - \text{etc.} \quad (551)$$

But by (93),

$$p' = -\epsilon_V' + t\eta_V' + \mu_1\gamma_1' + \mu_2\gamma_2' + \text{etc.},$$

$$p'' = -\epsilon_V'' + t\eta_V'' + \mu_1\gamma_1'' + \mu_2\gamma_2'' + \text{etc.}$$

Let us also write for brevity

$$W = [\epsilon] - t[\eta] - \mu_1[m_1] - \mu_2[m_2] - \text{etc.} \quad (552)$$

(It will be observed that the value of W is entirely determined by the nature of the physical system considered, and that the notion of the dividing surface does not in any way enter into its definition.) We shall then have

$$\sigma s = W + v'(p' - p''), \quad (553)$$

or, substituting for s and v' their values in terms of r,

$$4\pi r^2 \sigma = W + \tfrac{4}{3}\pi r^3(p' - p''), \quad (554)$$

and eliminating σ by (550),

$$\tfrac{2}{3}\pi r^3(p' - p'') = W, \quad (555)$$

$$r = \left(\frac{3W}{2\pi(p' - p'')}\right)^{\tfrac{1}{3}}. \quad (556)$$

If we eliminate r instead of σ, we have

$$\frac{16\pi\sigma^3}{3(p' - p'')^2} = W, \quad (557)$$

$$\sigma = \left(\frac{3W(p' - p'')^2}{16\pi}\right)^{\tfrac{1}{3}}. \quad (558)$$

Now, if we first suppose the difference of the pressures in the homogeneous masses to be very small, so that the surface of discontinuity is nearly plane, since without any important loss of generality we may regard σ as positive (for if σ is not positive when $p' = p''$, the surface when plane would not be stable in regard to position, as it certainly is, in every actual case, when the proper conditions are fulfilled with respect to its perimeter), we see by (550) that the pressure in the interior mass must be the greater; i.e., we may regard σ, $p' - p''$, and r as all positive. By (555), the value of W will also be positive. But it is evident from equation (552), which defines W, that the value of this quantity is necessarily real, in any possible case of equilibrium, and can only become infinite when r becomes infinite and $p' = p''$. Hence, by (556) and (558), as $p' - p''$ increases from very small values, W, r, and σ have single, real, and positive values until they simultaneously reach the value zero. Within this limit, our method is evidently applicable; beyond this limit, if such exist, it will hardly be profitable to seek to interpret the equations. But it must be remembered that the vanishing of the radius of the somewhat arbitrarily determined *dividing surface* may not necessarily involve the vanishing of the physical heterogeneity. It is evident, however (see pp. 225–227), that the globule must become insensible in magnitude before r can vanish.

It may easily be shown that the quantity denoted by W is the work which would be required to form (by a reversible process) the heterogeneous globule in the interior of a very large mass having initially the uniform phase of the exterior mass. For this work is equal to the increment of energy of the system when the globule is formed without change of the entropy or volume of the whole system or of the quantities of the several components. Now $[\eta]$, $[m_1]$, $[m_2]$, etc. denote the increments of entropy and of the components in the space where the globule is formed. Hence these quantities with the negative sign will be equal to the increments of entropy and of the components in the rest of the system. And hence, by equation (86),

$$-t[\eta] - \mu_1[m_1] - \mu_2[m_2] - \text{etc.}$$

will denote the increment of energy in all the system except where the globule is formed. But $[\epsilon]$ denotes the increment of energy in that part of the system. Therefore, by (552), W denotes the total increment of energy in the circumstances supposed, or the work required for the formation of the globule.

The conclusions which may be drawn from these considerations with respect to the stability of the homogeneous mass of the pressure p'' (supposed less than p', the pressure belonging to a different phase of the same temperature and potentials) are very obvious. Within those limits within which the method used has been justified, the mass in question must be regarded as in strictness stable with respect to the growth of a globule of the kind considered, since W, the work required for the formation of such a globule of a certain size (viz., that which would be in equilibrium with the surrounding mass), will always be positive. Nor can smaller globules be formed, for they can neither be in equilibrium with the surrounding mass, being too small, nor grow to the size of that to which W relates. If, however, by any external agency such a globular mass (of the size necessary for equilibrium) were formed, the equilibrium has already (page 243) been shown to be unstable, and with the least excess in size, the interior mass would tend to increase without limit except that depending on the magnitude of the exterior mass. We may therefore regard the quantity W as affording a kind of measure of the *stability* of the phase to which p'' relates. In equation (557) the value of W is given in terms of σ and $p'-p''$. If the three fundamental equations which give σ, p', and p'' in terms of the temperature and the potentials were known, we might regard the stability (W) as known in terms of the same variables. It will be observed that when $p'=p''$ the value of W is infinite. If $p'-p''$ increases without greater changes of the phases than are necessary for such increase, W will vary at first very nearly inversely as the square of $p'-p''$. If $p'-p''$ continues to increase, it may perhaps occur that W reaches the value zero ; but until this occurs the phase is certainly stable with respect to the kind of change considered. Another kind of change is conceivable, which initially is small in degree but may be great in its extent in space. Stability in this respect or *stability in respect to continuous changes of phase* has already been discussed (see page 105), and its limits determined. These limits depend entirely upon the fundamental equation of the homogeneous mass of which the stability is in question. But with respect to the kind of changes here considered, which are initially small in extent but great in degree, it does not appear how we can fix the limits of stability with the same precision. But it is safe to say that if there is such a limit it must be at or beyond the limit at which σ vanishes. This latter limit is determined entirely by the fundamental equation of the surface of discontinuity between the phase of which the stability is in question and that of which the possible formation is in question. We have already seen that when σ vanishes, the radius of the dividing surface and the work W vanish with it. If the fault in the homogeneity of the mass vanishes at the same time (it evidently cannot vanish sooner), the phase becomes unstable at this limit. But if the fault in the homogeneity of the physical mass does not vanish with r, σ and W,—and no sufficient reason appears why this should not be considered as the general case,—although the amount of work necessary to upset the equilibrium of the phase is infinitesimal, this is not enough to make the phase unstable.

It appears therefore that W is a somewhat one-sided measure of stability.

It must be remembered in this connection that the fundamental equation of a surface of discontinuity can hardly be regarded as capable of experimental determination, except for plane surfaces (see pp. 231-233), although the relation for spherical surfaces is in the nature of things entirely determined, at least so far as the phases are separately capable of existence. Yet the foregoing discussion yields the following practical results. It has been shown that the real stability of a phase extends in general beyond that limit (discussed on pages 103-105), which may be called the limit of practical stability, at which the phase can exist in contact with another at a plane surface, and a formula has been deduced to express the degree of stability in such cases as measured by the amount of work necessary to upset the equilibrium of the phase when supposed to extend indefinitely in space. It has also been shown to be entirely consistent with the principles established that this stability should have limits, and the manner in which the general equations would accommodate themselves to this case has been pointed out.

By equation (553), which may be written

$$W = \sigma s - (p' - p'')v', \tag{559}$$

we see that the work W consists of two parts, of which one is always positive, and is expressed by the product of the superficial tension and the area of the surface of tension, and the other is always negative, and is numerically equal to the product of the difference of pressure by the volume of the interior mass. We may regard the first part as expressing the work spent in forming the surface of tension, and the second part the work gained in forming the interior mass.* Moreover, the second of these quantities, if we neglect its

---

* To make the physical significance of the above more clear, we may suppose the two processes to be performed separately in the following manner. We may suppose a large mass of the same phase as that which has the volume v' to exist initially in the interior of the other. Of course, it must be surrounded by a resisting envelop, on account of the difference of the pressures. We may, however, suppose this envelop permeable to all the component substances, although not of such properties that a mass can form on the exterior like that within. We may allow the envelop to yield to the internal pressure until its contents are increased by v' without materially affecting its superficial area. If this be done sufficiently slowly, the phase of the mass within will remain constant. (See page 84.) A homogeneous mass of the volume v' and of the desired phase has thus been produced, and the work gained is evidently $(p' - p'')v'$.

Let us suppose that a small aperture is now opened and closed in the envelop so as to let out exactly the volume v' of the mass within, the envelop being pressed inwards in another place so as to diminish its contents by this amount. During the extrusion of the drop and until the orifice is entirely closed, the surface of the drop must adhere to the edge of the orifice, but not elsewhere to the outside surface of the envelop. The work done in forming the surface of the drop will evidently be σs or $\frac{1}{2}(p' - p'')v'$. Of

sign, is always equal to two-thirds of the first, as appears from equation (550) and the geometrical relation $v' = \frac{1}{3}rs$. We may therefore write

$$W = \frac{1}{3}\sigma s = \frac{1}{2}(p' - p'')v'. \tag{560}$$

On the Possible Formation at the Surface where two different Homogeneous Fluids meet of a Fluid of different Phase from either.

Let A, B, and C be three different fluid phases of matter, which satisfy all the conditions necessary for equilibrium when they meet at plane surfaces. The components of A and B may be the same or different, but C must have no components except such as belong to A or B. Let us suppose masses of the phases A and B to be separated by a very thin sheet of the phase C. This sheet will not necessarily be plane, but the sum of its principal curvatures must be zero. We may treat such a system as consisting simply of masses of the phases A and B with a certain surface of discontinuity, for in our previous discussion there has been nothing to limit the thickness or the nature of the film separating homogeneous masses, except that its thickness has generally been supposed to be small in comparison with its radii of curvature. The value of the superficial tension for such a film will be $\sigma_{AC} + \sigma_{BC}$, if we denote by these symbols the tensions of the surfaces of contact of the phases A and C, and B and C, respectively. This not only appears from evident mechanical considerations, but may also be easily verified by equations (502) and (93), the first of which may be regarded as defining the quantity σ. This value will not be affected by diminishing the thickness of the film, until the limit is reached at which the interior of the film ceases to have the properties of matter in mass. Now if $\sigma_{AC} + \sigma_{BC}$ is greater than σ_{AB}, the tension of the ordinary surface between A and B, such a film will be at least practically unstable. (See page 240.) We cannot suppose that $\sigma_{AB} > \sigma_{AC} + \sigma_{BC}$, for this would make the ordinary surface between A and B unstable and difficult to realize. If $\sigma_{AB} = \sigma_{AC} + \sigma_{BC}$, we may assume, in general, that this relation is not accidental, and that the ordinary surface of contact for A and B is of the kind which we have described.

Let us now suppose the phases A and B to vary, so as still to satisfy the conditions of equilibrium at plane contact, but so that the pressure of the phase C determined by the temperature and potentials

---

this work, the amount $(p' - p'')v'$ will be expended in pressing the envelop inward, and the rest in opening and closing the orifice. Both the opening and the closing will be resisted by the capillary tension. If the orifice is circular, it must have, when widest open, the radius determined by equation (550).

of A and B shall become less than the pressure of A and B. A system consisting of the phases A and B will be entirely stable with respect to the formation of any phase like C. (This case is not quite identical with that considered on page 104, since the system in question contains two different phases, but the principles involved are entirely the same.)

With respect to variations of the phases A and B in the opposite direction we must consider two cases separately. It will be convenient to denote the pressures of the three phases by p_A, p_B, p_C, and to regard these quantities as functions of the temperature and potentials.

If $\sigma_{AB} = \sigma_{AC} + \sigma_{BC}$ for values of the temperature and potentials which make $p_A = p_B = p_C$, it will not be possible to alter the temperature and potentials at the surface of contact of the phases A and B so that $p_A = p_B$, and $p_C > p_A$, for the relation of the temperature and potentials necessary for the equality of the three pressures will be preserved by the increase of the mass of the phase C. Such variations of the phases A and B might be brought about in separate masses, but if these were brought into contact, there would be an immediate formation of a mass of the phase C, with reduction of the phases of the adjacent masses to such as satisfy the conditions of equilibrium with that phase.

But if $\sigma_{AB} < \sigma_{AC} + \sigma_{BC}$, we can vary the temperature and potentials so that $p_A = p_B$, and $p_C > p_A$, and it will not be possible for a sheet of the phase of C to form *immediately*, i.e., while the pressure of C is sensibly equal to that of A and B; for mechanical work equal to $\sigma_{AC} + \sigma_{BC} - \sigma_{AB}$ per unit of surface might be obtained by bringing the system into its original condition, and therefore produced without any external expenditure, unless it be that of heat at the temperature of the system, which is evidently incapable of producing the work. The stability of the system in respect to such a change must therefore extend beyond the point where the pressure of C commences to be greater than that of A and B. We arrive at the same result if we use the expression (520) as a test of stability. Since this expression has a finite positive value when the pressures of the phases are all equal, the ordinary surface of discontinuity must be stable, and it must require a finite change in the circumstances of the case to make it become unstable.*

---

*It is true that such a case as we are now considering is formally excluded in the discussion referred to, which relates to a plane surface, and in which the system is supposed thoroughly stable with respect to the possible formation of any different homogeneous masses. Yet the reader will easily convince himself that the criterion (520) is perfectly valid in this case with respect to the possible formation of a thin sheet of the phase C, which, as we have seen, may be treated simply as a different kind of surface of discontinuity.

In the preceding paragraph it is shown that the surface of contact of phases A and B is stable under certain circumstances, with respect to the formation of a thin sheet of the phase C. To complete the demonstration of the stability of the surface with respect to the formation of the phase C, it is necessary to show that this phase cannot be formed at the surface in lentiform masses. This is the more necessary, since it is in this manner, if at all, that the phase is likely to be formed, for an incipient sheet of phase C would evidently be unstable when $\sigma_{AB} < \sigma_{AC} + \sigma_{BC}$, and would immediately break up into lentiform masses.

Fig. 10.

It will be convenient to consider first a lentiform mass of phase C in equilibrium between masses of phases A and B which meet in a plane surface. Let figure 10 represent a section of such a system through the centers of the spherical surfaces, the mass of phase A lying on the left of DEH'FG, and that of phase B on the right of DEH''FG. Let the line joining the centers cut the spherical surfaces in H' and H'', and the plane of the surface of contact of A and B in I. Let the radii of EH'F and EH''F be denoted by r', r'', and the segments IH', IH'', by x', x''. Also let IE, the radius of the circle in which the spherical surfaces intersect, be denoted by R. By a suitable application of the general condition of equilibrium we may easily obtain the equation

$$\sigma_{AC}\frac{r'-x'}{r'} + \sigma_{BC}\frac{r''-x''}{r''} = \sigma_{AB}, \qquad (561)$$

which signifies that the components parallel to EF of the tension σ_{AC} and σ_{BC} are together equal to σ_{AB}. If we denote by W the amount of work which must be expended in order to form such a lentiform mass as we are considering between masses of indefinite extent having the phases A and B, we may write

$$W = M - N, \qquad (562)$$

where M denotes the work expended in replacing the surface between A and B by the surfaces between A and C and B and C, and N denotes the work gained in replacing the masses of phases A and B by the mass of phase C. Then

$$M = \sigma_{AC}\,s_{AC} + \sigma_{BC}\,s_{BC} - \sigma_{AB}\,s_{AB}, \qquad (563)$$

where s_{AC}, s_{BC}, s_{AB} denote the areas of the three surfaces concerned; and

$$N = V'(p_C - p_A) + V''(p_C - p_B), \qquad (564)$$

where V' and V'' denote the volumes of the masses of the phases A and B which are replaced. Now by (500),

$$p_0 - p_A = \frac{2\,\sigma_{AC}}{r'}, \text{ and } p_0 - p_B = \frac{2\,\sigma_{BC}}{r''}. \tag{565}$$

We have also the geometrical relations

$$\left. \begin{aligned} V' &= \tfrac{2}{3}\pi r'^2 x' - \tfrac{1}{3}\pi R^2 (r' - x'), \\ V'' &= \tfrac{2}{3}\pi r''^2 x'' - \tfrac{1}{3}\pi R^2 (r'' - x''). \end{aligned} \right\} \tag{566}$$

By substitution we obtain

$$\begin{aligned} N = \tfrac{4}{3}\pi\,\sigma_{AC}\,r'x' &- \tfrac{2}{3}\pi R^2\,\sigma_{AC}\,\frac{r' - x'}{r'} \\ &+ \tfrac{4}{3}\pi\,\sigma_{BC}\,r''x'' - \tfrac{2}{3}\pi R^2\,\sigma_{BC}\,\frac{r'' - x''}{r''}, \end{aligned} \tag{567}$$

and by (561),

$$N = \tfrac{4}{3}\pi\,\sigma_{AC}\,r'x' + \tfrac{4}{3}\pi\,\sigma_{BC}\,r''x'' - \tfrac{2}{3}\pi\,R^2\,\sigma_{AB}. \tag{568}$$

Since

$$2\pi r'x' = s_{AC}, \quad 2\pi r''x'' = s_{BC}, \quad \pi R^2 = s_{AB},$$

we may write

$$N = \tfrac{2}{3}(\sigma_{AC}\,s_{AC} + \sigma_{BC}\,s_{BC} - \sigma_{AB}\,s_{AB}). \tag{569}$$

(The reader will observe that the ratio of M and N is the same as that of the corresponding quantities in the case of the spherical mass treated on pages 252–258.) We have therefore

$$W = \tfrac{1}{3}(\sigma_{AC}\,s_{AC} + \sigma_{BC}\,s_{BC} - \sigma_{AB}\,s_{AB}). \tag{570}$$

This value is positive so long as

$$\sigma_{AC} + \sigma_{BC} > \sigma_{AB},$$

since

$$s_{AC} > s_{AB}, \text{ and } s_{BC} > s_{AB}.$$

But at the limit, when

$$\sigma_{AC} + \sigma_{BC} = \sigma_{AB},$$

we see by (561) that

$$s_{AC} = s_{AB}, \text{ and } s_{BC} = s_{AB},$$

and therefore

$$W = 0.$$

It should however be observed that in the immediate vicinity of the circle in which the three surfaces of discontinuity intersect, the physical state of each of these surfaces must be affected by the vicinity of the others. We cannot, therefore, rely upon the formula (570) except when the dimensions of the lentiform mass are of sensible magnitude.

We may conclude that after we pass the limit at which p_0 becomes greater than p_A and p_B (supposed equal) lentiform masses of phase C will not be formed until either $\sigma_{AB} = \sigma_{AC} + \sigma_{BC}$, or $p_0 - p_A$ becomes so great that the lentiform mass which would be in equilibrium is one of insensible magnitude.   {The diminution of the radii with increasing values of $p_C - p_A$ is indicated by equation (565).}   Hence, no mass of phase C will be formed until one of these limits is reached.   Although the demonstration relates to a *plane* surface between A and B, the result must be applicable whenever the radii of curvature have a sensible magnitude, since the effect of such curvature may be disregarded when the lentiform mass is sufficiently small.

The equilibrium of the lentiform mass of phase C is easily proved to be unstable, so that the quantity W affords a kind of measure of the stability of plane surfaces of contact of the phases A and B.*

Essentially the same principles apply to the more general problem in which the phases A and B have moderately different pressures, so that their surfaces of contact must be curved, but the radii of curvature have a sensible magnitude.

In order that a thin film of the phase C may be in equilibrium between masses of the phases A and B, the following equations must be satisfied:—

$$\sigma_{AC}(c_1 + c_2) = p_A - p_C,$$

$$\sigma_{BC}(c_1 + c_2) = p_C - p_B,$$

where c_1 and c_2 denote the principal curvatures of the film, the centers of positive curvature lying in the mass having the phase A. Eliminating $c_1 + c_2$, we have

$$\sigma_{BC}(p_A - p_C) = \sigma_{AC}(p_C - p_B),$$

or
$$p_C = \frac{\sigma_{BC}\, p_A + \sigma_{AC}\, p_B}{\sigma_{BC} + \sigma_{AC}}. \tag{571}$$

It is evident that if p_C has a value greater than that determined by this equation, such a film will develop into a larger mass; if p_C has a less value, such a film will tend to diminish.   Hence, when

$$p_C < \frac{\sigma_{BC}\, p_A + \sigma_{AC}\, p_B}{\sigma_{BC} + \sigma_{AC}}, \tag{572}$$

the phases A and B have a stable surface of contact.

---

*If we represent phases by the position of points in such a manner that coexistent phases (in the sense in which the term is used on page 96) are represented by the same point, and allow ourselves, for brevity, to speak of the phases as having the positions of the points by which they are represented, we may say that three coexistent phases are situated where three series of pairs of coexistent phases meet or intersect.   If the three phases are all fluid, or when the effects of solidity may be disregarded, two cases are to be distinguished.   Either the three series of coexistent phases all intersect,—this is when each of the three surface tensions is less than the sum of the two others,—or one of the series terminates where the two others intersect,—this is where one surface tension is equal to the sum of the others.   The series of coexistent phases will be represented by lines or surfaces, according as the phases have one or two independently variable components.   Similar relations exist when the number of components is greater, except that they are not capable of geometrical representation without some limitation, as that of constant temperature or pressure or certain constant potentials.

Again, if more than one kind of surface of discontinuity is possible between A and B, for any given values of the temperature and potentials, it will be impossible for that having the greater tension to displace the other, at the temperature and with the potentials considered. Hence, when p_C has the value determined by equation (571), and consequently $\sigma_{AC} + \sigma_{BC}$ is one value of the tension for the surface between A and B, it is impossible that the ordinary tension of the surface σ_{AB} should be greater than this. If $\sigma_{AB} = \sigma_{AC} + \sigma_{BC}$, when equation (571) is satisfied, we may presume that a thin film of the phase C actually exists at the surface between A and B, and that a variation of the phases such as would make p_C greater than the second member of (571) cannot be brought about at that surface, as it would be prevented by the formation of a larger mass of the phase C. But if $\sigma_{AB} < \sigma_{AC} + \sigma_{BC}$ when equation (571) is satisfied, this equation does not mark the limit of the stability of the surface between A and B, for the temperature or potentials must receive a finite change before the film of phase C, or (as we shall see in the following paragraph) a lentiform mass of that phase, can be formed.

The work which must be expended in order to form on the surface between indefinitely large masses of phases A and B a lentiform mass of phase C in equilibrium, may evidently be represented by the formula

$$W = \sigma_{AC} S_{AC} + \sigma_{BC} S_{BC} - \sigma_{AB} S_{AB}$$
$$- p_C V_C + p_A V_A + p_B V_B, \tag{573}$$

where S_{AC}, S_{BC} denote the areas of the surfaces formed between A and C, and B and C; S_{AB} the diminution of the area of the surface between A and B; V_C the volume formed of the phase C; and V_A, V_B the diminution of the volumes of the phases A and B. Let us now suppose σ_{AC}, σ_{BC}, σ_{AB}, p_A, p_B to remain constant and the external boundary of the surface between A and B to remain fixed, while p_C increases and the surfaces of tension receive such alterations as are necessary for equilibrium. It is not necessary that this should be physically possible in the actual system; we may suppose the changes to take place, for the sake of argument, although involving changes in the fundamental equations of the masses and surfaces considered. Then, regarding W simply as an abbreviation for the second member of the preceding equation, we have

$$dW = \sigma_{AC} dS_{AC} + \sigma_{BC} dS_{BC} - \sigma_{AB} dS_{AB}$$
$$- p_C dV_C + p_A dV_A + p_B dV_B - V_C dp_C. \tag{574}$$

But the conditions of equilibrium require that

$$\sigma_{AC} dS_{AC} + \sigma_{BC} dS_{BC} - \sigma_{AB} dS_{AB}$$
$$- p_C dV_C + p_A dV_A + p_B dV_B = 0. \tag{575}$$

Hence,
$$dW = -V_C \, dp_C. \qquad (576)$$

Now it is evident that V_C will diminish as p_C increases. Let us integrate the last equation supposing p_C to increase from its original value until V_C vanishes. This will give

$$W'' - W' = \text{a negative quantity}, \qquad (577)$$

where W' and W'' denote the initial and final values of W. But $W'' = 0$. Hence W' is positive. But this is the value of W in the original system containing the lentiform mass, and expresses the work necessary to form the mass between the phases A and B. It is therefore impossible that such a mass should form on a surface between these phases. We must however observe the same limitation as in the less general case already discussed,—that $p_C - p_A$, $p_C - p_B$ must not be so great that the dimensions of the lentiform mass are of insensible magnitude. It may also be observed that the value of these differences may be so small that there will not be room on the surface between the masses of phases A and B for a mass of phase C sufficiently large for equilibrium. In this case we may consider a mass of phase C which is in equilibrium upon the surface between A and B in virtue of a *constraint* applied to the line in which the three surfaces of discontinuity intersect, which will not allow this line to become longer, although not preventing it from becoming shorter. We may prove that the value of W is positive by such an integration as we have used before.

Substitution of Pressures for Potentials in Fundamental Equations for Surfaces.

The fundamental equation of a surface which gives the value of the tension in terms of the temperature and potentials seems best adapted to the purposes of theoretical discussion, especially when the number of components is large or undetermined. But the experimental determination of the fundamental equations, or the application of any result indicated by theory to actual cases, will be facilitated by the use of other quantities in place of the potentials, which shall be capable of more direct measurement, and of which the numerical expression (when the necessary measurements have been made) shall depend upon less complex considerations. The numerical value of a potential depends not only upon the system of units employed, but also upon the arbitrary constants involved in the definition of the energy and entropy of the substance to which the potential relates, or, it may be, of the elementary substances of which that substance is formed. (See page 96.) This fact and the want of means of direct measurement may give a certain vagueness to the idea of the potentials, and render the equations which involve them less fitted to give a clear idea of physical relations.

Now the fundamental equation of each of the homogeneous masses which are separated by any surface of discontinuity affords a relation between the pressure in that mass and the temperature and potentials. We are therefore able to eliminate one or two potentials from the fundamental equation of a surface by introducing the pressures in the adjacent masses. Again, when one of these masses is a gas-mixture which satisfies Dalton's law as given on page 155, the potential for each simple gas may be expressed in terms of the temperature and the partial pressure belonging to that gas. By the introduction of these partial pressures we may eliminate as many potentials from the fundamental equation of the surface as there are simple gases in the gas-mixture.

An equation obtained by such substitutions may be regarded as a fundamental equation for the surface of discontinuity to which it relates, for when the fundamental equations of the adjacent masses are known, the equation in question is evidently equivalent to an equation between the tension, temperature, and potentials, and we must regard the knowledge of the properties of the adjacent masses as an indispensable preliminary, or an essential part, of a complete knowledge of any surface of discontinuity. It is evident, however, that from these fundamental equations involving pressures instead of potentials we cannot obtain by differentiation (without the use of the fundamental equations of the homogeneous masses) precisely the same relations as by the differentiation of the equations between the tensions, temperatures, and potentials. It will be interesting to inquire, at least in the more important cases, what relations may be obtained by differentiation from the fundamental equations just described alone.

If there is but one component, the fundamental equations of the two homogeneous masses afford one relation more than is necessary for the elimination of the potential. It may be convenient to regard the tension as a function of the temperature and the difference of the pressures. Now we have by (508) and (98)

$$d\sigma = -\eta_8 \, dt - \Gamma \, d\mu_1,$$

$$d(p' - p'') = (\eta_V{}' - \eta_V{}'')dt + (\gamma' - \gamma'')d\mu_1.$$

Hence we derive the equation

$$d\sigma = -\left(\eta_8 - \frac{\Gamma}{\gamma' - \gamma''}(\eta_V{}' - \eta_V{}'')\right)dt - \frac{\Gamma}{\gamma' - \gamma''}d(p' - p''), \quad (578)$$

which indicates the differential coefficients of σ with respect to t and $p' - p''$. For surfaces which may be regarded as nearly plane, it is evident that $\dfrac{\Gamma}{\gamma'-\gamma''}$ represents the distance from the surface of tension to a dividing surface located so as to make the superficial density of the single component vanish (being positive, when the latter surface is on the side specified by the double accents), and that the coefficient of dt (without the negative sign) represents the superficial density of entropy as determined by the latter dividing surface, i.e., the quantity denoted by $\eta_{S(1)}$ on page 235.

When there are two components, neither of which is confined to the surface of discontinuity, we may regard the tension as a function of the temperature and the pressures in the two homogeneous masses. The values of the differential coefficients of the tension with respect to these variables may be represented in a simple form if we choose such substances for the components that in the particular state considered each mass shall consist of a single component. This will always be possible when the composition of the two masses is not identical, and will evidently not affect the values of the differential coefficients. We then have

$$d\sigma = -\eta_S\, dt - \Gamma_{,}\, d\mu_{,} - \Gamma_{,,}\, d\mu_{,,},$$
$$dp' = \eta_V'\, dt + \gamma'\, d\mu_{,},$$
$$dp'' = \eta_V''\, dt + \gamma''\, d\mu_{,,},$$

where the marks $_{,}$ and $_{,,}$ are used instead of the usual $_1$ and $_2$ to indicate the identity of the component specified with the substance of the homogeneous masses specified by $'$ and $''$. Eliminating $d\mu_{,}$ and $d\mu_{,,}$ we obtain

$$d\sigma = -\left(\eta_S - \frac{\Gamma_{,}}{\gamma'}\eta_V' - \frac{\Gamma_{,,}}{\gamma''}\eta_V''\right) dt - \frac{\Gamma_{,}}{\gamma'}dp' - \frac{\Gamma_{,,}}{\gamma''}dp''. \tag{579}$$

We may generally neglect the difference of p' and p'', and write

$$d\sigma = -\left(\eta_S - \frac{\Gamma_{,}}{\gamma'}\eta_V' - \frac{\Gamma_{,,}}{\gamma''}\eta_V''\right) dt - \left(\frac{\Gamma_{,}}{\gamma'} + \frac{\Gamma_{,,}}{\gamma''}\right) dp. \tag{580}$$

The equation thus modified is strictly to be regarded as the equation for a plane surface. It is evident that $\dfrac{\Gamma_{,}}{\gamma'}$ and $\dfrac{\Gamma_{,,}}{\gamma''}$ represent the distances from the surface of tension of the two surfaces of which one would make $\Gamma_{,}$ vanish, and the other $\Gamma_{,,}$, that $\dfrac{\Gamma_{,}}{\gamma'} + \dfrac{\Gamma_{,,}}{\gamma''}$ represents the distance between these two surfaces, or the *diminution of volume* due to a unit of the surface of discontinuity, and that the coefficient of dt (without the negative sign) represents the excess of entropy in a system consisting of a unit of the surface of discontinuity with a part of each of the adjacent masses above that which the same matter would have if it existed in two homogeneous masses of the same phases but without any surface of discontinuity. (A mass thus existing without any surface of discontinuity must of course be entirely surrounded by matter of the same phase.)*

The form in which the values of $\left(\dfrac{d\sigma}{dt}\right)_p$ and $\left(\dfrac{d\sigma}{dp}\right)_t$ are given in equation (580) is adapted to give a clear idea of the relations of these quantities to the particular state of the system for which they are to be determined, but not to show how they vary with the state of the system. For this purpose it will be convenient to have the values of these differential coefficients expressed with reference to ordinary components. Let these be specified as usual by $_1$ and $_2$. If we eliminate $d\mu_1$ and $d\mu_2$ from the equations

$$-d\sigma = \eta_8\, dt + \Gamma_1\, d\mu_1 + \Gamma_2\, d\mu_2,$$

$$dp = \eta_V'\, dt + \gamma_1'\, d\mu_1 + \gamma_2'\, d\mu_2,$$

$$dp = \eta_V''\, dt + \gamma_1''\, d\mu_1 + \gamma_2''\, d\mu_2,$$

we obtain

$$d\sigma = \frac{B}{A}\, dt + \frac{C}{A}\, dp, \qquad (581)$$

---

* If we set

$$V = -\frac{\Gamma_{\prime}}{\gamma'} - \frac{\Gamma_{\prime\prime}}{\gamma''}, \qquad (a)$$

$$H_8 = \eta_8 - \frac{\Gamma_{\prime}}{\gamma'}\, \eta_V' - \frac{\Gamma_{\prime\prime}}{\gamma''}\, \eta_V'', \qquad (b)$$

and in like manner

$$E_8 = \epsilon_8 - \frac{\Gamma_{\prime}}{\gamma'}\, \epsilon_V' - \frac{\Gamma_{\prime\prime}}{\gamma''}\, \epsilon_V'', \qquad (c)$$

we may easily obtain, by means of equations (93) and (507),

$$E_8 = t H_8 + \sigma - p V. \qquad (d)$$

Now equation (580) may be written

$$d\sigma = -H_8\, dt + V\, dp. \qquad (e)$$

Differentiating (d), and comparing the result with (e), we obtain

$$dE_8 = t\, dH_8 - p\, dV. \qquad (f)$$

The quantities E_8 and H_8 might be called the superficial densities of energy and entropy quite as properly as those which we denote by ϵ_8 and η_8. In fact, when the composition of both of the homogeneous masses is invariable, the quantities E_8 and H_8^- are much more simple in their definition than ϵ_8 and η_8, and would probably be more naturally suggested by the terms *superficial density of energy* and *of entropy*. It would also be natural in this case to regard the quantities of the homogeneous masses as determined by the total quantities of matter, and not by the surface of tension or any other dividing surface. But such a nomenclature and method could not readily be extended so as to treat cases of more than two components with entire generality.

In the treatment of surfaces of discontinuity in this paper, the definitions and nomenclature which have been adopted will be strictly adhered to. The object of this note is to suggest to the reader how a different method might be used in some cases with advantage, and to show the precise relations between the quantities which are used in this paper and others which might be confounded with them, and which may be made more prominent when the subject is treated differently.

where

$$A = \gamma_1{}''\gamma_2{}' - \gamma_1{}'\gamma_2{}'', \qquad (582)$$

$$B = \begin{vmatrix} \eta_8 & \Gamma_1 & \Gamma_2 \\ \eta_V{}' & \gamma_1{}' & \gamma_2{}' \\ \eta_V{}'' & \gamma_1{}'' & \gamma_2{}'' \end{vmatrix}, \qquad (583)$$

$$C = \Gamma_1(\gamma_2{}'' - \gamma_2{}') + \Gamma_2(\gamma_1{}' - \gamma_1{}''). \qquad (584)$$

It will be observed that A vanishes when the composition of the two homogeneous masses is identical, while B and C do not, in general, and that the value of A is negative or positive according as the mass specified by $'$ contains the component specified by $_1$ in a greater or less proportion than the other mass. Hence, the values both of $\left(\dfrac{d\sigma}{dt}\right)_p$ and of $\left(\dfrac{d\sigma}{dp}\right)_t$ become infinite when the difference in the composition of the masses vanishes, and change sign when the greater proportion of a component passes from one mass to the other. This might be inferred from the statements on page 99 respecting co-existent phases which are identical in composition, from which it appears that when two coexistent phases have nearly the same composition, a small variation of the temperature or pressure of the coexistent phases will cause a relatively very great variation in the composition of the phases. The same relations are indicated by the graphical method represented in figure 6 on page 125.

With regard to gas-mixtures which conform to Dalton's law, we shall only consider the fundamental equation for plane surfaces, and shall suppose that there is not more than one component in the liquid which does not appear in the gas-mixture. We have already seen that in limiting the fundamental equation to plane surfaces we can get rid of one potential by choosing such a dividing surface that the superficial density of one of the components vanishes. Let this be done with respect to the component peculiar to the liquid, if such there is; if there is no such component, let it be done with respect to one of the gaseous components. Let the remaining potentials be eliminated by means of the fundamental equations of the simple gases. We may thus obtain an equation between the superficial tension, the temperature, and the several pressures of the simple gases in the gas-mixture or all but one of these pressures. Now, if we eliminate $d\mu_2$, $d\mu_3$, etc. from the equations

$$d\sigma = -\eta_{8(1)}dt - \Gamma_{2(1)}d\mu_2 - \Gamma_{3(1)}d\mu_3 - \text{etc.,}$$

$$dp_2 = \eta_{V2}dt + \gamma_2 d\mu_2,$$

$$dp_3 = \eta_{V3}dt + \gamma_3 d\mu_3,$$

$$\text{etc.,}$$

where the suffix $_1$ relates to the component of which the surface-density has been made to vanish, and γ_2, γ_3, etc. denote the densities of the gases specified in the gas-mixture, and p_2, p_3, etc., η_{V2}, η_{V3}, etc. the pressures and the densities of entropy due to these several gases, we obtain

$$d\sigma = -\left(\eta_{S(1)} - \frac{\Gamma_{2(1)}}{\gamma_2}\eta_{V2} - \frac{\Gamma_{3(1)}}{\gamma_3}\eta_{V3} - \text{etc.}\right)dt$$

$$-\frac{\Gamma_{2(1)}}{\gamma_2}dp_2 - \frac{\Gamma_{3(1)}}{\gamma_3}dp_3 - \text{etc.} \tag{585}$$

This equation affords values of the differential coefficients of σ with respect to t, p_2, p_3, etc., which may be set equal to those obtained by differentiating the equation between these variables.

Thermal and Mechanical Relations pertaining to the Extension of a Surface of Discontinuity.

The fundamental equation of a surface of discontinuity with one or two component substances, besides its statical applications, is of use to determine the heat absorbed when the surface is extended under certain conditions.

Let us first consider the case in which there is only a single component substance. We may treat the surface as plane, and place the dividing surface so that the surface density of the single component vanishes. (See page 234.) If we suppose the area of the surface to be increased by unity without change of temperature or of the quantities of liquid and vapor, the entropy of the whole will be increased by $\eta_{S(1)}$. Therefore, if we denote by Q the quantity of heat which must be added to satisfy the conditions, we shall have

$$Q = t\eta_{S(1)}, \tag{586}$$

and by (514),

$$Q = -t\frac{d\sigma}{dt} = -\frac{d\sigma}{d\log t}, \tag{587}$$

It will be observed that the condition of constant quantities of liquid and vapor as determined by the dividing surface which we have adopted is equivalent to the condition that the total volume shall remain constant.

Again, if the surface is extended without application of heat, while the pressure in the liquid and vapor remains constant, the temperature will evidently be maintained constant by condensation of the vapor. If we denote by M the mass of vapor condensed per unit of surface formed, and by η_M' and η_M'' the entropies of the liquid and vapor per unit of mass, the condition of no addition of heat will require that

$$M(\eta_M'' - \eta_M') = \eta_{S(1)}. \tag{588}$$

The increase of the volume of liquid will be

$$\frac{\eta_{S(1)}}{\gamma'(\eta_M{}'' - \eta_M{}')},\tag{589}$$

and the diminution of the volume of vapor

$$\frac{\eta_{S(1)}}{\gamma''(\eta_M{}'' - \eta_M{}')}.\tag{590}$$

Hence, for the work done (per unit of surface formed) by the external bodies which maintain the pressure, we shall have

$$W = \frac{p\,\eta_{S(1)}}{\eta_M{}'' - \eta_M{}'}\left(\frac{1}{\gamma''} - \frac{1}{\gamma'}\right),\tag{591}$$

and, by (514) and (131),

$$W = -p\frac{d\sigma}{dt}\frac{dt}{dp} = -p\frac{d\sigma}{dp} = -\frac{d\sigma}{d\log p}.\tag{592}$$

The work expended directly in extending the film will of course be equal to σ.

Let us now consider the case in which there are two component substances, neither of which is confined to the surface. Since we cannot make the superficial density of both these substances vanish by any dividing surface, it will be best to regard the surface of tension as the dividing surface. We may, however, simplify the formula by choosing such substances for components that each homogeneous mass shall consist of a single component. Quantities relating to these components will be distinguished as on page 266. If the surface is extended until its area is increased by unity, while heat is added at the surface so as to keep the temperature constant, and the pressure of the homogeneous masses is also kept constant, the phase of these masses will necessarily remain unchanged, but the quantity of one will be diminished by $\Gamma_{,}$, and that of the other by $\Gamma_{,,}$. Their entropies will therefore be diminished by $\dfrac{\Gamma_{,}}{\gamma'}\eta_V{}'$ and $\dfrac{\Gamma_{,,}}{\gamma''}\eta_V{}''$, respectively. Hence, since the surface receives the increment of entropy η_S, the total quantity of entropy will be increased by

$$\eta_S - \frac{\Gamma_{,}}{\gamma'}\eta_V{}' - \frac{\Gamma_{,,}}{\gamma''}\eta_V{}'',$$

which by equation (580) is equal to

$$-\left(\frac{d\sigma}{dt}\right)_p.$$

Therefore, for the quantity of heat Q imparted to the surface, we shall have

$$Q = -t\left(\frac{d\sigma}{dt}\right)_p = -\left(\frac{d\sigma}{d\log t}\right)_p.\tag{593}$$

We must notice the difference between this formula and (587). In (593) the quantity of heat Q is determined by the condition that the temperature and pressures shall remain constant. In (587) these conditions are equivalent and insufficient to determine the quantity of heat. The additional condition by which Q is determined may be most simply expressed by saying that the total volume must remain constant. Again, the differential coefficient in (593) is defined by considering p as constant; in the differential coefficient in (587) p cannot be considered as constant, and no condition is necessary to give the expression a definite value. Yet, notwithstanding the difference of the two cases, it is quite possible to give a single demonstration which shall be applicable to both. This may be done by considering a cycle of operations after the method employed by Sir William Thomson, who first pointed out these relations.*

The diminution of volume (per unit of surface formed) will be

$$V = \frac{\Gamma_,}{\gamma'} + \frac{\Gamma_{\prime\prime}}{\gamma''} = -\left(\frac{d\sigma}{dp}\right)_t;\tag{594}$$

and the work done (per unit of surface formed) by the external bodies which maintain the pressure constant will be

$$W = -p\left(\frac{d\sigma}{dp}\right)_t = -\left(\frac{d\sigma}{d\log p}\right)_t.\tag{595}$$

Compare equation (592).

The values of Q and W may also be expressed in terms of quantities relating to the ordinary components. By substitution in (593) and (595) of the values of the differential coefficients which are given by (581), we obtain

$$Q = -t\frac{B}{A}, \qquad W = -p\frac{C}{A},\tag{596}$$

where A, B, and C represent the expressions indicated by (582)–(584). It will be observed that the values of Q and W are in general infinite for the surface of discontinuity between coexistent phases which differ infinitesimally in composition, and change sign with the quantity A. When the phases are absolutely identical in composition, it is not in general possible to counteract the effect of extension of the surface of discontinuity by any supply of heat. For the matter at the surface will not in general have the same composition as the homogeneous masses, and the matter required for the increased surface cannot be obtained from these masses without altering their phase. The infinite values of Q and W are explained by the fact that when the phases are nearly identical in composition, the extension of the surface of

---

* See *Proc. Roy. Soc.*, vol. ix, p. 255 (June, 1858); or *Phil. Mag.*, ser. 4, vol. xvii, p. 61.

discontinuity is accompanied by the vaporization or condensation of a very large mass, according as the liquid or the vapor is the richer in that component which is necessary for the formation of the surface of discontinuity.

If, instead of considering the amount of heat necessary to keep the phases from altering while the surface of discontinuity is extended, we consider the variation of temperature caused by the extension of the surface while the pressure remains constant, it appears that this variation of temperature changes sign with $\gamma_1'' \gamma_2' - \gamma_1' \gamma_2''$, but vanishes with this quantity, i.e., vanishes when the composition of the phases becomes the same. This may be inferred from the statements on page 99, or from a consideration of the figure on page 125. When the composition of the homogeneous masses is initially absolutely identical, the effect on the temperature of a finite extension or contraction of the surface of discontinuity will be the same,—either of the two will lower or raise the temperature according as the temperature is a maximum or minimum for constant pressure.

The effect of the extension of a surface of discontinuity which is most easily verified by experiment is the effect upon the tension before complete equilibrium has been reëstablished throughout the adjacent masses. A fresh surface between coexistent phases may be regarded in this connection as an extreme case of a recently extended surface. When sufficient time has elapsed after the extension of a surface originally in equilibrium between coexistent phases, the superficial tension will evidently have sensibly its original value, unless there are substances at the surface which are either not found at all in the adjacent masses, or are found only in quantities comparable to those in which they exist at the surface. But a surface newly formed or extended may have a very different tension.

This will not be the case, however, when there is only a single component substance, since all the processes necessary for equilibrium are confined to a film of insensible thickness, and will require no appreciable time for their completion.

When there are two components, neither of which is confined to the surface of discontinuity, the reëstablishment of equilibrium after the extension of the surface does not necessitate any processes reaching into the interior of the masses except the transmission of heat between the surface of discontinuity and the interior of the masses. It appears from equation (593) that if the tension of the surface diminishes with a rise of temperature, heat must be supplied to the surface to maintain the temperature uniform when the surface is extended, i.e., the effect of extending the surface is to cool it ; but if the tension of any surface increases with the temperature, the effect of extending the surface will be to raise its temperature. In either case, it will be observed, the immediate effect of extending the surface is to increase its tension. A contraction of the surface will of course have the opposite effect. But the time necessary for the reëstablishment of sensible thermal equilibrium after extension or contraction of the surface must in most cases be very short.

In regard to the formation or extension of a surface between two coexistent phases of more than two components, there are two extreme cases which it is desirable to notice. When the superficial density of each of the components is exceedingly small compared with its density in either of the homogeneous masses, the matter (as well as the heat) necessary for the formation or extension of the normal surface can be taken from the immediate vicinity of the surface without sensibly changing the properties of the masses from which it is taken. But if any one of these superficial densities has a considerable value, while the density of the same component is very small in each of the homogeneous masses, both absolutely and relatively to the densities of the other components, the matter necessary for the formation or extension of the normal surface must come from a considerable distance. Especially if we consider that a small difference of density of such a component in one of the homogeneous masses will probably make a considerable difference in the value of the corresponding potential {see eq. (217)}, and that a small difference in the value of the potential will make a considerable difference in the tension {see eq. (508)}, it will be evident that in this case a considerable time will be necessary after the formation of a fresh surface or the extension of an old one for the reëstablishment of the normal value of the superficial tension. In intermediate cases, the reëstablishment of the normal tension will take place with different degrees of rapidity.

But whatever the number of component substances, provided that it is greater than one, and whether the reëstablishment of equilibrium is slow or rapid, extension of the surface will generally produce increase and contraction decrease of the tension. It would evidently be inconsistent with stability that the opposite effects should be produced. In general, therefore, a fresh surface between coexistent phases has a greater tension than an old one.* By the use of fresh surfaces, in experiments in capillarity, we may sometimes avoid the effect of minute quantities of foreign substances, which may be

---

* When, however, homogeneous masses which have not coexistent phases are brought into contact, the superficial tension may increase with the course of time. The superficial tension of a drop of alcohol and water placed in a large room will increase as the potential for alcohol is equalized throughout the room, and is diminished in the vicinity of the surface of discontinuity.

present without our knowledge or desire, in the fluids which meet at the surface investigated.

When the establishment of equilibrium is rapid, the variation of the tension from its normal value will be manifested especially during the extension or contraction of the surface, the phenomenon resembling that of viscosity, except that the variations of tension arising from variations in the densities at and about the surface will be the same in all directions, while the variations of tension due to any property of the surface really analogous to viscosity would be greatest in the direction of the most rapid extension.

We may here notice the different action of traces in the homogeneous masses of those substances which increase the tension and of those which diminish it. When the volume-densities of a component are very small, its surface-density may have a considerable positive value, but can only have a very minute negative one.* For the value when negative cannot exceed (numerically) the product of the greater volume-density by the thickness of the non-homogeneous film. Each of these quantities is exceedingly small. The surface-density when positive is of the same order of magnitude as the thickness of the non-homogeneous film, but is not necessarily small compared with other surface-densities because the volume-densities of the same substance in the adjacent masses are small. Now the potential of a substance which forms a very small part of a homogeneous mass certainly increases, and probably very rapidly, as the proportion of that component is increased. {See (171) and (217).} The pressure, temperature, and the other potentials, will not be sensibly affected. {See (98).} But the effect on the tension of this increase of the potential will be proportional to the surface-density, and will be to diminish the tension when the surface-density is positive. {See (508).} It is therefore quite possible that a very small trace of a substance in the homogeneous masses should greatly diminish the tension, but not possible that such a trace should greatly increase it.†

---

* It is here supposed that we have chosen for components such substances as are incapable of resolution into other components which are independently variable in the homogeneous masses. In a mixture of alcohol and water, for example, the components must be pure alcohol and pure water.

† From the experiments of M. E. Duclaux (*Annales de Chimie et de Physique*, ser. 4, vol. xxi, p. 383), it appears that one per cent. of alcohol in water will diminish the superficial tension to ·933, the value for pure water being unity. The experiments do not extend to pure alcohol, but the difference of the tensions for mixtures of alcohol and water containing 10 and 20 per cent. water is comparatively small, the tensions being ·322 and ·336 respectively.

According to the same authority (page 427 of the volume cited), one 3200th part of Castile soap will reduce the superficial tension of water by one-fourth ; one 800th part of soap by one-half. These determinations, as well as those relating to alcohol and

Impermeable Films.

We have so far supposed, in treating of surfaces of discontinuity, that they afford no obstacle to the passage of any of the component substances from either of the homogeneous masses to the other. The case, however, must be considered, in which there is a film of matter at the surface of discontinuity which is impermeable to some or all of the components of the contiguous masses. Such may be the case, for example, when a film of oil is spread on a surface of water, even when the film is too thin to exhibit the properties of the oil in mass. In such cases, if there is communication between the contiguous masses through other parts of the system to which they belong, such that the components in question can pass freely from one mass to the other, the impossibility of a direct passage through the film may be regarded as an immaterial circumstance, so far as states of equilibrium are concerned, and our formulæ will require no change. But when there is no such indirect communication, the potential for any component for which the film is impermeable may have entirely different values on opposite sides of the film, and the case evidently requires a modification of our usual method.

A single consideration will suggest the proper treatment of such cases. If a certain component which is found on both sides of a film cannot pass from either side to the other, the fact that the part of the component which is on one side is the same kind of matter with the part on the other side may be disregarded. All the general relations must hold true, which would hold if they were really different substances. We may therefore write μ_1 for the potential of the component on one side of the film, and μ_2 for the potential of the same substance (to be treated as if it were a different substance) on the other side; m_1^s for the excess of the quantity of the substance on the first side of the film above the quantity which would be on that side of the dividing surface (whether this is determined by the surface of tension or otherwise) if the density of the substance were the same near the dividing surface as at a distance, and m_2^s for a similar quantity relating to the other side of the film and dividing

---

water, are made by the method of drops, the weight of the drops of different liquids (from the same pipette) being regarded as proportional to their superficial tensions.

M. Athanase Dupré has determined the superficial tensions of solutions of soap by different methods. A statical method gives for one part of common soap in 5000 of water a superficial tension about one-half as great as for pure water, but if the tension be measured on a jet close to the orifice, the value (for the same solution) is sensibly identical with that of pure water. He explains these different values of the superficial tension of the same solution as well as the great effect on the superficial tension which a very small quantity of soap or other trifling impurity may produce, by the tendency of the soap or other substance to form a film on the surface of the liquid. (See *Annales de Chimie et de Physique*, ser. 4, vol. vii, p. 409, and vol. ix, p. 379.)

surface. On the same principle, we may use Γ_1 and Γ_2 to denote the values of m_1^s and m_2^s per unit of surface, and m_1', m_2'', γ_1', γ_2'' to denote the quantities of the substance and its densities in the two homogeneous masses.

With such a notation, which may be extended to cases in which the film is impermeable to any number of components, the equations relating to the surface and the contiguous masses will evidently have the same form as if the substances specified by the different suffixes were all really different. The superficial tension will be a function of μ_1 and μ_2, with the temperature and the potentials for the other components, and $-\Gamma_1$, $-\Gamma_2$ will be equal to its differential coefficients with respect to μ_1 and μ_2. In a word, all the general relations which have been demonstrated may be applied to this case, if we remember always to treat the component as a different substance according as it is found on one side or the other of the impermeable film.

When there is free passage for the component specified by the suffixes $_1$ and $_2$ through other parts of the system (or through any flaws in the film), we shall have in case of equilibrium $\mu_1 = \mu_2$. If we wish to obtain the fundamental equation for the surface when satisfying this condition, without reference to other possible states of the surface, we may set a single symbol for μ_1 and μ_2 in the more general form of the fundamental equation. Cases may occur of an impermeability which is not absolute, but which renders the transmission of some of the components exceedingly slow. In such cases, it may be necessary to distinguish at least two different fundamental equations, one relating to a state of approximate equilibrium which may be quickly established, and another relating to the ultimate state of complete equilibrium. The latter may be derived from the former by such substitutions as that just indicated.

The Conditions of Internal Equilibrium for a System of Hetero-
geneous Fluid Masses without neglect of the Influence of the
Surfaces of Discontinuity or of Gravity.

Let us now seek the complete value of the variation of the energy of a system of heterogeneous fluid masses, in which the influence of gravity and of the surfaces of discontinuity shall be included, and deduce from it the conditions of internal equilibrium for such a system. In accordance with the method which has been developed, the intrinsic energy (i.e. the part of the energy which is independent of gravity), the entropy, and the quantities of the several components must each be divided into two parts, one of which we regard as belonging to the surfaces which divide approximately homogeneous masses, and the other as belonging to these masses. The elements of intrinsic energy, entropy, etc., relating to an element of surface Ds will be denoted by $D\epsilon^s$, $D\eta^s$, Dm_1^s, Dm_2^s, etc., and those relating to an element of volume Dv, by $D\epsilon^v$, $D\eta^v$, Dm_1^v, Dm_2^v, etc. We shall also use Dm^s or $\Gamma\, Ds$ and Dm^v or $\gamma\, Dv$ to denote the total quantities of matter relating to the elements Ds and Dv respectively. That is,

$$Dm^s = \Gamma\, Ds = Dm_1^s + Dm_2^s + \text{etc.,} \tag{597}$$

$$Dm^v = \gamma\, Dv = Dm_1^v + Dm_2^v + \text{etc.} \tag{598}$$

The part of the energy which is due to gravity must also be divided into two parts, one of which relates to the elements Dm^s, and the other to the elements Dm^v. The complete value of the variation of the energy of the system will be represented by the expression

$$\delta \int D\epsilon^v + \delta \int D\epsilon^s + \delta \int gz\, Dm^v + \delta \int gz\, Dm^s, \tag{599}$$

in which g denotes the force of gravity, and z the height of the element above a fixed horizontal plane.

It will be convenient to limit ourselves at first to the consideration of reversible variations. This will exclude the formation of new masses or surfaces. We may therefore regard any infinitesimal variation in the state of the system as consisting of infinitesimal variations of the quantities relating to its several elements, and bring the sign of variation in the preceding formula after the sign of integration. If we then substitute for $\delta D\epsilon^v$, $\delta D\epsilon^s$, δDm^v, δDm^s, the values given by equations (13), (497), (597), (598), we shall have for the condition of equilibrium with respect to reversible variations of the internal state of the system

$$\int t\, \delta D\eta^v - \int p\, \delta Dv + \int \mu_1\, \delta Dm_1^v + \int \mu_2\, \delta Dm_2^v + \text{etc.}$$
$$+ \int t\, \delta D\eta^s + \int \sigma\, \delta Ds + \int \mu_1\, \delta Dm_1^s + \int \mu_2\, \delta Dm_2^s + \text{etc.}$$
$$+ \int g\, \delta z\, Dm^v + \int gz\, \delta Dm_1^v + \int gz\, \delta Dm_2^v + \text{etc.}$$
$$+ \int g\, \delta z\, Dm^s + \int gz\, \delta Dm_1^s + \int gz\, \delta Dm_2^s + \text{etc.} = 0. \tag{600}$$

Since equation (497) relates to surfaces of discontinuity which are initially in equilibrium, it might seem that this condition, although always necessary for equilibrium, may not always be sufficient. It is evident, however, from the form of the condition, that it includes the particular conditions of equilibrium relating to every possible deformation of the system, or reversible variation in the distribution of entropy or of the several components. It therefore includes all the relations between the different parts of the system which are necessary for equilibrium, so far as reversible variations are concerned. (The necessary relations between the various quantities relating to each element of the masses and surfaces are expressed by the fundamental equation of the mass or surface concerned, or may be immediately derived from it. See pp. 85–89 and 229–231.)

The variations in (600) are subject to the conditions which arise from the nature of the system and from the supposition that the changes in the system are not such as to affect external bodies. This supposition is necessary, unless we are to consider the variations in the state of the external bodies, and is evidently allowable in seeking the conditions of equilibrium which relate to the interior of the system.* But before we consider the equations of condition in detail, we may divide the condition of equilibrium (600) into the three conditions

$$\int t \, \delta D\eta^V + \int t \, \delta D\eta^S = 0, \tag{601}$$

$$-\int p \, \delta Dv + \int \sigma \, \delta Ds + \int g \delta z \, Dm^V + \int g \delta z \, Dm^S = 0, \tag{602}$$

$$\int \mu_1 \, \delta Dm_1^V + \int \mu_1 \, \delta Dm_1^S + \int gz \, \delta Dm_1^V + \int gz \, \delta Dm_1^S$$
$$+ \int \mu_2 \, \delta Dm_2^V + \int \mu_2 \, \delta Dm_2^S + \int gz \, \delta Dm_2^V + \int gz \, \delta Dm_2^S$$
$$+ \text{etc.} = 0. \tag{603}$$

For the variations which occur in any one of the three are evidently independent of those which occur in the other two, and the equations of condition will relate to one or another of these conditions separately.

The variations in condition (601) are subject to the condition that the entropy of the whole system shall remain constant. This may be expressed by the equation

$$\int \delta D\eta^V + \int \delta D\eta^S = 0. \tag{604}$$

To satisfy the condition thus limited it is necessary and sufficient that

$$t = \text{const.} \tag{605}$$

throughout the whole system, which is the condition of thermal equilibrium.

The conditions of mechanical equilibrium, or those that relate to the possible deformation of the system, are contained in (602), which may also be written

$$-\int p \, \delta Dv + \int \sigma \, \delta Ds + \int g\gamma \, \delta z \, Dv + \int g\Gamma \, \delta z \, Ds = 0. \tag{606}$$

---

* We have sometimes given a physical expression to a supposition of this kind, in problems in which the peculiar condition of matter in the vicinity of surfaces of discontinuity was to be neglected, by regarding the system as surrounded by a rigid and impermeable envelop. But the more exact treatment which we are now to give the problem of equilibrium would require us to take account of the influence of the envelop on the immediately adjacent matter. Since this involves the consideration of surfaces of discontinuity between solids and fluids, and we wish to limit ourselves at present to the consideration of the equilibrium of fluid masses, we shall give up the conception of an impermeable envelop, and regard the system as bounded simply by an imaginary surface, which is not a surface of discontinuity. The variations of the system must be such as do not deform the surface, nor affect the matter external to it.

It will be observed that this condition has the same form as if the different fluids were separated by heavy and elastic membranes without rigidity and having at every point a tension uniform in all directions in the plane of the surface. The variations in this formula, beside their necessary geometrical relations, are subject to the conditions that the external surface of the system, and the lines in which the surfaces of discontinuity meet it, are fixed. The formula may be reduced by any of the usual methods, so as to give the particular conditions of mechanical equilibrium. Perhaps the following method will lead as directly as any to the desired result.

It will be observed the quantities affected by δ in (606) relate exclusively to the position and size of the elements of volume and surface into which the system is divided, and that the variations δp and $\delta \sigma$ do not enter into the formula either explicitly or implicitly. The equations of condition which concern this formula also relate exclusively to the variations of the system of geometrical elements, and do not contain either δp or $\delta \sigma$. Hence, in determining whether the first member of the formula has the value zero for every possible variation of the system of geometrical elements, we may assign to δp and $\delta \sigma$ any values whatever which may simplify the solution of the problem, without inquiring whether such values are physically possible.

Now when the system is in its initial state, the pressure p, in each of the parts into which the system is divided by the surfaces of tension, is a function of the co-ordinates which determine the position of the element Dv, to which the pressure relates. In the varied state of the system, the element Dv will in general have a different position. Let the variation δp be determined solely by the change in position of the element Dv. This may be expressed by the equation

$$\delta p = \frac{dp}{dx}\,\delta x + \frac{dp}{dy}\,\delta y + \frac{dp}{dz}\,\delta z, \tag{607}$$

in which $\dfrac{dp}{dx}$, $\dfrac{dp}{dy}$, $\dfrac{dp}{dz}$ are determined by the function mentioned, and δx, δy, δz by the variation of the position of the element Dv.

Again, in the initial state of the system the tension σ, in each of the different surfaces of discontinuity, is a function of two co-ordinates ω_1, ω_2, which determine the position of the element Ds. In the varied state of the system, this element will in general have a different position. The change of position may be resolved into a component lying in the surface and another normal to it. Let the variation $\delta \sigma$ be determined solely by the first of these components of the motion of Ds. This may be expressed by the equation

$$\delta \sigma = \frac{d\sigma}{d\omega_1}\,\delta \omega_1 + \frac{d\sigma}{d\omega_2}\,\delta \omega_2, \tag{608}$$

in which $\dfrac{d\sigma}{d\omega_1}$, $\dfrac{d\sigma}{d\omega_2}$ are determined by the function mentioned, and $\delta\omega_1$, $\delta\omega_2$, by the component of the motion of Ds which lies in the plane of the surface.

With this understanding, which is also to apply to δp and $\delta\sigma$ when contained implicitly in any expression, we shall proceed to the reduction of the condition (606).

With respect to any one of the volumes into which the system is divided by the surfaces of discontinuity, we may write

$$\int p\,\delta\,Dv = \delta\int p\,Dv - \int \delta p\,Dv.$$

But it is evident that

$$\delta\int p\,Dv = \int p\,\delta N\,Ds,$$

where the second integral relates to the surfaces of discontinuity bounding the volume considered, and δN denotes the normal component of the motion of an element of the surface, measured outward. Hence,

$$\int p\,\delta\,Dv = \int p\,\delta N\,Ds - \int \delta p\,Dv.$$

Since this equation is true of each separate volume into which the system is divided, we may write for the whole system

$$\int p\,\delta\,Dv = \int (p'-p'')\delta N\,Ds - \int \delta p\,Dv, \tag{609}$$

where p' and p'' denote the pressures on opposite sides of the element Ds, and δN is measured toward the side specified by double accents.

Again, for each of the surfaces of discontinuity, taken separately,

$$\int \sigma\,\delta\,Ds = \delta\int \sigma\,Ds - \int \delta\sigma\,Ds,$$

and

$$\delta\int \sigma\,Ds = \int \sigma(c_1+c_2)\delta N\,Ds + \int \sigma\,\delta T\,Dl,$$

where c_1 and c_2 denote the principal curvatures of the surface (positive, when the centers are on the side opposite to that toward which δN is measured), Dl an element of the perimeter of the surface, and δT the component of the motion of this element which lies in the plane of the surface and is perpendicular to the perimeter (positive, when it extends the surface). Hence we have for the whole system

$$\int \sigma\,\delta\,Ds = \int \sigma(c_1+c_2)\,\delta N\,Ds + \int \Sigma(\sigma\,\delta T)\,Dl - \int \delta\sigma\,Ds, \tag{610}$$

where the integration of the elements Dl extends to all the lines in which the surfaces of discontinuity meet, and the symbol Σ denotes a summation with respect to the several surfaces which meet in such a line.

By equations (609) and (610), the general condition of mechanical equilibrium is reduced to the form

$$-\int (p'-p'')\,\delta N\,Ds + \int \delta p\,Dv + \int \sigma(c_1+c_2)\,\delta N\,Ds$$
$$+ \int \Sigma(\sigma\,\delta T)\,Dl - \int \delta\sigma\,Ds + \int g\gamma\,\delta z\,Dv + \int g\Gamma\,\delta z\,Ds = 0.$$

Arranging and combining terms, we have

$$\int (g\gamma\, \delta z + \delta p)\, Dv + \int [(p'' - p')\, \delta N + \sigma(c_1 + c_2)\, \delta N + g\Gamma\, \delta z - \delta\sigma]\, Ds$$
$$+ \int \Sigma(\sigma\, \delta T)\, Dl = 0. \tag{611}$$

To satisfy this condition, it is evidently necessary that the coefficients of Dv, Ds, and Dl shall vanish throughout the system.

In order that the coefficient of Dv shall vanish, it is necessary and sufficient that in each of the masses into which the system is divided by the surfaces of tension, p shall be a function of z alone, such that

$$\frac{dp}{dz} = -g\gamma. \tag{612}$$

In order that the coefficient of Ds shall vanish in all cases, it is necessary and sufficient that it shall vanish for normal and for tangential movements of the surface. For normal movements we may write

$$\delta\sigma = 0, \quad \text{and} \quad \delta z = \cos\theta\, \delta N,$$

where θ denotes the angle which the normal makes with a vertical line. The first condition therefore gives the equation

$$p' - p'' = \sigma(c_1 + c_2) + g\Gamma\cos\theta, \tag{613}$$

which must hold true at every point in every surface of discontinuity. The condition with respect to tangential movements shows that in each surface of tension σ is a function of z alone, such that

$$\frac{d\sigma}{dz} = g\Gamma. \tag{614}$$

In order that the coefficient of Dl in (611) shall vanish, we must have, for every point in every line in which surfaces of discontinuity meet, and for any infinitesimal displacement of the line,

$$\Sigma(\sigma\, \delta T) = 0. \tag{615}$$

This condition evidently expresses the same relations between the tensions of the surfaces meeting in the line and the directions of perpendiculars to the line drawn in the planes of the various surfaces, which hold for the magnitudes and directions of forces in equilibrium in a plane.

In condition (603), the variations which relate to any component are to be regarded as having the value zero in any part of the system in which that substance is not an actual component.* The same is true

---

*The term *actual component* has been defined for homogeneous masses on page 64, and the definition may be extended to surfaces of discontinuity. It will be observed that if a substance is an actual component of either of the masses separated by a surface of discontinuity, it must be regarded as an actual component for that surface, as well as when it occurs at the surface but not in either of the contiguous masses.

with respect to the equations of condition, which are of the form

$$\left.\begin{array}{l} \int \delta Dm_1^\gamma + \int \delta Dm_1^s = 0, \\ \int \delta Dm_2^\gamma + \int \delta Dm_2^s = 0, \\ \text{etc.} \end{array}\right\} \tag{616}$$

(It is here supposed that the various components are independent, i.e., that none can be formed out of others, and that the parts of the system in which any component actually occurs are not entirely separated by parts in which it does not occur.) To satisfy the condition (603), subject to these equations of condition, it is necessary and sufficient that the conditions

$$\left.\begin{array}{l} \mu_1 + gz = M_1, \\ \mu_2 + gz = M_2, \\ \text{etc.,} \end{array}\right\} \tag{617}$$

(M_1, M_2, etc. denoting constants,) shall each hold true in those parts of the system in which the substance specified is an actual component. We may here add the condition of equilibrium relative to the possible absorption of any substance (to be specified by the suffix $_a$) by parts of the system of which it is not an actual component, viz., that the expression $\mu_a + gz$ must not have a less value in such parts of the system than in a contiguous part in which the substance is an actual component.

From equation (613) with (605) and (617) we may easily obtain the differential equation of a surface of tension (in the geometrical sense of the term), when p', p'', and σ are known in terms of the temperature and potentials. For $c_1 + c_2$ and θ may be expressed in terms of the first and second differential coefficients of z with respect to the horizontal co-ordinates, and p', p'', σ, and Γ in terms of the temperature and potentials. But the temperature is constant, and for each of the potentials we may substitute—gz increased by a constant. We thus obtain an equation in which the only variables are z and its first and second differential coefficients with respect to the horizontal co-ordinates. But it will rarely be necessary to use so exact a method. Within moderate differences of level, we may regard γ', γ'', and σ as constant. We may then integrate the equation {derived from (612)}

$$d(p' - p'') = g(\gamma'' - \gamma') \, dz,$$

which will give

$$p' - p'' = g(\gamma'' - \gamma')z, \tag{618}$$

where z is to be measured from the horizontal plane for which $p' = p''$. Substituting this value in (613), and neglecting the term containing Γ, we have

$$c_1 + c_2 = \frac{g(\gamma'' - \gamma')}{\sigma} z, \tag{619}$$

where the coefficient of z is to be regarded as constant. Now the value of z cannot be very large, in any surface of sensible dimensions, unless $\gamma'' - \gamma'$ is very small. We may therefore consider this equation as practically exact, unless the densities of the contiguous masses are very nearly equal. If we substitute for the sum of the curvatures its value in terms of the differential coefficients of z with respect to the horizontal rectangular co-ordinates, x and y, we have

$$\frac{\left(1+\dfrac{dz^2}{dy^2}\right)\dfrac{d^2z}{dx^2} - 2\dfrac{dz}{dx}\dfrac{dz}{dy}\dfrac{d^2z}{dx\,dy} + \left(1+\dfrac{dz^2}{dx^2}\right)\dfrac{d^2z}{dy^2}}{\left(1+\dfrac{dz^2}{dx^2}+\dfrac{dz^2}{dy^2}\right)^{\frac{3}{2}}} = \frac{g(\gamma''-\gamma')}{\sigma}z. \quad (620)$$

With regard to the sign of the root in the denominator of the fraction, it is to be observed that, if we always take the positive value of the root, the value of the whole fraction will be positive or negative according as the greater concavity is turned upward or downward. But we wish the value of the fraction to be positive when the greater concavity is turned toward the mass specified by a single accent. We should therefore take the positive or negative value of the root according as this mass is above or below the surface.

The particular conditions of equilibrium which are given in the last paragraph but one may be regarded in general as the conditions of chemical equilibrium between the different parts of the system, since they relate to the separate components.* But such a designation is not entirely appropriate unless the number of components is greater than one. In no case are the conditions of mechanical equilibrium entirely independent of those which relate to temperature and the potentials. For the conditions (612) and (614) may be regarded as consequences of (605) and (617) in virtue of the necessary relations (98) and (508). †

The mechanical conditions of equilibrium, however, have an especial importance, since we may always regard them as satisfied in any liquid (and not decidedly viscous) mass in which no sensible motions are observable. In such a mass, when isolated, the attainment of mechanical equilibrium will take place very soon; thermal and chemical equilibrium will follow more slowly. The thermal equilibrium will generally require less time for its approximate attainment than the chemical; but the processes by which the latter is produced will generally cause certain inequalities of temperature until a state of complete equilibrium is reached.

---

* Concerning another kind of conditions of chemical equilibrium, which relate to the molecular arrangement of the components, and not to their sensible distribution in space, see pages 138–144.

† Compare page 146, where a similar problem is treated without regard to the influence of the surfaces of discontinuity.

When a surface of discontinuity has more components than one which do not occur in the contiguous masses, the adjustment of the potentials for these components in accordance with equations (617) may take place very slowly, or not at all, for want of sufficient mobility in the components of the surface. But when this surface has only one component which does not occur in the contiguous masses, and the temperature and potentials in these masses satisfy the conditions of equilibrium, the potential for the component peculiar to the surface will very quickly conform to the law expressed in (617), since this is a necessary consequence of the condition of mechanical equilibrium (614) in connection with the conditions relating to temperature and the potentials which we have supposed to be satisfied. The necessary distribution of the substance peculiar to the surface will be brought about by expansions and contractions of the surface. If the surface meets a third mass containing this component and no other which is foreign to the masses divided by the surface, the potential for this component in the surface will of course be determined by that in the mass which it meets.

The particular conditions of mechanical equilibrium (612)–(615), which may be regarded as expressing the relations which must subsist between contiguous portions of a fluid system in a state of mechanical equilibrium, are serviceable in determining whether a given system is or is not in such a state. But the mechanical theorems which relate to finite parts of the system, although they may be deduced from these conditions by integration, may generally be more easily obtained by a suitable application of the general condition of mechanical equilibrium (606), or by the application of ordinary mechanical principles to the system regarded as subject to the forces indicated by this equation.

It will be observed that the conditions of equilibrium relating to temperature and the potentials are not affected by the surfaces of discontinuity. {Compare (228) and (234).}* Since a phase cannot vary continuously without variations of the temperature or the potentials, it follows from these conditions that the phase at any point in a fluid system which has the same independently variable components throughout, and is in equilibrium under the influence of gravity, must be one of a certain number of phases which are completely determined by the phase at any given point and the difference of level of the two points considered. If the phases

---

* If the fluid system is divided into separate masses by solid diaphragms which are permeable to all the components of the fluids independently, the conditions of equilibrium of the fluids relating to temperature and the potentials will not be affected. (Compare page 84.) The propositions which follow in the above paragraph may be extended to this case.

throughout the fluid system satisfy the general condition of practical stability for phases existing in large masses (viz., that the pressure shall be the least consistent with the temperature and potentials), they will be entirely determined by the phase at any given point and the differences of level. (Compare page 149, where the subject is treated without regard to the influence of the surfaces of discontinuity.)

Conditions of equilibrium relating to irreversible changes.—The conditions of equilibrium relating to the absorption, by any part of the system, of substances which are not actual components of that part have been given on page 282. Those relating to the formation of new masses and surfaces are included in the conditions of stability relating to such changes, and are not always distinguishable from them. They are evidently independent of the action of gravity. We have already discussed the conditions of stability with respect to the formation of new fluid masses within a homogeneous fluid and at the surface when two such masses meet (see pages 252–264), as well as the condition relating to the possibility of a change in the nature of a surface of discontinuity. (See pages 237–240, where the surface considered is plane, but the result may easily be extended to curved surfaces.) We shall hereafter consider, in some of the more important cases, the conditions of stability with respect to the formation of new masses and surfaces which are peculiar to lines in which several surfaces of discontinuity meet, and points in which several such lines meet.

Conditions of stability relating to the whole system.—Besides the conditions of stability relating to very small parts of a system, which are substantially independent of the action of gravity, and are discussed elsewhere, there are other conditions, which relate to the whole system or to considerable parts of it. To determine the question of the stability of a given fluid system under the influence of gravity, when all the conditions of equilibrium are satisfied as well as those conditions of stability which relate to small parts of the system taken separately, we may use the method described on page 249, the demonstration of which (pages 247, 248) will not require any essential modification on account of gravity.

When the variations of temperature and of the quantities M_1, M_2, etc. {see (617)} involved in the changes considered are so small that they may be neglected, the condition of stability takes a very simple form, as we have already seen to be the case with respect to a system uninfluenced by gravity. (See page 251.)

We have to consider a varied state of the system in which the total entropy and the total quantities of the various components are unchanged, and all variations vanish at the exterior of the system,—

in which, moreover, the conditions of equilibrium relating to temperature and the potentials are satisfied, and the relations expressed by the fundamental equations of the masses and surfaces are to be regarded as satisfied, although the state of the system is not one of complete equilibrium. Let us imagine the state of the system to vary continuously in the course of time in accordance with these conditions and use the symbol d to denote the simultaneous changes which take place at any instant. If we denote the total energy of the system by E, the value of dE may be expanded like that of δE in (599) and (600), and then reduced (since the values of t, $\mu_1 + gz$, $\mu_2 + gz$, etc., are uniform throughout the system, and the total entropy and total quantities of the several components are constant) to the form

$$dE = -\int p\, dDv + \int g\, dz\, Dm^{\mathrm{v}} + \int \sigma\, dDs + \int g\, dz\, Dm^{\mathrm{s}}$$
$$= -\int p\, dDv + \int g\, \gamma\, dz\, Dv + \int \sigma\, dDs + \int g\, \Gamma\, dz\, Ds, \qquad (621)$$

where the integrations relate to the elements expressed by the symbol D. The value of p at any point in any of the various masses, and that of σ at any point in any of the various surfaces of discontinuity are entirely determined by the temperature and potentials at the point considered. If the variations of t and M_1, M_2, etc. are to be neglected, the variations of p and σ will be determined solely by the change in position of the point considered. Therefore, by (612) and (614),

$$dp = -g\gamma\, dz, \quad d\sigma = g\Gamma\, dz\,;$$

and

$$dE = -\int p\, dDv - \int dp\, Dv + \int \sigma\, dDs + \int d\sigma\, Ds$$
$$= -d\int p\, Dv + d\int \sigma\, Ds. \qquad (622)$$

If we now integrate with respect to d, commencing at the given state of the system, we obtain

$$\Delta E = -\Delta\int p\, Dv + \Delta\int \sigma\, Ds, \qquad (623)$$

where Δ denotes the value of a quantity in a varied state of the system diminished by its value in the given state. This is true for finite variations, and is therefore true for infinitesimal variations without neglect of the infinitesimals of the higher orders. The condition of stability is therefore that

$$\Delta\int p\, Dv - \Delta\int \sigma\, Ds < 0, \qquad (624)$$

or that the quantity

$$\int p\, Dv - \int \sigma\, Ds \qquad (625)$$

has a maximum value, the values of p and σ, for each different mass or surface, being regarded as determined functions of z. (In ordinary cases σ may be regarded as constant in each surface of discontinuity, and p as a linear function of z in each different mass.) It may easily be shown (compare page 252) that this condition is always *sufficient* for stability with reference to motion of surfaces of discontinuity, even when the variations of t, M_1, M_2, etc. cannot be neglected in the determination of the *necessary* condition of stability with respect to such changes.

On the Possibility of the Formation of a New Surface of Discontinuity where several Surfaces of Discontinuity meet.

When more than three surfaces of discontinuity between homogeneous masses meet along a line, we may conceive of a new surface being formed between any two of the masses which do not meet in a surface in the original state of the system. The condition of stability with respect to the formation of such a surface may be easily obtained by the consideration of the limit between stability and instability, as exemplified by a system which is in equilibrium when a very small surface of the kind is formed.

To fix our ideas, let us suppose that there are four homogeneous masses A, B, C, and D, which meet one another in four surfaces, which we may call A-B, B-C, C-D, and D-A, these surfaces all meeting along a line L. This is indicated in figure 11 by a section of the

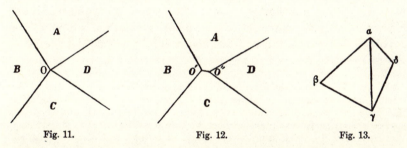

Fig. 11.   Fig. 12.   Fig. 13.

surfaces cutting the line L at right angles at a point O. In an infinitesimal variation of the state of the system, we may conceive of a small surface being formed between A and C (to be called A-C), so that the section of the surfaces of discontinuity by the same plane takes the form indicated in figure 12. Let us suppose that the condition of equilibrium (615) is satisfied both for the line L in which the surfaces of discontinuity meet in the original state of the system, and for the two such lines (which we may call L' and L'') in the varied state of the system, at least at the points O' and O'' where they are cut by the plane of section. We may therefore form a quadrilateral of which the sides $\alpha\beta$, $\beta\gamma$, $\gamma\delta$, $\delta\alpha$ are equal in numerical value to the tensions of the several surfaces A-B, B-C, C-D, D-A, and are parallel to the normals to these surfaces at the point O in the original state of the system. In like manner, for the varied state of the system we can construct two triangles having similar relations to the surfaces of discontinuity meeting at O′ and O″. But the directions of the normals to the surfaces A-B and B-C at O′ and to C-D and D-A at O″ in the varied state of the system differ infinitely little from the directions of the corresponding normals at O in the initial state. We may therefore regard $\alpha\beta$, $\beta\gamma$ as two sides of the triangle representing the surfaces meeting at O′, and $\gamma\delta$, $\delta\alpha$ as two sides of the triangle representing the surfaces meeting at O″. Therefore, if we join $\alpha\gamma$, this line will represent the direction of the normal to the surface A-C, and the value of its tension. If the tension of a surface between such masses as A and C had been greater than that represented by $\alpha\gamma$, it is evident that the initial state of the system of surfaces (represented in figure 11) would have been stable with respect to the possible formation of any such surface. If the tension had been less, the state of the system would have been at least practically unstable. To determine whether it is unstable in the strict sense of the term, or whether or not it is properly to be regarded as in equilibrium, would require a more refined analysis than we have used.*

The result which we have obtained may be generalized as follows. When more than three surfaces of discontinuity in a fluid system meet in equilibrium along a line, with respect to the surfaces and masses immediately adjacent to any point of this line, we may form a polygon of which the angular points shall correspond in order to the different masses separated by the surfaces of discontinuity, and

---

* We may here remark that a nearer approximation in the theory of equilibrium and stability might be attained by taking special account, in our general equations, of the lines in which surfaces of discontinuity meet. These lines might be treated in a manner entirely analogous to that in which we have treated surfaces of discontinuity. We might recognize linear densities of energy, of entropy, and of the several substances which occur about the line, also a certain linear tension. With respect to these quantities and the temperature and potentials, relations would hold analogous to those which have been demonstrated for surfaces of discontinuity. (See pp. 229–231.) If the sum of the tensions of the lines L′ and L″, mentioned above, is greater than the tension of the line L, this line will be in strictness stable (although practically unstable) with respect to the formation of a surface between A and C, when the tension of such a surface is a little less than that represented by the diagonal $\alpha\gamma$.

The different use of the term *practically unstable* in different parts of this paper need not create confusion, since the general meaning of the term is in all cases the same. A system is called practically unstable when a very small (not necessarily indefinitely small) disturbance or variation in its condition will produce a considerable change. In the former part of this paper, in which the influence of surfaces of discontinuity was neglected, a system was regarded as practically unstable when such a result would be produced by a disturbance of the same order of magnitude as the quantities relating to surfaces of discontinuity which were neglected. But where surfaces of discontinuity are considered, a system is not regarded as practically unstable, unless the disturbance which will produce such a result is very small compared with the quantities relating to surfaces of discontinuity of any appreciable magnitude.

the sides to these surfaces, each side being perpendicular to the corresponding surface, and equal to its tension. With respect to the formation of new surfaces of discontinuity in the vicinity of the point especially considered, the system is stable, if every diagonal of the polygon is less, and practically unstable, if any diagonal is greater, than the tension which would belong to the surface of discontinuity between the corresponding masses. In the limiting case, when the diagonal is exactly equal to the tension of the corresponding surface, the system may often be determined to be unstable by the application of the principle enunciated to an adjacent point of the line in which the surfaces of discontinuity meet. But when, in the polygons constructed for all points of the line, no diagonal is in any case greater than the tension of the corresponding surface, but a certain diagonal is equal to the tension in the polygons constructed for a finite portion of the line, farther investigations are necessary to determine the stability of the system. For this purpose, the method described on page 249 is evidently applicable.

A similar proposition may be enunciated in many cases with respect to a point about which the angular space is divided into solid angles by surfaces of discontinuity. If these surfaces are in equilibrium, we can always form a closed solid figure without reentrant angles of which the angular points shall correspond to the several masses, the edges to the surfaces of discontinuity, and the sides to the lines in which these edges meet, the edges being perpendicular to the corresponding surfaces, and equal to their tensions, and the sides being perpendicular to the corresponding lines. Now if the solid angles in the physical system are such as may be subtended by the sides and bases of a triangular prism enclosing the vertical point, or can be derived from such by deformation, the figure representing the tensions will have the form of two triangular pyramids on opposite sides of the same base, and the system will be stable or practically unstable with respect to the formation of a surface between the masses which only meet in a point, according as the tension of a surface between such masses is greater or less than the diagonal joining the corresponding angular points of the solid representing the tensions. This will easily appear on consideration of the case in which a very small surface between the masses would be in equilibrium.

The Conditions of Stability for Fluids relating to the Formation of a New Phase at a Line in which Three Surfaces of Discontinuity meet.

With regard to the formation of new phases there are particular conditions of stability which relate to lines in which several surfaces

of discontinuity meet. We may limit ourselves to the case in which
there are three such surfaces, this being the only one of frequent
occurrence, and may treat them as meeting in a straight line. It
will be convenient to commence by considering the equilibrium of a
system in which such a line is replaced by a filament of a different
phase.

Let us suppose that three homogeneous fluid masses, A, B, and C
are separated by cylindrical (or plane) surfaces, A-B, B-C, C-A, which
at first meet in a straight line, each of the surface-tensions $\sigma_{AB}, \sigma_{BC}, \sigma_{CA}$
being less than the sum of the other two. Let us suppose that the
system is then modified by the introduction of a fourth fluid mass D,
which is placed between A, B, and C, and is separated from them by
cylindrical surfaces D-A, D-B, D-C meeting A-B, B-C, and C-A in
straight lines. The general form of the surfaces is shown by figure 14,
in which the full lines represent a section perpendicular to all the
surfaces. The system thus modified is to be in equilibrium, as well
as the original system, the position of the surfaces A-B, B-C, C-A
being unchanged. That the last condition is consistent with equilibrium will appear from the following mechanical considerations.

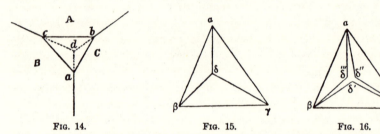

FIG. 14.   FIG. 15.   FIG. 16.

Let v_D denote the volume of the mass D per unit of length or the area
of the curvilinear triangle abc. Equilibrium is evidently possible for
any values of the surface tensions (if only $\sigma_{AB}, \sigma_{BC}, \sigma_{CA}$ satisfy the condition mentioned above, and the tensions of the three surfaces meet-
ing at each of the edges of D satisfy a similar condition) with any
value (not too large) of v_D, if the edges of D are constrained to remain
in the original surfaces A-B, B-C, and C-A, or these surfaces extended,
if necessary, without change of curvature. (In certain cases one of
the surfaces D-A, D-B, D-C may disappear and D will be bounded
by only two cylindrical surfaces.) We may therefore regard the
system as maintained in equilibrium by forces applied to the edges
of D and acting at right angles to A-B, B-C, C-A. The same forces
would keep the system in equilibrium if D were rigid. They must
therefore have a zero resultant, since the nature of the mass D is immaterial when it is rigid, and no forces external to the system would
be required to keep a corresponding part of the original system in

equilibrium. But it is evident from the points of application and directions of these forces that they cannot have a zero resultant unless each force is zero. We may therefore introduce a fourth mass D without disturbing the parts which remain of the surfaces A-B, B-C, C-D.

It will be observed that all the angles at a, b, c, and d in figure 14 are entirely determined by the six surface-tensions σ_{AB}, σ_{BC}, σ_{CA}, σ_{DA}, σ_{DB}, σ_{DC}. (See (615).) The angles may be derived from the tensions by the following construction, which will also indicate some important properties. If we form a triangle $\alpha\beta\gamma$ (figure 15 or 16) having sides equal to σ_{AB}, σ_{BC}, σ_{CA}, the angles of the triangle will be supplements of the angles at d. To fix our ideas, we may suppose the sides of the triangle to be perpendicular to the surfaces at d. Upon $\beta\gamma$ we may then construct (as in figure 16) a triangle $\beta\gamma\delta'$ having sides equal to σ_{BC}, σ_{DC}, σ_{DB}, upon $\gamma\alpha$ a triangle $\gamma\alpha\delta''$ having sides equal to σ_{CA}, σ_{DA}, σ_{DC}, and upon $\alpha\beta$ a triangle $\alpha\beta\delta'''$ having sides equal to σ_{AB}, σ_{DB}, σ_{DA}. These triangles are to be on the same sides of the lines $\beta\gamma$, $\gamma\alpha$, $\alpha\beta$, respectively, as the triangle $\alpha\beta\gamma$. The angles of these triangles will be supplements of the angles of the surfaces of discontinuity at a, b, and c. Thus $\beta\gamma\delta' = dab$, and $\alpha\gamma\delta'' = dba$. Now if δ' and δ'' fall together in a single point δ within the triangle $\alpha\beta\gamma$, δ''' will fall in the same point, as in figure 15. In this case we shall have $\beta\gamma\delta + \alpha\gamma\delta = \alpha\gamma\beta$, and the three angles of the curvilinear triangle adb will be together equal to two right angles. The same will be true of the three angles of each of the triangles bdc, cda, and hence of the three angles of the triangle abc. But if δ', δ'', δ''' do not fall together in the same point within the triangle $\alpha\beta\gamma$, it is either possible to bring these points to coincide within the triangle by increasing some or all of the tensions σ_{DA}, σ_{DB}, σ_{DC}, or to effect the same result by diminishing some or all of these tensions. (This will easily appear when one of the points δ', δ'', δ''' falls within the triangle, if we let the two tensions which determine this point remain constant, and the third tension vary. When all the points δ', δ'', δ''' fall without the triangle $\alpha\beta\gamma$, we may suppose the greatest of the tensions σ_{DA}, σ_{DB}, σ_{DC}—the two greatest, when these are equal, and all three when they all are equal—to diminish until one of the points δ', δ'', δ''' is brought within the triangle $\alpha\beta\gamma$.) In the first case we may say that the tensions of the new surfaces are too small to be represented by the distances of an internal point from the vertices of the triangle representing the tensions of the original surfaces (or, for brevity, that they are too small to be represented as in figure 15); in the second case we may say that they are too great to be thus represented. In the first case, the sum of the angles in each of the triangles adb, bdc, cda is less than two right angles (compare figures 14 and 16);

in the second case, each pair of the triangles $\alpha\beta\delta'''$, $\beta\gamma\delta'$, $\gamma\alpha\delta''$ will overlap, at least when the tensions σ_{DA}, σ_{DB}, σ_{DC} are only a little too great to be represented as in figure 15, and the sum of the angles of each of the triangles adb, bdc, cda will be greater than two right angles.

Let us denote by v_A, v_B, v_C the portions of v_D which were originally occupied by the masses A, B, C, respectively, by s_{DA}, s_{DB}, s_{DC}, the areas of the surfaces specified per unit of length of the mass D, and by s_{AB}, s_{BC}, s_{CA}, the areas of the surfaces specified which were replaced by the mass D per unit of its length. In numerical value, v_A, v_B, v_C will be equal to the areas of the curvilinear triangles bcd, cad, abd; and s_{DA}, s_{DB}, s_{DC}, s_{AB}, s_{BC}, s_{CA} to the lengths of the lines bc, ca, ab, cd, ad, bd. Also let

$$W_s = \sigma_{DA}\,s_{DA} + \sigma_{DB}\,s_{DB} + \sigma_{DC}\,s_{DC} - \sigma_{AB}\,s_{AB} - \sigma_{BC}\,s_{BC} - \sigma_{CA}\,s_{CA}, \quad (626)$$

and
$$W_v = p_D v_D - p_A v_A - p_B v_B - p_C v_C. \quad (627)$$

The general condition of mechanical equilibrium for a system of homogeneous masses not influenced by gravity, when the exterior of the whole system is fixed, may be written

$$\Sigma(\sigma\,\delta s) - \Sigma(p\,\delta v) = 0. \quad (628)$$

(See (606).) If we apply this both to the original system consisting of the masses A, B, and C, and to the system modified by the introduction of the mass D, and take the difference of the results, supposing the deformation of the system to be the same in each case, we shall have

$$\sigma_{DA}\,\delta s_{DA} + \sigma_{DB}\,\delta s_{DB} + \sigma_{DC}\,\delta s_{DC} - \sigma_{AB}\,\delta s_{AB} - \sigma_{BC}\,\delta s_{BC}$$
$$- \sigma_{CA}\,\delta s_{CA} - p_D\,\delta v_D + p_A\,\delta v_A + p_B\,\delta v_B + p_C\,\delta v_C = 0. \quad (629)$$

In view of this relation, if we differentiate (626) and (627) regarding all quantities except the pressures as variable, we obtain

$$dW_s - dW_v = s_{DA}\,d\sigma_{DA} + s_{DB}\,d\sigma_{DB} + s_{DC}\,d\sigma_{DC}$$
$$- s_{AB}\,d\sigma_{AB} - s_{BC}\,d\sigma_{BC} - s_{CA}\,d\sigma_{CA}. \quad (630)$$

Let us now suppose the system to vary in size, remaining always similar to itself in form, and that the tensions diminish in the same ratio as lines, while the pressures remain constant. Such changes will evidently not impair the equilibrium. Since all the quantities s_{DA}, σ_{DA}, s_{DB}, σ_{DB}, etc. vary in the same ratio,

$$s_{DA}\,d\sigma_{DA} = \tfrac{1}{2}d(\sigma_{DA}\,s_{DA}), \quad s_{DB}\,d\sigma_{DB} = \tfrac{1}{2}d(\sigma_{DB}\,s_{DB}), \quad \text{etc.} \quad (631)$$

We have therefore by integration of (630)

$$W_s - W_v = \tfrac{1}{2}(\sigma_{DA}\,s_{DA} + \sigma_{DB}\,s_{DB} + \sigma_{DC}\,s_{DC} - \sigma_{AB}\,s_{AB} - \sigma_{BC}\,s_{BC} - \sigma_{CA}\,s_{CA}), (632)$$

whence, by (626),

$$W_s = 2W_v. \quad (633)$$

The condition of stability for the system when the pressures and tensions are regarded as constant, and the position of the surfaces A-B, B-C, C-A as fixed, is that $W_S - W_V$ shall be a minimum under the same conditions. (See (549).) Now for any constant values of the tensions and of p_A, p_B, p_C, we may make v_D so small that when it varies, the system remaining in equilibrium (which will in general require a variation of p_D), we may neglect the curvatures of the lines da, db, dc, and regard the figure $abcd$ as remaining similar to itself. For the *total curvature* (i.e., the curvature measured in degrees) of each of the lines ab, bc, ca may be regarded as constant, being equal to the constant difference of the sum of the angles of one of the curvilinear triangles adb, bdc, cda and two right angles. Therefore, when v_D is very small, and the system is so deformed that equilibrium would be preserved if p_D had the proper variation, but this pressure as well as the others and all the tensions remain constant, W_S will vary as the lines in the figure $abcd$, and W_V as the square of these lines. Therefore, for such deformations,

$$W_V \propto W_S^2.$$

This shows that the system cannot be stable for constant pressures and tensions when v_D is small and W_V is positive, since $W_S - W_V$ will not be a minimum. It also shows that the system is stable when W_V is negative. For, to determine whether $W_S - W_V$ is a minimum for constant values of the pressures and tensions, it will evidently be sufficient to consider such varied forms of the system as give the least value to $W_S - W_V$ for any value of v_D in connection with the constant pressures and tensions. And it may easily be shown that such forms of the system are those which would preserve equilibrium if p_D had the proper value.

These results will enable us to determine the most important questions relating to the stability of a line along which three homogeneous fluids A, B, C meet, with respect to the formation of a different fluid D. The components of D must of course be such as are found in the surrounding bodies. We shall regard p_D and $\sigma_{DA}, \sigma_{DB}, \sigma_{DC}$ as determined by that phase of D which satisfies the conditions of equilibrium with the other bodies relating to temperature and the potentials. These quantities are therefore determinable, by means of the fundamental equations of the mass D and of the surfaces D-A, D-B, D-C, from the temperature and potentials of the given system.

Let us first consider the case in which the tensions, thus determined, can be represented as in figure 15, and p_D has a value consistent with the equilibrium of a small mass such as we have been considering. It appears from the preceding discussion that when v_D is sufficiently small the figure $abcd$ may be regarded as rectilinear, and that its angles are entirely determined by its tensions. Hence the ratios of v_A, v_B, v_C, v_D, for sufficiently small values of v_D, are determined by the tensions alone, and for convenience in calculating these ratios, we may suppose p_A, p_B, p_C to be equal, which will make the figure $abcd$ absolutely rectilinear, and make p_D equal to the other pressures, since it is supposed that this quantity has the value necessary for equilibrium. We may obtain a simple expression for the ratios of v_A, v_B, v_C, v_D in terms of the tensions in the following manner. We shall write [DBC], [DCA], etc., to denote the areas of triangles having sides equal to the tensions of the surfaces between the masses specified.

$$v_A : v_B :: \text{triangle } bdc : \text{triangle } adc$$
$$:: bc \sin bcd : ac \sin acd$$
$$:: \sin bac \sin bcd : \sin abc \sin acd$$
$$:: \sin \gamma\delta\beta \sin \delta a\beta : \sin \gamma\delta a \sin \delta\beta a$$
$$:: \sin \gamma\delta\beta \; \delta\beta : \sin \gamma\delta a \; \delta a$$
$$:: \text{triangle } \gamma\delta\beta : \text{triangle } \gamma\delta a$$
$$:: [DBC] : [DCA].$$

Hence,

$$v_A : v_B : v_C : v_D :: [DBC] : [DCA] : [DAB] : [ABC], \tag{634}$$

where

$$\tfrac{1}{4}\sqrt{[(\sigma_{AB}+\sigma_{BC}+\sigma_{CA})(\sigma_{AB}+\sigma_{BC}-\sigma_{CA})(\sigma_{BC}+\sigma_{CA}-\sigma_{AB})(\sigma_{CA}+\sigma_{AB}-\sigma_{BC})]}$$

may be written for [ABC], and analogous expressions for the other symbols, the sign $\sqrt{}$ denoting the positive root of the necessarily positive expression which follows. This proportion will hold true in any case of equilibrium, when the tensions satisfy the condition mentioned and v_D is sufficiently small. Now if $p_A = p_B = p_C$, p_D will have the same value, and we shall have by (627) $W_V = 0$, and by (633) $W_S = 0$. But when v_D is very small, the value of W_S is entirely determined by the tensions and v_D. Therefore, whenever the tensions satisfy the condition supposed, and v_D is very small (whether p_A, p_B, p_C are equal or unequal),

$$0 = W_S = W_V = p_D v_D - p_A v_A - p_B v_B - p_C v_C, \tag{635}$$

which with (634) gives

$$p_D = \frac{[DBC]\, p_A + [DCA]\, p_B + [DAB]\, p_C}{[DBC] + [DCA] + [DAB]}. \tag{636}$$

Since this is the only value of p_D for which equilibrium is possible when the tensions satisfy the condition supposed and v_D is small, it follows that when p_D has a less value, the line where the fluids

A, B, C meet is stable with respect to the formation of the fluid D. When p_D has a greater value, if such a line can exist at all, it must be at least practically unstable, i.e., if only a very small mass of the fluid D should be formed it would tend to increase.

Let us next consider the case in which the tensions of the new surfaces are too small to be represented as in figure 15. If the pressures and tensions are consistent with equilibrium for any very small value of v_D, the angles of each of the curvilinear triangles adb, bdc, cda will be together less than two right angles, and the lines ab, bc, ca will be convex toward the mass D. For given values of the pressures and tensions, it will be easy to determine the magnitude of v_D. For the tensions will give the total curvatures (in degrees) of the lines ab, bc, ca; and the pressures will give the radii of curvature. These lines are thus completely determined. In order that v_D shall be very small it is evidently necessary that p_D shall be less than the other pressures. Yet if the tensions of the new surfaces are only a very little too small to be represented as in figure 15, v_D may be quite small when the value of p_D is only a little less than that given by equation (636). In any case, when the tensions of the new surfaces are too small to be represented as in figure 15, and v_D is small, W_V is negative, and the equilibrium of the mass D is stable. Moreover, $W_S - W_V$, which represents the work necessary to form the mass D with its surfaces in place of the other masses and surfaces, is negative.

With respect to the stability of a line in which the surfaces A-B, B-C, C-A meet, when the tensions of the new surfaces are too small to be represented as in figure 15, we first observe that when the pressures and tensions are such as to make v_D moderately small but not so small as to be neglected (this will be when p_D is somewhat smaller than the second member of (636),—more or less smaller according as the tensions differ more or less from such as are represented in figure 15), the equilibrium of such a line as that supposed (if it is capable of existing at all) is at least practically unstable. For greater values of p_D (with the same values of the other pressures and the tensions) the same will be true. For somewhat smaller values of p_D, the mass of the phase D which will be formed will be so small, that we may neglect this mass and regard the surfaces A-B, B-C, C-A as meeting in a line in stable equilibrium. For still smaller values of p_D, we may likewise regard the surfaces A-B, B-C, C-A as capable of meeting in stable equilibrium. It may be observed that when v_D, as determined by our equations, becomes quite insensible, the conception of a small mass D having the properties deducible from our equations ceases to be accurate, since the matter in the vicinity of a line where these surfaces of discontinuity meet must be in a peculiar state of equilibrium not recognized by our equations.* But this cannot affect the validity of our conclusion with respect to the stability of the line in question.

The case remains to be considered in which the tensions of the new surfaces are too great to be represented as in figure 15. Let us suppose that they are not very much too great to be thus represented. When the pressures are such as to make v_D moderately small (in case of equilibrium) but not so small that the mass D to which it relates ceases to have the properties of matter in mass (this will be when p_D is somewhat greater than the second member of (636),—more or less greater according as the tensions differ more or less from such as are represented in figure 15), the line where the surfaces A-B, B-C, C-A meet will be in stable equilibrium with respect to the formation of such a mass as we have considered, since $W_S - W_V$ will be positive. The same will be true for less values of p_D. For greater values of p_D, the value of $W_S - W_V$, which measures the stability with respect to the kind of change considered, diminishes. It does not vanish, according to our equations, for finite values of p_D. But these equations are not to be trusted beyond the limit at which the mass D ceases to be of sensible magnitude.

But when the tensions are such as we now suppose, we must also consider the possible formation of a mass D within a closed figure in which the surfaces D-A, D-B, D-C meet together (with the surfaces A-B, B-C, C-A) in two opposite points. If such a figure is to be in equilibrium, the six tensions must be such as can be represented by the six distances of four points in space (see pages 288, 289),—a condition which evidently agrees with the supposition which we have made. If we denote by w_V the work gained in forming the mass D (of such size and form as to be in equilibrium) in place of the other masses, and by w_S the work expended in forming the new surfaces in place of the old, it may easily be shown by a method similar to that used on page 292 that $w_S = \frac{3}{2}w_V$. From this we obtain $w_S - w_V = \frac{1}{2}w_V$. This is evidently positive when p_D is greater than the other pressures. But it diminishes with increase of p_D, as easily appears from the

---

* See note on page 288. We may here add that the linear tension there mentioned may have a negative value. This would be the case with respect to a line in which three surfaces of discontinuity are regarded as meeting, but where nevertheless there really exists in stable equilibrium a filament of different phase from the three surrounding masses. The value of the linear tension for the supposed line, would be nearly equal to the value of $W_S - W_V$ for the actually existing filament. (For the exact value of the linear tension it would be necessary to add the sum of the linear tensions of the three edges of the filament.) We may regard two soap-bubbles adhering together as an example of this case. The reader will easily convince himself that in an exact treatment of the equilibrium of such a double bubble we must recognize a certain negative tension in the line of intersection of the three surfaces of discontinuity.

equivalent expression $\frac{1}{3}w_8$. Hence the line of intersection of the surfaces of discontinuity A-B, B-C, C-A is stable for values of p_D greater than the other pressures (and therefore for all values of p_D) so long as our method is to be regarded as accurate, which will be so long as the mass D which would be in equilibrium has a sensible size.

In certain cases in which the tensions of the new surfaces are much too large to be represented as in figure 15, the reasoning of the two last paragraphs will cease to be applicable. These are cases in which the six tensions cannot be represented by the sides of a tetrahedron. It is not necessary to discuss these cases, which are distinguished by the different shape which the mass D would take if it should be formed, since it is evident that they can constitute no exception to the results which we have obtained. For an increase of the values of σ_{DA}, σ_{DB}, σ_{DC} cannot favor the formation of D, and hence cannot impair the stability of the line considered, as deduced from our equations. Nor can an increase of these tensions essentially affect the fact that the stability thus demonstrated may fail to be realized when p_D is considerably greater than the other pressures, since the *a priori* demonstration of the stability of any one of the surfaces A-B, B-C, C-A, taken singly, is subject to the limitation mentioned. (See pages 261, 262.)

The Condition of Stability for Fluids relating to the Formation of a New Phase at a Point where the Vertices of Four Different Masses meet.

Let four different fluid masses A, B, C, D meet about a point, so as to form the six surfaces of discontinuity A-B, B-C, C-A, D-A, D-B, D-C, which meet in the four lines A-B-C, B-C-D, C-D-A, D-A-B, these lines meeting in the vertical point. Let us suppose the system stable in other respects, and consider the conditions of stability for the vertical point with respect to the possible formation of a different fluid mass E.

If the system can be in equilibrium when the vertical point has been replaced by a mass E against which the four masses A, B, C, D abut, being truncated at their vertices, it is evident that E will have four vertices, at each of which six surfaces of discontinuity meet. (Thus at one vertex there will be the surfaces formed by A, B, C, and E.) The tensions of each set of six surfaces (like those of the six surfaces formed by A, B, C, and D) must therefore be such that they can be represented by the six edges of a tetrahedron. When the tensions do not satisfy these relations, there will be no particular condition of stability for the point about which A, B, C, and D meet, since if a mass E should be formed, it would distribute itself along some of the lines or surfaces which meet at the vertical point, and it

is therefore sufficient to consider the stability of these lines and surfaces. We shall suppose that the relations mentioned are satisfied.

If we denote by W_V the work gained in forming the mass E (of
such size and form as to be in equilibrium) in place of the portions
of the other masses which are suppressed, and by W_S the work expended in forming the new surfaces in place of the old, it may easily
be shown by a method similar to that used on page 292 that

$$W_S = \tfrac{3}{2} W_V, \tag{637}$$

whence
$$W_S - W_V = \tfrac{1}{2} W_V; \tag{638}$$

also, that when the volume E is small, the equilibrium of E will be
stable or unstable according as W_S and W_V are negative or positive.

A critical relation for the tensions is that which makes equilibrium
possible for the system of the five masses A, B, C, D, E, when all
the surfaces are plane. The ten tensions may then be represented in
magnitude and direction by the ten distances of five points in space
a, β, γ, δ, ϵ, viz., the tension of A-B and the direction of its normal
by the line $a\beta$, etc. The point ϵ will lie within the tetrahedron
formed by the other points. If we write v_E for the volume of E, and
v_A, v_B, v_C, v_D for the volumes of the parts of the other masses which
are suppressed to make room for E, we have evidently

$$W_V = p_E v_E - p_A v_A - p_B v_B - p_C v_C - p_D v_D. \tag{639}$$

Hence, when all the surfaces are plane, $W_V = 0$, and $W_S = 0$. Now
equilibrium is always possible for a given small value of v_E with any
given values of the tensions and of p_A, p_B, p_C, p_D. When the tensions
satisfy the critical relation, $W_S = 0$, if $p_A = p_B = p_C = p_D$. But when
v_E is small and constant, the value of W_S must be independent of
p_A, p_B, p_C, p_D, since the angles of the surfaces are determined by the
tensions and their curvatures may be neglected. Hence, $W_S = 0$, and
$W_V = 0$, when the critical relation is satisfied and v_E small. This gives

$$p_E = \frac{v_A p_A + v_B p_B + v_C p_C + v_D p_D}{v_E}. \tag{640}$$

In calculating the ratios of v_A, v_B, v_C, v_D, v_E, we may suppose all the
surfaces to be plane. Then E will have the form of a tetrahedron,
the vertices of which may be called a, b, c, d (each vertex being
named after the mass which is not found there), and v_A, v_B, v_C, v_D will
be the volumes of the tetrahedra into which it may be divided
by planes passing through its edges and an interior point e. The
volumes of these tetrahedra are proportional to those of the five
tetrahedra of the figure $a\beta\gamma\delta\epsilon$, as will easily appear if we recollect
that the line ab is common to the surfaces C-D, D-E, E-C, and there-
fore perpendicular to the surface common to the lines $\gamma\delta$, $\delta\epsilon$, $\epsilon\gamma$, i.e.
to the surface $\gamma\delta\epsilon$, and so in other cases (it will be observed that
γ, δ, and ϵ are the letters which do not correspond to a or b); also

that the surface abc is the surface D-E and therefore perpendicular to $\delta\epsilon$, etc. Let tetr abcd, trian abc, etc. denote the volume of the tetrahedron or the area of the triangle specified, sin (ab, bc), sin (abc, dbc), sin (abc, ad), etc. the sines of the angles made by the lines and surfaces specified, and [BCDE], [CDEA], etc. the volumes of tetrahedra having edges equal to the tensions of the surfaces between the masses specified. Then, since we may express the volume of a tetrahedron either by $\frac{1}{3}$ of the product of one side, an edge leading to the opposite vertex, and the sine of the angle which these make, or by $\frac{2}{3}$ of the product of two sides divided by the common edge and multiplied by the sine of the included angle,

$$v_A : v_B :: \text{tetr bcde} : \text{tetr acde}$$
$$:: \text{bc} \sin(\text{bc, cde}) : \text{ac} \sin(\text{ac, cde})$$
$$:: \sin(\text{ba, ac}) \sin(\text{bc, cde}) : \sin(\text{ab, bc}) \sin(\text{ac, cde})$$
$$:: \sin(\gamma\delta\epsilon, \beta\delta\epsilon) \sin(a\delta\epsilon, a\beta) : \sin(\gamma\delta\epsilon, a\delta\epsilon) \sin(\beta\delta\epsilon, a\beta)$$
$$:: \frac{\text{tetr } \gamma\beta\delta\epsilon \text{ tetr } \beta a\delta\epsilon}{\text{trian } \beta\delta\epsilon \text{ trian } a\delta\epsilon} : \frac{\text{tetr } \gamma a\delta\epsilon \text{ tetr } a\beta\delta\epsilon}{\text{trian } a\delta\epsilon \text{ trian } \beta\delta\epsilon}$$
$$:: \text{tetr } \gamma\beta\delta\epsilon : \text{tetr } \gamma a\delta\epsilon$$
$$:: [\text{BCDE}] : [\text{CDEA}].$$

Hence,

$$v_A : v_B : v_C : v_D :: [\text{BCDE}] : [\text{CDEA}] : [\text{DEAB}] : [\text{EABC}], \quad (641)$$

and (640) may be written

$$p_E = \frac{[\text{BCDE}]p_A + [\text{CDEA}]p_B + [\text{DEAB}]p_C + [\text{EABC}]p_D}{[\text{BCDE}] + [\text{CDEA}] + [\text{DEAB}] + [\text{EABC}]}. \quad (642)$$

If the value of p_E is less than this, when the tensions satisfy the critical relation, the point where vertices of the masses A, B, C, D meet is stable with respect to the formation of any mass of the nature of E. But if the value of p_E is greater, either the masses A, B, C, D cannot meet at a point in equilibrium, or the equilibrium will be at least practically unstable.

When the tensions of the new surfaces are too small to satisfy the critical relation with the other tensions, these surfaces will be convex toward E; when their tensions are too great for that relation, the surfaces will be concave toward E. In the first case, W_V is negative, and the equilibrium of the five masses A, B, C, D, E is stable, but the equilibrium of the four masses A, B, C, D meeting at a point is impossible or at least practically unstable. This is subject to the limitation that when p_E is sufficiently small the mass E which will form will be so small that it may be neglected. This will only be the case when p_E is smaller—in general considerably smaller—than the second member of (642). In the second case, the equilibrium of the five masses A, B, C, D, E will be unstable, but the equilibrium of the four masses A, B, C, D will be stable unless v_E (calculated for the case of the five masses) is of insensible magnitude. This will only be the case when p_E is greater—in general considerably greater— than the second member of (642).

Liquid Films.

When a fluid exists in the form of a thin film between other fluids, the great inequality of its extension in different directions will give rise to certain peculiar properties, even when its thickness is sufficient for its interior to have the properties of matter in mass. The frequent occurrence of such films, and the remarkable properties which they exhibit, entitle them to particular consideration. To fix our ideas, we shall suppose that the film is liquid and that the contiguous fluids are gaseous. The reader will observe our results are not dependent, so far as their general character is concerned, upon this supposition.

Let us imagine the film to be divided by surfaces perpendicular to its sides into small portions of which all the dimensions are of the same order of magnitude as the thickness of the film,—such portions to be called *elements of the film*,—it is evident that far less time will in general be required for the attainment of approximate equilibrium between the different parts of any such element and the other fluids which are immediately contiguous, than for the attainment of equilibrium between all the different elements of the film. There will accordingly be a time, commencing shortly after the formation of the film, in which its separate elements may be regarded as satisfying the conditions of internal equilibrium, and of equilibrium with the contiguous gases, while they may not satisfy all the conditions of equilibrium with each other. It is when the changes due to this want of complete equilibrium take place so slowly that the film appears to be at rest, except so far as it accommodates itself to any change in the external conditions to which it is subjected, that the characteristic properties of the film are most striking and most sharply defined.

Let us therefore consider the properties which will belong to a film sufficiently thick for its interior to have the properties of matter in mass, in virtue of the approximate equilibrium of all its elements taken separately, when the matter contained in each element is regarded as invariable, with the exception of certain substances which are components of the contiguous gas-masses and have their potentials thereby determined. The occurrence of a film which precisely satisfies these conditions may be exceptional, but the discussion of this somewhat ideal case will enable us to understand the principal laws which determine the behavior of liquid films in general.

Let us first consider the properties which will belong to each element of the film under the conditions mentioned. Let us suppose the element extended, while the temperature and the potentials which are determined by the contiguous gas-masses are unchanged. If the film has no components except those of which the potentials are maintained constant, there will be no variation of tension in its surfaces. The same will be true when the film has only one component of which the potential is not maintained constant, provided that this is a component of the interior of the film and not of its surface alone. If we regard the thickness of the film as determined by *dividing surfaces* which make the surface-density of this component vanish, the thickness will vary inversely as the area of the element of the film, but no change will be produced in the nature or the tension of its surfaces. If, however, the single component of which the potential is not maintained constant is confined to the surfaces of the film, an extension of the element will generally produce a decrease in the potential of this component, and an increase of tension. This will certainly be true in those cases in which the component shows a tendency to distribute itself with a uniform superficial density.

When the film has two or more components of which the potentials are not maintained constant by the contiguous gas-masses, they will not in general exist in the same proportion in the interior of the film as on its surfaces, but those components which diminish the tensions will be found in greater proportion on the surfaces. When the film is extended, there will therefore not be enough of these substances to keep up the same volume- and surface-densities as before, and the deficiency will cause a certain increase of tension. The value of the *elasticity of the film* (i.e., the infinitesimal increase of the united tensions of its surfaces divided by the infinitesimal increase of area in a unit of surface) may be calculated from the quantities which specify the nature of the film, when the fundamental equations of the interior mass, of the contiguous gas-masses, and of the two surfaces of discontinuity are known. We may illustrate this by a simple example.

Let us suppose that the two surfaces of a plane film are entirely alike, that the contiguous gas-masses are identical in phase, and that they determine the potentials of all the components of the film except two. Let us call these components S_1 and S_2, the latter denoting that which occurs in greater proportion on the surface than in the interior of the film. Let us denote by γ_1 and γ_2 the densities of these components in the interior of the film, by λ the thickness of the film determined by such dividing surfaces as make the surface-density of S_1 vanish (see page 234), by $\Gamma_{2(1)}$ the surface-density of the other component as determined by the same surfaces, by σ and s the tension and area of one of these surfaces, and by E the elasticity of the film when extended under the supposition that the total quantities of S_1 and S_2 in the part of the film extended are invariable, as also the temperature and the potentials of the other components. From the definition of E we have

$$2\,d\sigma = E\frac{ds}{s}, \tag{643}$$

and from the conditions of the extension of the film

$$\frac{ds}{s} = -\frac{d(\lambda\gamma_1)}{\lambda\gamma_1} = -\frac{d(\lambda\gamma_2 + 2\Gamma_{2(1)})}{\lambda\gamma_2 + 2\Gamma_{2(1)}}. \tag{644}$$

Hence we obtain

$$\lambda\gamma_1\frac{ds}{s} = -\gamma_1 d\lambda - \lambda d\gamma_1,$$

$$(\lambda\gamma_2 + 2\Gamma_{2(1)})\frac{ds}{s} = -\gamma_2 d\lambda - \lambda d\gamma_2 - 2\,d\Gamma_{2(1)};$$

and eliminating $d\lambda$,

$$2\gamma_1\Gamma_{2(1)}\frac{ds}{s} = -\lambda\gamma_1 d\gamma_2 + \lambda\gamma_2 d\gamma_1 - 2\gamma_1 d\Gamma_{2(1)}. \tag{645}$$

If we set

$$r = \frac{\gamma_2}{\gamma_1}, \tag{646}$$

we have

$$dr = \frac{\gamma_1 d\gamma_2 - \gamma_2 d\gamma_1}{\gamma_1^{\,2}}, \tag{647}$$

and

$$2\Gamma_{2(1)}\frac{ds}{s} = -\lambda\gamma_1 dr - 2\,d\Gamma_{2(1)}. \tag{648}$$

With this equation we may eliminate ds from (643). We may also eliminate $d\sigma$ by the necessary relation (see (514))

$$d\sigma = -\Gamma_{2(1)}d\mu_2.$$

This will give

$$4\Gamma_{2(1)}^{\,2}\,d\mu_2 = E(\lambda\gamma_1 dr + 2\,d\Gamma_{2(1)}), \tag{649}$$

or

$$\frac{4\Gamma_{2(1)}^{\,2}}{E} = \lambda\gamma_1\frac{dr}{d\mu_2} + 2\frac{d\Gamma_{2(1)}}{d\mu_2}, \tag{650}$$

where the differential coefficients are to be determined on the conditions that the temperature and all the potentials except μ_1 and μ_2 are constant, and that the pressure in the interior of the film shall remain equal to that in the contiguous gas-masses. The latter condition may be expressed by the equation

$$(\gamma_1 - \gamma_1')d\mu_1 + (\gamma_2 - \gamma_2')d\mu_2 = 0, \tag{651}$$

in which γ_1' and γ_2' denote the densities of S_1 and S_2 in the contiguous gas-masses. (See (98).) When the tension of the surfaces of the film and the pressures in its interior and in the contiguous gas-masses are known in terms of the temperature and potentials, equation (650) will give the value of E in terms of the same variables together with λ.

If we write G_1 and G_2 for the total quantities of S_1 and S_2 per unit of area of the film, we have

$$G_1 = \lambda \gamma_1, \tag{652}$$

$$G_2 = \lambda \gamma_2 + 2\Gamma_{2(1)}. \tag{653}$$

Therefore,

$$G_2 = G_1 r + 2\Gamma_{2(1)},$$

$$\left(\frac{dG_2}{d\mu_2}\right)_{G_1} = \lambda \gamma_1 \frac{dr}{d\mu_2} + 2\frac{d\Gamma_{2(1)}}{d\mu_2}, \tag{654}$$

where the differential coefficients in the second member are to be determined as in (650), and that in the first member with the additional condition that G_1 is constant. Therefore,

$$\frac{4\Gamma_{2(1)}^2}{E} = \left(\frac{dG_2}{d\mu_2}\right)_{G_1},$$

and

$$E = 4\Gamma_{2(1)}^2 \left(\frac{d\mu_2}{dG_2}\right)_{G_1}, \tag{655}$$

the last differential coefficient being determined by the same conditions as that in the preceding equation. It will be observed that the value of E will be positive in any ordinary case.

These equations give the elasticity of any element of the film when the temperature and the potentials for the substances which are found in the contiguous gas-masses are regarded as constant, and the potentials for the other components, μ_1 and μ_2, have had time to equalize themselves throughout the element considered. The increase of tension immediately after a rapid extension will be greater than that given by these equations.

The existence of this elasticity, which has thus been established from *a priori* considerations, is clearly indicated by the phenomena which liquid films present. Yet it is not to be demonstrated simply by comparing the tensions of films of different thickness, even when they are made from the same liquid, for difference of thickness does not necessarily involve any difference of tension. When the phases within the films as well as without are the same, and the surfaces of the films are also the same, there will be no difference of tension. Nor will the tension of the same film be altered, if a part of the interior drains away in the course of time, without affecting the surfaces. In case the thickness of the film is reduced by evaporation, the tension may be either increased or diminished. (The evaporation of the substance S_1, in the case we have just considered, would diminish the tension.) Yet it may easily be shown that extension increases the tension of a film and contraction diminishes it. When a plane film is held vertically, the tension of the upper portions must evidently be greater than that of the lower. The tensions in every part of the film may be reduced to equality by turning it into a horizontal position. By restoring the original position we may restore the original tensions, or nearly so. It is evident that the same element of the film is capable of supporting very unequal tensions. Nor can this be always attributed to viscosity of the film. For in many cases, if we hold the film nearly horizontal, and elevate first one side and then another, the lighter portions of the film will dart from one side to the other, so as to show a very striking mobility in the film. The differences of tension which cause these rapid movements are only a very small fraction of the difference of tension in the upper and lower portions of the film when held vertically.

If we account for the power of an element of the film to support an increase of tension by viscosity, it will be necessary to suppose that the viscosity offers a resistance to a deformation of the film in which its surface is enlarged and its thickness diminished, which is enormously great in comparison with the resistance to a deformation in which the film is extended in the direction of one tangent and contracted in the direction of another, while its thickness and the areas of its surfaces remain constant. This is not to be readily admitted as a physical explanation, although to a certain extent the phenomena resemble those which would be caused by such a singular viscosity. (See page 274.) The only natural explanation of the phenomena is that the extension of an element of the film, which is the immediate result of an increase of external force applied to its perimeter, causes an increase of its tension, by which it is brought into true equilibrium with the external forces.

The phenomena to which we have referred are such as are apparent to a very cursory observation. In the following experiment, which is described by M. Plateau,* an increased tension is manifested in a film while contracting after a previous extension. The warmth of a finger brought near to a bubble of soap-water with glycerine, which is thin enough to show colors, causes a spot to appear indicating a diminution of thickness. When the finger is removed, the spot returns to its original color. This indicates a contraction, which would be resisted by any viscosity of the film, and can only be due to an excess of tension in the portion stretched, on the return of its original temperature.

We have so far supposed that the film is thick enough for its

---

* *Statique expérimentale et théorique des liquides soumis aux seules forces moléculaires*, vol. i, p. 294.

interior to have the properties of matter in mass. Its properties are then entirely determined by those of the three phases and the two surfaces of discontinuity. From these we can also determine, in part at least, the properties of a film at the limit at which the interior ceases to have the properties of matter in mass. The elasticity of the film, which increases with its thinness, cannot of course vanish at that limit, so that the film cannot become unstable with respect to extension and contraction of its elements immediately after passing that limit.

Yet a certain kind of instability will probably arise, which we may here notice, although it relates to changes in which the condition of the invariability of the quantities of certain components in an element of the film is not satisfied. With respect to variations in the distribution of its components, a film will in general be stable, when its interior has the properties of matter in mass, with the single exception of variations affecting its thickness without any change of phase or of the nature of the surfaces. With respect to this kind of change, which may be brought about by a current in the interior of the film, the equilibrium is neutral. But when the interior ceases to have the properties of matter in mass, it is to be supposed that the equilibrium will generally become unstable in this respect. For it is not likely that the neutral equilibrium will be unaffected by such a change of circumstances, and since the film certainly becomes unstable when it is sufficiently reduced in thickness, it is most natural to suppose that the first effect of diminishing the thickness will be in the direction of instability rather than in that of stability. (We are here considering liquid films between gaseous masses. In certain other cases, the opposite supposition might be more natural, as in respect to a film of water between mercury and air, which would certainly become stable when sufficiently reduced in thickness.)

Let us now return to our former suppositions—that the film is thick enough for the interior to have the properties of matter in mass, and that the matter in each element is invariable, except with respect to those substances which have their potentials determined by the contiguous gas-masses—and consider what conditions are necessary for equilibrium in such a case.

In consequence of the supposed equilibrium of its several elements, such a film may be treated as a simple surface of discontinuity between the contiguous gas-masses (which may be similar or different), whenever its radius of curvature is very large in comparison with its thickness,—a condition which we shall always suppose to be fulfilled. With respect to the film considered in this light, the mechanical conditions of equilibrium will always be satisfied, or very nearly so, as soon as a state of approximate rest is attained, except in those cases in which the film exhibits a decided viscosity. That is, the relations (613), (614), (615) will hold true, when by σ we understand the tension of the film regarded as a simple surface of discontinuity (this is equivalent to the sum of the tensions of the two surfaces of the film), and by Γ its mass per unit of area diminished by the mass of gas which would occupy the same space if the film should be suppressed and the gases should meet at its surface of tension. This *surface of tension of the film* will evidently divide the distance between the surfaces of tension for the two surfaces of the film taken separately, in the inverse ratio of their tensions. For practical purposes, we may regard Γ simply as the mass of the film per unit of area. It will be observed that the terms containing Γ in (613) and (614) are not to be neglected in our present application of these equations.

But the mechanical conditions of equilibrium for the film regarded as an approximately homogeneous mass in the form of a thin sheet bounded by two surfaces of discontinuity are not necessarily satisfied when the film is in a state of apparent rest. In fact, these conditions cannot be satisfied (in any place where the force of gravity has an appreciable intensity) unless the film is horizontal. For the pressure in the interior of the film cannot satisfy simultaneously condition (612), which requires it to vary rapidly with the height z, and condition (613) applied separately to the different surfaces, which makes it a certain mean between the pressures in the adjacent gas-masses. Nor can these conditions be deduced from the general condition of mechanical equilibrium (606) or (611), without supposing that the interior of the film is free to move independently of the surfaces, which is contrary to what we have supposed.

Moreover, the potentials of the various components of the film will not in general satisfy conditions (617), and cannot (when the temperature is uniform) unless the film is horizontal. For if these conditions were satisfied, equation (612) would follow as a consequence. (See page 283.)

We may here remark that such a film as we are considering cannot form any exception to the principle indicated on page 284,—that when a surface of discontinuity which satisfies the conditions of mechanical equilibrium has only one component which is not found in the contiguous masses, and these masses satisfy all the conditions of equilibrium, the potential for the component mentioned must satisfy the law expressed in (617), as a consequence of the condition of mechanical equilibrium (614). Therefore, as we have just seen that it is impossible that all the potentials in a liquid film which is not horizontal should conform to (617) when the temperature is uniform, it follows that if a liquid film exhibits any persistence which is not due to viscosity, or to a horizontal position, or to differences of temperature, it must have more than one component of which the potential is not determined by the contiguous gas-masses in accordance with (617).

The difficulties of the quantitative experimental verification of the properties which have been described would be very great, even in cases in which the conditions we have imagined were entirely fulfilled. Yet the general effect of any divergence from these conditions will be easily perceived, and when allowance is made for such divergence, the general behavior of liquid films will be seen to agree with the requirements of theory.

The formation of a liquid film takes place most symmetrically when a bubble of air rises to the top of a mass of the liquid. The motion of the liquid, as it is displaced by the bubble, is evidently such as to stretch the two surfaces in which the liquid meets the air, where these surfaces approach one another. This will cause an increase of tension, which will tend to restrain the extension of the surfaces. The extent to which this effect is produced will vary with the nature of the liquid. Let us suppose that the case is one in which the liquid contains one or more components which, although constituting but a very small part of its mass, greatly reduce its tension. Such components will exist in excess on the surfaces of the liquid. In this case the restraint upon the extension of the surfaces will be considerable, and as the bubble of air rises above the general level of the liquid, the motion of the latter will consist largely of a running out from between the two surfaces. But this running out of the liquid will be greatly retarded by its viscosity as soon as it is reduced to the thickness of a film, and the effect of the extension of the surfaces in increasing their tension will become greater and more permanent as the quantity of liquid diminishes which is available for supplying the substances which go to form the increased surfaces.

We may form a rough estimate of the amount of motion which is possible for the interior of a liquid film, relatively to its exterior, by calculating the descent of water between parallel vertical planes at which the motion of the water is reduced to zero. If we use the coefficient of viscosity as determined by Helmholtz and Piotrowski,[*] we obtain

$$V = 581 \, D^2, \qquad (656)$$

where V denotes the mean velocity of the water (i.e., that velocity

---

[*] *Sitzungsberichte der Wiener Akademie* (*mathemat.-naturwiss. Classe*), B. xl, S. 607. The calculation of formula (656) and that of the factor ($\frac{2}{3}$) applied to the formula of Poiseuille, to adapt it to a current between plane surfaces, have been made by means of the general equations of the motion of a viscous liquid as given in the memoir referred to.

which, if it were uniform throughout the whole space between the fixed planes, would give the same discharge of water as the actual variable velocity) expressed in millimeters per second, and D denotes the distance in millimeters between the fixed planes, which is supposed to be very small in proportion to their other dimensions. This is for the temperature of 24·5° C. For the same temperature, the experiments of Poiseuille * give

$$V = 337\,D^2$$

for the descent of water in long capillary tubes, which is equivalent to

$$V = 899\,D^2 \tag{657}$$

for descent between parallel planes. The numerical coefficient in this equation differs considerably from that in (656), which is derived from experiments of an entirely different nature, but we may at least conclude that in a film of a liquid which has a viscosity and specific gravity not very different from those of water at the temperature mentioned the mean velocity of the interior relatively to the surfaces will not probably exceed $1000\,D^2$. This is a velocity of 1^{mm} per second for a thickness of $\cdot01^{mm}$, $\cdot06^{mm}$ per *minute* for a thickness of $\cdot001$ (which corresponds to the red of the fifth order in a film of water), and $\cdot036^{mm}$ per *hour* for a thickness of $\cdot0001^{mm}$ (which corresponds to the white of the first order). Such an internal current is evidently consistent with great persistence of the film, especially in those cases in which the film can exist in a state of the greatest tenuity. On the other hand, the above equations give so large a value of V for thicknesses of 1^{mm} or $\cdot1^{mm}$, that the film can evidently be formed without carrying up any great weight of liquid, and any such thicknesses as these can have only a momentary existence.

A little consideration will show that the phenomenon is essentially of the same nature when films are formed in any other way, as by dipping a ring or the mouth of a cup in the liquid and then withdrawing it. When the film is formed in the mouth of a pipe, it may sometimes be extended so as to form a large bubble. Since the elasticity (i.e., the increase of the tension with extension) is greater in the thinner parts, the thicker parts will be most extended, and the effect of this process (so far as it is not modified by gravity) will be to diminish the ratio of the greatest to the least thickness of the film. During this extension, as well as at other times, the increased elasticity due to imperfect communication of heat, etc., will serve to protect the bubble from fracture by shocks received from the air or the pipe. If the bubble is now laid upon a suitable support, the condition (613) will be realized almost instantly. The bubble will

---

* *Ibid.*, p. 653; or *Mémoires des Savants Étrangers*, vol. ix, p. 532.

then tend toward conformity with condition (614), the lighter portions rising to the top, more or less slowly, according to the viscosity of the film. The resulting difference of thickness between the upper and the lower parts of the bubble is due partly to the greater tension to which the upper parts are subject, and partly to a difference in the matter of which they are composed. When the film has only two components of which the potentials are not determined by the contiguous atmosphere, the laws which govern the arrangement of the elements of the film may be very simply expressed. If we call these components S_1 and S_2, the latter denoting (as on page 301) that which exists in excess at the surface, one element of the film will tend toward the same level with another, or a higher, or a lower level, according as the quantity of S_2 bears the same ratio to the quantity of S_1 in the first element as in the second, or a greater, or a less ratio.

When a film, however formed, satisfies both the conditions (613) and (614), its thickness being sufficient for its interior to have the properties of matter in mass, the interior will still be subject to the slow current which we have already described, if it is truly fluid, however great its viscosity may be. It seems probable, however, that this process is often totally arrested by a certain gelatinous consistency of the mass in question, in virtue of which, although practically fluid in its behavior with reference to ordinary stresses, it may have the properties of a solid with respect to such very small stresses as those which are caused by gravity in the interior of a very thin film which satisfies the conditions (613) and (614).

However this may be, there is another cause which is often more potent in producing changes in a film, when the conditions just mentioned are approximately satisfied, than the action of gravity on its interior. This will be seen if we turn our attention to the edge where the film is terminated. At such an edge we generally find a liquid mass, continuous in phase with the interior of the film, which is bounded by concave surfaces, and in which the pressure is therefore less than in the interior of the film. This liquid mass therefore exerts a strong suction upon the interior of the film, by which its thickness is rapidly reduced. This effect is best seen when a film which has been formed in a ring is held in a vertical position. Unless the film is very viscous, its diminished thickness near the edge causes a rapid upward current on each side, while the central portion slowly descends. Also at the bottom of the film, where the edge is nearly horizontal, portions which have become thinned escape from their position of unstable equilibrium beneath heavier portions, and pass upwards, traversing the central portion of the film until they find a position of stable equilibrium. By these processes, the whole film is rapidly reduced in thickness.

The energy of the suction which produces these effects may be inferred from the following considerations. The pressure in the slender liquid mass which encircles the film is of course variable, being greater in the lower portions than in the upper, but it is everywhere less than the pressure of the atmosphere. Let us take a point where the pressure is less than that of the atmosphere by an amount represented by a column of the liquid one centimeter in height. (It is probable that much greater differences of pressure occur.) At a point near by in the interior of the film the pressure is that of the atmosphere. Now if the difference of pressure of these two points were distributed uniformly through the space of one centimeter, the intensity of its action would be exactly equal to that of gravity. But since the change of pressure must take place very suddenly (in a small fraction of a millimeter), its effect in producing a current in a limited space must be enormously great compared with that of gravity.

Since the process just described is connected with the descent of the liquid in the mass encircling the film, we may regard it as another example of the downward tendency of the interior of the film. There is a third way in which this descent may take place, when the principal component of the interior is volatile, viz., through the air. Thus, in the case of a film of soap-water, if we suppose the atmosphere to be of such humidity that the potential for water at a level mid-way between the top and bottom of the film has the same value in the atmosphere as in the film, it may easily be shown that evaporation will take place in the upper portions and condensation in the lower. These processes, if the atmosphere were otherwise undisturbed, would occasion currents of diffusion and other currents, the general effect of which would be to carry the moisture downward. Such a precise adjustment would be hardly attainable, and the processes described would not be so rapid as to have a practical importance.

But when the potential for water in the atmosphere differs considerably from that in the film, as in the case of a film of soap-water in a dry atmosphere, or a film of soap-water with glycerine in a moist atmosphere, the effect of evaporation or condensation is not to be neglected. In the first case, the diminution of the thickness of the film will be accelerated, in the second, retarded. In the case of the film containing glycerine, it should be observed that the water condensed cannot in all respects replace the fluid carried down by the internal current but that the two processes together will tend to wash out the glycerine from the film.

But when a component which greatly diminishes the tension of the film, although forming but a small fraction of its mass (therefore existing in excess at the surface), is volatile, the effect of evaporation and condensation may be considerable, even when the mean value of the potential for that component is the same in the film as in the surrounding atmosphere. To illustrate this, let us take the simple case of two components S_1 and S_2, as before. (See page 301.) It appears from equation (508) that the potentials must vary in the film with the height z, since the tension does, and from (98) that these variations must (very nearly) satisfy the relation

$$\gamma_1 \frac{d\mu_1}{dz} + \gamma_2 \frac{d\mu_2}{dz} = 0, \tag{658}$$

γ_1 and γ_2 denoting the densities of S_1 and S_2 in the interior of the film. The variation of the potential of S_2 as we pass from one level to another is therefore as much more rapid than that of S_1, as its density in the interior of the film is less. If then the resistances restraining the evaporation, transmission through the atmosphere, and condensation of the two substances are the same, these processes will go on much more rapidly with respect to S_2. It will be observed that the values of $\frac{d\mu_1}{dz}$ and $\frac{d\mu_2}{dz}$ will have opposite signs, the tendency of S_1 being to pass down through the atmosphere, and that of S_2 to pass up. Moreover, it may easily be shown that the evaporation or condensation of S_2 will produce a very much greater effect than the evaporation or condensation of the same quantity of S_1. These effects are really of the same kind. For if condensation of S_2 takes place at the top of the film, it will cause a diminution of tension, and thus occasion an extension of this part of the film, by which its thickness will be reduced, as it would be by evaporation of S_1. We may infer that it is a general condition of the persistence of liquid films, that the substance which causes the diminution of tension in the lower parts of the film must not be volatile.

But apart from any action of the atmosphere, we have seen that a film which is truly fluid in its interior is in general subject to a continual diminution of thickness by the internal currents due to gravity and the suction at its edge. Sooner or later, the interior will somewhere cease to have the properties of matter in mass. The film will then probably become unstable with respect to a flux of the interior (see page 305), the thinnest parts tending to become still more thin (apart from any external cause) very much as if there were an attraction between the surfaces of the film, insensible at greater distances, but becoming sensible when the thickness of the film is sufficiently reduced. We should expect this to determine the rupture of the film, and such is doubtless the case with most liquids. In a film of soap-water, however, the rupture does not take place, and the processes which go on can be watched.   It is apparent even to a very superficial observation that a film of which the tint is approaching the black exhibits a remarkable instability.   The continuous change of tint is interrupted by the breaking out and rapid extension of black spots. That in the formation of these black spots a separation of different substances takes place, and not simply an extension of a part of the film, is shown by the fact that the film is made thicker at the edge of these spots.

This is very distinctly seen in a plane vertical film, when a single black spot breaks out and spreads rapidly over a considerable area which was before of a nearly uniform tint approaching the black.   The edge of the black spot as it spreads is marked as it were by a string of bright beads, which unite together on touching, and thus becoming larger, glide down across the bands of color below.   Under favorable circumstances, there is often quite a shower of these bright spots. They are evidently small spots very much thicker—apparently many times thicker—than the part of the film out of which they are formed. Now if the formation of the black spots were due to a simple extension of the film, it is evident that no such appearance would be presented.   The thickening of the edge of the film cannot be accounted for by *contraction*.   For an extension of the upper portion of the film and contraction of the lower and thicker portion, with descent of the intervening portions, would be far less resisted by viscosity, and far more favored by gravity than such extensions and contractions as would produce the appearances described.   But the rapid formation of a thin spot by an internal current would cause an accumulation at the edge of the spot of the material forming the interior of the film, and necessitate a thickening of the film in that place.

That which is most difficult to account for in the formation of the black spots is the arrest of the process by which the film grows thinner.   It seems most natural to account for this, *if possible*, by passive resistance to motion due to a very viscous or gelatinous condition of the film.   For it does not seem likely that the film, after becoming unstable by the flux of matter from its interior, would become stable (without the support of such resistance) by a continuance of the same process.   On the other hand, gelatinous properties are very marked in soap-water which contains somewhat more soap than is best for the formation of films, and it is entirely natural that, even when such properties are wanting in the interior of a mass or thick film of a liquid, they may still exist in the immediate vicinity of the surface (where we know that the soap or some of its components exists in excess), or throughout a film which is so thin that the interior has ceased to have the properties of matter in mass.* But these considerations do not amount to any *a priori* probability of an arrest of the tendency toward an internal current between adjacent elements of a black spot which may differ slightly in thickness, in time to prevent rupture of the film. For, in a thick film, the increase of the tension with the extension, which is necessary for its stability with respect to extension, is connected with an excess of the soap (or of some of its components) at the surface as compared with the interior of the film. With respect to the black spots, although the interior has ceased to have the properties of matter in mass, and any quantitative determinations derived from the surfaces of a mass of the liquid will not be applicable, it is natural to account for the stability with reference to extension by supposing that the same general difference of composition still exists. If therefore we account for the arrest of internal currents by the increasing density of soap or some of its components in the interior of the film, we must still suppose that the characteristic difference of composition in the interior and surface of the film has not been obliterated.

The preceding discussion relates to liquid films between masses of gas. Similar considerations will apply to liquid films between other liquids or between a liquid and a gas, and to films of gas between masses of liquid. The latter may be formed by gently depositing a liquid drop upon the surface of a mass of the same or a different liquid. This may be done (with suitable liquids) so that the continuity of the air separating the liquid drop and mass is not broken, but a film of air is formed, which, if the liquids are similar, is a counterpart of the liquid film which is formed by a bubble of air rising to the top of a mass of the liquid. (If the bubble has the same volume as the drop, the films will have precisely the same form, as well as the rest of the surfaces which bound the bubble and the drop.) Sometimes, when the weight and momentum of the drop carry it through the surface of the mass on which it falls, it appears surrounded by a complete spherical film of air, which is the counterpart on a small scale of a soap-bubble hovering in air.† Since, however, the substance to which the necessary differences of

---

* The experiments of M. Plateau (chapter VII of the work already cited) show that this is the case to a very remarkable degree with respect to a solution of saponine. With respect to soap-water, however, they do not indicate any greater superficial viscosity than belongs to pure water. But the resistance to an internal current, such as we are considering, is not necessarily measured by the resistance to such motions as those of the experiments referred to.

† These spherical air-films are easily formed in soap-water. They are distinguishable from ordinary air-bubbles by their general behavior and by their appearance. The two concentric spherical surfaces are distinctly seen, the diameter of one appearing to be about three-quarters as large as that of the other. This is of course an optical illusion, depending upon the index of refraction of the liquid.

tension in the film are mainly due is a component of the liquid masses on each side of the air film, the necessary differences of the potential of this substance cannot be permanently maintained, and these films have little persistence compared with films of soap-water in air. In this respect, the case of these air-films is analogous to that of liquid films in an atmosphere containing substances by which their tension is greatly reduced. Compare pages 310, 311.

Surfaces of Discontinuity between Solids and Fluids.

We have hitherto treated of surfaces of discontinuity on the supposition that the contiguous masses are fluid. This is by far the most simple case for any rigorous treatment, since the masses are necessarily isotropic both in nature and in their state of strain. In this case, moreover, the mobility of the masses allows a satisfactory experimental verification of the mechanical conditions of equilibrium. On the other hand, the rigidity of solids is in general so great, that any tendency of the surfaces of discontinuity to variation in area or form may be neglected in comparison with the forces which are produced in the interior of the solids by any sensible strains, so that it is not generally necessary to take account of the surfaces of discontinuity in determining the state of strain of solid masses. But we must take account of the nature of the surfaces of discontinuity between solids and fluids with reference to the tendency toward solidification or dissolution at such surfaces, and also with reference to the tendencies of different fluids to spread over the surfaces of solids.

Let us therefore consider a surface of discontinuity between a fluid and a solid, the latter being either isotropic or of a continuous crystalline structure, and subject to any kind of stress compatible with a state of mechanical equilibrium with the fluid. We shall not exclude the case in which substances foreign to the contiguous masses are present in small quantities at the surface of discontinuity, but we shall suppose that the nature of this surface (i.e., of the non-homogeneous film between the approximately homogeneous masses) is entirely determined by the nature and state of the masses which it separates, and the quantities of the foreign substances which may be present. The notions of the *dividing surface*, and of the *superficial densities* of energy, entropy, and the several components, which we have used with respect to surfaces of discontinuity between fluids (see pages 219 and 224), will evidently apply without modification to the present case. We shall use the suffix $_1$ with reference to the substance of the solid, and shall suppose the dividing surface to be determined so as to make the superficial density of this substance vanish. The superficial densities of energy, of entropy, and of the

other component substances may then be denoted by our usual
symbols (see page 235),

$$\epsilon_{8(1)}, \quad \eta_{8(1)}, \quad \Gamma_{2(1)}, \quad \Gamma_{3(1)}, \quad \text{etc.}$$

Let the quantity σ be defined by the equation

$$\sigma = \epsilon_{8(1)} - t\eta_{8(1)} - \mu_2\Gamma_{2(1)} - \mu_3\Gamma_{3(1)} - \text{etc.}, \qquad (659)$$

in which t denotes the temperature, and μ_2, μ_3, etc. the potentials
for the substances specified at the surface of discontinuity.

As in the case of two fluid masses (see page 257), we may regard
σ as expressing the work spent in forming a unit of the surface of
discontinuity—under certain conditions, which we need not here
specify—but it cannot properly be regarded as expressing the tension
of the surface. The latter quantity depends upon the work spent in
stretching the surface, while the quantity σ depends upon the work
spent in *forming* the surface. With respect to perfectly fluid masses,
these processes are not distinguishable, unless the surface of discontinuity has components which are not found in the contiguous masses,
and even in this case (since the surface must be supposed to be formed
out of matter supplied at the same potentials which belong to the
matter in the surface) the work spent in increasing the surface
infinitesimally by stretching is identical with that which must be
spent in forming an equal infinitesimal amount of new surface. But
when one of the masses is solid, and its states of strain are to be
distinguished, there is no such equivalence between the stretching of
the surface and the forming of new surface.*

---

* This will appear more distinctly if we consider a particular case. Let us consider
a thin plane sheet of a crystal in a vacuum (which may be regarded as a limiting case
of a very attenuated fluid), and let us suppose that the two surfaces of the sheet are
alike. By applying the proper forces to the edges of the sheet, we can make all stress
vanish in its interior. The *tensions* of the two surfaces are in equilibrium with these
forces, and are measured by them. But the tensions of the surfaces, thus determined,
may evidently have different values in different directions, and are entirely different
from the quantity which we denote by σ, which represents the work required to form
a unit of the surface by any reversible process, and is not connected with any idea of
direction.

In certain cases, however, it appears probable that the values of σ and of the
superficial tension will not greatly differ. This is especially true of the numerous
bodies which, although generally (and for many purposes properly) regarded as solids,
are really very viscous fluids. Even when a body exhibits no fluid properties at its
actual temperature, if its surface has been formed at a higher temperature, at which
the body was fluid, and the change from the fluid to the solid state has been by
insensible gradations, we may suppose that the value of σ coincided with the superficial
tension until the body was decidedly solid, and that they will only differ so far as they
may be differently affected by subsequent variations of temperature and of the stresses
applied to the solid. Moreover, when an amorphous solid is in a state of equilibrium
with a solvent, although it may have no fluid properties in its interior, it seems not
improbable that the particles at its surface, which have a greater degree of mobility,
may so arrange themselves that the value of σ will coincide with the superficial tension,
as in the case of fluids.

With these preliminary notions, we now proceed to discuss the condition of equilibrium which relates to the dissolving of a solid at the surface where it meets a fluid, when the thermal and mechanical conditions of equilibrium are satisfied. It will be necessary for us to consider the case of isotropic and of crystallized bodies separately, since in the former the value of σ is independent of the direction of the surface, except so far as it may be influenced by the state of strain of the solid, while in the latter the value of σ varies greatly with the direction of the surface with respect to the axes of crystallization, and in such a manner as to have a large number of sharply defined minima.* This may be inferred from the phenomena which crystalline bodies present, as will appear more distinctly in the following discussion. Accordingly, while a variation in the direction of an element of the surface may be neglected (with respect to its effect on the value of σ) in the case of isotropic solids, it is quite otherwise with crystals. Also, while the surfaces of equilibrium between fluids and soluble isotropic solids are without discontinuities of direction, being in general curved, a crystal in a state of equilibrium with a fluid in which it can dissolve is bounded in general by a broken surface consisting of sensibly plane portions.

For isotropic solids, the conditions of equilibrium may be deduced as follows. If we suppose that the solid is unchanged, except that an infinitesimal portion is dissolved at the surface where it meets the fluid, and that the fluid is considerable in quantity and remains homogeneous, the increment of energy in the vicinity of the surface will be represented by the expression

$$\int [\epsilon_V{}' - \epsilon_V{}'' + (c_1 + c_2)\epsilon_{S(1)}]\, \delta N\, Ds$$

where Ds denotes an element of the surface, δN the variation in its position (measured normally, and regarded as *negative* when the solid is dissolved), c_1 and c_2 its principal curvatures (positive when their centers lie on the same side as the solid), $\epsilon_{S(1)}$ the surface-density of energy, $\epsilon_V{}'$ and $\epsilon_V{}''$ the volume-densities of energy in the solid and fluid respectively, and the sign of integration relates to the elements Ds. In like manner, the increments of entropy and of the quantities of the several components in the vicinity of the surface will be

$$\int [\eta_V{}' - \eta_V{}'' + (c_1 + c_2)\eta_{S(1)}]\, \delta N\, Ds,$$
$$\int [\gamma_1{}' - \gamma_1{}'']\, \delta N\, Ds,$$
$$\int [-\gamma_2{}'' + (c_1 + c_2)\Gamma_{2(1)}]\, \delta N\, Ds,$$
$$\text{etc.}$$

The entropy and the matter of different kinds representd by these

---

* The differential coefficients of σ with respect to the direction-cosines of the surface appear to be discontinuous functions of the latter quantities.

expressions we may suppose to be derived from the fluid mass. These expressions, therefore, with a change of sign, will represent the increments of entropy and of the quantities of the components in the whole space occupied by the fluid except that which is immediately contiguous to the solid. Since this space may be regarded as constant, the increment of energy in this space may be obtained (according to equation (12)) by multiplying the above expression relating to entropy by $-t$, and those relating to the components by $-\mu_1''$, $-\mu_2$, etc.,* and taking the sum. If to this we add the above expression for the increment of energy near the surface, we obtain the increment of energy for the whole system. Now by (93) we have

$$p'' = -\epsilon_V'' + t\eta_V'' + \mu_1''\gamma_1'' + \mu_2\gamma_2'' + \text{etc.}$$

By this equation and (659), our expression for the total increment of energy in the system may be reduced to the form

$$\int [\epsilon_V' - t\eta_V' - \mu_1''\gamma_1' + p'' + (c_1 + c_2)\sigma]\, \delta N\, Ds. \tag{660}$$

In order that this shall vanish for any values of δN, it is necessary that the coefficient of $\delta N\, Ds$ shall vanish. This gives for the condition of equilibrium

$$\mu_1'' = \frac{\epsilon_V' - t\eta_V' + p'' + (c_1 + c_2)\sigma}{\gamma_1'}. \tag{661}$$

This equation is identical with (387), with the exception of the term containing σ, which vanishes when the surface is plane.†

We may also observe that when the solid has no stresses except an isotropic pressure, if the quantity represented by σ is equal to the true tension of the surface, $p'' + (c_1 + c_2)\sigma$ will represent the pressure in the interior of the solid, and the second member of the equation will represent (see equation (93)) the value of the potential in the solid for the substance of which it consists. In this case, therefore, the equation reduces to

$$\mu_1'' = \mu_1',$$

that is, it expresses the equality of the potentials for the substance of

---

*The potential μ_1'' is marked by double accents in order to indicate that its value is to be determined in the fluid mass, and to distinguish it from the potential μ_1' relating to the solid mass (when this is in a state of isotropic stress), which, as we shall see, may not always have the same value. The other potentials μ_2, etc., have the same values as in (659), and consist of two classes, one of which relates to substances which are components of the fluid mass (these might be marked by the double accents), and the other relates to substances found only at the surface of discontinuity. The expressions to be multiplied by the potentials of this latter class all have the value zero.

†In equation (387), the density of the solid is denoted by Γ, which is therefore equivalent to γ_1' in (661).

the solid in the two masses—the same condition which would subsist if both masses were fluid.

Moreover, the compressibility of all solids is so small that, although σ may not represent the true tension of the surface, nor $p'' + (c_1 + c_2)\sigma$ the true pressure in the solid when its stresses are isotropic, the quantities ϵ_V' and η_V' if calculated for the pressure $p'' + (c_1 + c_2)\sigma$ with the actual temperature will have sensibly the same values as if calculated for the true pressure of the solid. Hence, the second member of equation (661), when the stresses of the solid are sensibly isotropic, is sensibly equal to the potential of the same body at the same temperature but with the pressure $p'' + (c_1 + c_2)\sigma$, and the condition of equilibrium with respect to dissolving for a solid of isotropic stresses may be expressed with sufficient accuracy by saying that the potential for the substance of the solid in the fluid must have this value. In like manner, when the solid is not in a state of isotropic stress, the difference of the two pressures in question will not sensibly affect the values of ϵ_V' and η_V', and the value of the second member of the equation may be calculated as if $p'' + (c_1 + c_2)\sigma$ represented the true pressure in the solid in the direction of the normal to the surface. Therefore, if we had taken for granted that the quantity σ represents the tension of a surface between a solid and a fluid, as it does when both masses are fluid, this assumption would not have led us into any practical error in determining the value of the potential μ_1'' which is necessary for equilibrium. On the other hand, if in the case of any amorphous body the value of σ differs notably from the true surface-tension, the latter quantity substituted for σ in (661) will make the second member of the equation equal to the true value of μ_1', when the stresses are isotropic, but this will not be equal to the value of μ_1'' in case of equilibrium, unless $c_1 + c_2 = 0$.

When the stresses in the solid are not isotropic, equation (661) may be regarded as expressing the condition of equilibrium with respect to the dissolving of the solid, and is to be distinguished from the condition of equilibrium with respect to an increase of solid matter, since the new matter would doubtless be deposited in a state of isotropic stress. (The case would of course be different with crystalline bodies, which are not considered here.) The value of μ_1'' necessary for equilibrium with respect to the formation of new matter is a little less than that necessary for equilibrium with respect to the dissolving of the solid. In regard to the actual behavior of the solid and fluid, all that the theory enables us to predict with certainty is that the solid will not dissolve if the value of the potential μ_1'' is greater than that given by the equation for the solid with its distorting stresses, and that new matter will not be formed if the value of μ_1'' is less than the same equation would give for the case of the solid with isotropic stresses.* It seems probable, however, that if the fluid in contact with the solid is not renewed, the system will generally find a state of equilibrium in which the outermost portion of the solid will be in a state of isotropic stress. If at first the solid should dissolve, this would supersaturate the fluid, perhaps until a state is reached satisfying the condition of equilibrium with the stressed solid, and then, if not before, a deposition of solid matter in a state of isotropic stress would be likely to commence and go on until the fluid is reduced to a state of equilibrium with this new solid matter.

The action of gravity will not affect the nature of the condition of equilibrium for any single point at which the fluid meets the solid, but it will cause the values of p'' and μ_1'' in (661) to vary according to the laws expressed by (612) and (617). If we suppose that the outer part of the solid is in a state of isotropic stress, which is the most important case, since it is the only one in which the equilibrium is in every sense stable, we have seen that the condition (661) is at least sensibly equivalent to this:—that the potential for the substance of the solid which would belong to the solid mass at the temperature t and the pressure $p'' + (c_1 + c_2)\sigma$ must be equal to μ_1''. Or, if we denote by (p') the pressure belonging to solid with the temperature t and the potential equal to μ_1'', the condition may be expressed in the form

$$(p') = p'' + (c_1 + c_2)\sigma. \tag{662}$$

Now if we write γ'' for the total density of the fluid, we have by (612)

$$dp'' = -g\gamma'' dz.$$

By (98)

$$d(p') = \gamma_1' d\mu_1'',$$

and by (617)

$$d\mu_1'' = -g\,dz\,;$$

whence

$$d(p') = -g\,\gamma_1' dz.$$

Accordingly we have

$$d(p') - dp'' = g(\gamma'' - \gamma_1')dz,$$

and

$$(p') - p'' = g(\gamma'' - \gamma_1')z,$$

z being measured from the horizontal plane for which $(p') = p''$. Substituting this value in (662), we obtain

$$c_1 + c_2 = \frac{g\,(\gamma'' - \gamma_1')}{\sigma}z, \tag{663}$$

---

* The possibility that the new solid matter might differ in composition from the original solid is here left out of account. This point has been discussed on pages 79–82, but without reference to the state of strain of the solid or the influence of the curvature of the surface of discontinuity. The statement made above may be generalized so as to hold true of the formation of new solid matter of any kind on the surface as follows:—that new solid matter of any kind will not be formed upon the surface (with more than insensible thickness), if the second member of (661) calculated for such new matter is greater than the potential in the fluid for such matter.

precisely as if both masses were fluid, and σ denoted the tension of their common surface, and (p') the true pressure in the mass specified. (Compare (619).)

The obstacles to an exact experimental realization of these relations are very great, principally from the want of absolute uniformity in the internal structure of amorphous solids, and on account of the passive resistances to the processes which are necessary to bring about a state satisfying the conditions of theoretical equilibrium, but it may be easy to verify the general tendency toward diminution of surface, which is implied in the foregoing equations.*

Let us apply the same method to the case in which the solid is a crystal. The surface between the solid and fluid will now consist of plane portions, the directions of which may be regarded as invariable. If the crystal grows on one side a distance δN, without other change, the increment of energy in the vicinity of the surface will be

$$(\epsilon_V{}' - \epsilon_V{}'')s\,\delta N + \Sigma'(\epsilon_{S(1)}{}'\,l'\cos\mathrm{ec}\,\omega' - \epsilon_{S(1)}{}'\,l'\cot\omega')\delta N,$$

---

*It seems probable that a tendency of this kind plays an important part in some of the phenomena which have been observed with respect to the freezing together of pieces of ice. (See especially Professor Faraday's "Note on Regelation" in the *Proceedings of the Royal Society*, vol. x, p. 440; or in the *Philosophical Magazine*, 4th ser., vol. xxi, p. 146.) Although this is a body of crystalline structure, and the action which takes place is doubtless influenced to a certain extent by the directions of the axes of crystallization, yet since the phenomena have not been observed to depend upon the orientation of the pieces of ice we may conclude that the effect, so far as its general character is concerned, is such as might take place with an isotropic body. In other words, for the purposes of a general explanation of the phenomena we may neglect the differences in the values of σ_{IW} (the suffixes are used to indicate that the symbol relates to the surface between ice and water) for different orientations of the axes of crystallization, and also neglect the influence of the surface of discontinuity with respect to crystalline structure, which must be formed by the freezing together of the two masses of ice when the axes of crystallization in the two masses are not similarly directed. In reality, this surface—or the necessity of the formation of such a surface if the pieces of ice freeze together—must exert an influence adverse to their union, measured by a quantity σ_{II}, which is determined for this surface by the same principles as when one of two contiguous masses is fluid, and varies with the orientations of the two systems of crystallographic axes relatively to each other and to the surface. But under the circumstances of the experiment, since we may neglect the possibility of the two systems of axes having precisely the same directions, this influence is probably of a tolerably constant character, and is evidently not sufficient to alter the general nature of the result. In order wholly to prevent the tendency of pieces of ice to freeze together, when meeting in water with curved surfaces and without pressure, it would be necessary that $\sigma_{II} \geqq 2\sigma_{IW}$, except so far as the case is modified by passive resistances to change, and by the inequality in the values of σ_{II} and σ_{IW} for different directions of the axes of crystallization.

It will be observed that this view of the phenomena is in harmony with the opinion of Professor Faraday. With respect to the union of pieces of ice as an indirect consequence of pressure, see page 198 of volume xi of the *Proceedings of the Royal Society*; or the *Philosophical Magazine*, 4th ser., vol. xxiii, p. 407.

where ϵ_V' and ϵ_V'' denote the volume-densities of energy in the crystal and fluid respectively, s the area of the side on which the crystal grows, $\epsilon_{S(1)}$ the surface-density of energy on that side, $\epsilon_{S(1)}'$ the surface-density of energy on an adjacent side, ω' the external angle of these two sides, l' their common edge, and the symbol Σ' a summation with respect to the different sides adjacent to the first. The increments of entropy and of the quantities of the several components will be represented by analogous formulæ, and if we deduce as on pages 316, 317 the expression for the increase of energy in the whole system due to the growth of the crystal without change of the total entropy or volume, and set this expression equal to zero, we shall obtain for the condition of equilibrium

$$(\epsilon_V' - t\eta_V' - \mu_1''\gamma_1' + p'')s\,\delta N + \Sigma'(\sigma'l'\operatorname{cosec}\omega' - \sigma l'\cot\omega')\delta N = 0, \quad (664)$$

where σ and σ' relate respectively to the same sides as $\epsilon_{S(1)}$ and $\epsilon_{S(1)}'$ in the preceding formula. This gives

$$\mu_1'' = \frac{\epsilon_V' - t\eta_V' + p''}{\gamma_1'} + \frac{\Sigma'(\sigma'l'\operatorname{cosec}\omega' - \sigma l'\cot\omega')}{s\gamma_1'}. \quad (665)$$

It will be observed that unless the side especially considered is small or narrow, we may neglect the second fraction in this equation, which will then give the same value of μ_1'' as equation (387), or as equation (661) applied to a plane surface.

Since a similar equation must hold true with respect to every other side of the crystal of which the equilibrium is not affected by meeting some other body, the condition of equilibrium for the crystalline form (when unaffected by gravity) is that the expression

$$\frac{\Sigma'(\sigma'l'\operatorname{cosec}\omega' - \sigma l'\cot\omega')}{s} \quad (666)$$

shall have the same value for each side of the crystal. (By the value of this expression for any side of the crystal is meant its value when σ and s are determined by that side and the other quantities by the surrounding sides in succession in connection with the first side.) This condition will not be affected by a change in the size of a crystal while its proportions remain the same. But the tendencies of similar crystals toward the form required by this condition, as measured by the inequalities in the composition or the temperature of the surrounding fluid which would counterbalance them, will be inversely as the linear dimensions of the crystals, as appears from the preceding equation.

If we write v for the volume of a crystal, and $\Sigma(\sigma s)$ for the sum of the areas of all its sides multiplied each by the corresponding value of σ, the numerator and denominator of the fraction (666), multiplied each by δN, may be represented by $\delta\Sigma(\sigma s)$ and δv

respectively. The value of the fraction is therefore equal to that of the differential coefficient

$$\frac{d\Sigma(\sigma s)}{dv}$$

as determined by the displacement of a particular side while the other sides are fixed. The condition of equilibrium for the form of a crystal (when the influence of gravity may be neglected) is that the value of this differential coefficient must be independent of the particular side which is supposed to be displaced. For a constant volume of the crystal, $\Sigma(\sigma s)$ has therefore a minimum value when the condition of equilibrium is satisfied, as may easily be proved more directly.

When there are no foreign substances at the surfaces of the crystal, and the surrounding fluid is indefinitely extended, the quantity $\Sigma(\sigma s)$ represents the work required to form the surfaces of the crystal, and the coefficient of $s\,\delta N$ in (664) with its sign reversed represents the work gained in forming a mass of volume unity like the crystal but regarded as without surfaces. We may denote the work required to form the crystal by

$$W_S - W_V,$$

W_S denoting the work required to form the surfaces $\{$i.e., $\Sigma(\sigma s)\}$, and W_V the work gained in forming the mass as distinguished from the surfaces. Equation (664) may then be written

$$-\delta W_V + \Sigma(\sigma\,\delta s) = 0. \tag{667}$$

Now (664) would evidently continue to hold true if the crystal were diminished in size, remaining similar to itself in form and in nature, if the values of σ in all the sides were supposed to diminish in the same ratio as the linear dimensions of the crystal. The variation of W_S would then be determined by the relation

$$dW_S = d\Sigma(\sigma s) = \tfrac{3}{2}\Sigma(\sigma\,ds),$$

and that of W_V by (667). Hence,

$$dW_S = \tfrac{3}{2}dW_V,$$

and, since W_S and W_V vanish together,

$$W_S = \tfrac{3}{2}W_V,$$

$$W_S - W_V = \tfrac{1}{3}W_S = \tfrac{1}{2}W_V, \tag{668}$$

—the same relation which we have before seen to subsist with respect to a spherical mass of fluid as well as in other cases. (See pages 257, 261, 298.)

The equilibrium of the crystal is unstable with respect to variations in size when the surrounding fluid is indefinitely extended, but it may be made stable by limiting the quantity of the fluid.

To take account of the influence of gravity, we must give to μ_1'' and p'' in (665) their average values in the side considered. These coincide (when the fluid is in a state of internal equilibrium) with their values at the center of gravity of the side. The values of γ_1', ϵ_V', η_V' may be regarded as constant, so far as the influence of gravity is concerned. Now since by (612) and (617)

$$dp'' = -g\gamma'' dz,$$

and

$$d\mu_1'' = -g\, dz,$$

we have

$$d(\gamma_1'\mu_1'' - p'') = g(\gamma'' - \gamma_1')dz.$$

Comparing (664), we see that the upper or the lower faces of the crystal will have the greater tendency to grow (other things being equal), according as the crystal is lighter or heavier than the fluid. When the densities of the two masses are equal, the effect of gravity on the form of the crystal may be neglected.

In the preceding paragraph the fluid is regarded as in a state of internal equilibrium. If we suppose the composition and temperature of the fluid to be uniform, the condition which will make the effect of gravity vanish will be that

$$\frac{d(\gamma_1'\mu_1'' - p'')}{dz} = 0,$$

when the value of the differential coefficient is determined in accordance with this supposition. This condition reduces to

$$\left(\frac{d\mu_1}{dp}\right)_{t,\,m}'' = \frac{1}{\gamma_1''},*$$

which, by equation (92), is equivalent to

$$\left(\frac{dv}{dm_1}\right)_{t,\,p,\,m}'' = \frac{1}{\gamma_1'}. \tag{669}$$

The tendency of a crystal to grow will be greater in the upper or lower parts of the fluid, according as the growth of a crystal at constant temperature and pressure will produce expansion or contraction.

Again, we may suppose the composition of the fluid and its entropy per unit of mass to be uniform. The temperature will then vary with the pressure, that is, with z. We may also suppose the temperature of different crystals or different parts of the same crystal to be determined by the fluid in contact with them. These conditions express a state which may perhaps be realized when the fluid is gently stirred. Owing to the differences of temperature we cannot regard ϵ_V' and η_V'

---

* A suffixed m is used to represent all the symbols m_1, m_2, etc., except such as may occur in the differential coefficient.

in (664) as constant, but we may regard their variations as subject to the relation $d\epsilon_V{}' = t\, d\eta_V{}'$. Therefore, if we make $\eta_V{}' = 0$ for the mean temperature of the fluid (which involves no real loss of generality), we may treat $\epsilon_V{}' - t\eta_V{}'$ as constant. We shall then have for the condition that the effect of gravity shall vanish

$$\frac{d(\gamma_1{}'\mu_1{}'' - p'')}{dz} = 0,$$

which signifies in the present case that

$$\left(\frac{d\mu_1}{dp}\right)''_{\eta,\,m} = \frac{1}{\gamma_1{}''},$$

or, by (90),

$$\left(\frac{dv}{dm_1}\right)''_{\eta,\,p,\,m} = \frac{1}{\gamma_1{}''}. \tag{670}$$

Since the entropy of the crystal is zero, this equation expresses that the dissolving of a small crystal in a considerable quantity of the fluid will produce neither expansion nor contraction, when the pressure is maintained constant and no heat is supplied or taken away.

The manner in which crystals actually grow or dissolve is often principally determined by other differences of phase in the surrounding fluid than those which have been considered in the preceding paragraph. This is especially the case when the crystal is growing or dissolving rapidly. When the great mass of the fluid is considerably supersaturated, the action of the crystal keeps the part immediately contiguous to it nearer the state of exact saturation. The farthest projecting parts of the crystal will therefore be most exposed to the action of the supersaturated fluid, and will grow most rapidly. The same parts of a crystal will dissolve most rapidly in a fluid considerably below saturation.*

But even when the fluid is supersaturated only so much as is necessary in order that the crystal shall grow at all, it is not to be expected that the form in which $\Sigma(\sigma s)$ has a minimum value (or such a modification of that form as may be due to gravity or to the influence of the body supporting the crystal) will always be the ultimate result. For we cannot imagine a body of the internal structure and external form of a crystal to grow or dissolve by an entirely continuous process, or by a process in the same sense continuous as condensation or evaporation between a liquid and gas, or the corresponding processes between an amorphous solid and a fluid. The process is rather to be regarded as periodic, and the formula (664)

---

* See O. Lehmann, "Ueber das Wachsthum der Krystalle," *Zeitschrift für Krystallographie und Mineralogie*, Bd. i, S. 453; or the review of the paper in Wiedemann's *Beiblätter*, Bd. ii, S. 1.

cannot properly represent the true value of the quantities intended unless δN is equal to the distance between two successive layers of molecules in the crystal, or a multiple of that distance. Since this can hardly be treated as an infinitesimal, we can only conclude with certainty that sensible changes cannot take place for which the expression (664) would have a positive value.*

---

*That it is necessary that certain relations shall be precisely satisfied in order that equilibrium may subsist between a liquid and gas with respect to evaporation, is explained (see Clausius, "Ueber die Art der Bewegung, welche wir Wärme nennen," *Pogg. Ann.*, Bd. c, S. 353; or *Abhand. über die mechan. Wärmetheorie*, XIV) by supposing that a passage of individual molecules from the one mass to the other is continually taking place, so that the slightest circumstance may give the preponderance to the passage of matter in either direction. The same supposition may be applied, at least in many cases, to the equilibrium between amorphous solids and fluids. Also in the case of crystals in equilibrium with fluids, there may be a passage of individual molecules from one mass to the other, so as to cause insensible fluctuations in the mass of the solid. If these fluctuations are such as to cause the occasional deposit or removal of a whole layer of particles, the least cause would be sufficient to make the probability of one kind of change prevail over that of the other, and it would be necessary for equilibrium that the theoretical conditions deduced above should be precisely satisfied. But this supposition seems quite improbable, except with respect to a very small side.

The following view of the molecular state of a crystal when in equilibrium with respect to growth or dissolution appears as probable as any. Since the molecules at the corners and edges of a perfect crystal would be less firmly held in their places than those in the middle of a side, we may suppose that when the condition of theoretical equilibrium (665) is satisfied several of the outermost layers of molecules on each side of the crystal are incomplete toward the edges. The boundaries of these imperfect layers probably fluctuate, as individual molecules attach themselves to the crystal or detach themselves, but not so that a layer is entirely removed (on any side of considerable size), to be restored again simply by the irregularities of the motions of the individual molecules. Single molecules or small groups of molecules may indeed attach themselves to the side of the crystal but they will speedily be dislodged, and if any molecules are thrown out from the middle of a surface, these deficiencies will also soon be made good; nor will the frequency of these occurrences be such as greatly to affect the general smoothness of the surfaces, except near the edges where the surfaces fall off somewhat, as before described. Now a continued growth on any side of a crystal is impossible unless new layers can be formed. This will require a value of μ_1'' which may exceed that given by equation (665) by a finite quantity. Since the difficulty in the formation of a new layer is at or near the commencement of the formation, the necessary value of μ_1'' may be independent of the area of the side, except when the side is very small. The value of μ_1'' which is necessary for the growth of the crystal will however be different for different kinds of surfaces, and probably will generally be greatest for the surfaces for which σ is least.

On the whole, it seems not improbable that the form of very minute crystals in equilibrium with solvents is principally determined by equation (665), (i.e., by the condition that $\Sigma(\sigma s)$ shall be a minimum for the volume of the crystal except so far as the case is modified by gravity or the contact of other bodies), but as they grow larger (in a solvent no more supersaturated than is necessary to make them grow at all), the deposition of new matter on the different surfaces will be determined more by the nature (orientation) of the surfaces and less by their size and relations to the surrounding surfaces. As a final result, a large crystal, thus formed, will generally be bounded by those surfaces alone on which the deposit of new matter takes place least readily, with small, perhaps insensible truncations. If one kind of surfaces

Let us now examine the special condition of equilibrium which relates to a line at which three different masses meet, when one or more of these masses is solid. If we apply the method of pages 316, 317 to a system containing such a line, it is evident that we shall obtain in the expression corresponding to (660), beside the integral relating to the surfaces, a term of the form

$$\int \Sigma(\sigma\, \delta T)\, Dl$$

to be interpreted as the similar term in (611), except so far as the definition of σ has been modified in its extension to solid masses. In order that this term shall be incapable of a negative value it is necessary that at every point of the line

$$\Sigma(\sigma\, \delta T) \geqq 0 \tag{671}$$

for any *possible* displacement of the line. Those displacements are to be regarded as possible which are not prevented by the solidity of the masses, when the interior of every solid mass is regarded as incapable of motion. At the surfaces between solid and fluid masses, the processes of solidification and dissolution will be possible in some cases, and impossible in others.

The simplest case is when two masses are fluid and the third is solid and insoluble. Let us denote the solid by S, the fluids by A and B, and the angles filled by these fluids by α and β respectively. If the surface of the solid is continuous at the line where it meets the two fluids, the condition of equilibrium reduces to

$$\sigma_{AB} \cos \alpha = \sigma_{BS} - \sigma_{AS}. \tag{672}$$

If the line where these masses meet is at an edge of the solid, the condition of equilibrium is that

$$
\left.
\begin{aligned}
\sigma_{AB} \cos \alpha &\leqq \sigma_{BS} - \sigma_{AS}, \\
\text{and} \qquad \sigma_{AB} \cos \beta &\leqq \sigma_{AS} - \sigma_{BS};
\end{aligned}
\right\} \tag{673}
$$

which reduces to the preceding when $\alpha + \beta = \pi$. Since the displacement of the line can take place by a purely mechanical process, this

---

satisfying this condition cannot form a closed figure, the crystal will be bounded by two or three kinds of surfaces determined by the same condition. The kinds of surface thus determined will probably generally be those for which σ has the least values. But the relative development of the different kinds of sides, even if unmodified by gravity or the contact of other bodies, will not be such as to make $\Sigma(\sigma s)$ a minimum. The growth of the crystal will finally be confined to sides of a single kind.

It does not appear that any part of the operation of removing a layer of molecules presents any especial difficulty so marked as that of commencing a new layer; yet the values of μ_1'' which will just allow the different stages of the process to go on must be slightly different, and therefore, for the continued dissolving of the crystal the value of μ_1'' must be less (by a finite quantity) than that given by equation (665). It seems probable that this would be especially true of those sides for which σ has the least values. The effect of dissolving a crystal (even when it is done as slowly as possible) is therefore to produce a form which probably differs from that of theoretical equilibrium in a direction opposite to that of a growing crystal.

experimental verification
esses of solidification and
a displacement of the line
f three fluids, since the
of matter are enormously
ering to the solid are not
by extensions and con-
n the case of fluid masses.
is arbitrary to a greater
in which a single foreign
is uniformly distributed,
stributed as to make the
esence of these substances
a more irregular manner.
ntinuity which meet in a
uid in which it is soluble,
le, and the particular con-
y modified. If the soluble
be treated by the method
equilibrium relating to the
ble from those relating to
of the line will involve a
tal which is terminated at
increment of energy in the
nvolving any variation in
f two parts, of which one
the system, and the other

es of discontinuity. This
ges diminishing the value

oned here. The fact that a fiber
nder water will become attached
solid body be brought into such
, the water will generally rise up
er a surface of some extent. The

, and to water, respectively. In
to the wool, if we neglect the

r, and to ice, respectively. See
, vol. xxi, p. 151.

General Relations.—For any constant state of strain of the surface of the solid we may write

$$d\epsilon_{s(1)} = t\,d\eta_{s(1)} + \mu_2\,d\Gamma_{2(1)} + \mu_3\,d\Gamma_{3(1)} + \text{etc.,} \qquad (674)$$

since this relation is implied in the definition of the quantities involved. From this and (659) we obtain

$$d\sigma = -\eta_{s(1)}\,dt - \Gamma_{2(1)}\,d\mu_2 - \Gamma_{3(1)}\,d\mu_3 - \text{etc.,} \qquad (675)$$

which is subject, in strictness, to the same limitation—that the state of strain of the surface of the solid remains the same. But this limitation may in most cases be neglected. (If the quantity σ represented the true tension of the surface, as in the case of a surface between fluids, the limitation would be wholly unnecessary.)

Another method and notation.—We have so far supposed that we have to do with a non-homogeneous film of matter between two homogeneous (or very nearly homogeneous) masses, and that the nature and state of this film is in all respects determined by the nature and state of these masses together with the quantities of the foreign substances which may be present in the film. (See page 314.) Problems relating to processes of solidification and dissolution seem hardly capable of a satisfactory solution, except on this supposition, which appears in general allowable with respect to the surfaces produced by these processes. But in considering the equilibrium of fluids at the surface of an unchangeable solid, such a limitation is neither necessary nor convenient. The following method of treating the subject will be found more simple and at the same time more general.

Let us suppose the superficial density of energy to be determined by the excess of energy in the vicinity of the surface over that which would belong to the solid, if (with the same temperature and state of strain) it were bounded by a vacuum in place of the fluid, and to the fluid, if it extended with a uniform volume-density of energy just up to the surface of the solid, or, if in any case this does not sufficiently define a surface, to a surface determined in some definite way by the exterior particles of the solid. Let us use the symbol (ϵ_s) to denote the superficial energy *thus defined*. Let us suppose a superficial density of entropy to be determined in a manner entirely analogous, and be denoted by (η_s). In like manner also, for all the components of the fluid, and for all foreign fluid substances which may be present at the surface, let the superficial densities be determined, and denoted by (Γ_2), (Γ_3), etc. These *superficial densities of the fluid components* relate solely to the matter which is fluid or movable. All matter which is immovably attached to the solid mass is to be regarded as a part of the same. Moreover, let ς be defined by the equation

$$\varsigma = (\epsilon_s) - t(\eta_s) - \mu_2(\Gamma_2) - \mu_3(\Gamma_3) - \text{etc.} \qquad (676)$$

These quantities will satisfy the following general relations :—

$$d(\epsilon_8) = t \, d(\eta_8) + \mu_2 \, d(\Gamma_2) + \mu_3 \, d(\Gamma_3) + \text{etc.,} \qquad (677)$$

$$d\varsigma = -(\eta_8)dt - (\Gamma_2)d\mu_2 - (\Gamma_3)d\mu_3 = \text{etc.} \qquad (678)$$

In strictness, these relations are subject to the same limitation as (674) and (675). But this limitation may generally be neglected. In fact, the values of ς, (ϵ_8), etc. must in general be much less affected by variations in the state of strain of the surface of the solid than those of σ, $\epsilon_{8(1)}$, etc.

The quantity ς evidently represents the tendency to contraction in that portion of the surface of the fluid which is in contact with the solid. It may be called *the superficial tension of the fluid in contact with the solid*. Its value may be either positive or negative.

It will be observed for the same solid surface and for the same temperature but for different fluids the values of σ (in all cases to which the definition of this quantity is applicable) will differ from those of ς by a constant, viz., the value of σ for the solid surface in a vacuum.

For the condition of equilibrium of two different fluids at a line on the surface of the solid, we may easily obtain

$$\sigma_{AB} \cos a = \varsigma_{BS} - \varsigma_{AS}, \qquad (679)$$

the suffixes, etc., being used as in (672), and the condition being subject to the same modification when the fluids meet at an edge of the solid.

It must also be regarded as a condition of theoretical equilibrium at the line considered (subject, like (679), to limitation on account of passive resistances to motion), that if there are any foreign substances in the surfaces A-S and B-S, the potentials for these substances shall have the same value on both sides of the line; or, if any such substance is found only on one side of the line, that the potential for that substance must not have a less value on the other side; and that the potentials for the components of the mass A, for example, must have the same values in the surface B-C as in the mass A, or, if they are not actual components of the surface B-C, a value not less than in A. Hence, we cannot determine the difference of the surface-tensions of two fluids in contact with the same solid, by bringing them together upon the surface of the solid, unless these conditions are satisfied, as well as those which are necessary to prevent the mixing of the fluid masses.

The investigation on pages 276–282 of the conditions of equilibrium for a fluid system under the influence of gravity may easily be extended to the case in which the system is bounded by or includes solid masses, when these can be treated as rigid and incapable of dissolution. The general condition of mechanical equilibrium would be of the form

$$-\int p \, \delta Dv + \int g\gamma \, \delta z \, Dv + \int \sigma \, \delta Ds + \int g\Gamma \, \delta z \, Ds$$
$$+\int g \, \delta z \, Dm + \int \varsigma \, \delta Ds + \int g(\Gamma) \delta z \, Ds = 0, \qquad (680)$$

where the first four integrals relate to the fluid masses and the surfaces which divide them, and have the same signification as in equation (606), the fifth integral relates to the movable solid masses, and the sixth and seventh to the surfaces between the solids and fluids, (Γ) denoting the sum of the quantities (Γ₂), (Γ₃), etc. It should be observed that at the surface where a fluid meets a solid δz and δz, which indicate respectively the displacements of the solid and the fluid, may have different values, but the components of these displacements which are normal to the surface must be equal.

From this equation, among other particular conditions of equilibrium, we may derive the following :—

$$d\varsigma = g(\Gamma)dz \qquad (681)$$

(compare (614)), which expresses the law governing the distribution of a thin fluid film on the surface of a solid, when there are no passive resistances to its motion.

By applying equation (680) to the case of a vertical cylindrical tube containing two different fluids, we may easily obtain the well-known theorem that the product of the perimeter of the internal surface by the difference $\varsigma' - \varsigma''$ of the superficial tensions of the upper and lower fluids in contact with the tube is equal to the excess of weight of the matter in the tube above that which would be there, if the boundary between the fluids were in the horizontal plane at which their pressures would be equal. In this theorem, we may either include or exclude the weight of a film of fluid matter adhering to the tube. The proposition is usually applied to the column of fluid *in mass* between the horizontal plane for which $p' = p''$ and the actual boundary between the two fluids. The superficial tensions ς' and ς'' are then to be measured in the vicinity of this column. But we may also include the weight of a film adhering to the internal surface of the tube. For example, in the case of water in equilibrium with its own vapor in a tube, the weight of all the water-substance in the tube above the plane $p' = p''$, diminished by that of the water-vapor which would fill the same space, is equal to the perimeter multiplied by the difference in the values of ς at the top of the tube and at the plane $p' = p''$. If the height of the tube is infinite, the value of ς at the top vanishes, and the weight of the film of water adhering to the tube and of the mass of liquid water above the plane $p' = p''$ diminished by the weight of vapor which would fill the same space is equal in numerical value but of opposite sign to the product of the perimeter of the internal surface of the tube multiplied by ς'', the superficial tension of liquid water in contact with the tube at the pressure at which the water and its vapor would be in equilibrium at a plane surface. In this sense, the total weight of water which can be supported by the tube per unit of the perimeter of its surface is directly measured by the value of $-\varsigma$ for water in contact with the tube.

Modification of the Conditions of Equilibrium by Electromotive Force.—Theory of a Perfect Electro-Chemical Apparatus.

We know by experience that in certain fluids (electrolytic conductors) there is a connection between the fluxes of the component substances and that of electricity. The quantitative relation between these fluxes may be expressed by an equation of the form

$$De = \frac{Dm_a}{a_a} + \frac{Dm_b}{a_b} + \text{etc.} - \frac{Dm_g}{a_g} - \frac{Dm_h}{a_h} - \text{etc.,} \qquad (682)$$

where De, Dm_a, etc. denote the infinitesimal quantities of electricity and of the components of the fluid which pass simultaneously through any same surface, which may be either at rest or in motion, and a_a, a_b, etc., a_g, a_h, etc. denote positive constants. We may evidently regard Dm_a, Dm_b, etc., Dm_g, Dm_h, etc. as independent of one another. For, if they were not so, one or more could be expressed in terms of the others, and we could reduce the equation to a shorter form in which all the terms of this kind would be independent.

Since the motion of the fluid as a whole will not involve any electrical current, the densities of the components specified by the suffixes must satisfy the relation

$$\frac{\gamma_a}{a_a} + \frac{\gamma_b}{a_b} + \text{etc.} = \frac{\gamma_g}{a_g} + \frac{\gamma_h}{a_h} + \text{etc.} \qquad (683)$$

These densities, therefore, are not independently variable, like the densities of the components which we have employed in other cases.

We may account for the relation (682) by supposing that electricity (positive or negative) is inseparably attached to the different kinds of molecules, so long as they remain in the interior of the fluid, in such a way that the quantities a_a, a_b, etc. of the substances specified are each charged with a unit of positive electricity, and the quantities a_g, a_h, etc. of the substances specified by these suffixes are each charged with a unit of negative electricity. The relation (683) is accounted for by the fact that the constants a_a, a_g, etc. are so small that the electrical charge of any sensible portion of the fluid varying sensibly from the law expressed in (683) would be enormously great, so that the formation of such a mass would be resisted by a very great force.

It will be observed that the choice of the substances which we regard as the components of the fluid is to some extent arbitrary, and that the same physical relations may be expressed by different equations of the form (682), in which the fluxes are expressed with reference to different sets of components. If the components chosen are such as represent what we believe to be the actual molecular constitution of the fluid, those of which the fluxes appear in the equation of the form (682) are called the *ions*, and the constants of the equation are called their *electro-chemical equivalents*. For our present purpose, which has nothing to do with any theories of molecular constitution, we may choose such a set of components as may be convenient, and call those *ions*, of which the fluxes appear in the equation of the form (682), without farther limitation.

Now, since the fluxes of the independently variable components of an electrolytic fluid do not necessitate any electrical currents, all the conditions of equilibrium which relate to the movements of these components will be the same as if the fluid were incapable of the electrolytic process. Therefore all the conditions of equilibrium which we have found without reference to electrical considerations, will apply to an electrolytic fluid and its independently variable components. But we have still to seek the remaining conditions of equilibrium, which relate to the possibility of electrolytic conduction.

For simplicity, we shall suppose that the fluid is without internal surfaces of discontinuity (and therefore homogeneous except so far as it may be slightly affected by gravity), and that it meets metallic conductors (*electrodes*) in different parts of its surface, being otherwise bounded by non-conductors. The only electrical currents which it is necessary to consider are those which enter the electrolyte at one electrode and leave it at another.

If all the conditions of equilibrium are fulfilled in a given state of the system, except those which relate to changes involving a flux of electricity, and we imagine the state of the system to be varied by the passage from one electrode to another of the quantity of electricity δe accompanied by the quantity δm_a of the component specified, without any flux of the other components or any variation in the total entropy, the total variation of energy in the system will be represented by the expression

$$(V'' - V')\delta e + (\mu_a'' - \mu_a')\delta m_a + (\Upsilon' - \Upsilon'')\delta m_a,$$

in which V', V'' denote the electrical potentials in pieces of the same kind of metal connected with the two electrodes, Υ', Υ'', the gravitational potentials at the two electrodes, and μ_a', μ_a'', the intrinsic potentials for the substance specified. The first term represents the increment of the potential energy of electricity, the second the increment of the intrinsic energy of the ponderable matter, and the third the increment of the energy due to gravitation.* But by (682)

$$\delta m_a = a_a \delta \epsilon.$$

It is therefore necessary for equilibrium that

$$V'' - V' + a_a(\mu_a'' - \mu_a' - \Upsilon'' + \Upsilon') = 0. \tag{684}$$

To extend this relation to all the electrodes we may write

$$V' + a_a(\mu_a' - \Upsilon') = V'' + a_a(\mu_a'' - \Upsilon'') = V''' + a_a(\mu_a''' - \Upsilon''') = \text{etc.} \tag{685}$$

For each of the other cations (specified by $_b$ etc.) there will be a similar condition, and for each of the anions a condition of the form

$$V' - a_g(\mu_g' - \Upsilon') = V'' - a_g(\mu_g'' - \Upsilon'') = V''' - a_g(\mu_g''' - \Upsilon''') = \text{etc.} \tag{686}$$

When the effect of gravity may be neglected, and there are but two electrodes, as in a galvanic or electrolytic cell, we have for any cation

$$V'' - V' = a_a(\mu_a' - \mu_a''), \tag{687}$$

and for any anion

$$V'' - V' = a_g(\mu_g'' - \mu_g'), \tag{688}$$

where $V'' - V'$ denotes the electromotive force of the combination. That is:—

When all the conditions of equilibrium are fulfilled in a galvanic or electrolytic cell, the electromotive force is equal to the difference in the values of the potential for any ion or apparent ion at the surfaces of the electrodes multiplied by the electro-chemical equivalent of that ion, the greater potential of an anion being at the same electrode as the greater electrical potential, and the reverse being true of a cation.

Let us apply this principle to different cases.

(I.) If the ion is an independently variable component of an electrode, or by itself constitutes an electrode, the potential for the ion (in any case of equilibrium which does not depend upon passive resistances to change) will have the same value within the electrode as on its surface, and will be determined by the composition of the electrode with its temperature and pressure. This might be illustrated by a cell with electrodes of mercury containing certain quantities of zinc in solution (or with one such electrode and the other of pure zinc) and an electrolytic fluid containing a salt of zinc, but not capable of dissolving the mercury.† We may regard

---

* It is here supposed that the gravitational potential may be regarded as constant for each electrode. When this is not the case the expression may be applied to small parts of the electrodes taken separately.

† If the electrolytic fluid dissolved the mercury as well as the zinc, equilibrium could only subsist when the electromotive force is zero, and the composition of the electrodes identical. For when the electrodes are formed of the two metals in different proportions, that which has the greater potential for zinc will have the less potential for mercury. (See equation (98).) This is inconsistent with equilibrium, according to the principle mentioned above, if both metals can act as cations.

a cell in which hydrogen acts as an ion between electrodes of palladium charged with hydrogen as another illustration of the same principle, but the solidity of the electrodes and the consequent resistance to the diffusion of the hydrogen within them (a process which cannot be assisted by convective currents as in a liquid mass) present considerable obstacles to the experimental verification of the relation.

(II.) Sometimes the ion is soluble (as an independently variable component) in the electrolytic fluid. Of course its condition in the fluid when thus dissolved must be entirely different from its condition when acting on an ion, in which case its quantity is not independently variable, as we have already seen. Its diffusion in the fluid in this state of solution is not necessarily connected with any electrical current, and in other relations its properties may be entirely changed. In any discussion of the internal properties of the fluid (with respect to its fundamental equation, for example), it would be necessary to treat it as a different substance. (See page 63.) But if the process by which the charge of electricity passes into the electrode, and the ion is dissolved in the electrolyte is *reversible*, we may evidently regard the potentials for the substance of the ion in (687) or (688) as relating to the substance thus dissolved in the electrolyte. In case of absolute equilibrium, the density of the substance thus dissolved would of course be uniform throughout the fluid (since it can move independently of any electrical current), so that by the strict application of our principle we only obtain the somewhat barren result that if any of the ions are soluble in the fluid without their electrical charges, the electromotive force must vanish in any case of absolute equilibrium not dependent upon passive resistances. Nevertheless, cases in which the ion is thus dissolved in the electrolytic fluid only to a very small extent, and its passage from one electrode to the other by ordinary diffusion is extremely slow, may be regarded as approximating to the case in which it is incapable of diffusion. In such cases, we may regard the relations (687), (688) as approximately valid, although the condition of equilibrium relating to the diffusion of the dissolved ion is not satisfied. This may be the case with hydrogen and oxygen as ions (or apparent ions) between electrodes of platinum in some of its forms.

(III.) The ion may appear in mass at the electrode. If it be a conductor of electricity, it may be regarded as forming an electrode, as soon as the deposit has become thick enough to have the properties of matter in mass. The case therefore will not be different from that first considered. When the ion is a non-conductor, a continuous thick deposit on the electrode would of course prevent the possibility of an electrical current. But the case in which the ion being a non-conductor is disengaged in masses contiguous to the electrode but not entirely covering it, is an important one. It may be illustrated by hydrogen appearing in bubbles at a cathode. In case of perfect equilibrium, independent of passive resistances, the potential of the ion in (687) or (688) may be determined in such a mass. Yet the circumstances are quite unfavorable for the establishment of perfect equilibrium, unless the ion is to some extent absorbed by the electrode or electrolytic fluid, or the electrode is fluid. For if the ion must pass *immediately* into the non-conducting mass, while the electricity passes into the electrode, it is evident that the only possible terminus of an electrolytic current is at the line where the electrode, the non-conducting mass, and the electrolytic fluid meet, so that the electrolytic process is necessarily greatly retarded, and an approximate ceasing of the current cannot be regarded as evidence that a state of approximate equilibrium has been reached. But even a slight degree of solubility of the ion in the electrolytic fluid or in the electrode may greatly diminish the resistance to the electrolytic process, and help toward producing that state of complete equilibrium which is supposed in the theorem we are discussing. And the mobility of the surface of a liquid electrode may act in the same way. When the ion is absorbed by the electrode, or by the electrolytic fluid, the case of course comes under the heads which we have already considered, yet the fact that the ion is set free in mass is important, since it is in such a mass that the determination of the value of the potential will generally be most easily made.

(IV.) When the ion is not absorbed either by the electrode or by the electrolytic fluid, and is not set free in mass, it may still be deposited on the surface of the electrode. Although this can take place only to a limited extent (without forming a body having the properties of matter in mass), yet the electro-chemical equivalents of all substances are so small that a very considerable flux of electricity may take place before the deposit will have the properties of matter in mass. Even when the ion appears in mass, or is absorbed by the electrode or electrolytic fluid, the non-homogeneous film between the electrolytic fluid and the electrode may contain an additional portion of it. Whether the ion is confined to the surface of the electrode or not, we may regard this as one of the cases in which we have to recognize a certain superficial density of substances at surfaces of discontinuity, the general theory of which we have already considered.

The deposit of the ion will affect the superficial tension of the electrode if it is liquid, or the closely related quantity which we have denoted by the same symbol σ (see pages 314–331) if the electrode is solid. The effect can of course be best observed in the case of a liquid electrode.   But whether the electrodes are liquid or solid, if the external electromotive force $V' - V''$ applied to an electrolytic combination is varied, when it is too weak to produce a lasting current, and the electrodes are thereby brought into a new state of polarization in which they make equilibrium with the altered value of the electromotive force, without change in the nature of the electrodes or of the electrolytic fluid, then by (508) or (675)

$$d\sigma' = -\Gamma_a' d\mu_a',$$
$$d\sigma'' = -\Gamma_a'' d\mu_a'';$$

and by (687),

$$d(V' - V'') = -a_a(d\mu_a' - d\mu_a'').$$

Hence

$$d(V' - V'') = \frac{a_a}{\Gamma_a'} d\sigma' - \frac{a_a}{\Gamma_a''} d\sigma''. \tag{689}$$

If we suppose that the state of polarization of only one of the electrodes is affected (as will be the case when its surface is very small compared with that of the other), we have

$$d\sigma' = \frac{\Gamma_a'}{a_a} d(V' - V''). \tag{690}$$

The superficial tension of one of the electrodes is then a function of the electromotive force.

This principle has been applied by M. Lippmann to the construction of the electrometer which bears his name.[*]   In applying equations (689) and (690) to dilute sulphuric acid between electrodes of mercury, as in a Lippmann's electrometer, we may suppose that the suffix refers to hydrogen.   It will be most convenient to suppose the *dividing surface* to be so placed as to make the surface-density of mercury zero.   (See page 234.)   The matter which exists in excess or deficiency at the surface may then be expressed by the surface-densities of sulphuric acid, of water, and of hydrogen.   The value of the last may be determined from equation (690).   According to M. Lippmann's determinations, it is negative when the surface is in its natural state (i.e., the state to which it tends when no external electromotive force is applied), since σ' increases with $V'' - V'$.   When $V'' - V'$ is equal to nine-tenths of the electromotive force of a Daniell's cell, the electrode to which V''' relates remaining in its natural state, the tension σ' of the surface of the other electrode has a maximum value, and there is no excess or deficiency of hydrogen at that surface.   This is the condition toward which a surface tends when it is extended while no flux of electricity takes place.   The flux of electricity per unit of new surface formed, which will maintain a surface in a

---

[*] See his memoir: "Relations entre les phénomènes électriques et capillaires," *Annales de Chimie et de Physique*, 5ᵉ série, t. v, p. 494.

constant condition while it is extended, is represented by $\dfrac{\Gamma_a{}'}{a_a}$ in numerical value, and its direction, when $\Gamma_a{}'$ is negative, is from the mercury into the acid.

We have so far supposed, in the main, that there are no passive resistances to change, except such as vanish with the rapidity of the processes which they resist. The actual condition of things with respect to passive resistances appears to be nearly as follows. There does not appear to be any passive resistance to the electrolytic process by which an ion is transferred from one electrode to another, except such as vanishes with the rapidity of the process. For, in any case of equilibrium, the smallest variation of the externally applied electromotive force appears to be sufficient to cause a (temporary) electrolytic current. But the case is not the same with respect to the molecular changes by which the ion passes into new combinations or relations, as when it enters into the mass of the electrodes, or separates itself in mass, or is dissolved (no longer with the properties of an ion) in the electrolytic fluid. In virtue of the passive resistance to these processes, the external electromotive force may often vary within wide limits, without creating any current by which the ion is transferred from one of the masses considered to the other. In other words, the value of $V' - V''$ may often differ greatly from that obtained from (687) or (688) when we determine the values of the potentials for the ion as in cases I, II, and III. We may, however, regard these equations as entirely valid, when the potentials for the ions are determined at the surface of the electrodes with reference to the ion in the condition in which it is brought there or taken away by an electrolytic current, without any attendant irreversible processes. But in a complete discussion of the properties of the surface of an electrode it may be necessary to distinguish (both in respect to surface-densities and to potentials) between the substance of the ion in this condition and the same substance in other conditions into which it cannot pass (directly) without irreversible processes. No such distinction, however, is necessary when the substance of the ion can pass at the surface of the electrode by reversible processes from any one of the conditions in which it appears to any other.

The formulæ (687), (688) afford as many equations as there are ions. These, however, amount to only one independent equation additional to those which relate to the independently variable components of the electrolytic fluid. This appears from the consideration that a flux of any cation may be combined with a flux of any anion in the same direction so as to involve no electrical current, and that this may be regarded as the flux of an independently variable component of the electrolytic fluid.

General Properties of a Perfect Electro-chemical Apparatus.

When an electrical current passes through a galvanic or electro-lytic cell, the state of the cell is altered. If no changes take place in the cell except during the passage of the current, and all changes which accompany the current can be reversed by reversing the current, the cell may be called a perfect electro-chemical apparatus. The electromotive force of the cell may be determined by the equations which have just been given. But some of the general relations to which such an apparatus is subject may be conveniently stated in a form in which the ions are not explicitly mentioned.

In the most general case, we may regard the cell as subject to external action of four different kinds. (1) The supply of electricity at one electrode and the withdrawal of the same quantity at the other. (2) The supply or withdrawal of a certain quantity of heat. (3) The action of gravity. (4) The motion of the surfaces enclosing the apparatus, as when its volume is increased by the liberation of gases.

The increase of the energy in the cell is necessarily equal to that which it receives from external sources. We may express this by the equation

$$d\epsilon = (V' - V'')de + dQ + dW_G + dW_P, \qquad (691)$$

in which $d\epsilon$ denotes the increment of the intrinsic energy of the cell, de the quantity of electricity which passes through it, V' and V'' the electrical potentials in masses of the same kind of metal connected with the anode and cathode respectively, dQ the heat received from external bodies, dW_G the work done by gravity, and dW_P the work done by the pressures which act on the external surface of the apparatus.

The conditions under which we suppose the processes to take place are such that the increase of the entropy of the apparatus is equal to the entropy which it receives from external sources. The only external source of entropy is the heat which is communicated to the cell by the surrounding bodies. If we write $d\eta$ for the increment of entropy in the cell, and t for the temperature, we have

$$d\eta = \frac{dQ}{t}. \qquad (692)$$

Eliminating dQ, we obtain

$$d\epsilon = (V' - V'')de + t\,d\eta + dW_G + dW_P, \qquad (693)$$

or

$$V'' - V' = -\frac{d\epsilon}{de} + t\frac{d\eta}{de} + \frac{dW_G}{de} + \frac{dW_P}{de}. \qquad (694)$$

It is worth while to notice that if we give up the condition of the reversibility of the processes, so that the cell is no longer supposed to be a perfect electro-chemical apparatus, the relation (691) will still subsist. But, if we still suppose, for simplicity, that all parts of the cell have the same temperature, which is *necessarily* the case with a perfect electro-chemical apparatus, we shall have, instead of (692),

$$d\eta \geqq \frac{dQ}{t},\qquad(695)$$

and instead of (693), (694)

$$(V'' - V')\,de \leqq -d\epsilon + t\,d\eta + dW_{\mathrm{G}} + dW_{\mathrm{P}}.\qquad(696)$$

The values of the several terms of the second member of (694) for a given cell, will vary with the external influences to which the cell is subjected. If the cell is enclosed (with the products of electrolysis) in a rigid envelop, the last term will vanish. The term relating to gravity is generally to be neglected. If no heat is supplied or withdrawn, the term containing $d\eta$ will vanish. But in the calculation of the electromotive force, which is the most important application of the equation, it is generally more convenient to suppose that the temperature remains constant.

The quantities expressed by the terms containing dQ and $d\eta$ in (691), (693), (694), and (696) are frequently neglected in the consideration of cells of which the temperature is supposed to remain constant. In other words, it is frequently assumed that neither heat nor cold is produced by the passage of an electrical current through a perfect electro-chemical combination (except that heat which may be indefinitely diminished by increasing the time in which a given quantity of electricity passes), and that only *heat* can be produced in *any* cell, unless it be by processes of a secondary nature, which are not immediately or necessarily connected with the process of electrolysis.

It does not appear that this assumption is justified by any sufficient reason. In fact, it is easy to find a case in which the electromotive force is determined entirely by the term $t\dfrac{d\eta}{de}$ in (694), all the other terms in the second member of the equation vanishing. This is true of a Grove's gas battery charged with hydrogen and nitrogen. In this case, the hydrogen passes over to the nitrogen,—a process which does not alter the energy of the cell, when maintained at a constant temperature. The work done by external pressures is evidently nothing, and that done by gravity is (or may be) nothing. Yet an electrical current is produced. The work done (or which may be done) by the current outside of the cell is the equivalent of the work (or of a part of the work) which might be gained by allowing the gases to mix in other ways. This is equal, as has been shown by

Lord Rayleigh,* to the work which may be gained by allowing each gas separately to expand at constant temperature from its initial volume to the volume occupied by the two gases together. The same work is equal, as appears from equations (278), (279) on page 156 (see also page 159), to the increase of the entropy of the system multiplied by the temperature.

It is possible to vary the construction of the cell in such a way that nitrogen or other neutral gas will not be necessary. Let the cell consist of a U-shaped tube of sufficient height, and have pure hydrogen at each pole under very unequal pressures (as of one and two atmospheres respectively) which are maintained constant by properly weighted pistons, sliding in the arms of the tube. The difference of the pressures in the gas-masses at the two electrodes must of course be balanced by the difference in the height of the two columns of acidulated water. It will hardly be doubted that such an apparatus would have an electromotive force acting in the direction of a current which would carry the hydrogen from the denser to the rarer mass. Certainly the gas could not be carried in the opposite direction by an external electromotive force without the expenditure of as much (electromotive) work as is equal to the mechanical work necessary to pump the gas from the one arm of the tube to the other. And if by any modification of the metallic electrodes (which remain unchanged by the passage of electricity) we could reduce the passive resistances to zero, so that the hydrogen could be carried reversibly from one mass to the other without finite variation of the electromotive force, the only possible value of the electromotive force would be represented by the expression $t\dfrac{d\eta}{de}$, as a very close approximation. It will be observed that although gravity plays an essential part in a cell of this kind by maintaining the difference of pressure in the masses of hydrogen, the electromotive force cannot possibly be ascribed to gravity, since the work done by gravity, when hydrogen passes from the denser to the rarer mass, is negative.

Again, it is entirely improbable that the electrical currents caused by differences in the concentration of solutions of salts (as in a cell containing sulphate of zinc between zinc electrodes, or sulphate of copper between copper electrodes, the solution of the salt being of unequal strength at the two electrodes), which have recently been investigated theoretically and experimentally by MM. Helmholtz and Moser,† are confined to cases in which the mixture of solutions of different degrees of concentration will produce heat. Yet in cases in which the mixture of more and less concentrated solutions is not

---

* *Philosophical Magazine*, vol. xlix, p. 311.

† *Annalen der Physik und Chemie*, Neue Folge, Band iii, February, 1878.

attended with evolution or absorption of heat, the electromotive force must vanish in a cell of the kind considered, if it is determined simply by the diminution of energy in the cell. And when the mixture produces cold, the same rule would make any electromotive force impossible except in the direction which would tend to increase the difference of concentration. Such conclusions would be quite irreconcilable with the theory of the phenomena given by Professor Helmholtz.

A more striking example of the necessity of taking account of the variations of entropy in the cell in *a priori* determinations of electromotive force is afforded by electrodes of zinc and mercury in a solution of sulphate of zinc. Since heat is absorbed when zinc is dissolved in mercury,* the energy of the cell is increased by a transfer of zinc to the mercury, when the temperature is maintained constant. Yet in this combination, the electromotive force acts in the direction of the current producing such a transfer.† The couple presents certain anomalies when a considerable quantity of zinc is united with the mercury. The electromotive force changes its direction, so that this case is usually cited as an illustration of the principle that the electromotive force is in the direction of the current which diminishes the energy of the cell, i.e., which produces or allows those changes which are accompanied by evolution of heat when they take place directly. But whatever may be the cause of the electromotive force which has been observed acting in the direction from the amalgam through the electrolyte to the zinc (a force which according to the determinations of M. Gaugain is only one twenty-fifth part of that which acts in the reverse direction when pure mercury takes the place of the amalgam), these anomalies can hardly affect the general conclusions with which alone we are here concerned. If the electrodes of a cell are pure zinc and an amalgam containing zinc not in excess of the amount which the mercury will dissolve at the temperature of the experiment without losing its fluidity, and if the only change (other than thermal) accompanying a current is a transfer of zinc from one electrode to the other,—conditions which may not have been satisfied in all the experiments recorded, but which it is allowable to suppose in a theoretical discussion, and which certainly will not be regarded as inconsistent with the fact that heat is absorbed when zinc is dissolved in mercury,—it is impossible that the electromotive force should be in the direction of a current transferring zinc from the amalgam to the electrode of pure zinc. For, since the zinc eliminated from the amalgam by the electrolytic process might be re-dissolved directly,

---

* J. Regnauld, *Comptes Rendus*, t. li, p. 778.

† Gaugain, *Comptes Rendus*, t. xlii, p. 430.

such a direction of the electromotive force would involve the possibility of obtaining an indefinite amount of electromotive work, and therefore of mechanical work, without other expenditure than that of heat at the constant temperature of the cell.

None of the cases which we have been considering involve combinations by definite proportions, and, except in the case of the cell with electrodes of mercury and zinc, the electromotive forces are quite small. It may perhaps be thought that with respect to those cells in which combinations take place by definite proportions the electromotive force may be calculated with substantial accuracy from the diminution of the energy, without regarding the variation of entropy. But the phenomena of chemical combination do not in general seem to indicate any possibility of obtaining from the combination of substances by any process whatever an amount of mechanical work which is equivalent to the heat produced by the direct union of the substances.

A kilogramme of hydrogen, for example, combining by combustion under the pressure of the atmosphere with eight kilogrammes of oxygen to form liquid water, yields an amount of heat which may be represented in round numbers by 34000 calories.* We may suppose that the gases are taken at the temperature of 0° C., and that the water is reduced to the same temperature. *But this heat cannot be obtained at any temperature desired.* A very high temperature has the effect of preventing to a greater or less extent, the combination of the elements. Thus, according to M. Sainte-Claire Deville,† the temperature obtained by the combustion of hydrogen and oxygen cannot much if at all exceed 2500° C., which implies that less than one-half of the hydrogen and oxygen present combine at that temperature. This relates to combustion under the pressure of the atmosphere. According to the determinations of Professor Bunsen‡ in regard to combustion in a confined space, only one-third of a mixture of hydrogen and oxygen will form a chemical compound at the temperature of 2850° C. and a pressure of ten atmospheres, and only a little more than one-half when the temperature is reduced by the addition of nitrogen to 2024° C., and the pressure to about three atmospheres *exclusive* of the part due to the nitrogen.

Now 10 calories at 2500° C. are to be regarded as reversibly convertible into one calorie at 4° C. together with the mechanical work representing the energy of 9 calories. If, therefore, all the 34000 calories obtainable from the union of hydrogen and oxygen under atmospheric pressure could be obtained at the temperature of

---

* See Rühlmann's *Handbuch der mechanischen Wärmetheorie*, Bd. ii, p. 290.

† *Comptes Rendus*, t. lvi, p. 199; and t. lxiv, 67.

‡ *Pogg. Ann.*, Bd. cxxxi (1867), p. 161.

2500° C., and no higher, we should estimate the electromotive work performed in a perfect electro-chemical apparatus in which these elements are combined or separated at ordinary temperatures and under atmospheric pressure as representing nine-tenths of the 34000 calories, and the heat evolved or absorbed in the apparatus as representing one-tenth of the 34000 calories.* This, of course, would give an electromotive force exactly nine-tenths as great as is obtained on the supposition that all the 34000 calories are convertible into electromotive or mechanical work. But, according to all indications, the estimate 2500° C. (for the temperature at which we may regard all the heat of combustion as obtainable) is far too high,† and we must regard the theoretical value of the electromotive force necessary to electrolyze water as considerably less than nine-tenths of the value obtained on the supposition that it is necessary for the electromotive agent to supply all the energy necessary for the process.

The case is essentially the same with respect to the electrolysis of hydrochloric acid, which is probably a more typical example of the process than the electrolysis of water. The phenomenon of dissociation is equally marked, and occurs at a much lower temperature, more than half of the gas being dissociated at 1400° C.‡ And the heat which is obtained by the combination of hydrochloric acid gas with water, especially with water which already contains a considerable quantity of the acid, is probably only to be obtained at temperatures comparatively low. This indicates that the theoretical value of the electromotive force necessary to electrolyze this acid (i.e., the electromotive force which would be necessary in a reversible electro-chemical apparatus) must be very much less than that which could perform in electromotive work the equivalent of all the heat evolved in the combination of hydrogen, chlorine and water to form the liquid submitted to electrolysis. This presumption, based upon

---

* These numbers are not subject to correction for the pressure of the atmosphere, since the 34000 calories relate to combustion under the same pressure.

† Unless the received ideas concerning the behavior of gases at high temperatures are quite erroneous, it is possible to indicate the general character of a process (involving at most only such difficulties as are neglected in theoretical discussions) by which water may be converted into separate masses of hydrogen and oxygen without other expenditure than that of an amount of heat equal to the difference of energy of the matter in the two states and supplied at a temperature far below 2500° C. The essential parts of the process would be (1) vaporizing the water and heating it to a temperature at which a considerable part will be dissociated, (2) the partial separation of the hydrogen and oxygen by filtration, and (3) the cooling of both gaseous masses until the vapor they contain is condensed. A little calculation will show that in a continuous process all the heat obtained in the operation of cooling the products of filtration could be utilized in heating fresh water.

‡ Sainte-Claire Deville, *Comptes Rendus*, t. lxiv, p. 67.

the phenomena exhibited in the direct combination of the substances, is corroborated by the experiments of M. Favre, who has observed an absorption of heat in the cell in which this acid was electrolyzed.[*] The electromotive work expended must therefore have been less than the increase of energy in the cell.

In both cases of composition in definite proportions which we have considered, the compound has more entropy than its elements, and the difference is by no means inconsiderable. This appears to be the rule rather than the exception with respect to compounds which have less energy than their elements. Yet it would be rash to assert that it is an invariable rule. And when one substance is substituted for another in a compound, we may expect great diversity in the relations of energy and entropy.

In some cases there is a striking correspondence between the electromotive force of a cell and the rate of diminution of its energy per unit of electricity transmitted, the temperature remaining constant. A Daniell's cell is a notable example of this correspondence. It may perhaps be regarded as a very significant case, since of all cells in common use, it has the most constant electromotive force, and most nearly approaches the condition of reversibility. If we apply our previous notation (compare (691)) with the substitution of finite for infinitesimal differences to the determinations of M. Favre,[†] estimating energy in calories, we have for each equivalent (32·6 kilogrammes) of zinc dissolved

$$(V'' - V')\Delta e = 24327^{\text{cal.}}, \quad \Delta \epsilon = -25394^{\text{cal.}}, \quad \Delta Q = -1067^{\text{cal.}}.$$

It will be observed that the electromotive work performed by the cell is about four per cent. less than the diminution of energy in the cell.[‡] The value of ΔQ, which, when negative, represents the heat evolved in the cell when the external resistance of the circuit is very great, was determined by direct measurement, and does not appear to have been corrected for the resistance of the cell. This correction would diminish the value of $-\Delta Q$, and increase that of $(V'' - V')\Delta e$, which was obtained by subtracting $-\Delta Q$ from $-\Delta \epsilon$.

It appears that under certain conditions neither heat nor cold is produced in a Grove's cell. For M. Favre has found that with different degrees of concentration of the nitric acid sometimes heat

---

* See *Mémoires des Savants Étrangers*, sér. 2, t. xxv, no. 1, p. 142 ; or *Comptes Rendus*, t. lxxiii, p. 973. The figures obtained by M. Favre will be given hereafter, in connection with others of the same nature.

† See *Mém. Savants Étrang.*, loc. cit., p. 90 ; or *Comptes Rendus*, t. lxix, p. 35, where the numbers are slightly different.

‡ A comparison of the experiments of different physicists has in some cases given a much closer correspondence. See Wiedemann's *Galvanismus*, etc., 2te Auflage, Bd. ii, §§ 1117, 1118.

and sometimes cold is produced.* When neither is produced, of course the electromotive force of the cell is exactly equal to its diminution of energy per unit of electricity transmitted. But such a coincidence is far less significant than the fact that an absorption of heat has been observed. With acid containing about seven equivalents of water $(HNO_6 + 7HO)$ $[HNO_3 + 3\tfrac{1}{2}H_2O]$, M. Favre has found

$$(V'' - V')\Delta e = 46781^{cal.}, \quad \Delta\epsilon = -41824^{cal.}, \quad \Delta Q = 4957^{cal.};$$

and with acid containing about one equivalent of water

$$(HNO_6 + HO)\ [HNO_3 + \tfrac{1}{2}H_2O],$$

$$(V'' - V')\Delta e = 49847^{cal.}, \quad \Delta\epsilon = -52714^{cal.}, \quad \Delta Q = -2867^{cal.}.$$

In the first example, it will be observed that the quantity of heat absorbed in the cell is not small, and that the electromotive force is nearly one-eighth greater than can be accounted for by the diminution of energy in the cell.

This absorption of heat in the cell he has observed in other cases, in which the chemical processes are much more simple.

For electrodes of cadmium and platinum in hydrochloric acid his experiments give†

$$(V'' - V')\Delta e = 9256^{cal.}, \qquad \Delta\epsilon = -8258^{cal.},$$
$$\Delta W_P = -290^{cal.}, \qquad \Delta Q = 1288^{cal.}.$$

In this case the electromotive force is nearly one-sixth greater than can be accounted for by the diminution of energy in the cell with the work done against the pressure of the atmosphere.

For electrodes of zinc and platinum in the same acid one series of experiments gives ‡

$$(V'' - V')\Delta e = 16950^{cal.}, \qquad \Delta\epsilon = -16189^{cal.},$$
$$\Delta W_P = -290^{cal.}, \qquad \Delta Q = 1051^{cal.};$$

and a later series,§

$$(V'' - V')\Delta e = 16738^{cal.}, \qquad \Delta\epsilon = -17702^{cal.},$$
$$\Delta W_P = -290^{cal.}, \qquad \Delta Q = -674^{cal.}.$$

In the electrolysis of hydrochloric acid in a cell with a porous partition, he has found ‖

$$(V' - V'')\Delta e = 34825^{cal.} \qquad \Delta Q = 2113^{cal.},$$

---

* *Mém Savants Étrang.*, loc. cit., p. 93; or *Comptes Rendus*, t. lxix, p. 37, and t. lxxiii, p. 893.

† *Comptes Rendus*, t. lxviii, p. 1305. The total heat obtained in the whole circuit (including the cell) when all the electromotive work is turned into heat, was ascertained by direct experiment. This quantity, 7968 calories, is evidently represented by $(V'' - V')\Delta e - \Delta Q$, also by $-\Delta\epsilon + \Delta W_P$. (See (691).) The value of $(V'' - V')\Delta e$ is obtained by adding ΔQ, and that of $-\Delta\epsilon$ by adding $-\Delta W_P$, which is easily estimated, being determined by the evolution of one kilogramme of hydrogen.

‡ *Ibid.*

§ *Mém. Savants Étrang.*, loc. cit., p. 145.

‖ *Ibid.*, p. 142.

whence

$$\Delta\epsilon - \Delta W_P = 36938.$$

We cannot assign a precise value to ΔW_P, since the quantity of chlorine which was evolved in the form of gas is not stated. But the value of $-\Delta W_P$ must lie between $290^{cal.}$ and $580^{cal.}$, probably nearer to the former.

The great difference in the results of the two series of experiments relating to electrodes of zinc and platinum in hydrochloric acid is most naturally explained by supposing some difference in the conditions of the experiment, as in the concentration of the acid, or in the extent to which the substitution of zinc for hydrogen took place.* That which it is important for us to observe in all these cases is that there are conditions under which heat is absorbed in a galvanic or electrolytic cell, so that the galvanic cell has a greater electromotive force than can be accounted for by the diminution of its energy, and the operation of electrolysis requires a less electromotive force than would be calculated from the increase of energy in the cell,—especially when the work done against the pressure of the atmosphere is taken into account.

It should be noticed that in all these experiments the quantity represented by ΔQ (which is the critical quantity with respect to the point at issue) was determined by direct measurement of the heat absorbed or evolved by the cell when placed alone in a calorimeter. The resistance of the circuit was made so great by a rheostat placed outside of the calorimeter that the resistance of the cell was regarded as insignificant in comparison, and no correction appears to have been made in any case for this resistance. With exception of the error due to this circumstance, which would in all cases diminish the heat absorbed in the cell (or increase the heat evolved), the probable error of ΔQ must be very small in comparison with that of $(V' - V'')\Delta e$, or with that of $\Delta\epsilon$, which were in general determined by the comparison of different calorimetrical measurements, involving very much greater quantities of heat.

In considering the numbers which have been cited, we should remember that when hydrogen is evolved as gas the process is in general very far from reversible. In a perfect electrochemical

---

*It should perhaps be stated that in his extended memoir published in 1877 in the *Mémoires des Savants Étrangers*, in which he has presumably collected those results of his experiments which he regards as most important and most accurate, M. Favre does not mention the absorption of heat in a cell of this kind, or in the similar cell in which cadmium takes the place of zinc. This may be taken to indicate a decided preference for the later experiments which showed an evolution of heat. Whatever the ground of this preference may have been, it can hardly destroy the significance of the absorption of heat, which was a matter of direct observation in repeated experiments. See *Comptes Rendus*, t. lxviii, p. 1305.

apparatus, the same changes in the cell would yield a much greater amount of electromotive work, or absorb a much less amount. In either case, the value of ΔQ would be much greater than in the imperfect apparatus, the difference being measured perhaps by thousands of calories.*

It often occurs in a galvanic or electrolytic cell that an ion which is set free at one of the electrodes appears in part as gas, and is in part absorbed by the electrolytic fluid, and in part absorbed by the electrode. In such cases, a slight variation in the circumstances, which would not sensibly affect the electromotive force, would cause all of the ion to be disposed of in one of the three ways mentioned, if the current were sufficiently weak. This would make a considerable difference in the variation of energy in the cell, and the electromotive force cannot certainly be calculated from the variation of energy alone in all these cases. The correction due to the work performed against the pressure of the atmosphere when the ion is set free as gas will not help us in reconciling these differences. It will appear on consideration that this correction will in general increase the discordance in the values of the electromotive force. Nor does it distinctly appear which of these cases is to be regarded

---

*Except in the case of the Grove's cell, in which the reactions are quite complicated, the absorption of heat is most marked in the electrolysis of hydrochloric acid. The latter case is interesting, since the experiments confirm the presumption afforded by the behavior of the substances in other circumstances. (See page 343.) In addition to the circumstances mentioned above tending to diminish the observed absorption of heat, the following, which are peculiar to this case, should be noticed.

The electrolysis was performed in a cell with a porous partition, in order to prevent the chlorine and hydrogen dissolved in the liquid from coming in contact with each other. It had appeared in a previous series of experiments (*Mém. Savants Étrang.*, loc. cit., p. 131 ; or *Comptes Rendus*, t. lxvi, p. 1231), that a very considerable amount of heat might be produced by the chemical union of the gases in solution. In a cell without partition, instead of an absorption, an evolution of heat took place, which sometimes exceeded 5000 calories. If, therefore, the partition did not perfectly perform its office, this could only cause a diminution in the value of ΔQ.

A large part at least of the chlorine appears to have been absorbed by the electrolytic fluid. It is probable that a slight difference in the circumstances of the experiment—a diminution of pressure, for example,—might have caused the greater part of the chlorine to be evolved as gas, without essentially affecting the electromotive force. The solution of chlorine in water presents some anomalies, and may be attended with complex reactions, but it appears to be always attended with a very considerable evolution of heat. (See Berthelot, *Comptes Rendus*, t. lxxvi, p. 1514.) If we regard the evolution of the chlorine in the form of gas as the normal process, we may suppose that the absorption of heat in the cell was greatly diminished by the retention of the chlorine in solution.

Under certain circumstances, oxygen is evolved in the electrolysis of dilute hydrochloric acid. It does not appear that this took place to any considerable extent in the experiments which we are considering. But so far as it may have occurred, we may regard it as a case of the electrolysis of water. The significance of the fact of the absorption of heat is not thereby affected.

as normal and which are to be rejected as involving secondary processes.*

If in any case secondary processes are excluded, we should expect it to be when the ion is identical in substance with the electrode upon which it is deposited, or from which it passes into the electrolyte. But even in this case we do not escape the difficulty of the different forms in which the substance may appear. If the temperature of the experiment is at the melting point of a metal which forms the ion and the electrode, a slight variation of temperature will cause the ion to be deposited in the solid or in the liquid state, or, if the current is in the opposite direction, to be taken up from a solid or from a liquid body. Since this will make a considerable difference in the variation of energy, we obtain different values for the electromotive force above and below the melting point of the metal, unless we also take account of the variations of entropy. Experiment does not indicate the existence of any such difference,† and when we take account of variations of entropy, as in equation (694), it is apparent that there ought not to be any, the terms $\dfrac{d\epsilon}{de}$ and $t\dfrac{d\eta}{de}$ being both affected by the same difference, viz., the heat of fusion of an electrochemical equivalent of the metal. In fact, if such a difference existed, it would be easy to devise arrangements by which the heat yielded by a metal in passing from the liquid to the solid state could be transformed into electromotive work (and therefore into mechanical work) without other expenditure.

The foregoing examples will be sufficient, it is believed, to show the necessity of regarding other considerations in determining the electromotive force of a galvanic or electrolytic cell than the variation of its energy alone (when its temperature is supposed to remain constant), or corrected only for the work which may be done by external

---

* It will be observed that in using the formulæ (694) and (696) we do not have to make any distinction between *primary* and *secondary* processes. The only limitation to the generality of these formulæ depends upon the *reversibility* of the processes, and this limitation does not apply to (696).

† M. Raoult has experimented with a galvanic element having an electrode of bismuth in contact with phosphoric acid containing phosphate of bismuth in solution. (See *Comptes Rendus*, t. lxviii, p. 643.) Since this metal absorbs in melting 12·64 calories per kilogramme or 885 calories per equivalent (70$^{\text{kil.}}$), while a Daniell's cell yields about 24000 calories of electromotive work per equivalent of metal, the solid or liquid state of the bismuth ought to make a difference of electromotive force represented by ·037 of a Daniell's cell, if the electromotive force depended simply upon the energy of the cell. But in M. Raoult's experiments no sudden change of electromotive force was manifested at the moment when the bismuth changed its state of aggregation. In fact, a change of temperature in the electrode from about fifteen degrees above to about fifteen degrees below the temperature of fusion only occasioned a variation of electromotive force equal to ·002 of a Daniell's cell.

Experiments upon lead and tin gave similar results.

pressures or by gravity. But the relations expressed by (693), (694), and (696) may be put in a briefer form.

If we set, as on page 89,

$$\psi = \epsilon - t\eta,$$

we have, for any constant temperature,

$$d\psi = d\epsilon - t\,d\eta;$$

and for any perfect electro-chemical apparatus, the temperature of which is maintained constant,

$$V'' - V' = -\frac{d\psi}{de} + \frac{dW_G}{de} + \frac{dW_P}{de};\qquad(697)$$

and for any cell whatever, when the temperature is maintained uniform and constant,

$$(V'' - V')de \leqq -d\psi + dW_G + dW_P.\qquad(698)$$

In a cell of any ordinary dimensions, the work done by gravity, as well as the inequalities of pressure in different parts of the cell may be neglected. If the pressure as well as the temperature is maintained uniform and constant, and we set, as on page 91,

$$\zeta = \epsilon - t\eta + pv,$$

where p denotes the pressure in the cell, and v its total volume (including the products of electrolysis), we have

$$d\zeta = d\epsilon - t\,d\eta + p\,dv,$$

and for a perfect electro-chemical apparatus,

$$V'' - V' = -\frac{d\zeta}{de},\qquad(699)$$

or for any cell,

$$(V'' - V')de \leqq -d\zeta.\qquad(700)$$

[SYNOPSIS.

SYNOPSIS OF SUBJECTS TREATED.

IV.

ON THE EQUILIBRIUM OF HETEROGENEOUS SUBSTANCES.

ABSTRACT OF THE PRECEDING PAPER BY THE AUTHOR.

[*American Journal of Science*, 3 ser., vol. XVI., pp. 441–458, Dec., 1878.]

IT is an inference naturally suggested by the general increase of entropy which accompanies the changes occurring in any isolated material system that when the entropy of the system has reached a maximum, the system will be in a state of equilibrium. Although this principle has by no means escaped the attention of physicists, its importance does not appear to have been duly appreciated. Little has been done to develop the principle as a foundation for the general theory of thermodynamic equilibrium.

The principle may be formulated as follows, constituting a criterion of equilibrium :—

I. *For the equilibrium of any isolated system it is necessary and sufficient that in all possible variations of the state of the system which do not alter its energy, the variation of its entropy shall either vanish or be negative.*

The following form, which is easily shown to be equivalent to the preceding, is often more convenient in application :—

II. *For the equilibrium of any isolated system it is necessary and sufficient that in all possible variations of the state of the system which do not alter its entropy, the variation of its energy shall either vanish or be positive.*

If we denote the energy and entropy of the system by ϵ and η respectively, the criterion of equilibrium may be expressed by either of the formulæ

$$(\delta\eta)_\epsilon \leqq 0, \tag{1}$$

$$(\delta\epsilon)_\eta \geqq 0. \tag{2}$$

Again, if we assume that the temperature of the system is uniform, and denote its absolute temperature by t, and set

$$\psi = \epsilon - t\eta, \tag{3}$$

the remaining conditions of equilibrium may be expressed by the formula

$$(\delta\psi)_t \geqq 0, \tag{4}$$

the suffixed letter, as in the preceding cases, indicating that the quantity which it represents is constant. This condition, in connection with that of uniform temperature, may be shown to be equivalent to (1) or (2). The difference of the values of ψ for two different states of the system which have the same temperature represents the work which would be expended in bringing the system from one state to the other by a reversible process and without change of temperature.

If the system is incapable of thermal changes, like the systems considered in theoretical mechanics, we may regard the entropy as having the constant value zero. Conditions (2) and (4) may then be written

$$\delta\epsilon \geqq 0, \qquad \delta\psi \geqq 0,$$

and are obviously identical in signification, since in this case $\psi = \epsilon$.

Conditions (2) and (4), as criteria of equilibrium, may therefore both be regarded as extensions of the criterion employed in ordinary statics to the more general case of a thermodynamic system. In fact, each of the quantities $-\epsilon$ and $-\psi$ (relating to a system without sensible motion) may be regarded as a kind of force-function for the system,—the former as the force-function *for constant entropy* (i.e., when only such states of the system are considered as have the same entropy), and the latter as the force-function *for constant temperature* (i.e., when only such states of the system are considered as have the same uniform temperature).

In the deduction of the particular conditions of equilibrium for any system, the general formula (4) has an evident advantage over (1) or (2) with respect to the brevity of the processes of reduction, since the limitation of constant temperature applies to every part of the system taken separately, and diminishes by one the number of independent variations in the state of these parts which we have to consider. Moreover, the transition from the systems considered in ordinary mechanics to thermodynamic systems is most naturally made by this formula, since it has always been customary to apply the principles of theoretical mechanics to real systems on the supposition (more or less distinctly conceived and expressed) that the temperature of the system remains constant, the mechanical properties of a thermodynamic system maintained at a constant temperature being such as might be imagined to belong to a purely mechanical system, and admitting of representation by a force-function, as follows directly from the fundamental laws of thermodynamics.

Notwithstanding these considerations, the author has preferred in general to use condition (2) as the criterion of equilibrium, believing that it would be useful to exhibit the conditions of equilibrium of thermodynamic systems in connection with those quantities which are most simple and most general in their definitions, and which appear most important in the general theory of such systems. The slightly different form in which the subject would develop itself, if condition (4) had been chosen as a point of departure instead of (2), is occasionally indicated.

Equilibrium of masses in contact.—The first problem to which the criterion is applied is the determination of the conditions of equilibrium for different masses in contact, when uninfluenced by gravity, electricity, distortion of the solid masses, or capillary tensions. The statement of the result is facilitated by the following definition.

If to any homogeneous mass in a state of hydrostatic stress we suppose an infinitesimal quantity of any substance to be added, the mass remaining homogeneous and its entropy and volume remaining unchanged, the increase of the energy of the mass divided by the quantity of the substance added is the *potential* for that substance in the mass considered.

In addition to equality of temperature and pressure in the masses in contact, it is necessary for equilibrium that the potential for every substance which is an independently variable component of any of the different masses shall have the same value in all of which it is such a component, so far as they are in contact with one another. But if a substance, without being an actual component of a certain mass in the given state of the system, is capable of being absorbed by it, it is sufficient if the value of the potential for that substance in that mass is not less than in any contiguous mass of which the substance is an actual component. We may regard these conditions as sufficient for equilibrium with respect to infinitesimal variations in the composition and thermodynamic state of the different masses in contact. There are certain other conditions which relate to the possible formation of masses entirely different in composition or state from any initially existing. These conditions are best regarded as determining the stability of the system, and will be mentioned under that head.

Anything which restricts the free movement of the component substances, or of the masses as such, may diminish the number of conditions which are necessary for equilibrium.

Equilibrium of osmotic forces.—If we suppose two fluid masses to be separated by a diaphragm which is permeable to some of the component substances and not to others, of the conditions of equilibrium which have just been mentioned, those will still subsist which relate to temperature and the potentials for the substances to which the diaphragm is permeable, but those relating to the potentials for the substances to which the diaphragm is impermeable will no longer be necessary. Whether the pressure must be the same in the two fluids will depend upon the rigidity of the diaphragm. Even when the diaphragm is permeable to all the components without restriction, equality of pressure in the two fluids is not always necessary for equilibrium.

Effect of gravity.—In a system subject to the action of gravity, the potential for each substance, instead of having a uniform value throughout the system, so far as the substance actually occurs as an independently variable component, will decrease uniformly with increasing height, the difference of its values at different levels being equal to the difference of level multiplied by the force of gravity.

Fundamental equations.—Let ϵ, η, v, t and p denote respectively the energy, entropy, volume, (absolute) temperature, and pressure of a homogeneous mass, which may be either fluid or solid, provided that it is subject only to hydrostatic pressures, and let m_1, m_2, ... m_n denote the quantities of its independently variable components, and μ_1, μ_2, ... μ_n the potentials for these components. It is easily shown that ϵ is a function of η, v, m_1, m_2, ... m_n, and that the complete value of $d\epsilon$ is given by the equation

$$d\epsilon = t\,d\eta - p\,dv + \mu_1 dm_1 + \mu_2 dm_2 \ldots + \mu_n dm_n. \tag{5}$$

Now if ϵ is known in terms of η, v, m_1, ... m_n, we can obtain by differentiation t, p, μ_1, ... μ_n in terms of the same variables. This will make $n+3$ independent known relations between the $2n+5$ variables, ϵ, η, v, m_1, m_2, ... m_n, t, p, μ_1, μ_2, ... μ_n. These are all that exist, for of these variables, $n+2$ are evidently independent. Now upon these relations depend a very large class of the properties of the compound considered,—we may say in general, all its thermal, mechanical, and chemical properties, so far as *active tendencies* are concerned, in cases in which the form of the mass does not require consideration. A single equation from which all these relations may be deduced may be called a fundamental equation. An equation between ϵ, η, v, m_1, m_2, ... m_n is a fundamental equation. But there are other equations which possess the same property.

If we suppose the quantity ψ to be determined for such a mass as we are considering by equation (3), we may obtain by differentiation and comparison with (5)

$$d\psi = -\eta\,dt - p\,dv + \mu_1 dm_1 + \mu_2 dm_2 \ldots + \mu_n dm_n. \tag{6}$$

If, then, ψ is known as a function of t, v, m_1, m_2, ... m_n, we can find η, p, μ_1, μ_2, ... μ_n in terms of the same variables. If we then substitute for ψ in our original equation its value taken from equation (3) we shall have again $n+3$ independent relations between the same $2n+5$ variables as before.

Let

$$\zeta = \epsilon - t\eta + pv, \tag{7}$$

then, by (5),

$$d\zeta = -\eta\,dt + v\,dp + \mu_1 dm_1 + \mu_2 dm_2 \ldots + \mu_n dm_n. \tag{8}$$

If, then, ζ is known as a function of $t, p, m_1, m_2, \ldots m_n$, we can find $\eta, v, \mu_1, \mu_2, \ldots \mu_n$ in terms of the same variables. By eliminating ζ, we may obtain again $n+3$ independent relations between the same $2n+5$ variables as at first.[*]

If we integrate (5), (6) and (8), supposing the quantity of the compound substance considered to vary from zero to any finite value, its nature and state remaining unchanged, we obtain

$$\epsilon = t\eta - pv + \mu_1 m_1 + \mu_2 m_2 \ldots + \mu_n m_n, \tag{9}$$

$$\psi = -pv + \mu_1 m_1 + \mu_2 m_2 \ldots, + \mu_n m_n, \tag{10}$$

$$\zeta = \mu_1 m_1 + \mu_2 m_2 \ldots + \mu_n m_n. \tag{11}$$

If we differentiate (9) in the most general manner, and compare the result with (5), we obtain

$$-v\,dp + \eta\,dt + m_1 d\mu_1 + m_2 d\mu_2 \ldots + m_n d\mu_n = 0, \tag{12}$$

or

$$dp = \frac{\eta}{v}dt + \frac{m_1}{v}d\mu_1 + \frac{m_2}{v}d\mu_2 \ldots + \frac{m_n}{v}d\mu_n = 0. \tag{13}$$

Hence, there is a relation between the $n+2$ quantities $t, p, \mu_1, \mu_2, \ldots \mu_n$, which, if known, will enable us to find in terms of these quantities all the ratios of the $n+2$ quantities $\eta, v, m_1, m_2, \ldots m_n$. With (9), this will make $n+3$ independent relations between the same $2n+5$ variables as at first.

Any equation, therefore, between the quantities

	$\epsilon,$	$\eta,$	$v,$	$m_1,$	$m_2, \ldots m_n,$
or	$\psi,$	$t,$	$v,$	$m_1,$	$m_2, \ldots m_n,$
or	$\zeta,$	$t,$	$p,$	$m_1,$	$m_2, \ldots m_n,$
or		$t,$	$p,$	$\mu_1,$	$\mu_2, \ldots \mu_n,$

is a fundamental equation, and any such is entirely equivalent to any other.

Coexistent phases.—In considering the different homogeneous bodies which can be formed out of any set of component substances, it is convenient to have a term which shall refer solely to the composition

---

[*] The properties of the quantities $-\psi$ and $-\zeta$ regarded as functions of the temperature and volume, and temperature and pressure, respectively, the composition of the body being regarded as invariable, have been discussed by M. Massieu in a memoir entitled "Sur les fonctions caractéristiques des divers fluides et sur la théorie des vapeurs" (*Mém. Savants Étrang.*, t. xxii). A brief sketch of his method in a form slightly different from that ultimately adopted is given in *Comptes Rendus*, t. lxix (1869), pp. 858 and 1057, and a report on his memoir by M. Bertrand in *Comptes Rendus*, t. lxxi, p. 257. M. Massieu appears to have been the first to solve the problem of representing all the properties of a body of invariable composition which are concerned in reversible processes by means of a single function.

and thermodynamic state of any such body without regard to its size or form. The word *phase* has been chosen for this purpose. Such bodies as differ in composition or state are called different phases of the matter considered, all bodies which differ only in size and form being regarded as different examples of the same phase. Phases which can exist together, the dividing surfaces being plane, in an equilibrium which does not depend upon passive resistances to change, are called *coexistent*.

The number of independent variations of which a system of co-existent phases is capable is $n+2-r$, where r denotes the number of phases, and n the number of independently variable components in the whole system. For the system of phases is completely specified by the temperature, the pressure, and the n potentials, and between these $n+2$ quantities there are r independent relations (one for each phase), which characterize the system of phases.

When the number of phases exceeds the number of components by unity, the system is capable of a single variation of phase. The pressure and all the potentials may be regarded as functions of the temperature. The determination of these functions depends upon the elimination of the proper quantities from the fundamental equations in p, t, μ_1, μ_2, etc. for the several members of the system. But without a knowledge of these fundamental equations, the values of the differential coefficients such as $\dfrac{dp}{dt}$ may be expressed in terms of the entropies and volumes of the different bodies and the quantities of their several components. For this end we have only to eliminate the differentials of the potentials from the different equations of the form (12) relating to the different bodies. In the simplest case, when there is but one component, we obtain the well-known formula

$$\frac{dp}{dt} = \frac{\eta' - \eta''}{v' - v''} = \frac{Q}{t(v'' - v')},$$

in which v', v'', η', η'' denote the volumes and entropies of a given quantity of the substance in the two phases, and Q the heat which it absorbs in passing from one phase to the other.

It is easily shown that if the temperature of two coexistent phases of two components is maintained constant, the pressure is in general a maximum or minimum when the composition of the phases is identical. In like manner, if the pressure of the phases is maintained constant, the temperature is in general a maximum or minimum when the composition of the phases is identical. The series of simultaneous values of t and p for which the composition of two coexistent phases is identical separates those simultaneous values of t and p for which no coexistent phases are possible from those for which there are two pairs of coexistent phases.

If the temperature of three coexistent phases of three components is maintained constant, the pressure is in general a maximum or minimum when the composition of one of the phases is such as can be produced by combining the other two. If the pressure is maintained constant, the temperature is in general a maximum or minimum when the same condition in regard to the composition of the phases is fulfilled.

Stability of fluids.—A criterion of the stability of a homogeneous fluid, or of a system of coexistent fluid phases, is afforded by the expression

$$\epsilon - t'\eta + p'v - \mu_1'm_1 - \mu_2'm_2 \ldots - \mu_n'm_n, \tag{14}$$

in which the values of the accented letters are to be determined by the phase or system of phases of which the stability is in question, and the values of the unaccented letters by any other phase of the same components, the possible formation of which is in question. We may call the former constants, and the latter variables. Now if the value of the expression, thus determined, is always positive for any possible values of the variables, the phase or system of phases will be stable with respect to the formation of any new phases of its components. But if the expression is capable of a negative value, the phase or system is at least *practically* unstable. By this is meant that, although, strictly speaking, an infinitely small disturbance or change may not be sufficient to destroy the equilibrium, yet a very small change in the initial state will be sufficient to do so. The presence of a small portion of matter in a phase for which the above expression has a negative value will in general be sufficient to produce this result. In the case of a system of phases, it is of course supposed that their contiguity is such that the formation of the new phase does not involve any transportation of matter through finite distances.

The preceding criterion affords a convenient point of departure in the discussion of the stability of homogeneous fluids. Of the other forms in which the criterion may be expressed, the following is perhaps the most useful :—

If the pressure of a fluid is greater than that of any other phase of its independent variable components which has the same temperature and potentials, the fluid is stable with respect to the formation of any other phase of these components; but if its pressure is not as great as that of some such phase, it will be practically unstable.

Stability of fluids with respect to continuous changes of phase.— In considering the changes which may take place in any mass, we have often to distinguish between infinitesimal changes in existing phases, and the formation of entirely new phases. A phase of a fluid may be stable with respect to the former kind of change, and unstable with respect to the latter. In this case, it may be capable of continued existence in virtue of properties which prevent the commencement of discontinuous changes. But a phase which is unstable with respect to continuous changes is evidently incapable of permanent existence on a large scale except in consequence of passive resistances to change. To obtain the conditions of stability with respect to continuous changes, we have only to limit the application of the variables in (14) to phases adjacent to the given phase. We obtain results of the following nature.

The stability of any phase with respect to continuous changes depends upon the same conditions with respect to the second and higher differential coefficients of the density of energy regarded as a function of the density of entropy and the densities of the several components, which would make the density of energy a minimum, if the necessary conditions with respect to the first differential coefficients were fulfilled.

Again, it is necessary and sufficient for the stability with respect to continuous changes of all the phases within any given limits, that within those limits the same conditions should be fulfilled with respect to the second and higher differential coefficients of the pressure regarded as a function of the temperature and the several potentials, which would make the pressure a minimum, if the necessary conditions with respect to the first differential coefficients were fulfilled.

The equation of the limits of stability with respect to continuous changes may be written

$$\left(\frac{d\mu_n}{d\gamma_n}\right)_{t,\,\mu_1,\,\ldots\,\mu_{n-1}} = 0, \text{ or } \left(\frac{d^2p}{d\mu_n{}^2}\right)_{t,\,\mu_1,\,\ldots\,\mu_{n-1}} = \infty, \qquad (15)$$

where γ_n denotes the density of the component specified or $m_n \div v$. It is in general immaterial to what component the suffix $_n$ is regarded as relating.

Critical phases.—The variations of two coexistent phases are sometimes limited by the vanishing of the difference between them. Phases at which this occurs are called *critical phases*. A critical phase, like any other, is capable of $n+1$ independent variations, n denoting the number of independently variable components. But when subject to the condition of remaining a critical phase, it is capable of only $n-1$ independent variations. There are therefore two independent equations which characterize critical phases. These may be written

$$\left(\frac{d\mu_n}{d\gamma_n}\right)_{t,\,\mu_1,\,\ldots\,\mu_{n-1}} = 0, \quad \left(\frac{d^2\mu_n}{d\gamma_n{}^2}\right)_{t,\,\mu_1,\,\ldots\,\mu_{n-1}} = 0. \qquad (16)$$

It will be observed that the first of these equations is identical with the equation of the limit of stability with respect to continuous

changes. In fact, stable critical phases are situated at that lim'
They are also situated at the limit of stability with respect to discontinuous changes. These limits are in general distinct, but touch
each other at critical phases.

Geometrical illustrations.—In an earlier paper,* the author has
described a method of representing the thermodynamic properties
of substances of invariable composition by means of surfaces. The
volume, entropy, and energy of a constant quantity of the substance
are represented by rectangular coordinates. This method corresponds
to the first kind of fundamental equation described above. Any
other kind of fundamental equation for a substance of invariable
composition will suggest an analogous geometrical method. In the
present paper, the method in which the coordinates represent temperature, pressure, and the potential, is briefly considered. But
when the composition of the body is variable, the fundamental
equation cannot be completely represented by any surface or finite
number of surfaces. In the case of three components, if we regard
the temperature and pressure as constant, as well as the total quantity
of matter, the relations between ζ, m_1, m_2, m_3 may be represented
by a surface in which the distances of a point from the three sides
of a triangular prism represent the quantities m_1, m_2, m_3, and the
distance of the point from the base of the prism represents the
quantity ζ. In the case of two components, analogous relations may
be represented by a plane curve. Such methods are especially useful
for illustrating the combinations and separations of the components,
and the changes in states of aggregation, which take place when the
substances are exposed in varying proportions to the temperature
and pressure considered.

Fundamental equations of ideal gases and gas-mixtures.—From
the physical properties which we attribute to ideal gases, it is easy
to deduce their fundamental equations. The fundamental equation
in ϵ, η, v, and m for an ideal gas is

$$c \log \frac{\epsilon - Em}{cm} = \frac{\eta}{m} - H + a \log \frac{m}{v}; \qquad (17)$$

that in ψ, t, v, and m is

$$\psi = Em + mt\left(c - H - c \log t + a \log \frac{m}{v}\right); \qquad (18)$$

that in p, t, and μ is

$$p = ae^{\frac{H-c-a}{a}} t^{\frac{c+a}{a}} e^{\frac{\mu - E}{at}}, \qquad (19)$$

where e denotes the base of the Naperian system of logarithms. As
for the other constants, c denotes the specific heat of the gas at

---

* [Page 33 of this volume.]

constant volume, a denotes the constant value of $pv \div mt$, E and H depend upon the zeros of energy and entropy. The two last equations may be abbreviated by the use of different constants. The properties of fundamental equations mentioned above may easily be verified in each case by differentiation.

The law of Dalton respecting a mixture of different gases affords a point of departure for the discussion of such mixtures and the establishment of their fundamental equations. It is found convenient to give the law the following form :—

The pressure in a mixture of different gases is equal to the sum of the pressures of the different gases as existing each by itself at the same temperature and with the same value of its potential.

A mixture of ideal gases which satisfies this law is called an *ideal gas-mixture.* Its fundamental equation in p, t, μ_1, μ_2, etc. is evidently of the form

$$p = \Sigma_1 \left(a_1 e^{\frac{H_1 - c_1 - a_1}{a_1}} \, t^{\frac{c_1 + a_1}{a_1}} \, e^{\frac{\mu_1 - E_1}{a_1 t}} \right), \tag{20}$$

where Σ_1 denotes summation with respect to the different components of the mixture. From this may be deduced other fundamental equations for ideal gas-mixtures. That in ψ, t, v, m_1, m_2, etc. is

$$\psi = \Sigma_1 \left(E_1 m_1 + m_1 t \left(c_1 - H_1 - c_1 \log t + a_1 \log \frac{m_1}{v} \right) \right). \tag{21}$$

Phases of dissipated energy of ideal gas-mixtures.—When the proximate components of a gas-mixture are so related that some of them can be formed out of others, although not necessarily in the gas-mixture itself at the temperatures considered, there are certain phases of the gas-mixture which deserve especial attention. These are the *phases of dissipated energy,* i.e., those phases in which the energy of the mass has the least value consistent with its entropy and volume. An atmosphere of such a phase could not furnish a source of mechanical power to any machine or chemical engine working within it, as other phases of the same matter might do. Nor can such phases be affected by any catalytic agent. A *perfect catalytic agent* would reduce any other phase of the gas-mixture to a phase of dissipated energy. The condition which will make the energy a minimum is that the potentials for the proximate components shall satisfy an equation similar to that which expresses the relation between the units of weight of these components. For example, if the components were hydrogen, oxygen and water, since one gram of hydrogen with eight grams of oxygen are chemically equivalent to nine grams of water, the potentials for these substances in a phase of dissipated energy must satisfy the relation

$$\mu_H + 8\mu_O = 9\mu_W.$$

Gas-mixtures with convertible components.—The theory of the phases of dissipated energy of an ideal gas-mixture derives an especial interest from its possible application to the case of those gas-mixtures in which the chemical composition and resolution of the components can take place in the gas-mixture itself, and actually does take place, so that the quantities of the proximate components are entirely determined by the quantities of a smaller number of ultimate components, with the temperature and pressure. These may be called *gas-mixtures with convertible components.* If the general laws of *ideal* gas-mixtures apply in any such case, it may easily be shown that the phases of dissipated energy are the only phases which can exist. We can form a fundamental equation which shall relate solely to these phases. For this end, we first form the equation in p, t, μ_1, μ_2, etc. for the gas-mixture, regarding its proximate components as *not* convertible. This equation will contain a potential for every proximate component of the gas-mixture. We then eliminate one (or more) of these potentials by means of the relations which exist between them in virtue of the convertibility of the components to which they relate, leaving the potentials which relate to those substances which naturally express the ultimate composition of the gas-mixture.

The validity of the results thus obtained depends upon the applicability of the laws of ideal gas-mixtures to cases in which chemical action takes place. Some of these laws are generally regarded as capable of such application, others are not so regarded. But it may be shown that in the very important case in which the components of a gas are convertible at certain temperatures, and not at others, the theory proposed may be established without other assumptions than such as are generally admitted.

It is, however, only by experiments upon gas-mixtures with convertible components, that the validity of any theory concerning them can be satisfactorily established.

The vapor of the peroxide of nitrogen appears to be a mixture of two different vapors, of one of which the molecular formula is double that of the other. If we suppose that the vapor conforms to the laws of an ideal gas-mixture in a state of dissipated energy, we may obtain an equation between the temperature, pressure, and density of the vapor, which exhibits a somewhat striking agreement with the results of experiment.

Equilibrium of stressed solids.—The second part of the paper[*] commences with a discussion of the conditions of internal and external equilibrium for solids in contact with fluids with regard to all possible states of strain of the solids. These conditions are deduced by

---

[*] [See footnote, p. 184.]

analytical processes from the general condition of equilibrium (2). The condition of equilibrium which relates to the dissolving of the solid at a surface where it meets a fluid may be expressed by the equation

$$\mu_1 = \frac{\epsilon - t\eta + pv}{m},\tag{22}$$

where ϵ, η, v, and m_1 denote respectively the energy, entropy, volume, and mass of the solid, if it is homogeneous in nature and state of strain,—otherwise, of any small portion which may be treated as thus homogeneous,—μ_1 the potential in the fluid for the substance of which the solid consists, p the pressure in the fluid and therefore one of the principal pressures in the solid, and t the temperature. It will be observed that when the pressure in the solid is isotropic, the second member of this equation will represent the potential in the solid for the substance of which it consists {see (9)}, and the condition reduces to the equality of the potential in the two masses, just as if it were a case of two fluids. But if the stresses in the solid are not isotropic, the value of the second member of the equation is not entirely determined by the nature and state of the solid, but has in general three different values (for the same solid at the same temperature, and in the same state of strain) corresponding to the three principal pressures in the solid. If a solid in the form of a right parallelopiped is subject to different pressures on its three pairs of opposite sides by fluids in which it is soluble, it is in general necessary for equilibrium that the composition of the fluids shall be different.

The *fundamental equations* which have been described above are limited, in their application to solids, to the case in which the stresses in the solid are isotropic. An example of a more general form of fundamental equation for a solid, is afforded by an equation between the energy and entropy of a given quantity of the solid, and the quantities which express its state of strain, or by an equation between ψ {see (3)} as determined for a given quantity of the solid, the temperature, and the quantities which express the state of strain.

Capillarity.—The solution of the problems which precede may be regarded as a first approximation, in which the peculiar state of thermodynamic equilibrium about the surfaces of discontinuity is neglected. To take account of the condition of things at these surfaces, the following method is used. Let us suppose that two homogeneous fluid masses are separated by a surface of discontinuity, i.e., by a very thin non-homogeneous film. Now we may imagine a state of things in which each of the homogeneous masses extends without variation of the densities of its several components, or of the densities of energy and entropy, quite up to a geometrical surface (to be called the dividing surface) at which the masses meet. We may suppose this surface to be sensibly coincident with the physical surface of discontinuity. Now if we compare the actual state of things with the supposed state, there will be in the former in the vicinity of the surface a certain (positive or negative) excess of energy, of entropy, and of each of the component substances. These quantities are denoted by ϵ^s, η^s, m_1^s, m_2^s, etc., and are treated as belonging to the surface. The ^{s} is used simply as a distinguishing mark, and must not be taken for an algebraic exponent.

It is shown that the conditions of equilibrium already obtained relating to the temperature and the potentials of the homogeneous masses, are not affected by the surfaces of discontinuity, and that the complete value of $\delta \epsilon^s$ is given by the equation

$$\delta \epsilon^s = t\, \delta \eta^s + \sigma\, \delta s + \mu_1 \delta m_1^s + \mu_2 \delta m_2^s + \text{etc.}, \qquad (23)$$

in which s denotes the area of the surface considered, t the temperature, μ_1, μ_2, etc., the potentials for the various components in the adjacent masses. It may be, however, that some of the components are found only at the surface of discontinuity, in which case the letter μ with the suffix relating to such a substance denotes, as the equation shows, the rate of increase of energy at the surface per unit of the substance added, when the entropy, the area of the surface, and the quantities of the other components are unchanged. The quantity σ we may regard as defined by the equation itself, or by the following, which is obtained by integration :—

$$\epsilon^s = t\eta^s + \sigma s + \mu_1 m_1^s + \mu_2 m_2^s + \text{etc.} \qquad (24)$$

There are terms relating to variations of the curvatures of the surface which might be added, but it is shown that we can give the dividing surface such a position as to make these terms vanish, and it is found convenient to regard its position as thus determined. It is always sensibly coincident with the physical surface of discontinuity. (Yet in treating of plane surfaces, this supposition in regard to the position of the dividing surface is unnecessary, and it is sometimes convenient to suppose that its position is determined by other considerations.)

With the aid of (23), the remaining condition of equilibrium for contiguous homogeneous masses is found, viz.,

$$\sigma(c_1 + c_2) = p' - p'', \qquad (25)$$

where p', p'' denote the pressures in the two masses, and c_1, c_2 the principal curvatures of the surface. Since this equation has the same form as if a tension equal to σ resided at the surface, the quantity σ is called (as is usual) the *superficial tension*, and the dividing surface in the particular position above mentioned is called the *surface of tension*.

By differentiation of (24) and comparison with (23), we obtain

$$d\sigma = -\eta_s dt - \Gamma_1 d\mu_1 - \Gamma_2 d\mu_2 - \text{etc.}, \qquad (26)$$

where η_8, Γ_1, Γ_2, etc. are written for $\dfrac{\eta^8}{8}$, $\dfrac{m_1^8}{8}$, $\dfrac{m_2^8}{8}$, etc., and denote the superficial densities of entropy and of the various substances. We may regard σ as a function of t, μ_1, μ_2, etc., from which if known η_8, Γ_1, Γ_2, etc. may be determined in terms of the same variables. An equation between σ, t, μ_1, μ_2, etc. may therefore be called a *fundamental equation for the surface of discontinuity*. The same may be said of an equation between ϵ^8, η^8, s, m_1^8, m_2^8, etc.

It is necessary for the stability of a surface of discontinuity that its tension shall be as small as that of any other surface which can exist between the same homogeneous masses with the same temperature and potentials. Besides this condition, which relates to the nature of the surface of discontinuity, there are other conditions of stability, which relate to the possible motion of such surfaces. One of these is that the tension shall be positive. The others are of a less simple nature, depending upon the extent and form of the surface of discontinuity, and in general upon the whole system of which it is a part. The most simple case of a system with a surface of discontinuity is that of two coexistent phases separated by a spherical surface, the outer mass being of indefinite extent. When the interior mass and the surface of discontinuity are formed entirely of substances which are components of the surrounding mass, the equilibrium is always unstable; in other cases, the equilibrium may be stable. Thus, the equilibrium of a drop of water in an atmosphere of vapor is unstable, but may be made stable by the addition of a little salt. The analytical conditions which determine the stability or instability of the system are easily found, when the temperature and potentials of the system are regarded as known, as well as the fundamental equations for the interior mass and the surface of discontinuity.

The study of surfaces of discontinuity throws considerable light upon the subject of the stability of such phases of fluids as have a less pressure than other phases of the same components with the same temperature and potentials. Let the pressure of the phase of which the stability is in question be denoted by p', and that of the other phase of the same temperature and potentials by p''. A spherical mass of the second phase and of a radius determined by the equation

$$2\,\sigma = (p'' - p')r, \qquad (27)$$

would be in equilibrium with a surrounding mass of the first phase. This equilibrium, as we have just seen, is unstable, when the surrounding mass is indefinitely extended. A spherical mass a little larger would tend to increase indefinitely. The work required to form such a spherical mass, by a reversible process, in the interior of an infinite mass of the other phase, is given by the equation

$$W = \sigma s - (p'' - p')v''. \qquad (28)$$

The term σs represents the work spent in forming the surface, and the term $(p''-p')v''$ the work gained in forming the interior mass. The second of these quantities is always equal to two-thirds of the first. The value of W is therefore positive, and the phase is in strictness stable, the quantity W affording a kind of measure of its stability. We may easily express the value of W in a form which does not involve any geometrical magnitudes, viz.,

$$W = \frac{16\pi\sigma^3}{3(p''-p')^2}, \tag{29}$$

where p'', p' and σ may be regarded as functions of the temperature and potentials. It will be seen that the stability, thus measured, is infinite for an infinitesimal difference of pressures, but decreases very rapidly as the difference of pressures increases. These conclusions are all, however, practically limited to the case in which the value of r, as determined by equation (27), is of sensible magnitude.

With respect to the somewhat similar problem of the stability of the surface of contact of two phases with respect to the formation of a new phase, the following results are obtained. Let the phases (supposed to have the same temperature and potentials) be denoted by A, B, and C; their pressures by p_A, p_B and p_C; and the tensions of the three possible surfaces by σ_{AB}, σ_{BC}, σ_{AC}. If p_C is less than

$$\frac{\sigma_{BC}\, p_A + \sigma_{AC}\, p_B}{\sigma_{BC} + \sigma_{AC}},$$

there will be no tendency toward the formation of the new phase at the surface between A and B. If the temperature or potentials are now varied until p_C is equal to the above expression, there are two cases to be distinguished. The tension σ_{AB} will be either equal to $\sigma_{AC} + \sigma_{BC}$ or less. (A greater value could only relate to an unstable and therefore unusual surface.) If $\sigma_{AB} = \sigma_{AC} + \sigma_{BC}$, a farther variation of the temperature or potentials, making p_C greater than the above expression, would cause the phase C to be formed at the surface between A and B. But if $\sigma_{AB} < \sigma_{AC} + \sigma_{BC}$, the surface between A and B would remain stable, but with rapidly diminishing stability, after p_C has passed the limit mentioned.

The conditions of stability for a line where several surfaces of discontinuity meet, with respect to the possible formation of a new surface, are capable of a very simple expression. If the surfaces A-B, B-C, C-D, D-A, separating the masses A, B, C, D, meet along a line, it is necessary for equilibrium that their tensions and directions at any point of the line should be such that a quadrilateral α, β, γ, δ may be formed with sides representing in direction and length the normals and tensions of the successive surfaces. For the stability of the system with reference to the possible formation of surfaces between A and C, or between B and D, it is farther necessary that the tensions σ_{AC} and σ_{BD} should be greater than the diagonals $\alpha\gamma$ and $\beta\delta$ respectively. The conditions of stability are entirely analogous in the case of a greater number of surfaces. For the conditions of stability relating to the formation of a new phase at a line in which three surfaces of discontinuity meet, or at a point where four different phases meet, the reader is referred to the original paper.

Liquid films.—When a fluid exists in the form of a very thin film between other fluids, the great inequality of its extension in different directions will give rise to certain peculiar properties, even when its thickness is sufficient for its interior to have the properties of matter in mass. The most important case is where the film is liquid and the contiguous fluids are gaseous. If we imagine the film to be divided into elements of the same order of magnitude as its thickness, each element extending through the film from side to side, it is evident that far less time will in general be required for the attainment of approximate equilibrium between the different parts of any such element and the contiguous gases than for the attainment of equilibrium between all the different elements of the film.

There will accordingly be a time, commencing shortly after the formation of the film, in which its separate elements may be regarded as satisfying the conditions of internal equilibrium, and of equilibrium with the contiguous gases, while they may not satisfy all the conditions of equilibrium with each other. It is when the changes due to this want of complete equilibrium take place so slowly that the film appears to be at rest, except so far as it accommodates itself to any change in the external conditions to which it is subjected, that the characteristic properties of the film are most striking and most sharply defined. It is from this point of view that these bodies are discussed. They are regarded as satisfying a certain well-defined class of conditions of equilibrium, but as not satisfying at all certain other conditions which would be necessary for complete equilibrium, in consequence of which they are subject to gradual changes, which ultimately determine their rupture.

The elasticity of a film (i.e., the increase of its tension when extended) is easily accounted for. It follows from the general relations given above that when a film has more than one component, those components which diminish the tension will be found in greater proportion on the surfaces. When the film is extended, there will not be enough of these substances to keep up the same volume- and surface-densities as before, and the deficiency will cause a certain increase of tension. It does not follow that a thinner film has always a greater tension than a thicker formed of the same liquid. When the phases within the films as well as without are the same, and the surfaces of the films are also the same, there will be no difference of tension. Nor will the tension of the same film be altered, if a part of the interior drains away in the course of time, without affecting the surfaces. If the thickness of the film is reduced by evaporation, its tension may be either increased or diminished, according to the relative volatility of its different components.

Let us now suppose that the thickness of the film is reduced until the limit is reached at which the interior ceases to have the properties of matter in mass. The elasticity of the film, which determines its stability with respect to extension and contraction, does not vanish at this limit. But a certain kind of instability will generally arise, in virtue of which inequalities in the thickness of the film will tend to increase through currents in the interior of the film. This probably leads to the destruction of the film, in the case of most liquids. In a film of soap-water, the kind of instability described seems to be manifested in the breaking out of the black spots. But the sudden diminution in thickness which takes place in parts of the film is arrested by some unknown cause, possibly by viscous or gelatinous properties, so that the rupture of the film does not necessarily follow.

Electromotive force.—The conditions of equilibrium may be modified by electromotive force. Of such cases a galvanic or electrolytic cell may be regarded as the type. With respect to the potentials for the ions and the electrical potential the following relation may be noticed:—

When all the conditions of equilibrium are fulfilled in a galvanic or electrolytic cell, the electromotive force is equal to the difference in the values of the potential for any ion at the surfaces of the electrodes multiplied by the electro-chemical equivalent of that ion, the greater potential of an anion being at the same electrode as the greater electrical potential, and the reverse being true of a cation.

The relation which exists between the electromotive force of a *perfect electro-chemical apparatus* (i.e., a galvanic or electrolytic cell which satisfies the condition of reversibility), and the changes in the cell which accompany the passage of electricity, may be expressed by the equation

$$d\epsilon = (V' - V'')de + t\, d\eta + dW_G + dW_P, \qquad (30)$$

in which $d\epsilon$ denotes the increment of the intrinsic energy in the apparatus, $d\eta$ the increment of entropy, de the quantity of electricity which passes through it, V' and V'' the electrical potentials in pieces of the same kind of metal connected with the anode and cathode respectively, dW_G the work done by gravity, and dW_P the work done by the pressures which act on the external surface of the apparatus. The term dW_G may generally be neglected. The same is true of dW_P, when gases are not concerned. If no heat is supplied or withdrawn the term $t\,d\eta$ will vanish. But in the calculation of electromotive forces, which is the most important application of the equation, it is convenient and customary to suppose that the temperature is maintained constant. Now this term $t\,d\eta$, which represents the heat absorbed by the cell, is frequently neglected in the consideration of cells of which the temperature is supposed to remain constant. In other words, it is frequently assumed that neither heat or cold is produced by the passage of an electrical current through a perfect electro-chemical apparatus (except that heat which may be indefinitely diminished by increasing the time in which a given quantity of electricity passes), unless it be by processes of a secondary nature, which are not immediately or necessarily connected with the process of electrolysis.

That this assumption is incorrect is shown by the electromotive force of a gas battery charged with hydrogen and nitrogen, by the currents caused by differences in the concentration of the electrolyte, by electrodes of zinc and mercury in a solution of sulphate of zinc, by *a priori* considerations based on the phenomena exhibited in the direct combination of the elements of water or of hydrochloric acid, by the absorption of heat which M. Favre has in many cases observed in a galvanic or electrolytic cell, and by the fact that the solid or liquid state of an electrode (at its temperature of fusion) does not affect the electromotive force.

V.

ON THE VAPOR-DENSITIES OF PEROXIDE OF NITROGEN, FORMIC ACID, ACETIC ACID, AND PERCHLORIDE OF PHOSPHORUS.

[*American Journal of Science*, ser. 3, vol. XVIII, Oct.–Nov. 1879.]

THE relation between temperature, pressure, and volume, for the vapor of each of these substances differs widely from that expressed by the usual laws for the gaseous state,—the laws known most widely by the names of Mariotte, Gay-Lussac, and Avogadro. The *density* of each vapor, in the sense in which the term is usually employed in chemical treatises, i.e., its density taken relatively to air of the same temperature and pressure,* has not a constant value, but varies nearly in the ratio of one to two. And these variations are exhibited at pressures not exceeding that of the atmosphere and at temperatures comprised between zero and 200° or 300° of the centigrade scale.

Such anomalies have been explained by the supposition that the vapor consists of a mixture of two or three different kinds of gas or vapor, which have different densities. Thus it is supposed that the vapor of peroxide of nitrogen is a gas-mixture, the components of which are represented (in the newer chemical notation) by NO_2 and N_2O_4 respectively. The densities corresponding to these formulæ are 1·589 and 3·178. The density of the mixture should have a value intermediate between these numbers, which is substantially the case with the actual vapor. The case is similar with respect to the vapor of formic acid, which we may regard as a mixture of CH_2O_2 (density 1·589) and $C_2H_4O_4$ (density 3·178), and the vapor of acetic acid, which we may regard as a mixture of $C_2H_4O_2$ (density 2·073) and $C_4H_8O_4$ (density 4·146). In the case of perchloride of phosphorus, we must suppose the vapor to consist of three parts; PCl_5 (the proper perchloride, density 7·20), PCl_3 (the protochloride, density 4·98), and Cl_2 (chlorine, density 2·22). Since the chlorine and protochloride arise from the decomposition of the perchloride, there must be as many molecules of the type Cl_2 as of the type PCl_3. Now a gas-mixture containing an equal number

---

* The language of this paper will be conformed to this usage.

of molecules of PCl_3 and Cl_2 will have the density $\frac{1}{2}(4.98+2.22)$ or 3.60. It follows that, at least so far as the range of the possible values of its density is concerned, we may regard the vapor as a mixture in variable proportions of two kinds of gas having the densities 7.20 and 3.60 respectively. The observed values of the density accord with this supposition.

These hypotheses respecting the constitution of the vapors are corroborated, in the case of peroxide of nitrogen and perchloride of phosphorus, by other circumstances. The varying color of the first vapor may be accounted for by supposing that the molecules of the type N_2O_4 are colorless, while each molecule of the type NO_2 has a constant color. This supposition affords a simple relation between the density of the vapor and the depth of its color, which has been verified by experiment.[*]

The vapor of the perchloride of phosphorus shows with increasing temperature in an increasing degree the characteristic color of chlorine. The amount of the color appears to be such as is required by the hypothesis respecting the constitution of the vapor on the very probable supposition that the perchloride proper is colorless, but the case hardly admits of such exact numerical determinations as are possible with respect to the peroxide of nitrogen.[†] But since the products of dissociation are in this case dissimilar, they may be partially separated by diffusion through a neutral gas, the lighter chlorine diffusing more rapidly than the heavier protochloride. The fact of dissociation has in this way been proved by direct experiment.[‡]

In the case of acetic and formic acids, we have no other evidence than the variations of the densities in support of the hypothesis of the compound nature of the vapor, yet if these variations shall appear to follow the same law as those of the peroxide of nitrogen and the perchloride of phosphorus, it will be difficult to refer them to a different cause.

But however it may be with these acids, the peroxide of nitrogen and the perchloride of phosphorus evidently furnish us with the means of studying the laws of chemical equilibrium in gas-mixtures in which chemical change is possible and does in fact take place reversibly, with varying conditions of temperature and pressure. Or, if from any considerations we can deduce a general law

---

[*] Salet, "Sur la coloration du peroxyde d'azote," *Comptes Rendus*, t. lxvii, p. 488.

[†] H. Sainte-Claire Deville, "Sur les densités de vapeur," *Comptes Rendus*, t. lxii, p. 1157.

[‡] Wanklyn and Robinson, "On Diffusion of Vapours: a means of distinguishing between apparent and real Vapour-densities of Chemical Compounds," *Proc. Roy. Soc.*, vol. xii, p. 507.

determining the proportions of the component gases necessary for the equilibrium of such a mixture under any given conditions, these substances afford an appropriate test for such a law.

In a former paper* by the present writer, equations were proposed to express the relation between the temperature, the pressure or the volume, and the quantities of the components in such a gas-mixture as we are considering—a *gas-mixture of convertible components* in the language of that paper. Applied to the vapor of the peroxide of nitrogen, these equations led to a formula giving the density in terms of the temperature and pressure, which was shown to agree very closely with the experiments of Deville and Troost, and much less closely, but apparently within the limits of possible error, with the experiments of Playfair and Wanklyn. Since the publication of that paper, new determinations of the density have been published in different quarters, which render it possible to compare the equation with the results of experiment throughout a wider range of temperature and pressure. In the present paper, all experimental determinations of the density of this vapor which have come to the knowledge of the writer are cited, and compared with the values demanded by the formula, and a similar comparison of theory and experiment is made with respect to each of the other substances which have been mentioned.

The considerations from which these formulæ were deduced may be briefly stated as follows. It will be observed that they are based rather upon an extension of generally acknowledged principles to a new class of cases than upon the introduction of any new principle.

The energy of a gas-mixture may be represented by an expression of the form

$$m_1(c_1 t + E_1) + m_2(c_2 t + E_2) + \text{etc.,}$$

with as many terms as there are different kinds of gas in the mixture, m_1, m_2, etc. denoting the quantities (by weight) of the several component gases, c_1, c_2, etc., their several specific heats at constant volume, E_1, E_2, etc., other constants, and t the absolute temperature. In like manner the entropy of the gas-mixture is expressed by

$$m_1\left(H_1 + c_1 \log_N t - a_1 \log_N \frac{m_1}{v}\right) + m_2\left(H_2 + c_2 \log_N t - a_2 \log_N \frac{m_2}{v}\right) + \text{etc.,}$$

where v denotes the volume, and H_1, a_1, H_2, a_2, etc. denote constants relating to the component gases, a_1, a_2, etc. being inversely proportional to their several densities. The logarithms are Naperian.

---

* "On the Equilibrium of Heterogeneous Substances," this volume, page 55. The equations referred to are (313), (317), (319), and (320), on pages 171 and 172. The applicability of these equations to such cases as we are now considering is discussed under the heading "Gas-mixtures with Convertible Components," page 172.

These expressions for energy and entropy will undoubtedly apply to mixtures of different gases, whatever their chemical relations may be (with such limitations and with such a degree of approximation as belong to other laws of the gaseous state), when no chemical action can take place under the conditions considered. If we assume that they will apply to such cases as we are now considering, although chemical action is possible, and suppose the equilibrium of the mixture with respect to chemical change to be determined by the condition that its entropy has the greatest value consistent with its energy and its volume, we may easily obtain an equation between m_1, m_2, etc., t and v.*

The condition that the energy does not vary, gives

$$(m_1 c_1 + m_2 c_2 + \text{etc.})\, dt + (c_1 t + \mathrm{E}_1)\, dm_1 + (c_2 t + \mathrm{E}_2)\, dm_2 + \text{etc.} = 0. \quad (1)$$

The condition that the entropy is a maximum implies that its variation vanishes, when the energy and volume are constant. This gives

$$\frac{m_1 c_1 + m_2 c_2 + \text{etc.}}{t}\, dt + \left(\mathrm{H}_1 - a_1 + c_1 \log_{\mathrm{N}} t - a_1 \log_{\mathrm{N}} \frac{m_1}{v}\right) dm_1$$

$$+ \left(\mathrm{H}_2 - a_2 + c_2 \log_{\mathrm{N}} t - a_2 \log_{\mathrm{N}} \frac{m_2}{v}\right) dm_2 + \text{etc.} = 0. \quad (2)$$

Eliminating dt, we have

$$\left(\mathrm{H}_1 - a_1 - c_1 - \frac{\mathrm{E}_1}{t} + c_1 \log_{\mathrm{N}} t - a_1 \log_{\mathrm{N}} \frac{m_1}{v}\right) dm_1$$

$$+ \left(\mathrm{H}_2 - a_2 - c_2 - \frac{\mathrm{E}_2}{t} + c_2 \log_{\mathrm{N}} t - a_2 \log_{\mathrm{N}} \frac{m_2}{v}\right) dm_2 + \text{etc.} = 0. \quad (3)$$

If the case is like that of the peroxide of nitrogen, this equation will have two terms, of which the second may refer to the denser component of the gas-mixture. We shall then have $a_1 = 2a_2$, and $dm_1 = -dm_2$, and the equation will reduce to the form

$$\log \frac{m_2 v}{m_1^2} = -\mathrm{A} - \mathrm{B} \log t + \frac{\mathrm{C}}{t}, \quad (4)$$

where common logarithms have been substituted for Naperian, and A, B and C are constants. If in place of the quantities of the components we introduce the partial pressures, p_1, p_2, due to these components and measured in millimeters of mercury, by means of the relations

$$m_1 = \frac{p_1 v}{a_1 t} \qquad m_2 = \frac{p_2 v}{\frac{1}{2} a_1 t},$$

---

* For certain *a priori* considerations which give a degree of probability to these assumptions, the reader is referred to the paper already cited.

where a_1 denotes a constant, we have

$$\log \frac{p_2}{p_1{}^2} = -(A + \log 2a_1) - (1 + B)\log t + \frac{C}{t}$$

$$= -A' - B'\log t + \frac{C}{t}, \tag{5}$$

where A′ and B′ are new constants. Now if we denote by p the total pressure of the gas-mixture (in millimeters of mercury), by D its density (relative to air of the same temperature and pressure), and by D_1 the theoretical density of the rarer component, we shall have

$$p : p + p_2 :: D_1 : D.$$

This appears from the consideration that $p + p_2$ represents what the pressure would become, if without change of temperature or volume all the matter in the gas-mixture could take the form of the rarer component. Hence,

$$p_2 = p\frac{D - D_1}{D_1},$$

$$p_1 = p - p_2 = p\frac{2D_1 - D}{D_1},$$

and

$$\frac{p_2}{p_1{}^2} = \frac{D_1(D - D_1)}{p(2D_1 - D)^2}.$$

By substitution in (5) we obtain

$$\log \frac{D_1(D - D_1)}{(2D_1 - D)^2} = -A' - B'\log t + \frac{C}{t} + \log p. \tag{6}$$

By this formula, when the values of the constants are determined, we may calculate the density of the gas-mixture from its temperature and pressure. The value of D_1 may be obtained from the molecular formula of the rarer component. If we compare equations (3), (4) and (5), we see that

$$B' = B + 1, \qquad B = \frac{c_1 - c_2}{a_2}.$$

Now $c_1 - c_2$ is the difference of the specific heats at constant volume of NO_2 and N_2O_4. The general rule that the specific heat of a gas at constant volume and per unit of weight is independent of its *condensation*, would make $c_1 = c_2$, B = 0, and B′ = 1. It may easily be shown, with respect to any of the substances considered in this paper,* that unless the numerical value of B′ greatly exceeds unity, the term B′ log t may be neglected without serious error, if its omission is compensated in the values given to A′ and C. We may therefore cancel this term, and then determine the remaining constants by comparison of the formula with the results of experiment.

---

* For the case of peroxide of nitrogen, see pp. 180, 181 in the paper cited above.

In the case of a mixture of Cl_2, PCl_3 and PCl_5, equation (3) will have three terms distinguished by different suffixes. To fix our ideas, we may make these suffixes $_2$, $_3$ and $_5$, referring to Cl_2, PCl_3 and PCl_5 respectively. Since the constants a_2, a_3 and a_5 are inversely proportional to the densities of these gases,

$$a_2 dm_2 = a_3 dm_3 = -a_5 dm_5,$$

and we may substitute $\dfrac{1}{a_2}, \dfrac{1}{a_3}, \dfrac{-1}{a_5}$ for dm_2, dm_3 and dm_5 in equation (3), which is thus reduced to the form

$$\log \frac{m_5 v}{m_2 m_3} = -A - B \log t + \frac{C}{t}. \tag{7}$$

If we eliminate m_2, m_3, m_5 by means of the partial pressures p_2, p_3, p_5, we obtain

$$\log \frac{p_5}{4 p_2 p_3} = -A' - B' \log t + \frac{C}{t}, \tag{8}$$

when A', B', like A, B and C, are constants. If the chlorine and the protochloride are in such proportions as arise from the decomposition of the perchloride, $p_2 = p_3$ and $4 p_2 p_3 = (p_2 + p_3)^2$. In this case, therefore, we have

$$\log \frac{p_5}{(p_2 + p_3)^2} = -A' - B' \log t + \frac{C}{t}. \tag{9}$$

It will be seen that this equation is of the same form as equation (5), when p_5 in (9) is regarded as corresponding to p_2 in (5), and $p_2 + p_3$ in (9), which represents the pressure due to the products of decomposition, is regarded as corresponding to p_1 in (5), which has the same signification. It follows that equation (5), as well as (6), which is derived from it, may be regarded as applying to the vapor of perchloride of phosphorus, when the values of the constants are properly determined. This result might have been anticipated, but the longer course which we have taken has given us the more general equations, (7) and (8), which will apply to cases in which there is an excess of chlorine or of the protochloride.

If the gas-mixture considered, in addition to the components capable of chemical action, contains a neutral gas, the expressions for the energy and entropy of the gas-mixture should properly each contain a term relating to this neutral gas. This would make it necessary to add $c_n m_n$ to the coefficient of dt in (1), and $\dfrac{c_n m_n}{t}$ to the coefficient of dt in (2), the suffix $_n$ being used to mark the quantities relating to the neutral gas. But these quantities would disappear with the elimination of dt, and equation (3) and all the subsequent equations would require no modification, if only p and D are estimated (in accordance with usage) with exclusion of the pressure and weight due to the neutral gas. This result, which may be extended to any number of neutral gases, is simply an expression of Dalton's Law.

We now proceed to the comparison of the formulæ, especially of equation (6), with the results of experiment.

TABLE I.--PEROXIDE OF NITROGEN

Experiments at Atmospheric Pressure.

MITSCHERLICH,—R. MÜLLER,—DEVILLE and TROOST.

Temperature.	Pressure.	Density calculated by eq. (10).	Density observed. Deville & Troost.				Excess of observed density. Deville & Troost.			
			M—h. M—r.	I.	II.	III.	M—h. M—r.	I.	II.	III.
183·2	(760)	1·592				1·57				−·022
154·0	(760)	1·597				1·58				−·017
151·8	(760)	1·598		1·50				−·10		
135·0	(760)	1·607				1·60				−·007
121·8	(760)	1·622		1·64				+·02		
121·5	(760)	1·622				1·62				−·002
111·3	(760)	1·641				1·65				+·009
100·25	760	1·677	1·72				+·04			
100·1	(760)	1·676				1·68				+·004
100·0	(760)	1·677		1·71				+·03		
90·0	(760)	1·728				1·72				−·008
84·4	(760)	1·768		1·83				+·06		
80·6	(760)	1·801				1·80				−·001
79	748	1·814	1·84				+·03			
77·4	(760)	1·833			1·85				+·02	
70·0	(760)	1·920				1·92				·000
70	754·5	1·919	1·95				+·03			
68·8	(760)	1·937		1·99				+·05		
66·0	(760)	1·976			2·03				+·05	
60·2	(760)	2·067				2·08				+·013
55·0	(760)	2·157			2·20				+·04	
52	757	2·211	2·26				+·05			
49·7	(760)	2·255		2·34				+·09		
49·6	(760)	2·256				2·27				+·014
45·1	(760)	2·342			2·40				+·06	
39·8	(760)	2·443				2·46				+·017
35·4	(760)	2·524				2·53				+·006
35·2	(760)	2·528		2·66				+·13		
34·6	(760)	2·539			2·62				+·08	
32	748	2·582	2·65				+·07			
28·7	(760)	2·642			2·80				+·16	
28	751	2·652	2·70				+·05			
27·6	(760)	2·661			2·70				+·04	
26·7	(760)	2·676				2·65				−·026

Peroxide of nitrogen.—If we take the constants of the equation for this substance from the paper already cited,[*] we have

$$\log \frac{15 \cdot 89(D-1 \cdot 589)}{(3 \cdot 178-D)^2} = \frac{3118 \cdot 6}{t_c+273} + \log p - 12 \cdot 451, \qquad (10)$$

t_c denoting the temperature on the centigrade scale. The numbers 3·178 and 1·589 represent the theoretical densities of N_2O_4 and NO_2

---

[*] See equation (336) on page 177,—also the following equations in which the density is given in terms of the temperature and pressure. In comparing these equations, it must be observed that in (336) the pressures are measured in atmospheres, but in this paper in millimeters of mercury.

respectively. The two other constants were determined by the experiments of Deville and Troost.

The results of these and other experiments at atmospheric pressure, all made by Dumas' method, are exhibited in Table I. The first three columns give the temperature (centigrade), the pressure (in millimeters of mercury),* and the density calculated from the temperature and pressure by equation (10). The subsequent columns give the densities observed by different authorities, and the excess of the observed over the calculated densities. In the first column of observed densities, we have one observation by Mitscherlich† (at 100.25°) and five by R. Müller.‡ The three remaining columns contain each the results of a series of experiments by Deville and Troost.§ In each series the experiments were made with increasing temperatures, and with the same vessel, without refilling. It should be observed that the results of the three series are not regarded by their distinguished authors as of equal weight. It is expressly stated that the numbers in the two earlier series, and especially in the first, may be less exact. The last series agrees very closely with the formula. It was from this that the constants of the formula were determined. The experiments of series I and II, and those of Mitscherlich and Müller, give somewhat larger values, with a single exception, as is best seen in the columns which give the excess of the observed density. The differences between the different columns are far too regular to be attributed to the accidental errors of the individual observations, except in the case of the experiment at 151.8°, where some accident has evidently occurred either in the experiment itself or in the reduction of the result. Setting this observation aside, we must look for some constant cause for the other discrepancies between the different series.

We can hardly attribute these discrepancies to difference in the material employed, or to air or other foreign substance imperfectly expelled from the flask. For impurities which increase the density would make the divergence between the different series greatest when the densities are the least, whereas the divergences seem to vanish as the density approaches the limiting value. (A similar

---

* 760ᵐᵐ has been assumed as the pressure of the atmosphere in all cases in which the precise pressure is not recorded in the published account of the experiments. The figures inserted in the columns of pressures are in such cases enclosed in parentheses. The same course has been followed in the subsequent tables. With respect to the principal series of observations by Deville and Troost (series III), it is stated that the barometer varied between 747 and 764 millimeters. A difference of 13 millimeters in the pressure would in no case cause a difference of ·005 in the calculated densities. In this series, therefore, the errors due to this circumstance are not very serious.

† *Pogg. Ann.*, vol. xxix (1833), p. 220.

‡ *Lieb. Ann.*, vol. cxxii (1862), p. 15.

§ *Comptes Rendus*, vol. lxiv (1867), p. 237.

objection would apply to the supposition of any error in the determination of the weight of the flask when filled with air alone.) But if we should attribute the divergences to an impurity which diminishes the density (as air), we should be driven to the conclusion that the first series of Deville and Troost gives the most correct results, and that all the best attested numbers at temperatures below 90° are considerably in the wrong. It does not seem possible to account for these discrepancies by any causes which would apply to cases of normal or constant density. They are illustrations of the general fact that when the density varies rapidly with the temperature, determinations of density for the same temperature and pressure by different observers, or different determinations by the same observer, exhibit discordances which are entirely of a different order of magnitude from those which occur with substances of normal or constant densities, or which occur with the same substance at temperatures at which the density approaches a constant value. In some cases such results may be accounted for by carelessness on the part of the observers, not controlled by a comparison of the result with a value already known. But such an explanation is inadequate to explain the general fact, and evidently inadmissible in the present case.

It is probable that these discrepancies are in part attributable to a circumstance which has been noticed by M. Wurtz, in his account of his experiments upon the vapor-density of bromhydrate of amylene, in the following words:—"Le temps pendant lequel la vapeur est maintenue à la température où l'on détermine la densité n'est pas sans influence sur les nombres obtenus. C'est ce qui result des deux expériences faites à 225 degrés avec des produits identiques. Dans la première, la vapeur a été portée rapidement à 225 degrés. Dans la seconde elle a été maintenue pendant dix minutes à cette température. On voit que les nombres trouvés pour les densités ont été fort différents. (The numbers were 4·69 and 3·68 respectively.) Ce résultat ne doit point surprendre si l'on considère que le phénomène de décomposition de la vapeur doit absorber de la chaleur, et que les quantités de chaleur nécessaires pour produire et la dilatation et la décomposition ne sauraient être fournies instantanément."[*]

It is not difficult to form an estimate of the quantities of heat which come into play in such cases. With respect to peroxide of nitrogen, it was estimated in the paper already cited that the heat absorbed in the conversion of a unit of N_2O_4 into NO_2 under constant pressure is represented by 7181 a_2. (The heat is supposed to be measured in units of mechanical work.) Now the external

---

[*] *Comptes Rendus*, t. lx, p. 730.

work done by the conversion of a unit of N_2O_4 into NO_2 under constant pressure is a_2t. Therefore, the ratio of the heat absorbed to the external work done by the conversion of N_2O_4 into NO_2 is $7181 \div t$, or 23 at the temperature of 40° centigrade. Let us next consider how much more rapidly this vapor expands with increase of temperature at constant pressure than air. From the necessary relation

$$v = \frac{kmt}{pD},$$

where m denotes the weight of the vapor, and k a constant, we obtain

$$\left(\frac{dv}{dt}\right)_p = \frac{v}{t} - \frac{v}{D}\left(\frac{dD}{dt}\right)_p,$$

where the suffix $_p$ indicates that the differential coefficients are for constant pressure. The last term of this expression evidently denotes the part of the expansion which is due to the conversion of N_2O_4 into NO_2, and the preceding term the expansion which would take place if there were no such conversion, and which is identical with the expansion of the same volume of air under the same circumstances. The ratio of the two terms is $-\frac{t}{D}\left(\frac{dD}{dt}\right)_p$, the numerical value of which for the temperature of 40° is 2·42, as may be found by differentiating equation (10), or, with less precision, from the numbers in the third column of Table I. Let us now suppose that equal volumes of peroxide of nitrogen and of air at the temperature of 40° and the pressure of one atmosphere receive equal infinitesimal increments of temperature under constant pressure. The heat absorbed by the peroxide of nitrogen on account of the conversion of N_2O_4 into NO_2 is 23 times the external work due to the same cause, and this work is 2·42 times the external work done by the expansion of the air. But the heat absorbed by the air in expanding under constant pressure is well known to be 3·5 times the work done. Therefore the heat absorbed on account of the conversion of N_2O_4 into NO_2 is $(23 \times 2·42 \div 3·5 =)$ 15·9 times the heat absorbed by the air. To obtain the whole heat absorbed by the vapor we must add that which would be required if no conversion took place. At 40° the vapor of peroxide of nitrogen contains about 54 molecules of N_2O_4 to 46 of NO_2, as may easily be calculated from its density. The specific heat for constant pressure of a mixture in such proportions of gases of such molecular formulæ, if no chemical action could take place, would be about twice that of the same volume of air. Adding this to the heat absorbed by the chemical action we obtain the final result,—that at 40° and the pressure of the atmosphere the specific heat of peroxide of nitrogen at constant pressure is about eighteen times that of the same volume of air.*

But the greater amount of heat which is required to bring the vapor to the desired temperature is only one factor in the increased liability to error in cases of this kind. The expansion of peroxide of nitrogen for increase of temperature under constant pressure at 40° is 3·42 times that of air. If, then, in a determination of density, the vapor fails to reach the temperature of the bath, the error due to the difference of the temperature of the vapor and the bath, will be 3·42 times as great as would be caused by the same difference of temperatures in the case of any vapor or gas having a constant density. When we consider that we are liable not only to the same, but to a much greater difference of temperatures in a case like that of peroxide of nitrogen, when the exposure to the heat is of the same duration, it is evident that the common test of the exactness of a process for the determination of vapor-densities, by applying it to a case in which the density is nearly constant, is entirely insufficient.

That the experiments of the III^d series of Deville and Troost give numbers so regular and so much lower than the other experiments is probably to be attributed in part to the length of time of exposure to the heat of the experiment, which was half an hour in this series,—for the other series, the time is not given.

Another point should be considered in this connection. During the heating of the vapor in the bath, it is not immaterial whether the flask is open or closed. This will appear, if we compare the values of $\left(\dfrac{d\mathrm{D}}{dt}\right)_p$ and $\left(\dfrac{d\mathrm{D}}{dt}\right)_v$, the differential coefficients of the density with respect to the temperature on the suppositions respectively, of constant pressure, and of constant volume. For 40°, we have

$$\left(\frac{d\mathrm{D}}{dt}\right)_p = ·0189, \qquad \left(\frac{d\mathrm{D}}{dt}\right)_v = ·0163,$$

the first number being obtained immediately from equation (10) by differentiation, and the second by differentiation after substitution of $\dfrac{kmt}{v\mathrm{D}}$ for p. The ratio of these numbers evidently gives the proportion in which the chemical change takes place under the two suppositions. This shows that only about six-sevenths of the heat required for the chemical change can be supplied before opening the flask, and the remainder of this heat as well as that required for expansion must be supplied after the opening. The errors due

---

* Similar calculations from less precise data for the bromhydrate of amylene at 225° seem to indicate a specific heat as much as forty times as great as that of the same volume of air.

to this source may evidently be diminished by diminishing the intervals of temperature between the successive experiments in a series of this kind, and also by diminishing the opening made in the flask, which increases the time for which the flask may be left open without danger of the entrance of air. In the III^d series of experiments by Deville and Troost, the intervals of temperature did not exceed ten degrees (except after the density had nearly reached its limiting value), and the neck of the flask was drawn out into a very fine tube.

In Table II, which relates to experiments on the same substance at pressures less than that of the atmosphere, the principal series is that of Naumann,* which commences a few degrees below the lowest temperatures of Deville and Troost, and extends to $-6°$ centigrade, the pressures varying from 301 to 84 millimeters. These experiments were made by the method of Gay-Lussac. The numbers in the column of observed densities have been re-calculated from the more immediate results of the experiments, and are not in all cases identical with those given in Professor Naumann's paper. Every case of difference is marked with brackets. Instead of the numbers [2·66], [2·62], [2·85], [2·94], Naumann's paper has 2·57, 2·65, 2·84, 3·01, respectively. In some cases the temperatures and pressures of two experiments are so nearly the same that it would be allowable to average the results, at least in the column of excess of observed density. In such cases the numbers in this column have been united by a brace. The greatest difference between the observed and calculated densities is ·16, which occurs at the least pressure, 84 millimeters. In this experiment the weight of the substance employed is also less than in any other experiment. Under such circumstances, the liability to error is of course greatly increased. The average difference between the observed and calculated densities is ·063. Since these differences are almost uniformly positive and increase as the temperature diminishes, it is evident that they might be considerably diminished by slight changes in the constants of equation (10), without seriously impairing the agreement of that equation with the experiments of Deville and Troost. But it has not seemed necessary to re-calculate the formula, which, in its present form, will at least illustrate the degree of accuracy with which densities at low pressures and at temperatures below the boiling point of the liquid may be derived from experiments at atmospheric pressure above the boiling point. Moreover, the excess of observed density may be due in part to a circumstance mentioned by Professor Naumann, that the chemical action between the vapor and the

---

* *Berichte der deutschen chemischen Gesellschaft*, Jahrgang xi (1878), S. 2045.

mercury diminished the volume of the vapor, and thus increased the numbers obtained for the density.

TABLE II.—PEROXIDE OF NITROGEN.

Experiments at less than Atmospheric Pressure.

PLAYFAIR AND WANKLYN,—TROOST,—NAUMANN.

Tempera-ture.	Pressure.	Density calculated by eq. (10).	Density observed.			Excess of obs. density.		
			P. & W.	T.	N.	P. & W.	T.	N.
97·5	(301)	1·631	1·783			+ ·152		
27	35	1·90		1·6			− ·30	
27	16	1·77		1·59			− ·18	
24·5	(323)	2·524	2·52			− ·004		
22·5	136·5	2·34			2·35			+ ·01
22·5	101	2·26			2·28			+ ·02
21·5	161	2·41			2·38			− ·03⎱
20·8	153·5	2·41			2·46			+ ·05⎰
20	301	2·59			2·70			+ ·11
18·5	136	2·43			2·45			+ ·02
18	279	2·61			2·71			+ ·10
17·5	172	2·51			2·52			+ ·01⎱
16·8	172	2·53			2·55			+ ·02⎰
16·5	224	2·59			[2·66]			+ ·07⎱
16	228·5	2·61			[2·62]			+ ·01⎰
14·5	175	2·58			2·63			+ ·05
11·3	(159)	2·620	2·645			+ ·025		
11	190	2·66			2·76			+ ·10
10·5	163	2·64			2·73			+ ·09
4·2	(129)	2·710	2·588			− ·122		
4	172·5	2·77			2·85			+ ·08
2·5	145	2·76			[2·85]			+ ·09
1	138	2·78			2·84			+ ·06
− 1	153	2·83			2·87			+ ·04
− 3	84	2·76			2·92			+ ·16
− 5	123	2·85			2·98			+ ·13⎱
− 6	125·5	2·87			[2·94]			+ ·07⎰

The same table includes two experiments of Troost,[*] by Dumas' method, but at the very low pressures of 35^{mm} and 16^{mm}. In such experiments we cannot expect a close agreement with the formula, for the same error in the determination of the weight of the vapor, which would make a difference of ·01 in the density in experiments at atmospheric pressure, would make a difference of ·21 or ·47 in the circumstances of these experiments. In fact, the numbers obtained differ considerably from those demanded by the formula.

There remain four experiments by Playfair and Wanklyn[†] in which Dumas' method was varied by diluting the vapor with nitrogen. The numbers in the column of pressures represent the total pressure diminished by the pressure which the nitrogen alone would have exerted. They are not quite accurate, since the data given in the memoir cited only enable us to determine the ratios

---

[*] *Comptes Rendus*, t. lxxxvi (1878), p. 1395.

[†] *Trans. Roy. Soc. Edinb.*, vol. xxii (1861), p. 463.

of the total and the partial pressures. The numbers here given are obtained by setting the total pressure, which was that of the atmosphere at the time of the experiment, equal to 760^{mm}. The effect of this inaccuracy upon the calculated densities would be small. Two of these observations agree closely with the formula; and two show considerable divergence, but in opposite directions, and these are the two in which the quantities of peroxide of nitrogen were the smallest. The differences appear to be attributable rather to the difficulty of a precise determination of the quantities of nitrogen and of vapor, than to any effect of the one upon the other.

Special interest attaches to experiments at the same or nearly the same temperature but different pressures. For with experiments at the same temperature, the constants of the formula which are determined by observation are reduced to one, so that the verification of the formula by experiment cannot possibly be regarded as a case of interpolation. It is not necessary that the temperatures should be exactly the same, for it will be conceded that the formula represents the actual function well enough to answer for adjusting slight differences of temperature; but it is necessary that the range of pressures should be considerable in order that the differences of density should be large in proportion to the probable errors of observation. But the pressures must not be so low that accurate determinations become impossible.

In the experiments of Naumann we see some fair correspondences with the formula in respect to the influence of pressure, especially in the first four experiments of the list, where, if we average the results of the third and fourth experiments, as is evidently allowable, the observed values follow very closely the fluctuations of the calculated, extending from 2·26 to 2·41. In other cases the agreement is less satisfactory. The circumstance that the experiments at the two highest pressures (301 and 279^{mm}) give results exceeding the calculated values considerably more than any other experiments at adjacent temperatures may seem to indicate that the densities increase with the pressures more rapidly than the formula allows; but the differences are not too large to be ascribed to errors of observation, and the experiment at the lowest pressure (84^{mm}) also shows a large excess of observed density.

A much more critical test may be found in the comparison of Naumann's experiments with those of Deville and Troost, notwithstanding the interval of about 4° of temperature. The formula requires that a diminution of pressure from 760 to 101 millimeters shall reduce the density from 2·676 at 26·7° to 2·26 at 22·5°, notwithstanding the effect of the change of temperature. Experiment gives a reduction of density from 2·65 to 2·28, which is about one-ninth less. This is, it will be observed, a deviation from the formula in the opposite direction from that which the experiments of Naumann alone, or a comparison of the experiments of Troost with those of Deville and Troost, seemed to indicate. The experiment here compared with Naumann's belongs to the III[d] series of Deville and Troost. If instead of this experiment we should take an average of the experiments at lowest temperature in the II[d] and III[d] series, the agreement with the formula with respect to the effect of change of pressure would be almost perfect.

Formic acid.—In Table III, the determinations of Bineau are compared with the densities calculated by the formula

$$\log \frac{1·589\,(D-1·589)}{(3·178-D)^2} = \frac{3800}{t_C+273} + \log p - 12·641. \tag{11}$$

The observed densities are taken from the eighteenth volume of the third series of the *Annales de Chimie et de Physique* (1846), except in three cases, distinguished by parentheses, which are earlier determinations published in the nineteenth volume of the *Comptes Rendus* (1844). It may be added that the pressure (687) for the experiment at 108° is taken from Erdmann's *Journal für praktische Chemie* (vol. xl, p. 44), the impression being imperfect in the *Annales*, in the copies to which the writer has been able to refer, where the figures look much like 637. (The pressure 637 would make the calculated density 2·28.)

In the column which gives the excess of observed densities, the effect of nearness to the state of saturation is often very marked. Such cases are distinguished by an asterisk. The temperature of 99·5° is below the boiling point of formic acid, and the higher pressures employed at this temperature cannot be far from the pressure of saturated vapor. With respect to lower temperatures, we have the statement of Bineau that the pressure of saturated vapor is about 19[mm] at 13°, 20·5[mm] at 15°, 33·5[mm] at 22°, and 53·5[mm] at 32°. By interpolation between the *logarithms* of these pressures (in a single case, by *extrapolation*), we obtain the following result :—

Temperature,	-	-	-	10·5	12·5	16	18·5	22
Pressure of sat. vapor,		-		16·6	18·5	22	26·2	33·5
Pressure of experiment,		-		14·69	15·20	15·97	23·53	25·17

TABLE III.—FORMIC ACID.

EXPERIMENTS OF BINEAU.

Temperature.	Pressure.	Density calculated by eq. (11).	Density observed.	Excess of observed density.
216·0	690	1·60	1·61	+ ·01
184·0	750	1·64	1·68	+ ·04
125·5	687	2·03	2·05	+ ·02
125·5	645	2·02	2·03	+ ·01
124·5	670	2·04	2·06	+ ·02
124·5	640	2·03	2·04	+ ·01
118·0	655	2·13	(2·14)	(+ ·01)
118·0	650	2·13	2·13	·00
117·5	688	2·15	2·13	− ·02
115·5	649	2·17	2·20	+ ·03
115·5	640	2·16	2·16	·00
115	655	2·18	(2·13)	(− ·05)
111·5	690	2·25	2·22	− ·03
111·5	690	2·25	2·25	·00
111	608	2·22	(2·13)	(− ·09)
108	[687]	2·30	2·31	+ ·01
105·0	691	2·35	2·35	·00
105·0	650	2·34	2·33	− ·01
105·0	630	2·33	2·32	− ·01
101·0	693	2·42	2·44	+ ·02
101·0	650	2·40	2·41	+ ·01
99·5	690	2·44	2·52	+ ·08*
99·5	684	2·44	2·49	+ ·05
99·5	676	2·44	2·46	+ ·02
99·5	662	2·43	2·44	+ ·01
99·5	641	2·42	2·42	·00
99·5	619	2·41	2·41	·00
99·5	602	2·41	2·40	− ·01
99·5	557	2·39	2·34	− ·05
34·5	28·94	2·82	2·77	− ·05
31·5	3·04	2·40	2·60	+ ·20
30·5	8·83	2·67	2·69	+ ·02
30·0	18·28	2·81	2·76	− ·05
29·0	27·40	2·88	2·83	− ·05
24·5	17·39	2·88	2·86	− ·02
22·0	25·17	2·95	3·05	+ ·10*
20·0	16·67	2·93	2·94	+ ·01
20·0	7·99	2·84	2·85	+ ·01
20·0	2·72	2·64	2·80	+ ·16
18·5	23·53	2·98	3·23	+ ·25*
16·0	15·97	2·97	3·13	+ ·16*
15·5	2·61	2·72	2·86	+ ·14
15·0	7·60	2·90	2·93	+ ·03
12·5	15·20	3·00	3·14	+ ·14*
11·0	7·26	2·95	3·02	+ ·07
10·5	14·69	3·01	3·23	+ ·22*

Whether the large excess of observed density in these cases represents a property of the vapor, or an incipient condensation on the walls of the vessel which contains it, as has been supposed by eminent physicists in similar cases, we need not here discuss.

If we reject these cases of nearly saturated vapor, as well as the three earlier determinations, there remain 25 experiments at pressures somewhat less than one atmosphere in which the maximum difference between the observed and calculated densities is ·05, and the average difference ·016; nine experiments at pressures ranging from 29^{mm} to 7^{mm}, in which the maximum difference is ·07 and the average ·035; and three experiments at pressures of about 3^{mm}, in which the average difference is ·17. The extraordinary precision of the determinations at low pressures is doubtless due to the large scale on which the experiments were conducted. All the experiments at temperatures below 99° were made with a globe of the capacity of 5½ liters with a stem of suitable length to hold the barometric column.

The agreement is certainly as good as could be desired, and shows the accuracy of which the method of observation is capable. But in no part of the thermometric scale do we find so great a range of pressures as might be desired, without using pressures too low for accurate results, or observations which are to be rejected for other reasons.

Acetic acid.—For this substance the densities have been calculated by the formula

$$\log \frac{2\cdot073(D-2\cdot073)}{(4\cdot146-D)^2} = \frac{3520}{t_C+273} + \log p - 11\cdot349, \tag{12}$$

the constants 3520 and 11·349 being derived from the determinations of Cahours and Bineau, which with those of Horstmann and Troost are given in Table IV. The experiments of Cahours and Horstmann were made under atmospheric pressure, those of Horstmann[*] by the method of Bunsen, those of Cahours presumably by the method of Dumas. The numbers in the first column of the densities observed by Cahours are taken from the twentieth volume (1845) of the *Comptes Rendus*, except a few cases, distinguished by parentheses, which are taken from the preceding volume (1844). The numbers in the second column are taken from his *Leçons de chimie générale élémentaire*, 1856. These numbers seem to be based in part upon new experiments and in part upon a revision of the observations recorded in the *Comptes Rendus*, the calculations being carried out to another figure of decimals. They are therefore entitled to a greater weight than the numbers of the preceding column.

The agreement of the formula with the numbers given in the *Leçons de chimie* is very good, the greatest divergences being ·080 at 190° and ·062 at 180°. But at 190° the table in the *Comptes Rendus* agrees precisely with the formula, and at 171° (the next experiment) it shows a divergence in the opposite direction. The next divergences in the order of magnitude are − ·033, − ·036, − ·032

---

[*] *Lieb. Ann.*, suppl. vi, p. 65.

TABLE IV.—ACETIC ACID.

EXPERIMENTS OF CAHOURS,—HORSTMANN,—BINEAU,—TROOST.

Tempera-ture.	Pressure.	Density calculated by eq. (12).	Density observed. Cahours.			Excess of observed density. Cahours.		
			C. R.	Leçons.	Horst-mann.	C. R.	Leçons.	Horst-mann.
338	(760)	2·077	2·08			·00		
336	(760)	2·077		2·082			+ ·005	
327	(760)	2·078	2·08	2·085		·00	+ ·007	
321	(760)	2·079	2·08	2·083		·00	+ ·004	
308	(760)	2·081		2·085			+ ·004	
300	(760)	2·082	2·08			·00		
295	(760)	2·084		2·083			− ·001	
280	(760)	2·089	2·08			− ·01		
272	(760)	2·093		2·088			− ·005	
254·6	747·2	2·105			2·135			+ ·030
252	(760)	2·108		2·090			− ·018	
250	(760)	2·111	2·08			− ·03		
240	(760)	2·122		2·090			− ·032	
233·5	752·8	2·132			2·195			+ ·063
231	(760)	2·137	(2·12)	2·101		(− ·02)	− ·036	
230	(760)	2·139	2·09			− ·05		
219	(760)	2·165	2·17	2·132		+ ·01	− ·033	
200	(760)	2·239	2·22	2·248		− ·02	+ ·009	
190	(760)	2·298	2·30	2·378		·00	+ ·080	
181·7	749·7	2·359			2·419			+ ·060
180	(760)	2·376		2·438			+ ·062	
171	(760)	2·466	2·42			− ·05		
170	(760)	2·477		2·480			+ ·003	
165·0	754·1	2·534			2·647			+ ·113
162	(760)	2·575		2·583			+ ·008	
160·3	751·6	2·594			2·649			+ ·055
160	(760)	2·601	2·48			− ·12		
152	(760)	2·716	(2·72)	2·727		(·00)	+ ·011	
150	(760)	2·747	2·75			·00		
145	(760)	2·826	(2·75)			(− ·08)		
140	(760)	2·910	2·90	2·907		− ·01	− ·003	
134·3	748·8	3·001			3·108			+ ·107
131·3	754·1	3·055			3·070			+ ·015
130	(760)	3·082	3·12	3·105		+ ·04	+ ·023	
128·6	752·9	3·103			3·079			− ·024
125	(760)	3·168	3·20			+ ·03		
124	(760)	3·185		3·194			+ ·009	
			Bineau.		Troost.	Bineau.		Troost.
132	757	3·05	(2·86)			(− ·19)		
130	59·7	2·31			2·12			− ·19
130	30·6	2·21			2·10			− ·11
129	633	3·03	(2·88)			(− ·15)		
36·5	11·32	3·63	3·62			− ·01		
35·0	11·19	3·65	3·64			− ·01		
30·0	6·03	3·61	3·60			− ·01		
28·0	10·03	3·75	3·75			·00		
24·0	5·75	3·71	3·70			− ·01		
22·0	8·64	3·82	3·85			+ ·03		
22	2·70	3·59	3·56			− ·03		
21·0	4·06	3·70	3·72			+ ·02		
20·5	10·03	3·86	3·95			+ ·09		
20·0	8·55	3·84	3·88			+ ·04		
20·0	5·56	3·77	3·77			·00		
19·0	4·00	3·73	3·75			+ ·02		
19	2·60	3·65	3·66			+ ·01		
12·0	5·23	3·88	3·92			+ ·04		
12	2·44	3·77	3·80			+ ·03		
11·5	3·76	3·84	3·88			+ ·04		

at 219°, 231°, 240°, respectively. Here the table in the *Comptes Rendus* agrees substantially with that of the *Leçons*, but the experiments of Horstmann show a divergence in the opposite direction. In fact, the three columns of observed densities nowhere agree in the direction of their divergence from the formula.

The somewhat decided differences between the results of Horstmann and those of Cahours may be due in part to the different methods of observation, especially to the entirely different manner of applying the heat and measuring the temperature. But the higher values obtained by Horstmann cannot be accounted for by too short an exposure to the source of heat, for his experiments were made with decreasing temperatures.

The determinations of Bineau are taken from the same sources as those on formic acid, the earlier determinations being distinguished as before by parentheses. One of these (at 132°) was made by the method of Dumas, the other by that of Gay-Lussac. The smallness of the observed densities appears due to the presence of water. (An acidimetric test gave 295 parts of acid in 306.) The other experiments were made with the same apparatus which was used with formic acid and show even greater regularity in their results than the experiments with that substance. Only in one case is the influence of proximity to saturation seen, viz., at 20·5° and 10·03^{mm}, the pressure of saturated vapor at this temperature being about 12·7^{mm}.* In the remaining fifteen observations of this series, notwithstanding the very low pressures employed (from 2·44 to 11·32), the greatest difference between the observations and the formula is ·04, and the average difference ·02.

The two observations by Troost† were made by the method of Dumas, but at pressures very low for this method. The results obtained differ considerably from the formula, but not so much as in the case of his experiments at low pressure with peroxide of nitrogen.

Table V contains the experiments of Naumann‡ on acetic acid. These consist of ten series (distinguished by the letters A, B, C, etc.) of observations by Hoffmann's method.§ The temperatures of the observations in the different series are for the most part the same, so that for each temperature we have observations through a wide range of pressures. Within each compartment of the table are given

---

* This number is obtained from data given by Bineau by the same kind of interpolation which was used for formic acid.

† *Comptes Rendus*, vol. lxxxvi (1878), p. 1395.

‡ *Lieb. Ann.*, vol. clv, p. 325.

§ This is a modification of the method of Gay-Lussac, in which the heat is supplied by a vapor bath.

TABLE V.—ACETIC ACID.

EXPERIMENTS OF NAUMANN.

		TEMPERATURE.								
		78°	100°	110°	120°	130°	140°	150°	160°	185°
A	Pressure.		393·5	411	432	455	477	498·5		565
	D. calc.		3·39	3·23	3·06	2·90	2·75	2·61		2·28
	D. obs.		3·44	3·31	3·14	2·97	2·82	2·68		2·36
	Exc. of D. obs.		+·05	+·08	+·08	+·07	+·07	+·07		+·08
B	Pressure.		342·3	359·3	377·5	398·5	417·5	436·5		495
	D. calc.		3·35	3·18	3·02	2·85	2·70	2·57		2·26
	D. obs.		3·37	3·22	3·06	2·89	2·75	2·63		2·31
	Exc. of D. obs.		+·02	+·04	+·04	+·04	+·05	+·06		+·05
C	Pressure.		258							382
	D. calc.		3·26							2·22
	D. obs.		3·17							2·25
	Exc. of D. obs.		−·09							+·03
D	Pressure.		232		252	274	287·5	300		335
	D. calc.		3·23		2·87	2·72	2·58	2·46		2·21
	D. obs.		3·12		2·94	2·68	2·54	2·44		2·23
	Exc. of D. obs.		−·11		+·07	−·04	−·04	−·02		+·02
E	Pressure.	164	186	197	209	221	232	243	253	269
	D. calc.	3·53	3·15	2·97	2·81	2·65	2·52	2·41	2·32	2·18
	D. obs.	3·41	3·06	2·91	2·75	2·61	2·50	2·40	2·31	2·22
	Exc. of D. obs.	−·12	−·09	−·06	−·06	−·04	−·02	−·01	−·01	+·04
F	Pressure.	149	168			201				
	D. calc.	3·50	3·12			2·62				
	D. obs.	3·34	3·01			2·56				
	Exc. of D. obs.	−·16	−·11			−·06				
G	Pressure.	137	156	166·5	180	188	199	208·2		230
	D. calc.	3·48	3·09	2·92	2·75	2·60	2·47	2·37		2·17
	D. obs.	3·26	2·98	2·81	2·61	2·50	2·40	2·29		2·14
	Exc. of D. obs.	−·22	−·11	−·11	−·14	−·10	−·07	−·08		−·03
H	Pressure.	113	130	138·5	149	157·5	168·2	175		191·5
	D. calc.	3·42	3·03	2·85	2·69	2·55	2·43	2·33		2·15
	D. obs.	3·25	2·94	2·78	2·60	2·47	2·32	2·26		2·13
	Exc. of D. obs.	−·17	−·09	−·07	−·09	−·08	−·11	−·07		−·02
J	Pressure.	80	92	98·5	106	112·5	117·3		129·2	
	D. calc.	3·32	2·91	2·73	2·58	2·45	2·35		2·21	
	D. obs.	3·06	2·76	2·61	2·46	2·34	2·27		2·11	
	Exc. of D. obs.	−·26	−·15	−·12	−·12	−·11	−·08		−·10	
K	Pressure.	66	77·7	84	89·5	93	98	103		110·5
	D. calc.	3·26	2·85	2·68	2·53	2·40	2·31	2·24		2·12
	D. obs.	3·04	2·66	2·49	2·37	2·32	2·24	2·16		2·11
	Exc. of D. obs.	−·22	−·19	−·19	−·16	−·08	−·07	−·08		−·01

in order the pressure of an experiment, the density calculated by equation (12), the observed density, and the excess of observed density, the temperature of the experiment being given at the head of the column. These experiments, taken by themselves, seem to show an effect of pressure upon the density about one third greater than is indicated by the formula. But the divergences (of which the greatest is ·26 and the average ·085) are not large in view of the fact that the experiments were undertaken rather with the desire of obtaining a great number of observations with moderate labor, than with the intention of attaining the greatest possible accuracy.

The quantity of acid diminishes somewhat regularly from ·2084 grams in series A to ·0185 in series K. The volume, which was 154^{cc} in the experiment at 185° in series A, diminishes in the successive series, and in the same series with diminishing temperature, to 69·6^{cc} in the experiment at 78° in series K. It is worthy of notice that the greatest deviations from the formula occur where the liability to error is most serious with respect to pressure (which was measured without a cathetometer), to volume, and to the quantity of acid.

Far more serious than the absolute amount of these divergences, is the regularity which they exhibit. But it must be remembered that the observations are by no means entirely independent, and many sources of possible error, such as the calibration of the tube and the determination of the quantity of acid, might affect the results with considerable regularity.

Only to a slight degree can the divergences from the formula be accounted for by an insufficient exposure to the temperature of the experiment. The observations, except those at 78°, were made with increasing temperatures, and the greatest divergences from the formula are not in the positive direction. Yet the positive divergences occur where we should most expect to find them, if they were due to this cause, viz., in the series in which the greatest quantities of acid were used, and in cases in which the temperature seems to have been raised at once an unusual number of degrees. (See especially the observation at 120° in series D, and in general the observations at 185°, which exhibit if not a positive at least a diminution of negative excess.) In the observations at 78°, which were the last of each series, and therefore followed a fall of temperature from 185°, we find in some cases, especially in series G, H, and J, a negative divergence much greater than in the other determinations of the same series, and which appears to be referable to this circumstance.

In Table VI are exhibited the results of experiments by Playfair and Wanklyn,* in which the vapor of the acid was diluted with hydrogen or, in a single case (the experiment at 95·5°), by air. Columns I and II of the observed densities relate each to a series of observations by the method of Gay-Lussac, column III contains four independent determinations by the method of Dumas. The numbers in the column of pressures are, as in other similar cases, the partial pressures obtained by subtracting from the total pressure (which was never very much less than that of the atmosphere) that which would be exerted by the hydrogen or air alone.

The first observation of the first series gives the density 1·936, which is doubtless too small, since it is much less than the theoretical

---

* *Trans. Roy. Soc. Edinb.*, vol. xxii, p. 455.

limit 2·073. Since the greater part of the measurements from which this number was calculated were also used in reducing the other observations of the series, the error probably affects the other observations, and in a somewhat increased degree. This will account only for a part of the difference between the observations and the formula. The remaining part of the differences in this series, and the somewhat smaller differences in the next, may be due to the fact that the experiments of both series were conducted with descending temperatures. Yet the experiments of the third column, which were made by Dumas' method, do not exhibit any preponderance of positive values for the excess of observed density, but rather the opposite.

TABLE VI.—ACETIC ACID.

EXPERIMENTS OF PLAYFAIR AND WANKLYN.

Temperature.	Pressure.	Density calculated by eq. (12).	Density observed.			Excess of observed density.		
			I.	II.	III.	I.	II.	III.
212·5	322·8	2·124		2·060			− ·064	
194	326·0	2·168		2·055			− ·113	
186	254·4	2·173	1·936			− ·237		
182	319·4	2·213		2·108			− ·105	
166·5	289·5	2·293		2·350			+ ·057	
163	245·8	2·290	2·017			− ·273		
132	227·5	2·628	2·292			− ·336		
130·5	285·7	2·729		2·426			− ·303	
119	269·0	2·914		2·623			− ·291	
116·5	211·3	2·876	2·371			− ·505		
95·5	(123·8)	3·105			2·594			− ·511
86·5	(200·4)	3·432			3·172			− ·260
79·9	(83·3)	3·297			3·340			+ ·043
62·5	(46·2)	3·473			3·950			+ ·477

On the whole, these experiments furnish no decisive indication of any influence of the hydrogen or air upon the vapor. They may be thought to corroborate slightly the tendency observed in the experiments of Naumann and Troost toward lower densities than the formula gives at very low pressures. Yet where the experiments of Naumann show the greatest deficiency in observed density (at 78° and 80^{mm}), an experiment of Playfair and Wanklyn, at almost precisely the same temperature and pressure, gives a trifling excess of observed density, and at a little lower temperature and pressure, where we should expect from the experiments of Naumann that the deficiency would be still greater, an experiment of Playfair and Wanklyn shows a great excess of density.

By combining the experiments of Cahours, Naumann and Troost, we may obtain observations of density at 130° for a very wide range of pressures. For one atmosphere, we may regard the formula as coinciding with the average of the numbers given by Cahours. For pressures between three-quarters and one-half of an atmosphere the experiments of Naumann show an excess of density; at pressures below half an atmosphere the experiments both of Naumann and of Troost show a deficiency of density as compared with the formula. For an indefinite diminution of pressure, there can be little doubt that the real density, like the value given by the formula, approaches the theoretical value 2·073. The greatest excess in numbers obtained by experiment is ·07; the greatest deficiency is ·19, which occurs at 59·7^{mm}; the next in order of magnitude is ·11, which occurs more than once. These discrepancies are certainly such as may be accounted for by errors of observation. They do not appear to be greater than we might expect on the hypothesis of the entire correctness of the formula. On the other hand, the agreement is greater than we should expect, if we reject the theory on which the formula was obtained. It is about such as we might expect in a suitable formula of inter-polation with three constants, which have been determined by the values of the density for one atmosphere, for half an atmosphere, and for infinitesimal pressures. But we must regard the actual formula, in its application to this single temperature, as having only two constants, of which one is determined so as to make the formula give the theoretical value for infinitesimal pressures, and the other so as to make it agree with the experiments of Cahours at the pressure of one atmosphere.

An entirely different method has been employed by Horstmann* to determine the vapor-density of this substance. A current of dried air is forced through the liquid acid, which is heated to promote evaporation, and the mixture of air and vapor is cooled to any desired temperature, with deposition of the excess of acid, by passing upward through a spiral tube in a suitable bath. The acid is then separated from the air, and the quantity of each determined. It is assumed that the air is exactly saturated with vapor on leaving the coil, and that it has the temperature of the bath. If we know the pressure of saturated vapor for that temperature, and assume the validity of Dalton's law, it is easy to calculate the density of the vapor. For the pressure of the air is found by subtracting the pressure of the vapor from the total pressure (the experiments were so conducted that this was the same as the actual pressure of the atmosphere), and the ratio of the weights of the acid and the air obtained by analysis, divided by the ratio of their pressures, will give the ratio of their densities. The pressures of saturated vapor employed by Horstmann are those given by Landolt,† and differ greatly from the determina-tions of Regnault, in some cases being nearly twice as great,—a difference noticed but not explained by Landolt, who however gives

---

* *Berichte der deutschen chemischen Gesellschaft*, Jahrg. iii (1870), S. 78 ; and Jahrg. xi (1878), S. 1287.

† *Lieb. Ann.*, suppl. vi (1868), p. 157.

determinations (previously unpublished) of Wüllner, which somewhat exceed his own. (On the other hand, the observations of Bineau substantially agree with those of Regnault.)

If we compare the observations of Horstmann with the values given by equation (12), on the basis of Landolt's pressures, we find a very marked disagreement, as may be seen by the following numbers, which relate to the highest temperatures of Horstmann's experiments, where the disagreement is least:—

Temperature	-	-	63·1	62·9	59·9	51·1	49·0	48·7	44·6	41·4
Pressure (Land.)		-	110·0	109·2	97·0	69·0	63·4	63·0	53·1	46·6
Density calc. eq. (12)	-		3·67	3·67	3·69	3·75	3·77	3·77	3·79	3·81
Density obs.	-	-	3·19	3·11	3·12	3·16	2·89	2·98	2·75	2·62

It will be observed that while the values obtained from equation (12) increase with diminishing temperatures, the values obtained from Horstmann's experiments diminish. This diminution continues as far as the experiments go, until finally at 12° or 15° the densities are only one half as great as those obtained by Bineau, by direct experiment at the same temperatures and at somewhat less pressures, in a series of observations which bear every mark of a very exceptional precision. (Compare Tables VII and IV.) The explanation of this disagreement is doubtless to be found in the values of the pressures employed in the calculations, and it will be interesting to see how the results may be modified by the adoption of different pressures.

In determinations of the pressure of saturated vapors, too great values are so much more easily accounted for than errors in the opposite direction, especially when the pressures are small, that especial interest attaches to the lowest figures which are supported by a competent authority. The experiments of Regnault[*] were made with three different preparations of acetic acid, of which the second was once, and the third twice, purified by distillation over anhydrous phosphoric acid. Each distillation considerably diminished the pressure of the saturated vapor, the effect of the second distillation being about half that of the first. The numbers obtained with the third preparation are given in the following table with their logarithms, and the differences of the logarithms for one degree of temperature:—

Temperature.	Pressure.	log. pressure.	diff. per 1°.
9·71	6·42	·8075	
			·0239
12·12	7·33	·8651	
			·0272
14·33	8·42	·9253	
			·0161
14·87	8·59	·9340	
			·0252
17·23	9·85	·9934	
			·0251
19·84	11·455	1·0590	
			·0237
22·37	13·15	1·1189	
			·0232
25·28	15·36	1·1864	

[*] *Mém. Acad. Sciences*, vol. xxvi, p. 758. The experiments date from 1844.

The uniformity of the numbers in the last column shows the remarkable precision of the determinations. At the same time it is evident that the differences in these numbers are due principally to the errors of observation, so that numbers obtained by interpolation between the logarithms of the observed pressures will be somewhat better (on account of averaging of the errors) than the original determinations.

The values obtained by such an interpolation have been used for the comparison of Horstmann's experiments with the formula (12) which is given in Table VII. Unfortunately this comparison cannot be extended above 25°, which is the limit of Regnault's experiments. The first three columns of the table give the temperatures of Horstmann's experiments, the pressures corresponding to these temperatures according to the determinations of Landolt, and the density deduced from Horstmann's experiments by the use of these pressures. To

TABLE VII.—ACETIC ACID.

Determinations of Vapor-density by Distillation.

Temperature.	Pressure acc. to Landolt.	Density observed, Horstmann and Landolt.	Pressure acc. to Regnault.	Density calc. from Regnault's pressures by eq. (12).	Density observed, Horstmann and Regnault	Excess of observed density.	
						I.	II.
25·0	23·5	2·42	15·13	3·86	3·80	− ·06	
23·8	22·4	2·23	14·19	3·86	3·56		− ·30
22·6	21·6	2·29	13·31	3·87	3·76	− ·11	
21·5	20·4	2·24	12·54	3·87	3·68	− ·19	
20·4	19·2	2·05	11·81	3·88	3·37		− ·51
20·2	19·0	2·28	11·68	3·88	3·75	− ·13	
20·0	18·9	2·13	11·56	3·88	3·52		− ·36
17·4	16·8	2·09	9·95	3·89	3·56	− ·33	
15·6	15·6	1·98	8·96	3·90	3·48	− ·42	
15·3	15·3	1·95	8·81	3·90	3·42		− ·48
15·3	15·3	1·85	8·81	3·90	3·24		− ·66
14·7	15·1	1·78	8·54	3·91	3·18	− ·73	
12·7	13·7	1·96	7·60	3·91	3·56	− ·35	
12·4	13·5	1·89	7·46	3·92	3·45	− ·47	

these columns, which are taken from Horstmann's paper, are added the pressure derived from Regnault's observations by the logarithmic interpolation described above, the density calculated by equation (12) from these pressures and the temperatures of the first column, and the densities obtained by combining Horstmann's experiments with Regnault's pressures. This column is derived from the second, third and fourth, as follows. If w and W denote respectively the weights of vapor and of air which pass through the apparatus in the same time, P the height of the barometer, and p_L the pressure of saturated vapor as determined by Landolt, the densities obtained on the basis of Landolt's pressures, and given in the third column, are evidently represented by $\dfrac{w(\mathrm{P}-p_L)}{\mathrm{W}p_L}$. The numbers of the fifth column, which are represented in the same way by $\dfrac{w(\mathrm{P}-p_R)}{\mathrm{W}p_R}$, where p_R denotes the pressure as determined by Regnault's experiments, have been calculated by the present writer by multiplying the numbers of the third column by $\dfrac{p_{\mathrm{L}}(\mathrm{P}-p_{\mathrm{R}})}{p_{\mathrm{R}}(\mathrm{P}-p_{\mathrm{L}})}$.

As the height of the barometer in Horstmann's experiments is not given, it has been necessary to assume $\mathrm{P}=760$. The inaccuracy due to this circumstance is evidently trifling. The last two columns of the table, which relate to different series of experiments by Horstmann (a distinction not observed in other parts of the table), give the excess of the densities thus obtained from Horstmann's and Regnault's experiments above the values calculated from equation (12) with the use of Regnault's determinations of pressure.

The densities obtained by experiment are without exception less than those obtained from equation (12). At the highest temperatures, where the liability to error is the least, both in respect to the measurement of the pressure of saturated vapor and in respect to the analysis of the product of distillation, the results of experiment are most uniform, and most nearly approach the numbers required by the formula. At the lowest temperatures, the greatest observed density is about one-eleventh less than that required by the formula, the difference being about the same as between the highest and lowest observed values for the same temperature.

Since each successive purification of the substance employed by Regnault diminished the pressure of its vapor, it is not improbable that the pressures might have been still farther diminished by farther purification of the substance. The pressures which we have used are therefore liable to the suspicion of being too high, and it is quite possible that more accurate values of the pressure would still farther reduce the deficiency of observed density.

Perchloride of phosphorus.—For this substance, we have at atmospheric pressure a single determination of vapor-density by Mitscherlich,[*] and a series of determinations by Cahours;[†] at lower pressures we have determinations by Wurtz[‡] and by Troost and Hautefeuille.[§] In the experiments of Wurtz the pressure was reduced by mixing the vapor with air. In Table VIII all these determinations are compared with the formula

$$\log \frac{3\cdot6\,(\mathrm{D}-3\cdot6)}{(7\cdot2-\mathrm{D})^2} = \frac{5441}{t_{\mathrm{C}}+273} + \log p - 14\cdot353. \qquad (13)$$

The differences between the calculated and observed values are often large, in six cases exceeding ·30; but they exhibit in general that

---

[*] *Pogg. Ann.*, vol. xxix (1833), p. 221.

[†] *Comptes Rendus*, vol. xxi (1845), p. 625; and *Annales de Chimie et de Physique*, ser. 3, vol. xx (1847), p. 369.

[‡] *Comptes Rendus*, vol. lxxvi (1873), p. 601.   [§] *Ibid.*, vol. lxxxiii (1876), p. 977.

irregularity which is characteristic of errors of observation. We should expect large errors in the observed densities, on account of the difficulty of obtaining the substance in a state of purity, and because the large value of the density renders it very sensitive to the effect of impurities which diminish the density,—also because the specific heat of the vapor is great, as shown by the numerator of the fraction in the second member of (13),* and because the density varies very rapidly with the temperature as seen by the numbers in the third column of Table VIII.

TABLE VIII.—PERCHLORIDE OF PHOSPHORUS.

EXPERIMENTS OF MITSCHERLICH, CAHOURS, WURTZ, AND TROOST AND HAUTEFEUILLE.

Tempera-ture.	Pressure.	Density calculated by eq. (13).	Density observed.		Excess of observed density.	
			Mitscherlich.	Cahours.	Mitscherlich.	Cahours.
336	(760)	3·610		3·656		+ ·046
327	754	3·614		3·656		+ ·042
300	765	3·637		3·654		+ ·017
289	(760)	3·656		3·69		+ ·034
288	763	3·659		3·67		+ ·011
274	755	3·701		3·84		+ ·139
250	751	3·862		3·991		+ ·129
230	746	4·159		4·302		+ ·142
222	753	4·344	4·85		+ ·506	
208	(760)	4·752		4·73		− ·021
200	758	5·018		4·851		− ·167
190	758	5·368		4·987		− ·381
182	757	5·646		5·078		− ·568
			Wurtz.	T. & H.	Wurtz.	T. & H.
178·5	227·2	5·053		5·150		+ ·097
175·8	253·7	5·223		5·235		+ ·012
167·6	221·8	5·456		5·415		− ·041
154·7	221	5·926		5·619		− ·307
150·1	225	6·086		5·886		− ·200
148·6	244	6·169		5·964		− ·205
145	391	6·45	6·55		+ ·10	
145	311	6·37	6·70		+ ·33	
145	307	6·36	6·33		− ·03	
144·7	247	6·287		6·14		− ·147
137	281	6·53	6·48		− ·05	
137	269	6·51	6·54		+ ·03	
137	243	6·48	6·46		− ·02	
137	234	6·47	6·42		− ·05	
137	148	6·31	6·47		+ ·16	
129	191	6·59	6·18		− ·41	
129	170	6·56	6·63		+ ·07	
129	165	6·55	6·31		− ·24	

But at the two lowest temperatures of Cahours' experiments, the differences of the observed and calculated densities (·381 and ·568) are not only great, but exhibit, in connection with the adjacent numbers, a regularity which suggests a very different law from that of the

---

* Compare *Equilib. Het. Subs.*, this volume p. 180, and *supra* pp. 380-382.

formula. In fact, the densities obtained by Cahours at atmospheric pressure and those obtained by Troost and Hautefeuille at pressures a little less than one-third of an atmosphere seem to form a continuous series, notwithstanding the abrupt change of pressure. Yet it is difficult to admit that the density is independent of the pressure. So radical a difference between the behavior of this substance and that of the others which we have been considering requires unequivocal evidence. Now it is worthy of notice that the experiment at 182°, in which the greatest discrepancy is seen, is not given in the first record of the experiments, which was in the *Comptes Rendus* in 1845. It is given in the *Annales de Chimie et de Physique* in 1847, where it is called the first experiment. (The experiment at 336° is also omitted in the *Comptes Rendus* and that at 208° in the *Annales*,—otherwise the lists are the same.) If it was the first experiment in point of time, which is apparently the meaning, it was made before the publication in the *Comptes Rendus*, and we can only account for its omission by supposing that it was a preliminary experiment, in which its distinguished author did not feel sufficient confidence to include it at first with his other determinations, although he afterwards concluded to insert it. If we reject this observation as doubtful, the disagreement between the formula and observation appears to be within the limits of possible error, but additional experiments will be necessary to confirm the formula.*

Experiments have also been made by M. Wurtz in which the vapor of the perchloride of phosphorus was diluted with that of the protochloride.† These experiments may be used to test equation (8), which, when the values of its constants are determined by equation (13), reduces to the form

$$\log \frac{p_5}{p_2 p_3} = \frac{5441}{t_C + 273} - 13\cdot751, \tag{14}$$

where p_5, p_2, and p_3 denote the partial pressures due respectively to the PCl_5, the Cl_2, and the PCl_3, existing as such in the gas-mixture. Since these quantities cannot be the subjects of immediate observation, a farther transformation of the equation will be convenient. Let M_3, M_2 denote the quantities of the protochloride and of chlorine of which the mixture may be formed, and P_3, P_2 the pressure which

---

* Additional experiments on the density of this vapor have been made by M. Cahours, concerning which he says in 1866 : " Les déterminations qui je viens d'effectuer à 170 et 172 degrés (ce corps bout vers 160 à 165 degrés) m'ont donné des nombres qui, bien que notablement plus forts que ceux que j'ai obtenus antérieurement à 182 et 185 degrés, sont encore bien éloignés de celui que correspond à 4 volumes," *Comptes Rendus*, t. 63, p. 16. So far as the present writer has been able to ascertain, these determinations have not been published. The formula gives 6·025 for 170° and 5·973 for 172°, at atmospheric pressure. The number corresponding to four volumes is 7·20.

† *Comptes Rendus*, vol. lxxvi (1873), p. 601.

would belong to each of these if existing by itself with the same volume and temperature. These quantities will be connected by the equations

$$P_2 = \frac{kt M_2}{2 \cdot 22 v}, \qquad P_3 = \frac{kt M_3}{4 \cdot 98 v}, \tag{15}$$

where k denotes the same constant as on page 381. From the evident relations

$$P_2 = p_2 + p_5, \qquad P_3 = p_3 + p_5, \qquad p = p_2 + p_3 + p_5,$$

we obtain

$$p_5 = P_2 + P_3 - p, \qquad p_2 = p - P_3, \qquad p_3 = p - P_2;$$

and by substitution of these values in equation (14),

$$\log \frac{P_2 + P_3 - p}{(p - P_2)(p - P_3)} = \frac{5441}{t_C + 273} - 13 \cdot 751. \tag{16}$$

In view of the relations (15), this may be regarded as an equation between the pressure, the temperature, the volume, and the quantities of protochloride of phosphorus and chlorine into which the gas-mixture is resolvable.

TABLE IX.—PERCHLORIDE AND PROTOCHLORIDE OF PHOSPHORUS.

EXPERIMENTS ON THE MIXED VAPORS BY WURTZ.

No. of exp.	t_c	p (obs.)	π	δ	P_2	P_3	p calculated by eq. (16).	Excess of obs. value of p.
XII	173·29	756·1	423	6·68	392·4	725·5	760·7	− 4·6
X	165·4	748·4	413	6·80	390·1	725·5	747·9	+ ·5
VII	176·24	751·0	411	6·88	392·7	732·7	773·1	− 22·1
VIII	169·35	724·1	394	7·16	391·8	721·9	750·5	− 26·4
V	175·26	743·3	343	7·03	334·9	735·2	764·4	− 21·1
II	164·9	758·5	338	7·38	346·4	766·9	782·9	− 24·4
XI	175·75	760·0	318	7·00	309·2	751·2	776·8	− 16·8
IV	175·26	756·3	271	7·06	265·7	751·0	770·9	− 14·6
IX	160·47	753·5	214	7·44	221·1	760·6	766·8	− 13·3
I	165·4	760·0	194	7·25	195·3	761·3	768·5	− 8·5
VI	170·34	751·2	174	8·30	200·6	777·8	787·6	− 36·4
III	174·28	742·7	168	7·74	180·6	755·3	766·5	− 23·8

It is in this form that we shall apply the equation to the experiments of M. Wurtz, the results of which are exhibited in Table IX. The first column gives the number distinguishing each experiment in the original memoir; the second, the temperature; the third, the observed pressure (p) of the mixture of PCl_5, PCl_3, and Cl_2, which is the barometric pressure corrected for the small quantity of air remaining in the flask; the fourth, the pressure π due to the *possible perchloride*, found by subtracting the pressure due to the excess of protochloride (this pressure is calculated from the theoretical density of the protochloride) from the total pressure; the fifth, the density δ of the possible perchloride calculated from its pressure π with the temperature and volume. The numbers of these five columns are taken from the memoir cited, except that the correction of the barometric pressures has been applied by the present writer in accordance with the data furnished in that memoir. The two next columns contain the values of P_2 and P_3. These would naturally be calculated from M_2 and M_3 by equations (15). But since the values of M_2 and M_3 have not been given explicitly, those of P_2 and P_3 have been calculated from the recorded values of π and δ. Since the weight of the *possible perchloride* is $\dfrac{7 \cdot 2}{2 \cdot 22} M_2$, we have

$$\delta = \frac{7 \cdot 2 \, M_2 \, kt}{2 \cdot 22 \, v\pi} = \frac{7 \cdot 2}{\pi} P_3.$$

Moreover,

$$p - \pi = P_3 - P_2,$$

since both members of the equation express the pressure due to the excess of the protochloride. The values of P_2 and P_3 were obtained by these equations.

The eighth column of the table gives the values of p calculated from the preceding values of t_C, P_2, and P_3, by equation (16); and the last column, the difference of the observed and calculated values of p. The average difference is 18^{mm}, or a little more than two per cent., the observed pressure being almost uniformly less than the calculated value. This deficiency of pressure is doubtless to be accounted for by a fact which MM. Troost and Hautefeuille have noticed in this connection. The protochloride of phosphorus deviates quite appreciably from the laws of Mariotte, Gay-Lussac, and Avogadro, the product of the volume and pressure of a given quantity of vapor at 180° and the pressure of one atmosphere being 1·548 per cent. less than at the same temperature and the pressure of one-half an atmosphere.[*] Now we may assume as a general rule that when the product of volume and pressure of a gas is slightly less than the theoretical number (calculated by the laws of Mariotte, Gay-Lussac, and Avogadro) the difference for any same temperature is nearly proportional to the pressure.[†] It is therefore probable that between 160° and 180°, at pressures of about one atmosphere, the product of volume and pressure for protochloride of phosphorus is somewhat more than three per cent. less than the theoretical number. The experiments of Wurtz, as exhibited in Table IX, show that the pressure, and therefore the product of volume and pressure (we may evidently give the volume any constant value as unity), in a mixture consisting principally of the protochloride is on the average a little more than two per cent. less than is demanded by theory, the differences being greater when the proportion of the protochloride is

---

[*] Troost and Hautefeuille, *Comptes Rendus*, vol. lxxxiii (1876), p. 334.

[†] Andrews, "On the Gaseous State of Matter," *Phil. Trans.*, vol. clxvi (1876), p. 447.

greater. The deviation from the calculated values is therefore in the same direction and about such in quantity as we should expect.[*]

M. Wurtz has remarked that the average value of δ (the density of the *possible perchloride*) is nearly identical with the theoretical density of the perchloride, and appears inclined to attribute the variations from this value to the errors of experiment. Yet it appears very distinctly in Table IX, in which the experiments are arranged according to the value of π (the pressure due to the *possible perchloride*), that δ increases as π diminishes. The experiments of MM. Troost and Hautefeuille show that the coincidence remarked by M. Wurtz is due to the fact that on the average in these experiments the deficiency of the density of the possible perchloride (compared with the theoretical value) is counterbalanced by the excess of density of the protochloride. When $\pi > 400$, the effect of the deficiency in the density of the possible perchloride distinctly preponderates; when $\pi < 250$, the effect of the excess of density in the protochloride distinctly preponderates. But the magnitude of the differences concerned is not such as to invalidate the general conclusion established by the experiments of M. Wurtz, that the dissociation of the perchloride may be prevented (at least approximately) by mixing it with a large quantity of the protochloride.

Table for facilitating calculation.—The numerical solution of equations (10), (11), (12) and (13) for given values of t and p may be facilitated by the use of a table. If we set

$$\Delta = \frac{D}{D_1}, \tag{17}$$

$$L = \log \frac{1000\,D_1(D-D_1)}{(2D_1-D)^2} = \log \frac{1000(\Delta-1)}{(2-\Delta)^2}, \tag{18}$$

we have for peroxide of nitrogen,

$$L = \frac{3118\cdot6}{t_c+273} + \log p - 9\cdot451\,; \tag{19}$$

for formic acid,

$$L = \frac{3800}{t_c+273} + \log p - 9\cdot641\,; \tag{20}$$

for acetic acid,

$$L = \frac{3520}{t_c+273} + \log p - 8\cdot349\,; \tag{21}$$

---

[*] The deviation of the protochloride of phosphorus from the laws of ideal gases shows the impossibility of any *very close* agreement between such equations as have been deduced in this paper and the results of experiment in the case of gas-mixtures in which this substance is one of the components. With respect to the question whether future experiments on the vapor of the perchloride (alone, or with an excess of chlorine or of the protochloride) will reduce the disagreement between the calculated and observed values to such magnitudes as occur in the case of the protochloride alone, it would be rash to attempt to anticipate the result of experiment.

and for perchloride of phosphorus,

$$L = \frac{5441}{t_c + 273} + \log p - 11 \cdot 353. \tag{22}$$

By these equations the values of L are easily calculated. The values of Δ may then be obtained by inspection (with interpolation when necessary) of the following table. From Δ the value of D may be obtained by multiplying by D_1, viz., by 1·589 for peroxide of nitrogen or formic acid, by 2·073 for acetic acid, and by 3·6 for perchloride of phosphorus.*

TABLE X.

For the solution of the equation: $\log \dfrac{1000(\Delta - 1)}{(2 - \Delta)^2} = L.$

L	Δ	Diff.	L	Δ	Diff.	L	Δ	Diff.
·7	1·005		3·0	1·382		5·3	1·932	
·8	1·006	1	3·1	1·421	39	5·4	1·939	7
·9	1·008	2	3·2	1·461	40	5·5	1·945	6
1·0	1·010	2	3·3	1·500	39	5·6	1·951	6
1·1	1·012	2	3·4	1·537	37	5·7	1·956	5
1·2	1·015	3	3·5	1·574	37	5·8	1·961	5
1·3	1·019	4	3·6	1·609	35	5·9	1·965	4
1·4	1·024	5	3·7	1·642	33	6·0	1·969	4
1·5	1·030	6	3·8	1·673	31	6·1	1·972	3
1·6	1·037	7	3·9	1·703	30	6·2	1·975	3
1·7	1·046	9	4·0	1·730	27	6·3	1·978	3
1·8	1·056	10	4·1	1·755	25	6·4	1·980	2
1·9	1·069	13	4·2	1·778	23	6·5	1·982	2
2·0	1·084	15	4·3	1·800	22	6·6	1·984	2
2·1	1·102	18	4·4	1·819	19	6·7	1·986	2
2·2	1·122	20	4·5	1·837	18	6·8	1·987	1
2·3	1·146	24	4·6	1·854	17	6·9	1·989	2
2·4	1·172	26	4·7	1·868	14	7·0	1·990	1
2·5	1·202	30	4·8	1·882	14	7·2	1·992	
2·6	1·234	32	4·9	1·894	12	7·4	1·994	
2·7	1·268	34	5·0	1·905	11	7·6	1·995	
2·8	1·305	37	5·1	1·915	10	7·8	1·996	
2·9	1·343	38	5·2	1·924	9	8·0	1·997	
3·0	1·382	39	5·3	1·932	8	9·0	1·999	

The constants of these equations are of course subject to correction by future experiments, which must also decide the more general question—in what cases, and within what limits, and with what degree of approximation, the actual relations can be expressed by equations of such form. In the case of perchloride of phosphorus especially, the formula proposed requires confirmation.

---

* The value of Δ diminished by unity expresses the ratio of the number of the molecules of the more complex type to the whole number of molecules. Thus, if $\Delta = 1·20$, in the case of peroxide of nitrogen there are 20 molecules of the type N_2O_4 to 80 of the type NO_2, or in the case of perchloride of phosphorus there are 20 molecules of the type PCl_5 to 40 of the type PCl_3 and 40 of the type Cl_2. A consideration of the varying values of Δ is therefore more instructive than that of the values of D, and it would in some respects be better to make the comparison of theory and experiment with respect to the values of Δ.

ON AN ALLEGED EXCEPTION TO THE SECOND LAW OF THERMODYNAMICS.

[*Science*, vol. I, p. 160, Mar. 16, 1883.]

ACCORDING to the received doctrine of radiation, heat is transmitted with the same intensity in all directions and at all points within any space which is void of ponderable matter and entirely surrounded by stationary bodies of the same temperature. We may apply this principle to the arrangement recently proposed by Prof. H. T. Eddy * for transferring heat from a colder body A to a warmer B without expenditure of work.

In its simplest form the arrangement consists of parallel screens, which are placed between the bodies A and B, and have the form of very thin disks with certain apertures, and the property of totally reflecting heat. These disks, or screens, are supposed to be fixed on a common axis, and to revolve with a constant velocity. For the purposes of theoretical discussion, we may allow this velocity to be kept up without expenditure of work, since we may suppose the experiment to be made *in vacuo*. If the dimensions and velocity of the apparatus are such that the screens receive a considerable change of position during the time in which radiant heat traverses the distances between them, the apertures in the screens may be so placed that radiations can pass from A to B, but not from B to A. It is inferred that it is possible, by such means, to make heat pass from a colder to a warmer body without compensation.

In order to judge of the validity of this inference, let us suppose thermal equilibrium to subsist initially in the system, and inquire whether the motion of the screens will have any tendency to disturb that equilibrium. We suppose, then, that the screens, the bodies A and B, and the walls enclosing the space in which the experiment is made, have all the same temperature, and that the spaces between and around the screens and the bodies A and B are filled with the radiations which belong to that temperature, according to the principle cited above. Under such circumstances, it is evident that the presence of the screens, whether at rest or in motion, will not have

---

* *Journ. Frankl. Inst.*, March, 1883.

any influence upon the intensity of the radiations passing through the spaces between and around them; since the heat reflected by a screen in any direction is the exact equivalent of that which would proceed in the same direction (without reflection) if the screen were not there. So, also, the heat passing through any aperture in a screen is the exact equivalent of that which would be reflected in the same direction if there were no aperture. The quantities of radiant heat which fall upon the bodies A and B are therefore entirely unchanged by the presence and the motion of the screens, and their temperature cannot be affected.

We may conclude *a fortiori* that B will not grow warmer if A is colder than B, and none of the other bodies present are warmer than B.

Since the body A, for example, when the screens are in motion, does not receive radiations from every body to which it sends them, it is not without interest to inquire from what bodies it will receive its share of heat. This problem may be solved most readily by supposing the screens to move in the opposite direction, with the same velocity as before. One may easily convince himself that every body which receives radiant heat from A when the apparatus moves backward, will impart heat to A when the apparatus moves forward, and to exactly the same amount, if its temperature is the same as that of A.

VII.

ELECTROCHEMICAL THERMODYNAMICS.

*Two letters to the Secretary of the Electrolysis Committee of the
British Association for the Advancement of Science.*

[*Report Brit. Asso. Adv. Sci.*, 1886, pp. 388, 389; and 1888, pp. 343–346.]

New Haven, *January* 8, 1887.

Professor OLIVER J. LODGE,

Dear Sir,—Please accept my thanks for the proof copy of your
"Report on Electrolysis in its Physical and Chemical Bearings," which
I received a few days ago with the invitation, as I understand it, to
comment thereon.

I do not know that I have anything to say on the subjects more
specifically discussed in this report, but I hope I shall not do violence
to the spirit of your kind invitation or too much presume on your
patience if I shall say a few words on that part of the general subject
which you discussed with great clearness in your last report on
pages 745 ff. (Aberdeen). To be more readily understood, I shall
use your notation and terminology, and consider the most simple case
possible.

Suppose that two radicles unite in a galvanic cell during the
passage of a unit of electricity, and suppose that the same quantities
of the radicles would give θ_ϵ units of heat in uniting directly, that is,
without production of current; will the union of the radicles in the
galvanic cell give $J\theta_\epsilon$ units of electrical work? Certainly not, unless
the radicles can produce the heat at an infinitely high temperature,
which is not, so far as we know, the usual case. Suppose the highest
temperature at which the heat can be produced is t'', so that at this
temperature the union of the radicles with evolution of heat is a
reversible process; and let t' be the temperature of the cell, both
temperatures being measured on the absolute scale. Now θ_ϵ units
of heat at the temperature t'' are equivalent to $\theta_\epsilon \dfrac{t'}{t''}$ units of heat at
the temperature t', together with $J\theta_\epsilon \dfrac{t''-t'}{t''}$ units of mechanical or
electrical work. (I use the term "equivalent" *strictly* to denote

reciprocal convertibility, and not in the loose and often misleading sense in which we speak of heat and work as equivalent when there is only a one-sided convertibility.) Therefore the *rendement* of a perfect or reversible galvanic cell would be $J\theta\epsilon\dfrac{t''-t'}{t''}$ units of electrical work, with $\theta\epsilon\dfrac{t'}{t''}$ units of (reversible) heat, for each unit of electricity which passes.

You will observe that we have thus solved a very different problem from that which finds its answer in the Joule-Helmholtz-Thomson equation with term for reversible heat. That equation gives a relation between the E.M.F. and the reversible heat and certain other quantities, so that if we set up the cell and measure the reversible heat, we may determine the E.M.F. without direct measurement, or *vice versâ*. But the considerations just adduced enable us to predict both the electromotive force and the reversible heat without setting up the cell at all. Only in the case that the reversible heat is zero does this distinction vanish, and not then unless we have some way of knowing *à priori* that this is the case.

From this point of view it will appear, I think, that the production of reversible heat is by no means anything accidental, or superposed, or separable, but that it belongs to the very essence of the operation.

The thermochemical data on which such a prediction of E.M.F. and reversible heat is based must be something more than the heat of union of the radicles. They must give information on the more delicate question of the temperature at which that heat can be obtained. In the terminology of Clausius they must relate to entropy as well as to energy—a field of inquiry which has been far too much neglected.

Essentially the same view of the subject I have given in a form more general and more analytical, and, I fear, less easily intelligible, in the closing pages of a somewhat lengthy paper on the "Equilibrium of Heterogeneous Substances" (*Conn. Acad. Trans.*, vol. iii, 1878), of which I send you the Second Part, which contains the passage in question. My separate edition of the First Part has long been exhausted. The question whether the "reversible heat" is a negligible quantity is discussed somewhat at length on pages 510-519.* On page 503† is shown the connection between the electromotive force of a cell and the difference in the value of (what I call) the *potential for one of the ions* at the electrodes. The definition of the *potential for a material substance*, in the sense in which I use the term, will be found on page 443‡ of the synopsis from the *Am. Jour. Sci.*, vol. xvi, which I enclose. I cannot say that the term has been adopted by

---

* [This vol., pp. 339-347.]   † [*Ibid.*, p. 333.]   ‡ [*Ibid.*, p. 356.]

physicists. It has, however, received the unqualified commendation of Professor Maxwell (although not with reference to this particular application—see his lecture on the "Equilibrium of Heterogeneous Substances," in the science conferences at South Kensington, 1876); and I do not see how we can do very well without the idea in certain kinds of investigations.

Hoping that the importance of the subject will excuse the length of this letter,

I remain,

Yours faithfully,

J. WILLARD GIBBS.

New Haven, November 21, 1887.

Professor OLIVER J. LODGE,

Dear Sir,—As the letter which I wrote you some time since concerning the *rendement* of a perfect or reversible galvanic cell seems to have occasioned some discussion, I should like to express my views a little more fully.

It is easy to put the matter in the canonical form of a Carnot's cycle. Let a unit of electricity pass through the cell producing certain changes. We may suppose the cell brought back to its original condition by some reversible chemical process, involving a certain expenditure (positive or negative) of work and heat, but involving no electrical current nor any permanent changes in other bodies except the supply of this work and heat.

Now the first law of thermodynamics requires that the algebraic sum of all the work and heat (measured in "equivalent" units) supplied by external bodies during the passage of the electricity through the cell, and the subsequent processes by which the cell is restored to its original condition, shall be zero.

And the second law requires that the algebraic sum of all the heat received from external bodies, divided, each portion thereof, by the absolute temperature at which it is received, shall be zero.

Let us write W for the work and Q for the heat supplied by external bodies during the passage of the electricity, and [W], [Q] for the work and heat supplied in the subsequent processes.

Then

$$W + Q + [W] + [Q] = 0, \tag{1}$$

and

$$\frac{Q}{t'} + \int \frac{d[Q]}{t} = 0, \tag{2}$$

where t under the integral sign denotes the temperature at which the element of heat $d[Q]$ is supplied, and t' the temperature of the cell, which we may suppose constant.

Now the work W includes that required to carry a unit of electricity from the cathode having the potential V'' to the anode having the potential V'. (These potentials are to be measured in masses of the same kind of metal attached to the electrodes.) When there is any change of volume, a part of the work will be done by the atmosphere or other body enclosing the cell. Let this part be denoted by W_P. In some cases it may be necessary to add a term relating to gravity, but as such considerations are somewhat foreign to the essential nature of the problem which we are considering, we may set such cases aside. We have then

$$W = V' - V'' + W_P \qquad (3)$$

Combining these equations we obtain

$$V'' - V' = W_P + [W] + [Q] - t' \int \frac{d[Q]}{t}. \qquad (4)$$

It will be observed that this equation gives the electromotive force in terms of quantities which may be determined *without setting up the cell.*

Now $[W] + [Q]$ represents the increase of the intrinsic energy of the substances in the cell during the processes to which the brackets relate, and $\int \frac{d[Q]}{t}$ represents their increase of entropy during the same processes. The same expressions, therefore, with the contrary signs, will represent the increase of energy and entropy in the cell during the passage of the current. We may therefore write

$$V'' - V' = -\Delta \epsilon + t' \Delta \eta + W_P, \qquad (5)$$

where $\Delta \epsilon$ and $\Delta \eta$ denote respectively the increase of energy and entropy in the cell during the passage of a unit of electricity. This equation is identical in meaning, and nearly so in form, with equation (694) of the paper cited in my former letter, except that the latter contains the term relating to gravity. See *Trans. Conn. Acad.,* iii (1878), p. 509.* The matter is thus reduced to a question of energy and entropy. Thus, if we knew the energy and entropy of oxygen and hydrogen at the temperature and pressure at which they are disengaged in an electrolytic cell, and also the energy and entropy of the acidulated water from which they are set free (the latter, in strictness, as functions of the degree of concentration of the acid), we could at once determine the electromotive force for a reversible cell. This would be a limit below which the electromotive force required in an actual cell used electrolytically could not fall, and above which the electromotive force of any such cell used to produce a current (as in a Grove's gas battery) could not reach.

---

* [This volume, p. 338.]

Returning to equation (4), we may observe that if t under the integral sign has a constant value, say t'', the equation will reduce to

$$V'' - V' = \frac{t'' - t'}{t''}[Q] + [W] + W_P. \tag{6}$$

Such would be the case if we should suppose that at the temperature t'' the chemical processes to which the brackets relate take place reversibly with evolution or absorption of heat, and that the heat required to bring the substances from the temperature of the cell to the temperature t'', and that obtained in bringing them back again to the temperature of the cell, may be neglected as counterbalancing each other. This is the point of view of my former letter. I do not know that it is necessary to discuss the question whether any such case has a real existence. It appears to me that in supposing such a case we do not exceed the liberty usually allowed in theoretical discussions. But if this should appear doubtful, I would observe that the equation (6) must hold in all cases if we give a slightly different definition to t'', viz., if t'' be defined as a temperature determined so that

$$\frac{[Q]}{t''} = \int \frac{d[Q]}{t}. \tag{7}$$

The temperature t'', thus defined, will have an important physical meaning. For by means of perfect thermo-dynamic engines we may change a supply of heat $[Q]$ at the constant temperature t'' into a supply distributed among the various temperatures represented by t in the manner implied in the integral, or *vice versâ*. We may, therefore, while vastly complicating the experimental operations involved, obtain a theoretical result which may be very simply stated and discussed. For we now see that after the passage of the current we may (theoretically) by reversible processes bring back the cell to its original state simply by the expenditure of the heat $[Q]$ supplied at the temperature t'', with perhaps a certain amount of work represented by $[W]$, and that the electromotive force of the cell is determined by these quantities in the manner indicated by equation (6), which may sometimes be further simplified by the vanishing of $[W]$ and W_P.

If the current causes a separation of radicles, which are afterwards united with evolution of heat, $[Q]$ being in this case negative, t'' represents the highest temperature at which this heat can be obtained. I do not mean the highest at which any part of the heat can be obtained—that would be quite indefinite—but the highest at which the whole can be obtained. I should add that if the effect of the union of the radicles is obtained partly in work—$[W]$, and partly in heat—$[Q]$, we may vary the proportion of work and heat; and t''

will then vary directly as [Q]. But if the effect is obtained entirely in heat, t'' will have a perfectly definite value.

It is easy to show that these results are in complete accordance with Helmholtz's differential equation. We have only to differentiate the value which we have found for the electromotive force. For this purpose equation (5) is most suitable. It will be convenient to write E for the electromotive force $V' - V''$, and for the differences $\Delta\epsilon$, $\Delta\eta$ to write the fuller forms $\epsilon'' - \epsilon'$, $\eta'' - \eta'$, where the single and double accents distinguish the values before and after the passage of the current. We may also set $p(v' - v'')$ for W_P, where p is the pressure (supposed uniform) to which the cell is subjected, and $v'' - v'$ is the increase of volume due to the passage of the current. If we also omit the accent on the t, which is no longer required, the equation will read

$$E = \epsilon'' - \epsilon' - t(\eta'' - \eta') + p(v'' - v'). \tag{8}$$

If we suppose the temperature to vary, the pressure remaining constant, we have

$$dE = d\epsilon'' - d\epsilon' - t\,d\eta'' + t\,d\eta' - (\eta'' - \eta')\,dt + p\,dv'' - p\,dv'. \tag{9}$$

Now, the increase of energy $d\epsilon'$ is equal to the heat required to increase the temperature of the cell by dt diminished by the work done by the cell in expanding. Since $d\eta'$ is the heat imparted divided by the temperature, the heat imparted is $t\,d\eta'$, and the work is obviously $p\,dv'$. Hence

$$d\epsilon' = t\,d\eta' - p\,dv',$$

and in like manner

$$d\epsilon'' = t\,d\eta'' - p\,dv''.$$

If we substitute these values, the equation becomes

$$dE = (\eta' - \eta'')\,dt. \tag{10}$$

We have already seen that $\eta' - \eta''$ represents the integral $\int \dfrac{d[Q]}{t}$ of equations (2) and (4), which by equation (2) is equal to the reversible heat evolved, $-Q$, divided by the temperature of the cell, which we now call t. Substitution of this value gives

$$\frac{dE}{dt} = -\frac{Q}{t}, \tag{11}$$

which is Helmholtz's equation.

These results of the second law of thermodynamics are of course not to be applied to any real cells, except so far as they approach the condition of reversible action. They give, however, in many cases limits on one side of which the actual values must lie. Thus, if we set $\leqq$ for $=$ in equations (2), (4), (5), (6), and $\geqq$ for $=$ in (8), the formula will there hold true without the limitation of reversibility.

But we cannot get anything by differentiating an inequality, and it does not appear à priori which side of (10) is the greater when the condition of reversibility is not satisfied. The term $\frac{Q}{t}$ in (11) is certainly not greater than $\eta'' - \eta'$, for which it was substituted. But this does not determine which side of (11) is the greater in case of irreversibility. It is the same with Helmholtz's method of proof, which is quite different from that here given, but indicates nothing except so far as the condition of reversibility is fulfilled. (See *Sitzungsberichte Berl. Acad.*, 1882, pp. 24, 25.)

I fear that it is a poor requital for the kind wish which you expressed at Manchester, that I were present to explain and support my position, for me to impose so long a letter upon you. Trusting, however, in your forbearance, I remain, yours faithfully,

J. WILLARD GIBBS.

VIII.

SEMI-PERMEABLE FILMS AND OSMOTIC PRESSURE.

[*Nature*, vol. LV, pp. 461, 462, Mar. 18, 1897.]

LORD KELVIN'S very interesting problem concerning molecules which differ only in their power of passing a diaphragm (see *Nature* for January 21, p. 272), seems only to require for its solution the relation between density and pressure for the fluid at the temperature of the experiment, when this relation for small densities becomes that of an ideal gas; in other cases, a single numerical constant in addition to the relation between density and pressure is sufficient.

This will, perhaps, appear most readily if we imagine each of the vessels A and B connected with a vertical column of the fluid which it contains, these columns extending upwards until the state of an ideal gas is reached. The equilibrium which we suppose to subsist will not be disturbed by communications between the columns at as many levels as we choose, if these communications are always made through the same kind of semi-permeable diaphragm as that which separates the vessels A and B. It will be observed that the difference of level at which any same pressure is found in the two columns is a constant quantity, easily determined in the upper parts (where the fluids are in the ideal gaseous state) as a function of the composition of the fluid in the A-column, and giving at once the height above the vessel A, where in the A-column we find a pressure equal to that in the vessel B.

In fact, we have in either column

$$dp = -g\gamma\,dz,$$

where the letters denote respectively pressure, force of gravity, density, and vertical elevation. If we set

$$\frac{1}{\gamma} = F'(p),$$

we have

$$F'(p)\,dp = -g\,dz.$$

Integrating, with a different constant for each column, we get

$$F(p_A) = -g(z - C_A),$$
$$F(p_B) = -g(z - C_B),$$
$$F(p_A) - F(p_B) = g(C_A - C_B).$$

In the upper regions,

$$F'(p) = \frac{1}{\gamma} = \frac{at}{p};$$

$$\therefore F(p) = at \log p,$$

where t denotes temperature, and a the constant of the law of Boyle and Charles.   Hence,

$$at \log p_A - at \log p_B = g(C_A - C_B).$$

Moreover, if $1:n$ represents the constant ratio in which the S- and D-molecules are mixed in the A-column, we shall have in the upper regions, where the S-molecules have the same density in the two columns,

$$\gamma_A = (1+n)\gamma_B, \quad p_A = (1+n)p_B,$$

$$g(C_A - C_B) = at \log (1+n).$$

Therefore, at any height,

$$F(p_A) - F(p_B) = at \log (1+n).$$

This equation gives the required relation between the pressures in A and B and the composition of the fluid in A.   It agrees with van't Hoff's law, for when n is small the equation may be written

$$F'(p_A)(p_A - p_B) = atn$$

or

$$p_A - p_B = atn\gamma_A.$$

But we must not suppose, in any literal sense, that this difference of pressure represents the part of the pressure in A which is exerted by the D-molecules, for that would make the total pressure calculable by the law of Boyle and Charles.

To show that the case is substantially the same, at least for any one temperature, when the fluid is not volatile, we may suppose that we have many kinds of molecules, A, B, C, etc., which are identical in all properties except in regard to passing diaphragms.   Let us imagine a row of vertical cylinders or tubes closed at both ends.   Let the first contain A-molecules sufficient to give the pressure p' at a certain level.   Then let it be connected with the second cylinder through a diaphragm impermeable to B-molecules, freely permeable to all others. Let the second cylinder contain such quantities of A- and B-molecules as to be in equilibrium with the first cylinder, and to have a certain pressure p'' at the level of p' in the first cylinder.   At a higher level this second cylinder will have the pressure which we have called p'. There let it be connected with the third cylinder through a diaphragm impermeable to C-molecules, and to them alone.   Let this third cylinder contain such quantities of A-, B-, and C-molecules as to be in equilibrium with the second cylinder, and have the pressure p'' at the diaphragm ; and so on, the connections being so made, and the quantities of the several kinds of molecules so regulated, that the pressures at all the diaphragms shall have the same two values.

It is evident that the vertical distance between successive connections must be everywhere the same, say l; also, that at all the diaphragms, cn the side of the greater pressure, the proportion of molecules which can and which cannot pass the diaphragm must be the same. Let the ratio be $1:n$. If we write γ_A, γ_B, etc., for the densities of the several kinds of molecules, and γ for the total density, we have for the second cylinder

$$\frac{\gamma_A+\gamma_B}{\gamma_A}=1+n.$$

For the third cylinder we have this equation, and also

$$\frac{\gamma_A+\gamma_B+\gamma_C}{\gamma_A+\gamma_B}=1+n,$$

which gives

$$\frac{\gamma_A+\gamma_B+\gamma_C}{\gamma_A}=(1+n)^2.$$

In this way, we have for the rth cylinder

$$\frac{\gamma}{\gamma_A}=(1+n)^{r-1}.$$

Now the vertical distance between equal pressures in the first and rth cylinders, is

$$(r-1)l.$$

Now the equilibrium will not be destroyed if we connect all the cylinders with the first through diaphragms impermeable to all except A-molecules. And the last equation shows that as γ/γ_A increases geometrically, the vertical distance between any pressure in the column when this ratio of densities is found, and the same pressure in the first cylinder increases arithmetically. This distance, therefore, may be represented by $\log(\gamma/\gamma_A)$ multiplied by a constant. This is identical with our result for a volatile liquid, except that for that case we found the value of the constant to be at/g.

The following demonstration of van't Hoff's law, which is intended to apply to existing substances, requires only that the solutum, i.e., dissolved substance, should be capable of the ideal gaseous state, and that its molecules, as they occur in the gas, should not be broken up in the solution, nor united to one another in more complex molecules.

It will be convenient to use certain quantities which may be called the *potentials* of the solvent and of the solutum, the term being thus defined:—In any sensibly homogeneous mass, the *potential* of any independently variable component substance is the differential coefficient of the thermodynamic energy of the mass taken with respect to that component, the entropy and volume of the mass and the quantities of its other components remaining constant. The advantage of using such *potentials* in the theory of semi-permeable diaphragms consists partly in the convenient form of the conditions of equilibrium, the potential for any substance to which a diaphragm is freely permeable having the same value on both sides of the diaphragm, and partly in our ability to express van't Hoff's law as a relation between the quantities characterising the state of the solution, without reference to any experimental arrangement (see *Transactions of the Connecticut Academy*, vol. iii, pp. 116, 138, 148, 194) [this vol., pp. 63, 83, 92, 135].

Let there be three reservoirs, R', R'', R''', of which the first contains the solvent alone, maintained in a constant state of temperature and pressure, the second the solution, and the third the solutum alone. Let R' and R'' be connected through a diaphragm freely permeable to the solvent, but impermeable to the solutum, and let R'' and R''' be connected through a diaphragm impermeable to the solvent, but freely permeable to the solutum. We have then, if we write μ_1 and μ_2 for the potentials of the solvent and the solutum, and distinguish by accents quantities relating to the several reservoirs,

$$\mu_1'' = \mu_1' = \text{const.,} \quad \mu_2'' = \mu_2'''.$$

Now if the quantity of the solutum in the apparatus be varied, the ratio in which it is divided in equilibrium between the reservoirs R'' and R''' will be constant, so long as its densities in the two reservoirs, γ_2'', γ_2''', are small. For let us suppose that there is only a single molecule of the solutum. It will wander through R'' and R''', and in a time sufficiently long the parts of the time spent respectively in R'' and R''', which for convenience we may suppose of equal volume, will approach a constant ratio, say $1:B$. Now if we put in the apparatus a considerable number of molecules, they will divide themselves between R' and R'' sensibly in the ratio $1:B$, so long as they do not sensibly interfere with one another, i.e., so long as the number of molecules of the solutum which are within the spheres of action of other molecules of the solutum is a negligible part of the whole, both in R'' and R'''. With this limitation we have, therefore,

$$\gamma_2''' = B\gamma_2''.$$

Now in R''' let the solutum have the properties of an ideal gas, which give for any constant temperature (*ibid.* p. 212) [this vol., p. 152]

$$\mu_2''' = a_2 t \log \gamma_2''' + C,$$

where a_2 is the constant of the law of Boyle and Charles, and C another constant. Therefore,

$$\mu_2'' = a_2 t \log (B\gamma_2'') + C.$$

This equation, in which a single constant may evidently take the place of B and C, may be regarded as expressing the property of the solution implied in van't Hoff's law. For we have the general thermodynamic relation (*ibid.* p. 143) [this vol., p. 88].

$$v\,dp = \eta\,dt + m_1\,d\mu_1 + m_2\,d\mu_2,$$

where v and η denote the volume and entropy of the mass considered, and m_1 and m_2 the quantities of its components. Applied to this case, since t and μ_1 are constant, this becomes

$$dp'' = \gamma_2''\,d\mu_2''.$$

Substituting the value of $d\mu_2''$, derived from the last finite equation, we have

$$dp'' = a_2 t\,d\gamma_2'',$$

whence, integrating from $\gamma_2'' = 0$ and $p'' = p'$, we get

$$p'' - p' = a_2 t \gamma_2'',$$

which evidently expresses van't Hoff's law.

We may extend this proof to cases in which the solutum is not volatile by supposing that we give to its molecules mutually repulsive molecular forces, which, however, are entirely inoperative with respect to any other kind of molecules. In this way we may make the solutum capable of the ideal gaseous state. But the relations pertaining to the contents of R'' will not be affected by these new forces, since we suppose that only a negligible part of the molecules of the solutum are within the range of such forces. Therefore these relations cannot depend on the new forces, and must exist without them.

To give up the condition that the molecules of the solutum shall not be broken up in the solution, nor united to one another in more complex molecules, would involve the consideration of a good many cases, which it would be difficult to unite in a brief demonstration. The result, however, seems to be that the increase of pressure is to be estimated by Avogadro's law from the number of molecules in the solution which contain any part of the solutum, without reference to the quantity in each.   J. WILLARD GIBBS.

New Haven, Connecticut, *February* 18.

IX.

UNPUBLISHED FRAGMENTS.

[Being portions of a supplement to the "Equilibrium of Heterogeneous Substances" in preparation at the time of the author's death, and intended to accompany a proposed reprint of his thermodynamic papers.*]

[A list of subjects found with the manuscript and printed below appears to indicate the scope of the supplementary chapters as planned by Professor Gibbs. As will be observed, however, the author's unfinished manuscript, except for a number of disconnected notes, relates to only two of these subjects, the first and fourth in the list.]

On the values of potentials in liquids for small components. (Temperature coefficients.)

On the fundamental equations of molecules with latent differences.

On the fundamental equations for vanishing components.

On the equations of electric motion.

On the liquid state, $p = 0$.

On entropy as mixed-up-ness.

Geometrical illustrations.

On similarity in thermodynamics.

Cryohydrates.

---

*[See Preface.]

On the Values of Potentials in Liquids for Substances which form but a Small Part of the whole Mass.*

The value of a potential† for a volatile substance in a liquid may be measured in a coexistent gaseous phase,‡ and so far as the latter may be treated as an ideal gas or gas-mixture,§ the value of the potential will be given by the equation (276), [" Equilib. Het. Subs."] which may be briefly written

$$\mu = \text{func}\,(t) + at \log \gamma_{\text{gas}}, \qquad [1]$$

where μ is the potential of the volatile substance considered, either in the liquid or in the gas, t the absolute temperature, γ_{gas} the density of the volatile substance in the gas and a the constant of the law of Boyle and Charles. Since this last quantity is inversely proportional to the molecular weight we may set

$$a = \frac{A}{M},$$

where M denotes the molecular weight, and A an absolute constant (the constant of the law of Boyle, Charles, and Avogadro),‖ and write the equation in the form

$$\mu = \text{func}\,(t) + \frac{At}{M} \log \gamma_{\text{gas}}, \qquad [2]$$

in which the value of the potential depends explicitly on the molecular weight.

The validity of this equation, it is to be observed, is only limited by the applicability of the laws of ideal gases to the gaseous phase; there is no limitation in regard to the proportion of the substance in question to the whole liquid mass. Thus at 20° Cent. the equation may be determined by the potential for water or for alcohol in a mixture of the two substances in any proportions, since the vapor of the mixture may be regarded as an ideal gas-mixture. But at a temperature at which we approach the critical state, the same is not true without limitation, since the coexistent gaseous phase cannot be treated as an ideal gas-mixture. At the same temperature however, if we limit ourselves to cases in which the proportion of water does not exceed $\frac{1}{10}$ of one per cent., and suppose the density of the

---

*The object of this chapter is to show the relation of the doctrine of potentials to van't Hoff's Law (what form van't Hoff's Law takes from the standpoint of the potentials); and to the modern theory of dilute solutions as developed by van't Hoff and Arrhenius. "Equilib. Het. Subs." [this volume], pp. 135-138, 138-144, 164-165, 168-172, 172-184.

†For the definition of this term see p. 93, also pp. 92-96.

‡In some cases a semi-permeable membrane may be necessary. (Enlarge.) (Is the term *coexistent* right in this case?)

§ Definition. (Enlarge.)

‖ $\dfrac{pvM}{mt} = A, \quad pv = \dfrac{m}{M}At.$ Is *absolute* used correctly?

water-vapor, γ_{gas}, to be measured in a space containing only water-vapor and separated from the liquid by a diaphragm permeable to water and not to alcohol, then the above equation would probably be applicable, since then the water-vapor might probably be treated as an ideal gas. The same would be true (*mutatis mutandis*) of the potential for alcohol in a mixture of alcohol and water containing not more than $\frac{1}{10}$ of one per cent. of alcohol.*

This law, however, which makes the potential in a liquid depend upon the density of the substance in some other phase is manifestly not convenient for use. We may get over this difficulty most simply by the law of Henry according to which the ratio of the densities of a substance in coexistent liquid and gaseous phases is (in cases to which the law applies) constant. If γ be the density in the liquid phase and γ_{gas} in the gas, we have

$$\gamma_{gas} = c\gamma, \qquad [3]$$

and by substitution in equation [2] we have

$$\mu = \text{func}\,(t) + \frac{At}{M}\log c\gamma,$$

or

$$\mu = \text{func}\,(t) + \frac{At}{M}\log \gamma, \qquad [4]$$

where the function of the temperature has been increased by $\frac{At}{M}\log c$.

With this value of the potential, which is manifestly demonstrated only to be used so far as the law of Henry applies, in connection with the general equation (98), ["Equilib. Het. Subs."] viz.,

$$dp = \frac{\eta}{v}\,dt + \frac{m_1}{v}\,d\mu_1 + \frac{m_2}{v}\,d\mu_2 \cdots + \frac{m_n}{v}\,d\mu_n,$$

we may calculate the osmotic pressure, etc., etc., as we shall see more particularly hereafter.

    I. Osmotic pressure.

    II. Lowering freezing point.

    III. Diminishing pressure of other gas.

    IIIa. Effect on total pressure.

    IIII. Raising boiling point with one pressure.

    IIIIa. Raising boiling point with two pressures.

    V. Interpolation formula for mixtures of liquids.

In fact, when γ_D† is small, we have approximately

$$\gamma_D\,d\mu_D = \frac{At}{M_D}\,d\gamma_D = At\,d\frac{n_D}{v}, \qquad [5]$$

---

*Also the potentials of water and alcohol in a mixture may be measured in a vertical tube of sufficient height. [See p. 413.]

†[In the following discussion, D indicates the dissolved substance, or solutum, and S the solvent.]

where n_D denotes the number of molecules of the form (D). Hence we have for the solution

$$dp = \frac{\eta}{v} dt + \gamma_8 d\mu_8 + At d\frac{n_D}{v}.$$ [6]

If t is constant, and also μ_8,—a condition realized in equilibrium, when the solution is separated from the pure solvent by a diaphragm permeable to the solvent but not to the solutum,—the equation reduces to

$$dp = At d\frac{n_D}{v}.$$

Whence $$p - p' = \frac{At}{M_D} \gamma_D = At\frac{n_D}{v},$$ [7]

p' being the pressure where $\gamma_D = 0$, i.e., in the pure solvent. Here $p - p'$ is the so-called osmotic pressure, and $\frac{At}{M_D} \gamma_D$ is the pressure as calculated * by the laws of Boyle, Charles, and Avogadro for the solutum in the space occupied by the solution. The equation manifestly expresses van't Hoff's law.

For a coexistent solid phase of the solvent, with constant pressure, the general equation gives

$$0 = \eta dt + m_8 d\mu_8 + v At d\gamma_D$$

for the solution, and

$$0 = \eta' d\iota + m_1 d\mu_8$$

for the solid coexistent phase. Here t and μ_8 have necessarily the same values in the two equations, and we may suppose the quantity of one of the phases to be so chosen as to make the values of m_8 equal in the two equations. This gives

$$(\eta' - \eta)dt = v\frac{At}{M_D} d\gamma_D.$$ [8]

In integrating from $\gamma_D = 0$ to any small value of γ_D, we may treat the coefficients of dt and $d\gamma_D$ as having the same constant values as when $\gamma_D = 0$. This gives

$$(\eta' - \eta)\Delta t = v\frac{At}{M_D} \gamma_D = \frac{At}{M_D} m_D.$$

If we write Q_8 for $\frac{t(\eta - \eta')}{m_8}$ (the latent heat of melting for the unit of weight of the solvent), we get

$$-\Delta t = \frac{At}{M_D} m_D \frac{t}{Q_8 m_8},$$

or $$-\Delta t = \frac{\frac{m_D}{M_D}}{\frac{m_8}{M_8}} \frac{At^2}{Q_8 M_8} = \frac{m_D}{M_D} \frac{At^2}{Q_8 m_8},$$ [9]

---

* Not experimentally found.

$m_8 Q_8$ is the latent heat of so much of the solvent as occurs in the solution. (Or make $m_8 = 1$.)

Raoult makes $\Delta t \propto \dfrac{m_D}{M_D}$, with exceptions.

With a coexistent gaseous phase of the solvent (the solutum being not volatile), we have for the solution

$$dp = \gamma_8 \, d\mu_8 + \frac{At}{M_D} \, d\gamma_D,$$

and for the gaseous phase

$$dp = \gamma_8' \, d\mu_8.$$

Here, on account of the coexistence of the phases, p and μ_8 and dp and $d\mu_8$ have the same values. Hence

$$dp = \frac{\gamma_8}{\gamma_8'} \, dp + \frac{At}{M_D} \, d\gamma_D,$$

$$-dp \frac{(\gamma_8 - \gamma_8')}{\gamma_8'} = \frac{At}{M_D} \, d\gamma_D.$$

Say

$$-dp = \frac{\gamma_8'}{\gamma_8} \frac{At}{M_D} \, d\gamma_D,$$

$$p - P = \frac{\gamma_8'}{\gamma_8} \frac{At}{M_D} \, \gamma_D,^* \qquad\qquad [10]$$

or†

$$-\frac{dp}{p} = \frac{M_8}{\gamma_8} \frac{d\gamma_D}{M_D},$$

$$\frac{p - P}{p} = \frac{M_8}{\gamma_8} \frac{\gamma_D}{M_D}. \qquad\qquad [11]$$

M_D is the molecular weight [of solutum] in solution;

M_8 is the molecular weight [of solvent] in vapor.

But the foregoing equation suggests a generalization which is not confined to cases in which the law of Henry has been proved. The letter M in the equation has been defined as the molecular weight of the substance in the form of gas. Now the molecular weight which figures in the relation between the potential and the density of a substance in a liquid would naturally be the molecular weight of the substance as it exists in the liquid. It is therefore a natural supposition suggested by the equation that, in the case where Henry's law holds good, and consequently eq. [4], the molecular weight of the solutum is the same in the liquid and in the gaseous phase; that in

---

* [p is the vapor pressure of the pure solvent, P that of the solution.]

† $\left[\text{Assuming that the vapor behaves like an ideal gas, we have } \gamma_8' = \dfrac{pM_8}{At}.\right]$

case the law of Henry and eq. [4] do not hold, it may be on account of a difference in the molecular weight in the gas and the liquid, and that the eq. [4] may still hold if we give the proper value to M in that equation, viz., the molecular weight in the liquid.

But as these considerations, although natural, fall somewhat short of a rigorous demonstration, let us scrutinize the case more carefully. It is easy to give an *a priori* demonstration of Henry's law and equation [4] in cases in which there is only one molecular formula for the solutum in liquid and in gas, so long as the density both in liquid and in gas is so small that we may neglect the mutual action of the molecules of the solutum. In such a case the molecules of the solutum will be divided between the liquid and the gas in a (sensibly) constant ratio (the volume of the liquid and gas being kept constant), simply because every molecule, moving as if there were no others, would spend the same part of its time in the vapor and in the liquid as if the others were absent, and the number of the molecules being large, this would make the division sensibly constant. This proof will apply in cases in which the law of Henry can hardly be experimentally demonstrated, because the density of the solutum as gas is so small as to escape our power of measurement. Also in cases in which a semi-permeable diaphragm is necessary, an arrangement very convenient for theoretical demonstrations, but imperfectly realizable in practice. (Also in cases in [which a] difference of level is necessary, with or without diaphragm.) But in every case when the law of Henry is demonstrably untrue for dilute solutions, we may be sure that there is more than one value of the molecular weight of the solutum in the phases considered.

This theoretical proof will apply to cases in which experimental proof is impossible :

(1) When the density in gas is too small to measure.

(2) When the density in gas is too great, either the total density or the partial. (Diaphragm or vertical column.)

(3) When the liquid (or other phase) is sensitive to pressure and not in equilibrium with the gas.

Will the various theorems exist in these cases ?

If one or both appear in a larger molecular form, the densities of γ_M and γ_M' * are proportional and

$$\gamma_{2M} \propto \gamma_M^2, \quad \gamma_{2M}' \propto \gamma_M'^2, \\ \mu_{2M} = \mu_M = \mu_M' = \mu_{2M}', \tag{12}$$

hence one equation of form, $\mu_M = \dfrac{At}{M} \log \gamma_M$ proves all.

---

* [γ refers to the liquid, and γ' to the gaseous phase.]

Let us next consider the case in which the solutum appears with more than one molecular formula in the liquid or gas or both. Now there are two cases, that in which the quantities of the substance with the different molecular formulae are independently variable, and that in which they are not. In the [first] case there is no question. If, for example, hydrogen appears with the molecular formula H_2 and also in molecules with the molecular formula H_2O, these are to be treated as separate substances, and we have the two equations

$$\mu_{H_2} = \text{func } (t) + \frac{At}{M_{H_2}} \log \gamma_{H_2},$$

and

$$\mu_{H_2O} = \text{func } (t) + \frac{At}{M_{H_2O}} \log \gamma_{H_2O},$$

and also if free oxygen is present

$$\mu_{O_2} = \text{func } (t) + \frac{At}{M_{O_2}} \log \gamma_{O_2}.$$

But when the quantities of the substance associated in the different molecular combinations are not independently variable, then we have the equation

$$M_1\mu_1 + M_2\mu_2 = M_{12}\mu_{12}, \qquad [13]$$

which is exact and certain, and the considerations adduced on p. (*), which are not limited to gases, seem to show that in this case the equations of the form (†) all continue to subsist, but we have also the equation of form (‡).

It would therefore appear that we may regard the equation

$$\mu = \text{func } (t) + \frac{At}{M} \log \gamma$$

as expressing a general law of nature, where the letter M is the molecular weight corresponding to any molecular combination in the liquid and γ is the density of the matter which has that molecular formula, provided that the density γ is so small that of the molecules which it represents only a negligible fraction at any time are within the spheres of each other's attraction. It goes without saying that the law is approximative, as the last condition can only be satisfied approximately for any finite value of γ. (Need of verification on account of the unknown M.)

[The author's manuscript for the proposed supplement ends, so far at least as a connected treatment is concerned, at this point. The following notes are appended.]

---

* [Although left blank in the MS., this probably refers to p. 423.]
† [Probably equation [12].]   ‡ [Probably equation [13].]

In case of one molecular formula in liquid and none in gas, we may give the molecules repelling forces which will make the gas possible. (?) [See p. 417.]

Deduce Ostwald's law in more general form.

Deduce interpolation formula.

What use can we make of Latent Differences? μ_A, μ_{AA}, μ_B, μ_{BB}, μ_{AB} all conform to law, I think.

[On the Equations of Electric Motion.]

[A somewhat abbreviated copy of a letter written four years earlier (in May 1899) to Professor W. D. Bancroft of Cornell University had been placed by Professor Gibbs between the pages of the manuscript, and was evidently intended to serve as a basis for the chapter " On the equations of electric motion" mentioned in the list on page 418.

Through the courtesy of Professor Bancroft the original letter has been placed at the disposal of the editors and is here given in full. The major portion of this letter was incorporated by Professor Bancroft in an article entitled " Chemical Potential and Electromotive Force," published after the death of Professor Gibbs, in the Journal of Physical Chemistry, *vol. vii., p.* 416, *June* 1903.]

My dear Prof. Bancroft :

A working theory of galvanic cells requires (as you suggest) that we should be able to evaluate the (intrinsic or chemical) potentials involved, and your formula

$$d\mu = Rt \, d \log p,$$

is all right as you interpret it. I should perhaps prefer to write

$$\mu_D = B + \frac{At}{M_D} \log \gamma_D, \tag{1}$$

or

$$\gamma_D \, d\mu_D = \frac{At}{M_D} d\gamma_D, \tag{2}$$

for small values of γ_D, where γ_D is the density of a component (say the mass of the solutum divided by the volume of the solution), M_D its molecular weight (viz., for the kind of molecule which actually exists in the solution), A the constant of Avogadro's Law $\left(\frac{pv}{mt} = \frac{A}{M}\right)$, and B a quantity which depends upon the solvent and the solutum, as well as the temperature, but which may be regarded as independent of γ_D so long as this is small, and which is practically independent of the pressure in ordinary cases.

We may avoid 'hedging' in regard to B by using the differential equation (2). We may simply say that this equation holds for changes produced by varying the quantity of (D), when γ_D is small. It is not limited to changes in which t is constant, for the change in μ_D due to t appearing in (1) (both explicitly, and implicitly in B) becomes negligible when multiplied by the small quantity γ_D.

The formula contains the molecular weight M_D, and if all the solutum has not the same molecular formula, the γ_D must be understood as relating only to a single kind of molecule.

Thus if a salt $(_{12})$ is partly dissociated into the ions $(_1)$ and $(_2)$, we will have the three equations

$$\mu_1 = B_1 + \frac{At}{M_1} \log \gamma_1.$$

$$\mu_2 = B_2 + \frac{At}{M_2} \log \gamma_2,$$

$$\mu_{12} = B_{12} + \frac{At}{M_{12}} \log \gamma_{12}.$$

The three potentials are also connected by the relation

$$M_1 \mu_1 + M_2 \mu_2 = M_{12} \mu_{12},$$

which determines the amount of dissociation. We have, namely,

$$M_1 B_1 + M_2 B_2 - M_{12} B_{12} + At \log \frac{\gamma_1 \gamma_2}{\gamma_{12}} = 0,$$

which makes $\dfrac{\gamma_1 \gamma_2}{\gamma_{12}}$ constant, for constant temperature and solvent.

I may observe in passing that this relation, eq. (1) or (2), which is so fundamental in the modern theory of solutions, is somewhat vaguely indicated in my " Equilib. Het. Subs." (See [this volume] pp. 135-138, 156, and 164-165.) I say vaguely, because the coefficient of the logarithm is only given (in the general case) as constant for a given solvent and temperature. The generalization that this coefficient is in all cases of exactly the same form as for gases, even to the details which arise in cases of dissociation, is due to van't Hoff in connection with Arrhenius, who suggested that the " discords " are but " harmonies not understood," and that exceptions vanish when we use the true molecular weights. At all events, eq. (2) with (98) (*E.H.S.*) gives for a solvent (S) with one dissolved substance (D),

$$dp = \frac{\eta}{v} dt + \frac{m_S}{v} d\mu_S + \frac{At}{M_D} d\gamma_D.$$

If we integrate, keeping t constant and also μ_S (by connection with the pure solvent through a semi-permeable diaphragm), we have van't Hoff's Law,

$$p - p' = \frac{At}{M_D} \gamma_D.$$

In the above case of dissociation the formula would be

$$p - p_1 = At\left(\frac{\gamma_1}{M_1} + \frac{\gamma_2}{M_2} + \frac{\gamma_3}{M_3}\right).$$

For a coexistent solid phase of the solvent we have for constant pressure

$$0 = \eta\, dt + m\, d\mu_8 + v\frac{At}{M_D}\, d\gamma_D,$$

$$0 = \eta'\, dt + m\, d\mu_8,$$

m_8 being for convenience taken the same in both phases.

Then
$$(\eta' - \eta)\, dt = v\frac{At}{M_D}\, d\gamma_D.$$

In integrating for small values of γ_D we may treat the coefficients of dt and $d\gamma_D$ as constant. This gives

$$(\eta' - \eta)\Delta t = v\frac{At}{M_D}\gamma_D = At\frac{m_D}{M_D},$$

or if we write Q_8 for $\dfrac{t(\eta - \eta')}{m_8}$ (the latent heat of melting for the unit of weight of the solvent), we have

$$-\Delta t = \frac{At^2 m_D}{Q_8 M_D m_8}.$$

This may be written

$$-\Delta t\frac{m_8}{m_D}\frac{M_D}{M_8} = \frac{At^2}{Q_8 M_8}.$$

According to Raoult, the first member of this equation has a value nearly identical for all solvents and solutes (supposed definite compounds). This would make the second member the same for all liquids of "definite" composition, when we give M_8 the value for the molecule in the liquid state. I should think it more likely that these properties should hold for the two members of the equation

$$-\frac{\Delta t}{t}\frac{m_8}{m_D}\frac{M_D}{M_8} = \frac{At}{Q_8 M_8},$$

which are pure numbers (of no dimensions in physical units). In this form it has a certain analogy with van der Waals' law of "corresponding states."

With a coexistent vapor phase of the solvent, we have

$$v\, dp = m_8\, d\mu_8 + v\frac{At}{M_D}\, d\gamma_D,$$

$$v'\, dp = m_8\, d\mu_8,$$

$$(v - v')\, dp = v\frac{At}{M_D}\, d\gamma_D,$$

$$-dp = \frac{v}{v' - v}\frac{At}{M_D}\, d\gamma_D.$$

We may regard $\dfrac{v}{v'-v}$ as constant in integrating (for small γ_D), which gives

$$P - p = \frac{v}{v'-v}\, At \frac{\gamma_D}{M_D}.$$

Now

$$\frac{At}{v'-v} = \frac{At}{v'} = \frac{PM_S}{m_S} \text{ nearly, which gives}$$

$$\frac{P-p}{P} = \frac{m_D}{m_S}\frac{M_S}{M_D}, \text{ which is Raoult's Law.}$$

Raoult found values about 5 per cent. larger than this, which agrees very well with the fact that $\dfrac{At}{v'-v}$ is somewhat larger than $\dfrac{PM_S}{m_S}$. It is also to be observed that M_D relates to the molecules in the solution, but M_S to the molecules in the vapor. Or, with a coexistent vapor phase of the solutum (alone or mixed with other vapors or gases), we have

$$\mu = B + \frac{At}{M_D}\log \gamma_D,$$

$$\mu = B' + \frac{At}{M_D}\log \gamma_D',$$

$$\frac{B'-B}{At}M_D = \log \frac{\gamma_D}{\gamma_D''},$$

which makes $\dfrac{\gamma_D}{\gamma_D'}$ constant for the same solvent, solutum, and temperature, according to Henry's Law.

So for the galvanic cell which you first consider, I should write

$$V'' - V' = a_a(\mu' - \mu'') = a_a \frac{At}{M_a}\log \frac{\gamma_a'}{\gamma_a''},$$

γ_a', γ_a'' being the densities, supposed small, of the cation (a) in the two electrodes, which are supposed identical except for the dissolved (a). Here a_a has reference to the solution and M_a to the electrodes. It may be more convenient to divide a_a into the factors E_a, a_H, where a_H is the weight of hydrogen which carries the unit of electricity, and E_a the weight of (a) which carries the same quantity of electricity as the unit of weight of hydrogen. In other words E_a is Faraday's "electrochemical equivalent" and a_a is Maxwell's "electrochemical equivalent." This gives

$$V'' - V' = a_H At \frac{E_a}{M_a}\log \frac{\gamma_a'}{\gamma_a''},$$

where $a_H A$ is your R and $\dfrac{M_a}{E_a}$ your v, v'.*

---

* [The valence of the ion].

The meagreness of the results obtained in my *E.H.S.* in the matter of electrolysis has a deeper reason than the difficulty of the evaluation of the potentials.

In the first place, cases of true equilibrium (even for open circuit) are quite exceptional. Thus the single case of unequal concentration of the electrolyte cannot be one of equilibrium since the process of diffusion cannot be stopped. Cases in which equilibrium does not subsist were formally excluded by my subject, and indeed could not be satisfactorily treated without the introduction of new ideas quite foreign to those necessary for the treatment of equilibrium.

Again, the consideration of the electrical potential in the electrolyte, and especially the consideration of the difference of potential in electrolyte and electrode, involves the consideration of quantities of which we have no apparent means of physical measurement, while the difference of potential in "pieces of metal of the same kind attached to the electrodes" is exactly one of the things which we can and do measure.

Nevertheless, with some hedging in regard to the definition of the electrical potential, we may apply

$$V'' - V' = a_a(\mu_a' - \mu_a'')$$

to points in electrolyte (') and electrode ('').

This gives

$$V'' - V' = a_a\Big(B_a + \frac{At}{M_a}\log \gamma_a - \mu_a''\Big),$$

say,

$$V'' - V' = \frac{a_a At}{M_a}\log \frac{\gamma_a}{G}.$$

The G like the P of your formula seems to depend on the solvent, presumably varies with the temperature, but as Nernst remarks does not depend on the other ion associated with (a), so long as the solution is dilute.

The case of unequal concentration, or, in general, cases in which the electrolyte is not homogeneous, I should treat as follows; Let us suppose for convenience that the cell is in form of a rectangular parallelopiped with edge parallel to axis of x and cross section of unit area. The electrolyte is supposed homogeneous in planes parallel to the ends, which are formed by the electrodes.

Of course we should have equilibrium if proper forces could be applied to prevent the migration of the ions and also of the part of the solutum which is not dissociated. What would these forces be? For the molecules ($_{12}$) which are not dissociated, the force per unit of mass would be $\dfrac{d\mu_{12}}{dx}$. (The problem is practically the same as that discussed in *E.H.S.* [this volume], pp. 144 ff.) If the unit of mass of the cation $(_1)$ has the charge c_1, the force necessary to prevent its migration would be

$$\frac{d\mu_1}{dx} + c_1 \frac{dV}{dx}.$$

For an anion $(_2)$ the force would be

$$\frac{d\mu_2}{dx} - c_2 \frac{dV}{dx}. *$$

Now we may suppose that the same ion in different parts of a dilute solution will have velocities proportional to the forces which would be required to prevent its motion. We may therefore write for the velocity of the cation $(_1)$,

$$-\frac{k_1}{c_1}\left(\frac{d\mu_1}{dx} + c_1 \frac{dV}{dx}\right),$$

and for the flux of the cation $(_1)$,

$$\phi_1 = -\frac{k_1}{c_1}\gamma_1\left(\frac{d\mu_1}{dx} + c_1 \frac{dV}{dx}\right) = -At\frac{k_1}{c_1 M_1}\frac{d\gamma_1}{dx} - k_1 \frac{dV}{dx}\gamma_1; \qquad (3)$$

for the flux of the anion $(_2)$,

$$\phi_2 = -\frac{k_2}{c_2}\gamma_2\left(\frac{d\mu_2}{dx} - c_2 \frac{dV}{dx}\right) = -At\frac{k_2}{c_2 M_2}\frac{d\gamma_2}{dx} + k_2 \frac{dV}{dx}\gamma_2, \qquad (4)$$

where k_1, k_2 are constants ('migration velocities') depending on the solvent, the temperature, and the ion.† Now whatever the number of ions the flux of electricity is given by the equation

$$\phi = \Sigma \pm c_1 \phi_1, \ddagger$$

where the upper sign is for cations and the lower for anions, and the summation for all ions. This gives

$$\phi = At\Sigma \mp \frac{k_1}{M_1}\frac{d\gamma_1}{dx} - \frac{dV}{dx}\Sigma c_1 k_1 \gamma_1.$$

That is,

$$\phi \frac{dx}{\Sigma c_1 k_1 \gamma_1} = At \frac{\Sigma \mp \dfrac{k_1}{M_1} d\gamma_1}{\Sigma c_1 k_1 \gamma_1} - dV.$$

The form of this equation shows that since ϕ is the current, $\dfrac{dx}{\Sigma c_1 k_1 \gamma_1}$ is the "resistance" of an elementary slice of the cell, and the next term the (internal) electromotive force of that slice.

---

* [c_2 is a positive number equal numerically to the negative charge on unit mass of the anion.]

† [The positive direction for both these fluxes is the direction of increasing x.]

‡ [The sign of the charge is not included in c. Hence the double sign is necessary.]

Integrating from one point to another in the electrolyte,

$$\phi \int \frac{dx}{\Sigma c_1 k_1 \gamma_1} = At \int \frac{\Sigma \mp \frac{k_1}{M_1} d\gamma_1}{\Sigma c_1 k_1 \gamma_1} + V' - V''.$$

The evaluation of these integrals which denote the resistance and electromotive force for a finite part of the electrolyte depends on the distribution of the ions in the cell. For one salt with varying concentration,

$$\phi \frac{dx}{c_1 k_1 \gamma_1 + c_2 k_2 \gamma_2} = At \frac{-\frac{k_1}{M_1} d\gamma_1 + \frac{k_2}{M_2} d\gamma_2}{c_1 k_1 \gamma_1 + c_2 k_2 \gamma_2} - dV,$$

or, since $c_1 \gamma_1 = c_2 \gamma_2$ and $c_1 d\gamma_1 = c_2 d\gamma_2$,

$$\phi \frac{dx}{c_1 k_1 \gamma_1 + c_2 k_2 \gamma_2} = At \frac{-\frac{k_1}{c_1 M_1} + \frac{k_2}{c_2 M_2}}{k_1 + k_2} \frac{d\gamma_1}{\gamma_1} - dV,$$

$$\phi \int \frac{dx}{c_1 k_1 \gamma_1 + c_2 k_2 \gamma_2} = At \frac{-\frac{k_1}{c_1 M_1} + \frac{k_2}{c_2 M_2}}{k_1 + k_2} \log \frac{\gamma_1''}{\gamma_1'} + V' - V''.$$

The resistance depends on the concentration throughout the part of the cell considered, but the electromotive force depends only on the concentration at the terminal points (' and ").

For $c_1 M_1$ and $c_2 M_2$ we may write $\frac{v_1}{a_H}$ and $\frac{v_2}{a_H}$, where v_1 and v_2 are the " valencies " of the molecules. This gives

$$V'' - V' = a_H At \frac{\frac{k_1}{v_1} - \frac{k_2}{v_2}}{k_1 + k_2} \log \frac{\gamma_1'}{\gamma_1'''}, \text{ for } \phi = 0 \text{ (circuit open).}$$

I think this is identical with your equation (V) when your ions have the same valency.

Planck's problem is less simple.* We may regard it as relating to a tube connecting the two great reservoirs filled with different electrolytes of same concentration, i.e., $\Sigma_0 c_0 \gamma_0' = \Sigma_0 c_0 \gamma_0''$. I use $(_0)$ for any ion, $(_1)$ for any cation, $(_2)$ for any anion. [The accents (') and (") refer to the two reservoirs.]

The tube is supposed to have reached a stationary state and dissociation is complete. The number of ions is immaterial, but they all must have the same valency v.

Now by equations (3) and (4), since $c_0 M_0 = \frac{v}{a_H}$,

$$\phi_0 = -\frac{aAt}{v} k_0 \frac{d\gamma_0}{dx} \mp k_0 \frac{dV}{dx} \gamma_0,$$

---

*[Planck, *Wied. Ann.*, vol. xl (1890), p. 561.]

or, writing T for the constant $\dfrac{aAt}{v}$,

$$\phi_0 = -Tk_0\frac{d\gamma_0}{dx} \mp k_0\gamma_0\frac{dV}{dx},$$

$$\frac{c_0\phi_0}{k_0} = -Tc_0\frac{d\gamma_0}{dx} \mp c_0\gamma_0\frac{dV}{dx},$$

$$\Sigma_0\frac{c_0\phi_0}{k_0} = -T\Sigma_0\frac{c_0 d\gamma_0}{dx}.$$

[The terms $\mp\dfrac{dV}{dx}c_0\gamma_0$ disappear in the algebraic sum since $\Sigma c_1\gamma_1 = \Sigma c_2\gamma_2$. For a similar reason]

$$\Sigma_0 \pm \frac{c_0\phi_0}{k_0} = -\frac{dV}{dx}\Sigma_0 c_0\gamma_0.$$

The first equation makes $\dfrac{d\Sigma_0 c_0\gamma_0}{dx}$ constant throughout the tube, and since $\Sigma_0 c_0\gamma_0'' = \Sigma_0 c_0\gamma_0'$, $\Sigma c_0\gamma_0$ must be constant throughout the tube. The second equation then makes $\dfrac{dV}{dx}$ constant throughout the tube. Let $X = -\dfrac{dV}{dx}$.

Our original equation is

$$\phi_0 = -Tk_0\frac{d\gamma_0}{dx} \pm Xk_0\gamma_0.$$

Now with X constant this is easily integrated.

$$\frac{\phi_0}{Xk} = -\frac{T}{X}\frac{d\gamma_0}{dx} \pm \gamma_0,$$

$$\pm\frac{T}{X}\frac{d\gamma_0}{dx} = \gamma_0 \mp \frac{\phi_0}{Xk_0},$$

$$\frac{d\gamma_0}{\gamma_0 \mp \dfrac{\phi_0}{Xk_0}} = \pm\frac{X}{T}dx,$$

$$\log\left(\gamma_0 \mp \frac{\phi_0}{Xk_0}\right) = \pm\frac{X}{T}x + \log H_0,$$

$$\gamma_0 \mp \frac{\phi_0}{Xk_0} = H_0\, e^{\pm\frac{X}{T}x}.$$

To determine H_0 we have

$$\gamma_0'' - \gamma_0' = H_0\left(e^{\pm\frac{X}{T}x''} - e^{\pm\frac{X}{T}x'}\right).$$

If we put the origin of coordinates in the middle of the tube we have

$$x' = -x''.$$

Let $\qquad P = e^{\frac{X}{T}x''}, \qquad\qquad \gamma_0'' - \gamma_0' = \pm H_0(P - P^{-1}).$

Let $\qquad \Delta_0 = \gamma_0'' - \gamma_0', \qquad \gamma_0 \mp \dfrac{\phi_0}{Xk_0} = \pm \Delta_0 \dfrac{e^{\pm\frac{X}{T}x}}{P - P^{-1}},$

$$c_0 k_0 \gamma_0 \mp \frac{c_0\phi_0}{X} = \pm c_0 k_0 \Delta_0 \frac{e^{\pm\frac{X}{T}x}}{P - P^{-1}}.$$

The condition of no electrical current gives

$$\Sigma_0 c_0 k_0 \gamma_0 = \Sigma_0 \pm c_0 k_0 \Delta_0 \frac{e^{\pm\frac{X}{T}x}}{P - P^{-1}}.$$

Apply to both ends and add,

$$\Sigma_0 c_0 k_0 \gamma_0'' + \Sigma_0 c_0 k_0 \gamma_0' = \Sigma_0 \pm c_0 k_0 \Delta_0 \frac{e^{\pm\frac{X}{T}x''} + e^{\pm\frac{X}{T}x'}}{P - P^{-1}},$$

$$= \Sigma_0 \pm c_0 k_0 \Delta_0 \frac{P + P^{-1}}{P - P^{-1}}.$$

If we set, to abridge,

$$K_1' = \Sigma_1 c_1 k_1 \gamma_1', \quad K_1'' = \Sigma_1 c_1 k_1 \gamma_1'',$$
$$K_2' = \Sigma_2 c_2 k_2 \gamma_2', \quad K_2'' = \Sigma_2 c_2 k_2 \gamma_2''.$$

When the summations are for cations or anions *separately*, the last equation may be written

$$K_1'' + K_2'' + K_1' + K_2' = \frac{P + P^{-1}}{P - P^{-1}}(K_1'' - K_1' - K_2'' + K_2'),$$

which gives $\qquad\qquad P^2 = \dfrac{K_1'' + K_2'}{K_2'' + K_1'}.$

Now $\quad \log P = \dfrac{X}{T}x'' = \dfrac{-X}{T}x', \quad 2\log P = \dfrac{X(x'' - x')}{T} = \dfrac{V' - V''}{T},$

$$\frac{V' - V''}{T} = \log\frac{K_1'' + K_2'}{K_1' + K_2''}, \quad V'' - V' = \frac{A\,at}{v}\log\frac{K_1' + K_2''}{K_1'' + K_2'}.$$

K_1' is the part of the conductivity of the first electrolyte which is due to the cations.

If the first electrolyte contains only one cation ($_1$) and one anion ($_2$), and the second only one cation ($_3$) and one anion ($_4$), we have

$$V'' - V' = \frac{A\,at}{v}\log\frac{c_1 k_1 \gamma_1' + c_4 k_4 \gamma_4''}{c_3 k_3 \gamma_3'' + c_2 k_2 \gamma_2'},$$

or, since $\qquad c_1\gamma_1' = c_2\gamma_2' = c_3\gamma_3'' = c_4\gamma_4'',$

$$V'' - V' = \frac{A\,at}{v}\log\frac{k_1 + k_4}{k_3 + k_2},$$

like the formula which you quote.

I regret that I have been obliged to delay my writing so long. I presume that you would have preferred to have me reply more promptly and more briefly. But the matter did not seem to be capable of being dispatched in few words.

One might easily economize in letters in the formulae by referring densities (γ) and potentials (μ) to equivalent or molecular weights, as you have done, but I thought I was more sure to be understood with the notations which I have used. Moreover, since the molecular weight is often the doubtful point in the whole problem, there is a certain advantage in bringing it in explicitly rather than implicitly, so that we can see at a glance how a change in our assumptions in regard to the molecules will affect the measurable quantities.

Yours, very sincerely,

J. WILLARD GIBBS.